Petroleum Engineering

Petroleum Engineering

Edited by **Michael Dedini**

SYRAWOOD
PUBLISHING HOUSE
New York

Published by Syrawood Publishing House,
750 Third Avenue, 9th Floor,
New York, NY 10017, USA
www.syrawoodpublishinghouse.com

Petroleum Engineering
Edited by Michael Dedini

International Standard Book Number: 978-1-68286-370-1 (Hardback)

Printed in China.

Contents

Preface

Petroleum engineering is a prominent field that is concerned with the exploration and production of fuels like crude oil, natural gas and other related products. It is an interdisciplinary field that integrates concepts and applications of mineralogy, geology and other disciplines of engineering. The book elucidates various aspects of petroleum operations, petroleum geology and production engineering. The researches and case-studies compiled in this book are meant to provide an overview of the current practices and technological advancements in petroleum engineering across the globe. This book also aims to highlight some of the major impacts of petroleum engineering and mining on the environment. It will be helpful for students and researchers alike.

This book is a result of research of several months to collate the most relevant data in the field.

When I was approached with the idea of this book and the proposal to edit it, I was overwhelmed. It gave me an opportunity to reach out to all those who share a common interest with me in this field. I had 3 main parameters for editing this text:

1. Accuracy - The data and information provided in this book should be up-to-date and valuable to the readers.
2. Structure - The data must be presented in a structured format for easy understanding and better grasping of the readers.
3. Universal Approach - This book not only targets students but also experts and innovators in the field, thus my aim was to present topics which are of use to all.

Thus, it took me a couple of months to finish the editing of this book.

I would like to make a special mention of my publisher who considered me worthy of this opportunity and also supported me throughout the editing process. I would also like to thank the editing team at the back-end who extended their help whenever required.

Editor

A new method for determining the formation temperature from bottom-hole temperature logs

I. M. Kutasov[1] and L.V. Eppelbaum[2*]

[1]Pajarito Enterprises, 640 Alta Vista, Suite 124, Santa Fe, New Mexico 87505, USA.
[2]Department of Geophysics and Planetary Sciences, Raymond and Beverly Sackler Faculty of Exact Sciences, Tel Aviv University, Ramat Aviv 69978, Tel Aviv, Israel.

Earlier, the authors proposed a new method for prediction of formation temperatures in permafrost regions from temperature logs in deep wells, and developed working formulas to process field data. The application of the proposed method on predicting the undisturbed formation temperature does not depend: (a) on the well drilling history (vertical depth versus time and stops in mud circulation), (b) on the knowledge of thermal diffusivity of formations and well radius. This method is based on using only three values of shut-in temperatures. Our investigation indicates that for bottom-hole temperature (BHT) tests, where the thermal disturbance of formations is caused mainly by one short (3-24 hours) continuous drilling fluid circulation period, the developed earlier method can be utilized for processing results of BHT surveys. The results of bottom-hole temperature surveys in three wells and data for one simulated example were used to validate our proposal. We suggest that application of this procedure in oil and gas industry will increase the reliability of temperature field BHT tests.

Key words: Formation temperature, bottom-hole temperature logs, generalized Horner method, oil and gas industry.

INTRODUCTION

It is known that temperature field investigations play an important role by prospecting and exploitation of oil and gas deposits (Melton and Gardini, 1984; Deming and Chapman, 1988; Eppelbaum et al., 1996; Kutasov, 1999; McAleese, 2000; Andaverde et al., 2005; Waples et al., 2004; Zschocke, 2005; Verma et al., 2006; Pasquale et al., 2008, etc.).

The determination of physical properties of reservoir fluids, calculation of hydrocarbon volumes (estimation of oil and gas formation volume factors, gas solubility), predictions of the gas hydrate prone zones, well log interpretation, determination of heat flow density and evaluation of geothermal energy resources require knowledge of the undisturbed formation temperature. In most cases, bottom-hole temperature surveys are mainly used to determine the temperature of the earth's interior. The drilling process, however, greatly alters the temperature of formation immediately surrounding the well. The temperature change is affected by the duration of drilling fluid circulation, the temperature difference between the reservoir and the drilling fluid, the well radius, the thermal diffusivity of the reservoir and the drilling technology used. Given these factors, the exact determination of formation temperature at any depth requires a certain length of time in which the well is not in operation. In theory, this shut-in time is infinitely long to reach the original condition. There is, however, a practical limit to the time required for the difference in temperature between the well wall and surrounding reservoir to become a specified small value. The objective of this paper is to suggest a new approach for utilizing bottom-hole temperature logs in deep wells.

It should be noted that Horner method for obtaining the formation equilibrium temperature from the bottom-hole temperature is widely applied in oil and gas industry (Cao et al., 1988; Deming and Chapman, 1988; Kutasov, 1989; Nielsen et al., 1990; Kutasov, 1999; Mcaleese, 2000; Förster, 2001; Andaverde et al., 2005; Kutasov and Eppelbaum, 2005; Zschocke, 2005; Verma et al., 2006; Pasquale et al., 2008; Espinosa-Paredes et al., 2009). Earlier we proposed a new method for predicting the

*Corresponding author. E-mail: levap@post.tau.ac.il.

formation temperatures in permafrost regions from temperature logs in deep wells (Kutasov and Eppelbaum, 2003). The main features of the suggested method were the following:

1. In the permafrost section of the well, the starting point in the well thermal recovery is moved from the end of well completion to the moment of time when the refreezing of formations was completed. It is taken into account that the refreezing of thawed formations occurs in some temperature interval,
2. Below the permafrost base the starting point in the well thermal recovery is moved from the end of well completion to the moment of time when the first shut-in temperature log was taken.

The application of the proposed method of predicting the undisturbed formation temperature does not depend on the well drilling history (vertical depth versus time, stops in mud circulation). Specification of two parameters: thermal diffusivity of formations and well radius are not needed to compute the value of the undisturbed formation temperature (Kutasov and Eppelbaum, 2003). At conducting bottom-hole temperature logs, the thermal disturbance of formations (near the well's bottom) is caused mainly by one short continuous drilling fluid circulation period (prior to logging). The duration of this period usually consists of 3 - 24 h. As it is shown in this paper, the earlier suggested method (Kutasov and Eppelbaum, 2003) can be extended for determining the formation temperature from bottom-hole temperature surveys. Three field cases and one synthetic example are used to demonstrate the validity of this approach. The results of calculations are compared with those obtained after the Generalized Horner Method (GHM), where the values of well radius and formation diffusivity should be specified.

NEW METHOD (NM) AND WORKING FORMULA

Theoretically, the drilling process affects the temperature field of formations at very long radial distances. There is however, a practical limit to the distance – the radius of thermal influence (r_{in}), where for a given circulation period ($t = t_c$) the temperature $T(r_{in}, t_c)$ is "practically" equal to the geothermal temperature T_f. To avoid this uncertainty, it is essential that the parameter r_{in} must not depend on the temperature difference $T(r_{in}, t_c) - T_f$. For this reason, we used the thermal balance method to calculate the radius of thermal influence. The results of modelling, experimental works and field observations indicate that the following relation could be used for approximating the temperature distribution around the wellbore during drilling (Kutasov, 1968, 1999). Let us assume that three shut-in temperatures T_{s1}, T_{s2}, and T_{s3} are measured at a given depth. We can consider that the period of time $t_c^* = t_c + t_{s1}$ as a new "thermal disturbance" period. Then the "shut-in times" are

$$t_{s1}^* = t_{s2} - t_{s1}, \quad t_{s2}^* = t_{s3} - t_{s1}. \tag{1}$$

Now dimensionless temperature distribution at $t = t_{s1}$ (Figure 1)

$$\begin{Bmatrix} T_{cD}^*(r_D, t_{xD}) = 1, & 0 \le r_D \le 1, \\ T_{cD}^*(r_D, t_{xD}) = 1 - \ln r_D / \ln R_x, & 1 \le r_D \le R_x \\ T_{cD}^*(r_D, t_{xD}) = 0, & r_D > R_x. \end{Bmatrix}, \tag{2}$$

$$R_x = 1 + 2.184\sqrt{t_{xD}}, \qquad t_{xD} = \frac{a\, t_c^*}{r_{wx}^2}, \qquad R_x = \frac{r_{ix}}{r_{wx}}, \tag{3}$$

$$T_{cD}^*(r_D, t_{xD}) = \frac{T_c(r,t) - T_f}{T_{s1} - T_f}, \qquad r_D = \frac{r}{r_{wx}}, \tag{4}$$

Where; r_{wx} is the radius of a cylindrical source with a constant wall temperature (T_{s1}) during the thermal disturbance period (t_c^*), a is the thermal diffusivity of formations, t_c is the duration of the circulation period (at the bottom-hole), T_f is the undisturbed temperature of formations and r_{ix} is the radius of thermal influence. For the initial radial temperature distributions (equation 2) the dimensionless shut-in wellbore temperature (at $t_s > t_{s1}$) was presented earlier (Kritikos and Kutasov, 1988; Kutasov, 1999):

$$T_{sD}^* = \frac{T(0,t_s^*) - T_f}{T_{s1} - T_f} = 1 - \frac{Ei(-p^* R_x^2) - Ei(-p^*)}{2 \ln R_x}, \tag{5}$$

Where;

$$p^* = \frac{1}{4n^* t_{xD}}, \quad n^* = \frac{t_s^*}{t_c^*}. \tag{6}$$

By using measurements T_{s2}, T_{s3} and equation (5) we can eliminate the formation temperature T_f. After simple transformations we obtain (Kutasov and Eppelbaum, 2003)

$$\gamma = \frac{T_{s2} - T_{s1}}{T_{s3} - T_{s1}} = \frac{Ei(-p_1^* R_x^2) - Ei(-p_1^*)}{Ei(-p_2^* R_x^2) - Ei(-p_2^*)}, \tag{7}$$

Where;

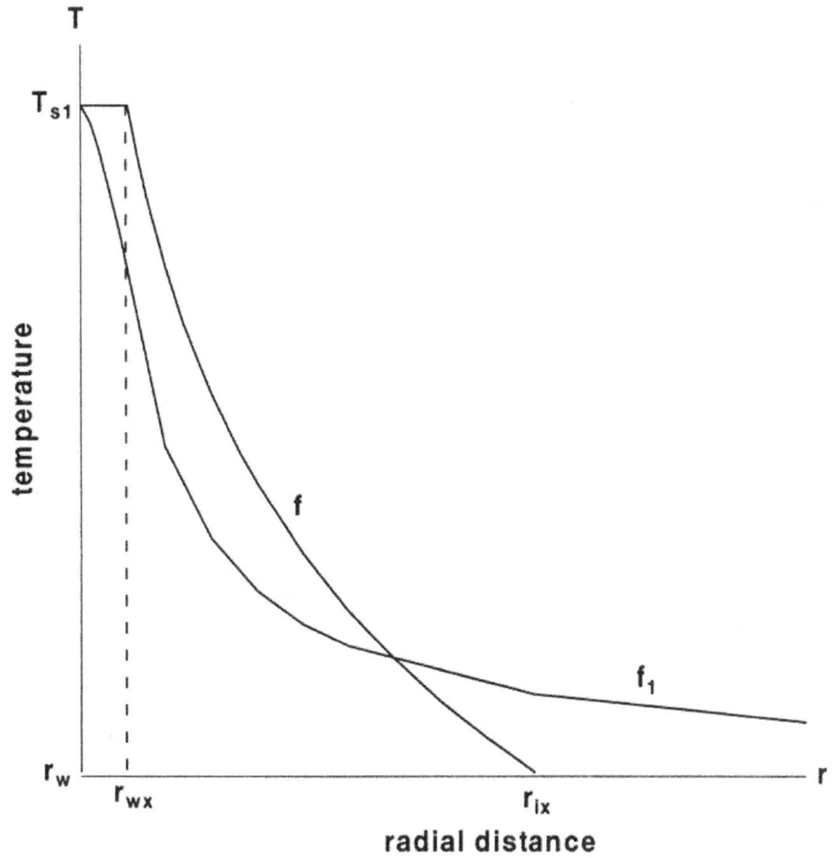

Figure 1. Actual (curve f_1) and assumed (curve f) radial temperature distributions at $t_s = t_{s1}$ – schematic curves.

$$p_1^* = \frac{1}{4 n_1^* t_{xD}}, \qquad n_1^* = \frac{t_{s1}^*}{t_c^*}, \qquad p_2^* = \frac{1}{4 n_2^* t_{xD}}, \qquad n_2^* = \frac{t_{s2}^*}{t_c^*}. \tag{8}$$

Substituting the value of R_x Formula (3) into Equation (7) we can obtain a formula for calculating the dimensionless disturbance time, t_{xD}, after this is possible to determine values of T_f, R_x, and $A = a_f / (r_{wx})^2$. Although the equation (7) is based on an analytical solution (Equation 6), we should mention several limitations in application of the suggested method. Firstly, the temperature ratio γ (Formula 7) should be determined with a high accuracy. This means that high accuracy of temperature measurements (T_{s1}, T_{s2}, T_{s3}) is needed. The temperature differences T_{s2} - T_{s1} and T_{s3} - T_{s1} should be significantly larger than the absolute accuracy of temperature measurements.

GENERALIZED HORNER METHOD (GHM)

Field investigations have shown that the bottom-hole circulating (without penetration) fluid temperature after some stabilization time can be considered constant (Figures 2 and 3). The solid curves in Figure 2 illustrate the calculated circulating mud temperatures (at a constant heat transfer coefficient) by using the Raymond (1969) model.

It was shown that by using the adjusted circulation time concept (Kutasov, 1987, 1989) a well with a constant borehole wall temperature can be substituted by a cylindrical source with a constant heat flow rate. Let us assume that at a given depth the fluid circulation started at the moment of time t = 0 and stopped at t = t_c. The corresponding values of the flow rates are

$$q(t = 0) = \infty, \qquad q(t = t_c) = q.$$

Using the adjusted circulation time concept and the principle of superposition for a well as a cylindrical source with a constant heat flow rate q = $q(t_c)$ which operates during the time t = $G{\cdot}t_c$ and shut-in thereafter, we obtained a working formula for processing field data (Kutasov and Eppelbaum, 2005):

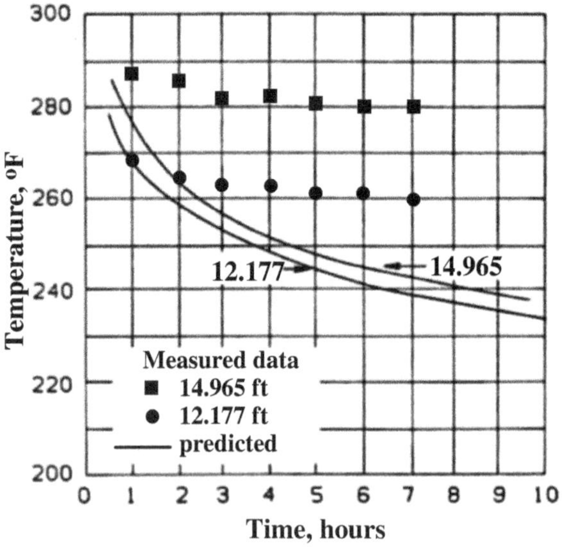

Figure 2. Comparison of measured and predicted circulating mud temperatures, Well 1 (Sump and Williams, 1973).

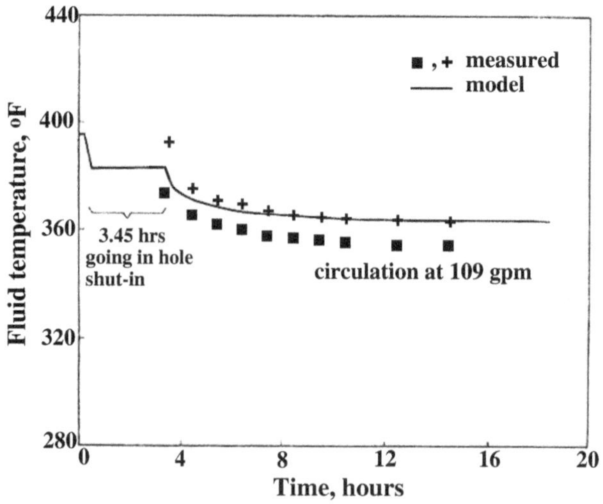

Figure 3. Circulating mud temperature at 23,669 ft (7214 m) – Mississippi well (Wooley et al., 1984).
Courtesy of Society of Petroleum Engineers.

$$T(r_w, t_s) = T_i + m \ln X, \quad m = \frac{q}{2\pi\lambda}, \tag{9}$$

$$X = \frac{1 + \left(c - \dfrac{1}{a + \sqrt{Gt_{cD} + t_{sD}}}\right)\sqrt{Gt_{cD} + t_{sD}}}{1 + \left(c - \dfrac{1}{a + \sqrt{t_{sD}}}\right)\sqrt{t_{sD}}}, \tag{10}$$

$$a = 2.7010505, \qquad c = 1.4986055,$$

$$\left\{\begin{array}{ll} G = 1 + \dfrac{1}{1 + AF}, & t_{cD} \le 10 \\ F = [\ln(1 + t_{cD})]^n, & n = 2/3 \quad A = 7/8 \end{array}\right\}, \tag{11}$$

$$G = \frac{\ln t_{cD} - \exp\left(-0.236\sqrt{t_{cD}}\right)}{\ln t_{cD} - 1}, \qquad t_{cD} > 10. \tag{12}$$

The correlation coefficient $G(t_{cD})$ varies in the narrow limits: $G(0) = 2$ and $G(\infty) = 1$ As can be seen from equation (9), the field data processing (semilog linear log) is similar to that of the Horner method. For this reason, we have given the name "Generalized Horner Method" to this procedure for determining the static temperature of formations. To calculate the ratio X, the thermal diffusivity of formations (a) should be determined with a reasonable accuracy. An example showing the effect of variation of this parameter on the accuracy of determining undisturbed formation temperature was presented in (Kutasov and Eppelbaum, 2005). It is easy to see that for large values of t_{cD} ($G \to 1$) and t_{sD} we obtain the well-known Horner equation (13).

$$T_s(r_w, t_s) = T_i + M \ln\left(\frac{t_s + t_c}{t_s}\right), \qquad M = \frac{m}{2} = \frac{q}{4\pi\lambda}. \tag{13}$$

FIELD EXAMPLES AND SYNTHETIC EXAMPLE

Field examples

We utilized the temperature data from Wells HL31 (depth 1,780 m) and HL30 (depth 1,394 m) (Andaverde et al., 2005). The third example is from Kelley Hot Springs geothermal reservoir, Moduc County, California (depth 1,035 m) (Roux et al., 1980).

Synthetic example

This example was taken from (Cao et al., 1980). We used in our calculations 8 points (from 15) and the assumed formation temperature was 120°C. The input data for 3 field cases and one synthetic example are presented in Tables 1 and 2.

Results of computations

Results of determinations of formation temperatures and some parameters by using the new method (equation (7)) are presented in Table 3. For comparison, we conducted calculations after formula (9) (Generalized Horner Method). The results of calculations are presented in Table 4. Comparing computed values of T_f (Table 5)

Table 1. Input data – Field cases.

Run number	t_{s1}, (h)	t_{s2}, (h)	t_{s3}, (h)	T_{s1}, (°C)	T_{s2}, (°C)	T_{s3}, (°C)
	Andaverde et al. (2005), HI31, t_c = 2.5 h					
1	6	12	18	186.3	206.9	219.0
2	6	12	24	186.3	206.9	231.9
3	6	12	30	186.3	206.9	239.8
4	6	12	36	186.3	206.9	247.7
5	12	18	24	206.9	219.0	231.9
6	12	18	30	206.9	219.0	239.8
7	12	18	36	206.9	219.0	247.7
8	18	24	30	219.0	231.9	239.8
9	18	24	36	219.0	231.9	247.7
10	24	30.	36	231.9	239.8	247.7
	Andaverde et al. (2005), HI30, t_c = 2.5 h					
11	6	12	18	178.6	198.9	211.4
12	6	12	24	178.6	198.9	225.8
13	6	12	30	178.6	198.9	235.1
14	6	12	36	178.6	198.9	240.4
15	6	12	42	178.6	198.9	247.1
16	12	18	24	198.9	211.4	225.8
17	12	18	30	198.9	211.4	235.1
18	12	18	36	198.9	211.4	240.4
19	12	18	42	198.9	211.4	247.1
20	18	24	30	211.4	225.8	235.1
21	18	24	36	211.4	225.8	240.4
22	18	24	42	211.4	225.8	247.1
23	24	30	36	225.8	235.1	240.4
24	24	30	42	225.8	235.1	247
	Roax et al. (1988), t_c = 12.0 h					
25	14.3	22.3	29.3	83.9	90.0	94.4

Table 2. Input data – synthetic example.

Run number	t_{s1}, (h)	T_{S2}, (h)	t_{s3}, (h)	T_{s1}, (°C)	T_{s2}, (°C)	T_{s3}, (°C)
	Cao et al. (1980), t_c = 5.0 h					
26	2	6	10	91.7	102.4	107.9
27	2	6	14	91.7	102.4	111.3
28	2	6	18	91.7	102.4	113.6
29	2	6	22	91.7	102.4	115.2
30	2	6	30	91.7	102.4	117.1
31	2	6	50	91.7	102.4	119.1
32	6	10	14	102.4	107.9	111.3
33	6	10	18	102.4	107.9	113.6
34	6	10	22	102.4	107.9	115.2
35	6	10	30	102.4	107.9	117.1
36	6	10	50	102.4	107.9	119.1
37	10	14	18	107.9	111.3	113.6
38	10	14	22	107.9	111.3	115.2
39	10	14	30	107.9	111.3	117.1
40	10	14	50	107.9	111.3	119.1

Table 2. Contd.

41	14	18	22	111.3	113.6	115.2
42	14	18	30	111.3	113.6	117.1
43	14	18	50	111.3	113.6	119.1
44	18	22	30	113.6	115.2	117.1
45	18	22	50	113.6	115.2	119.1
46	22	30	50	115.2	117.1	119.1

Table 3. Results of computations after Equations (3), (5) and (7)

t_{xD}	A,(1/h)	R_x	T_f, (°C)	t_{xD}	A (1/hr)	R_x	T_f, (°C)
Andaverde et al. (2005), HI31, t_c = 2.5 h				**Roax et al. (1988), t_c = 12.0 h**			
1.07	0.125	3.26	245.3	1.80	0.068	3.93	111.0
0.73	0.086	2.86	261.9	**Cao et al. (1980), t_c = 5.0 h**			
0.68	0.079	2.79	266.5	1.71	0.245	3.86	120.0
0.61	0.072	2.71	273.2	1.57	0.225	3.74	121.1
0.90	0.062	3.07	276.1	1.50	0.214	3.67	121.9
0.92	0.064	3.10	274.0	1.46	0.209	3.64	122.3
0.85	0.059	3.02	279.7	1.44	0.206	3.62	122.5
2.81	0.137	4.66	263.3	1.45	0.207	3.63	122.4
2.05	0.100	4.13	273.2	2.18	0.198	4.22	120.7
1.82	0.069	3.94	280.4	2.01	0.183	4.10	121.5
				1.94	0.176	4.04	122.0
Andaverde et al. (2005), HI30, t_c = 2.5 h				1.90	0.173	4.01	122.2
1.00	0.117	3.18	239.1	1.91	0.173	4.02	122.2
0.66	0.077	2.77	259.3	2.65	0.177	4.56	121.0
0.60	0.071	2.69	265.8	2.48	0.166	4.44	121.6
0.59	0.070	2.68	266.7	2.41	0.161	4.39	121.8
0.56	0.065	2.63	271.9	2.40	0.160	4.38	121.9
0.82	0.056	2.97	277.4	3.25	0.171	4.93	120.9
0.83	0.057	2.99	275.8	3.03	0.160	4.80	121.4
0.87	0.060	3.04	272.4	2.93	0.154	4.74	121.6
0.82	0.057	2.98	277.0	3.74	0.162	5.22	120.9
2.58	0.126	4.51	263.4	3.40	0.148	5.03	121.4
2.67	0.130	4.57	262.4	3.02	0.112	4.79	121.4
2.17	0.106	4.22	269.5				
4.16	0.157	5.45	257.4				
2.53	0.095	4.47	269.1				

shows that the suggested method provides determining the static formation temperatures with sufficient accuracy. As follows from these tables, in some cases the static formation temperatures estimated by two presented methods are very close.

The authors recommend the use of the proposed method when the temperature of the drilling fluid is not constant and stops in the mud circulation are documented. The authors will be glad to test this methodology at new results of thermal measurements in boreholes with the aim to obtain the reliable statistical data.

Conclusion

It is shown that the earlier proposed method for prediction of formation temperatures in permafrost regions from temperature logs in deep wells (Kutasov and Eppelbaum, 2003) can be applied to process results of bottom-hole temperature surveys in oil and gas exploration. Three field cases and one synthetic example are used to show the validity of this approach. The results of calculations are favourable compared with those obtained after the Generalized Horner Method (GHM). The authors hope to

Table 4. Results of determining the formation temperature after Formula (9). *R* is the squared averaged temperature deviation.

t_s (h)	T_s (°C)	T_s^* (°C)	ΔT (°C)	t_s (h)	T_s °C	T_s^* (°C)	ΔT (°C)
Andaverde et al. (2005), Hl31, T_f = 262.9 °C, t_c = 2.5 h, r_w = 0.106 m, a = 0.0036m²/h, R = 4.3 °C				**Roax et al. (1980), T_f = 111.3 °C, t_c = 12.0 h, r_w = 0.100 m, a = 0.0027 m²/h, R = 0.4 °C**			
6	183.6	179.8	3.8				
12	206.9	211.3	-4.4	14.3	83.9	83.7	0.2
18	219.0	225.1	-6.1	22.3	90.0	90.5	-0.5
24	231.9	233.0	-1.1	29.3	94.4	94.1	0.3
30	239.8	238.1	1.7				
36	247.7	241.7	6.0				
Andaverde et al. (2005), Hl30, T_f = 259.6 °C, t_c = 2.5 hr, r_w = 0.106 m, a = 0.0036m²/h, R = 4.3 °C				**Cao et al. (1988), T_f = 122.5 °C, t_c = 5.0 hr, r_w = 0.108 m, a = 0.0054 m²/h, R = 0.7 °C**			
6	178.6	172.5	6.1	2	91.7	90.7	1.0
2	198.9	205.5	-6.6	6	102.4	103.7	-1.3
8	211.4	220.0	-8.6	10	107.9	108.8	-0.9
4	225.8	228.2	-2.4	14	111.3	111.7	-0.4
30	235.1	233.6	1.5	18	113.6	113.5	0.1
36	240.4	237.3	3.1	22	115.2	114.8	0.4
42	247.1	240.1	7.0	30	117.1	116.5	0.6
				50	119.1	118.6	0.5

Table 5. Results of determining the formation temperature after Formulae (7) and (9). *R* is the squared averaged temperature deviation.

Case, Reference	Points	t_c (h)	t_s (h)	$T_f \pm R$, °C	
				NM	GHM
HL31	6	2.5	6 - 36	269.4 ± 10.1	262.9 ± 4.3
HL30	7	2.5	6 - 42	266.2 ± 9.7	259.6 ± 4.3
Roax et al. (1980	3	12.0	14.3 0- 29.3	111.0	111.3 ± 0.3
Cao et al. (1988)	8	5.0	2 – 50	121.6 ± 0.6	122.5 ± 0.4

apply this method on a set of thermal borehole data measurements with the aim to obtain the reliable statistics.

ACKNOWLEDGEMENTS

We want to thank Dr. Edgar R. Santoyo Gutierrez for providing us with some data related to Wells HL31 and HL30. The authors would like to thank two anonymous reviewers, who thoroughly reviewed the manuscript, and their critical comments and valuable suggestions were very helpful in preparing this paper.

REFERENCES

Andaverde J, Verma SP, Santoyo E (2005). Uncertainty estimates of static formation temperatures in boreholes and evaluation of regression models. Geophys. J. Int. 160: 1112-1122.

Cao S, Lerche I, Hermanrud C (1988). Formation temperature estimation by inversion of borehole measurements. Geophys. 53: 979-988.

Deming D, Chapman DS (1988). Heat flow in the Utah-Wyoming thrust belt from analysis of bottom-hole temperature data measured in oil and gas wells. J. Geophys. Res. 93: 13657-13672.

Eppelbaum LV, Modelevsky MM Jr., Pilchin AN (1996). Thermal investigation in petroleum geology: the experience of implication in the Dead Sea Rift zone, Israel. J. Petroleum Geol. 19(4): 425-444.

Espinosa-Paredesa G, Morales-Di´az A, Olea-Gonza´ lez U, Ambriz-Garcia JJ (2009). Application of a proportional-integral control for the estimation of static formation temperatures in oil wells. Mar. Petroleum Geol. 26: 259-268.

Förster A (2001) Analysis of borehole temperature data in the Northeast German Basin: continuous logs versus bottom-hole temperatures. Petroleum Geosci. 7(3): 241-254.

Kritikos WP, Kutasov IM (1988). Two-Point method for determination of undisturbed reservoir temperature. Form. Eval. 3(1): 222-226.

Kutasov IM (1968). Determination of Time Required for the Attainment of Temperature Equilibrium and Geothermal Gradient in Deep Boreholes. Freiberger Forshungshefte C238: 55-61

Kutasov IM (1987). Dimensionless temperature, cumulative heat flow and heat flow rate for a well with a constant bore-face temperature. Geothermics 16(2): 467-472.

Kutasov IM (1989). Application of the Horner method for a well produced at a constant bottomhole pressure. Form. Eval. 3: 90-92

Kutasov IM (1999). Applied Geothermics for Petroleum Engineers. Elsevier, Amsterdam.

Kutasov IM, Eppelbaum LV (2003). Prediction of formation temperatures in permafrost regions from temperature logs in deep wells –

field cases. Permafrost Periglacial Process.14: 247-258.
Kutasov IM, Eppelbaum LV (2005). Determination of formation temperature from bottom-hole temperature logs – a generalized Horner method. J. Geophys. Eng. 2: 90-96.
Mcaleese S (2000). Operational Aspects of Oil and Gas Well Testing (Handbook of Petroleum Exploration and Production), Elsevier, Amsterdam.
Melton CE, Giardini AA (1984). Petroleum formation and the thermal history of the earth's surface. J. Petroleum Geol. 7, No. 3: 303-312.
Nielsen SB, Balling N, Christiansen HS (1990) Formation temperatures determined from stochastic inversion of borehole observations. Geophys. J. Int. 101: 581-590.
Pasquale P, Chiozzi, Gola G, Verdoya M (2008). Depth-time correction of petroleum bottom-hole temperatures in the Po Plain, Italy. Geophys. 73(6): E187-E196.
Raymond LR (1969). Temperature Distribution in a Circulating Drilling Fluid. J. Petroleum Technol. 21: 333-341.
Roux B, Sanyal SK, Brown SL (1980). An Improved Approach to Estimating True Reservoir Temperature from Transient Temperature Data. SPE Paper 8888 Presented at the 1980 SPE California Regional Meeting, Los Angeles, 9-11 April.
Sump GD, Williams BB (1973). Prediction of Wellbore Temperatures During Mud Circulation and Cementing Operations. J. Eng. Ind. 11: 1083-1092.
Verma SP, Andaverde J, Santoyo E (2006). Statistical evaluation of methods for the calculation of static formation temperatures in geothermal and oil wells using an extension of the error propagation theory. J. Geochem. Explor. 89, Nos. 1-3: 398-404.
Waples DW, Pachco J, Vera A (2004). A method for correcting log-derived temperatures deep wells calibrated in the Gulf of Mexico. Petroleum Geosci. 10: 239-245.
Wooley GR, Giussani AP, Galate JW, Wederlich HF (III) (1984). Cementing Temperatures for Deep-Well Production Liners. SPE paper 13046 presented at the 59th Annual Technical Conference and Exhibition, Houston, Texas, 16-19 September.
Zschocke A (2005). Correction of non-equilibrated temperature logs and implications for geothermal investigations. J. Geophys. Eng. 2: 364-371.

2

A proactive approach for predicting and preventing wax deposition in production tubing strings: A Niger Delta experience

A. A. Sulaimon[1]*, G. K. Falade[1] and W. DeLandro[2]

[1]Department of Petroleum Engineering, University of Ibadan, Oyo State, Nigeria.
[2]Shell Petroleum Development Company, Nigeria.

A proactive approach for managing wax deposition problems during production operation is presented. A predictive thermodynamic model is employed to determine the onset condition(s) for wax precipitation in hydrocarbon mixtures. The parameters of a continuous distribution function are used to establish the nature of hydrocarbon mixtures while a fluid temperature distribution model enables the determination of the potential wax precipitating point in the production tubing. A field case study was conducted on 17 reservoirs in a field located in (Niger Delta, Nigeria). Results show that the thermodynamic model predicted Wax Appearance Temperature (WAT) decreases with reservoir depth. The WAT ranged between 21 and 46°C from the deepest (8,000 ft) to the shallowest (5,000 ft) reservoir. Field observation shows that wax formation is effectively prevented when the Wax-Inhibiting-Tool (WIT) is installed 100 - 500 ft below the tubing depth corresponding to the predicted WAT. Proper placement of the WIT has substantially reduced wax precipitation and potential problems that might have occurred due to wax deposition.

Key words: Thermodynamic model, wax appearance temperature, fluid temperature distribution, wax inhibiting tool, gamma distribution parameters.

INTRODUCTION

The identification, diagnosis, and alleviation of the problems caused by heavy organic compounds deposition are as old as the oil industry. The combined mass of major petroleum deposits are generally referred to as wax (Leontaritis, 2007) and past efforts by operators were limited to reactive or curative measures rather than preventive approaches to solving depositional problems. However, proper identification and characterization as well as accurate description of the behaviour of various deposits are essential for appropriate field development planning and design of effective mitigation strategies.

Current research efforts are geared towards developing strategies to predict, prevent and/or mitigate the formation of heavy organic compounds during petroleum production and processing (Baker, 2003; Weispfennig, 2006; Dalirsefat and Feyzi, 2007). The most effective proactive approach to forestall heavy organics' precipitation or flocculation is to develop efficient predictive tools or mathematical models that can predict the onset conditions for organic deposition. It is equally important to devise a means of establishing the location of the precipitation point or nucleation sites within the production and processing system using an appropriate fluid temperature distribution model (Sagar et al, 1991; Hassan and Kabir, 1994).

Studies (Erickson et al., 1993; Hammami and Raines, 1999) have shown that the wax appearance temperature (WAT) measured by cross polar microscopy (CPM) compare favourably with field measured temperatures at which wax deposition starts in the corresponding well. Similarly therefore, a good thermodynamic model's WAT prediction should agree fairly well with the field observations.

FIELD CASE STUDY

A major oil producing company in Nigeria's Niger Delta (ND) was faced with the challenges of managing wax

*Corresponding author. E-mail: aa.sulaimon@mail.ui.edu.ng, alisulal1@yahoo.com.

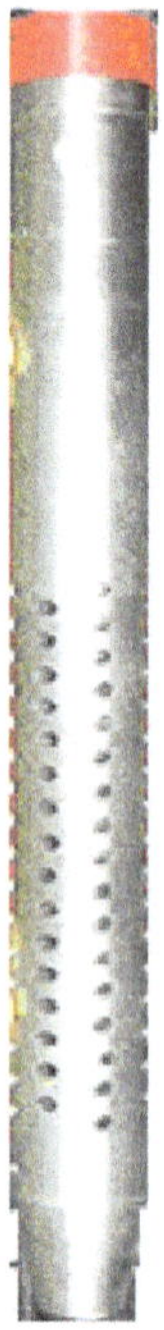

Figure 1. Wax inhibiting tool (WIT).

Table 1. ND field waxy strings.

Reservoir	Well(S)
D1.000X	Well 32S Well 5S Well 33S – Wax problems in flow-line
D2.000X	Well 12S Well 23L Well 25L Well 33L Well 37L Well 4T Well 51L Well 5L Well 28L – Closed in due to wax problem
D3.000C	Well 59T – Monthly pigging
D3.000X	Well 12L
D6.000N	Well 30S

deposition problems in its operations. A WIT (Figure 1) was designed to prevent or reduce wax deposition inside the production tubing or along the flow line. The equipment was tested in some wax-prone wells but with mixed success. It was subsequently discovered that the tool works better if installed very close to the wax nucleation point. Therefore, there was a need for an approach, which could predict the wax precipitation point in the production tubing, and a clear definition of the WIT depth of installation.

Challenges in ND field

ND field has a history of wax deposition problems. Fifteen wells producing from D1.000X, D2.000X, D3.000C, D3.000X, and D6.000N reservoir sands were identified as waxy wells but the wax problem is most severe in Wells 33S, 28L and 59T (Table 1). Well 28L was closed in due to wax problem, while Well 59T undergoes monthly pigging also as a result of wax deposition. In 2007, 1500BOPD was deferred due to wells closed in as a result of wax problems.

A number of reactive measures have been adopted in the past to resolve the problems caused by wax deposition in ND field. Some of these methods included pigging, solvent soaking, and wax cutting. These techniques had yielded little positive results as their application was characterized by sudden increase in production with subsequent gradual production loss. This called for a more effective mitigation strategy.

Figure 2 shows the high frequency of well interventions occasioned by wax cutting exercises followed by increase in production and subsequent gradual rate decline for a well in a different AG field. The periodic wax cutting operations was maintained throughout 1997 -1999 but could not be sustained from 2000 due to increase in host community hostilities. The WIT was arbitrarily installed (at any convenient point) after wax cutting and tubing wash in December 2003 while production had averaged 600 BOPD. Although, the well has been producing consistently with little decline, production could have been significantly improved if the WAT and the corresponding nucleation point had been predicted before setting the WIT at the most appropriate point. This is because the WIT performs more efficiently when set close to the predicted wax nucleation point.

In this work, we present an approach based on the application of three-phase thermodynamic model (Won, 1985; Sulaimon, 2009) and fluid temperature model (Sagar et al, 1991; Hassan and Kabir, 1994; Sulaimon, 2009) to respectively predict the WAT and generate fluid temperature profiles inside the production tubing string. The objective is to locate potential wax nucleation points inside the tubing, where a WIT can be installed to prevent wax formation.

Thermodynamic model

When vapour, liquid and solid phases coexist at equilibrium, the following thermodynamic equilibrium

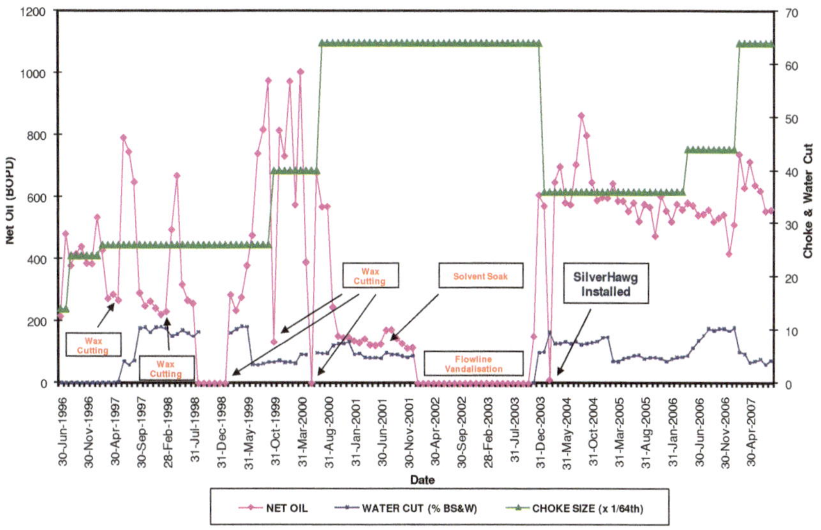

Figure 2. Typical intervention for well 39L in AG field.

condition must be satisfied:

$$f_i^V = f_i^L \quad (1)$$

$$f_n^V = f_n^L = f_n^S \quad (2)$$

Where

f_i = fugacity of component i
n = number of components in the mixture
V, L, S = Vapour, Liquid, Solid phases

A detailed description of the three phase thermodynamic model has been presented (Won, 1985; Sulaimon, 2009). However, Sulaimon (2009) has recently developed a new set of thermodynamic properties' correlations that afford better comprehensive estimation of the solid-liquid equilibrium constant. The thermodynamic model has been validated (Sulaimon, 2009) with experimental data from 32 major oil fields with an average absolute deviation (AAD) of 0.75%. This makes it a reliable tool for predicting the onset of wax precipitation in the Niger Delta.

Fluid temperature distribution model

Fluid and wellbore temperature profiles for a naturally flowing or gas lifted well are determined by modifying the Temperature Distribution Model (TDM) developed by Sagar et al. (1991). The Ramey's time function (Ramey, 1962) used in the model was replaced by the algebraic approximation developed by Hassan and Kabir (1994) which offers better dimensionless time solution.

MODEL APPLICATION PROCEDURES

A computer package which combines both the thermodynamic model and temperature distribution model (TDM) was used to predict the onset conditions and generate temperature profiles inside the wellbore respectively. The following are the steps involve in running the computer program:

1. Feed in the input parameters (hydrocarbon components' mole fractions and molecular weights) for the three-phase thermodynamic model.
2. The temperature value that corresponds to the least amount of 'precipitated wax (%)' is the cloud point or Wax Appearance Temperature (WAT).
3. Initialize the Temperature Distribution Model (TDM) subroutine using production and well parameters.
4. Generate earth (formation) temperatures and fluid's temperatures at different depth at specified regular intervals.
5. Create the temperature distribution profiles by plotting 'Depth (ft)' against 'Temperature (°F)'; trace the WAT from the horizontal axis to the fluid temperature profile and find the corresponding precipitation depth on the vertical axis.
6. From the well status diagram, establish the nearest possible point that is closest to the estimated precipitation depth for hanging the WIT.

Data gathering

To facilitate a comprehensive study and analysis of the problem of wax deposition potential in ND field, it is imperative to uniformly select representative candidate wells (identified waxy and non-waxy wells) from every reservoir in the field. This would afford a means of generating a broad database from the field and laboratory investigations of sampled wells. It is assumed that the initial fluid composition is the same at every drainage points (wells) within a given reservoir.

Originally, a well was selected from each of the 28 hydrocarbon-bearing blocks found in 21 reservoir sands. Table 2 shows the reservoir sands, blocks and all wells in the field from which the 29 candidates were chosen. The compositional data obtained from 17 wells (in bold letters) producing from different blocks were used as input data for the thermodynamic model.

Crude oil samples were collected from the 17 of the 29 selected candidate wells and analysed. The results of the laboratory analyses are shown in Table 3. Pour point is an important parameter in flow assurance studies and can provide an indication for partial or total plugging of flowlines. Therefore, due to their relatively high pour point values, the most susceptible wells to wax deposition problems are those producing from D4.000X, D8.000N and E2.000X reservoir sands. The observed variations in the measured pour points (Table 3) for wells producing from the same reservoir (D1.000X, D2.000X and D3.000X) are due to changes in fluid compositions and differences in well depth, bottom-hole temperatures and pressures.

DISCUSSION OF RESULTS

WAT predictions for ND oils

The validated thermodynamic model (Sulaimon, 2009) was deployed to study the wax deposition tendencies of wells in the ND field. Table 4 shows the WAT predictions for the 17 well fluids' samples collected from different reservoirs. The depths at the Oil-Water-contact (OWC) or Oil-Down-To (ODT) were used as the reference point.

Study results have revealed that wells producing from D8.000A, D9.000A, E2.000A, F2.700N, F4.100 and F4.200 reservoir sands are less likely to have wax deposition problems. This is because there were no wax precipitations at all temperatures and pressures after several computer runs. Nevertheless, similar computer runs performed using other reservoirs' fluids' analytical data (D2.000N, D2.000X, D4.000X, D3.000X, D3.000C, D4.000P, D6.000N, D6.000A, D9.000N, E3.800A, and E5.000A) show that they are all prone to wax deposition but at different WATs.

Fluid temperature distributions

To determine the point or depth where precipitation would start in the production tubing strings, the fluid temperature model was run for the 11 reservoirs' fluids identified to have high potential for wax deposition in the ND field. Analysis of the temperature profiles revealed that wax precipitation could occur in five conduits (Table 5). These include Well 6S (D2.000X), Well 19L (D4.000X), Well 25S (D2.000N), Well 57S (D4.000P), and Well 59T (D3.000C). The remaining six wax prone wells: Well 19L (D3.000X), Well 30S (D6.000N), Well 6L (D6.000A), Well 20L (D9.000N), Well 53L (E3.800A) and Well 45L (E5.000A), would most likely experience wax deposition in the surface equipment as also shown in Table 5.

Wax deposition tendency

D2.000N reservoir fluid

The D2.000N reservoir sand is the shallowest of all the reservoirs under study and Well 25S with completion interval within 5,814 - 5,823 ft (1,772.1 - 1,774.9 m) was used for analysis. Modelling results showed that wax precipitation would occur when the fluid's temperature drops below the predicted WAT of 107.6°F (42°C). This corresponds to a tubing depth of 1,490 ft (454.2 m) as illustrated on the temperature distribution profile for Well 25S (Figure 3). Thus, it suffices to install the wax-inhibiting tool (WIT) below this depth to prevent precipitation inside the production tubing. Therefore, a tubing depth of 2,000 ft (609.6 m) would be appropriate.

D2.000X reservoir fluid

The D2.000X reservoir is characterized by complex geologic features and has the highest concentration of wells (44 wells) in ND field. The fluid's compositional data for Well 6S was used as input data for the model runs. The well was completed with perforations through 7" casing with interval 5,906.6 - 5,912.5 ft (1,800.8 - 1,802.6 m) open to production.

Model predictions have indicated that the WAT is 111.2°F (44°C) and this corresponds to a tubing depth of 2,000 ft (609.6 m) as highlighted on the temperature profiles in production tubing (Figure 4). A placement depth of 2,500 ft (762.0 m) is recommended on the temperature profile.

It was observed that well 6S has been closed-in since April 1, 1999, due to wax problems as depicted by the sudden increase in the tubing head pressure from 280 - 450 psi with subsequent decline in oil production from 4,988 - 203 bbls (Figure 5). Therefore, it is necessary to identify the corresponding wax precipitation depths in all producing wells in D2.000X reservoir to ascertain probable wax nucleation points using the predicted WAT as a benchmark. Thereafter, the well status diagrams can be used to locate the most practicable depth for the WIT installation.

Table 2. Selection of candidate wells.

S/N	Reservoir	Block	All wells	Selected wells
1	D1.000	X	5S, 10S, 20S, 21S, 23S, 25S, 28S, 29S, 32S, 33S, 37S, 43T, 51S, 55S	25S
		N	25S	
2	D2.000	X	1L, 1T, 3L, 4T, 5L, 5S, 6S, 9S, 10L, 11T, 12S, 16T, 18S, 19S, 21C, 21L, 22S, 23L, 23S, 24S, 25L, 26S, 28L, 29L, 30L, 32L, 33L, 37L, 39C, 39T, 40T, 43T, 48S, 49L, 50L, 51L, 54T, 55L, 58T, 60T, 61T, 63T, 64L, 64S	6S
		C	59T	59T
3	D3.000	X	5L, 12L, 13C, 13S, 19L(1), 21T, 22L, 23L, 25L, 26L, 48L, 65S, 66T	19L
		P	57S	57S
4	D4.000	X	19L(2), 31C, 36S, 38S	19L(2)
		N	20S	20S
5	D5.000X	X	13T, 20L, 26L, 36L, 38L, 57S	57S
		A	2C, 6L, 7C, 8S, 9L, 10L, 12L, 13L, 36L, 45S	6L
6	D6.000	N	30S	30S
		P	57L	57L
7	D7.000	A	14S	14S
		N	24L	24L
8	D8.000	A	7S, 56S	7S
		N	30L	30L
9	D9.000	A	7L, 53S	7L
		N	20L, 44S	20L
11	E2.000	A	53L	53L
		X	18L, 20L, 24L, 34S, 65L	65L
12	E3.000	A	18L, 56L	56L
13	E3.800	A	31S	31S
14	E4.000	A	31L	31L
15	E5.000	A	45L	45L
16	E6.000	N	30S, 34L	30S
17	E9.400	N	14L	14L(1)
18	F2.000	N	31L	31L
19	F2.700	N	14L, 31L	14L(2)
20	F4.100	N	47S	47S
21	F4.200	N	47L	47L

Table 3. Laboratory results.

Sand	Well	TAN (MgKOH/g)	Dry S.G @ 15/15°C	API gravity	Kinematic viscosity @ 40°C	Pour point	
						°C	°F
D1.000 X	5S	0.4	0.9182	22.6	24.7	-7	19.4
D2.000 X	1T	0.3	0.9224	21.9	27.3	-3.9	24.98
	4T	0.6	0.9248	21.5	28.9	-1.1	30.02
	12S	0.3	0.9564	16.5	26.1	-7	19.4
	18S	0.5	0.9463	18	26.2	-7	19.4
	43T	0.3	0.9216	22	22.6	-1.1	30.02

Table 3. Contd.

D3.000 X	12L	0.3	0.8904	27.4	7.7	-1.1	30.02
	22L	0.2	0.9058	24.7	13.4	-7	19.4
D3.000 C	59T	0.2	0.8834	28.7	10.5	1.7	35.06
D4.000 X	36S	0.2	0.8696	31.2	22.3	4.4	39.92
D8.000 N	30L	0.3	0.8571	33.6	4.3	4.4	39.92
E2.000 X	65L	0.1	0.8267	39.7	2.8	4.4	39.92
E3.000 A	56L	0.1	0.8329	38.4	3.2	-1.1	30.02
E3.800 A	31S	0.1	0.827	39.6	3.2	1.7	35.06

Table 4. ND field WAT predictions.

S/N	Well	Sand	OWC/ODT (ft-ss)	Predicted WAT	
				(°F)	(°C)
1	25S	D2.000N	5814	107.6	42.0
2	06S	D2.000X	5900	111.2	44.0
3	19L2	D4.000X	6185	114.8	46.0
4	19L1	D3.000X	6190	91.4	33.0
5	59T	D3.000C	6254	114.8	46.0
6	57S	D4.000P	6420	98.6	37.0
7	30S	D6.000N	6457	102.2	39.0
8	06L	D6.000A	6560	89.6	32.0
9	07S	D8.000A	6720	No wax	No wax
10	20L	D9.000N	6723	100.4	38.0
11	07L	D9.000A	6780	No wax	No wax
12	53L	E2.000A	7132	No wax	No wax
13	31S	E3.800A	7657	69.8	21.0
14	45L	E5.000A	8073	80.6	27.0
15	31L	F2.700N	9619	No wax	No wax
16	47S	F4.100N	9811	No wax	No wax
17	47L	F4.200N	9851	No wax	No wax

Table 5. Characterization of hydrocarbon mixtures in ND field.

S/N	Sand	Well no.	API gravity	Gamma distribution parameters			Fluid type
				Alpha (α)	Beta (β)	Variance (η)	
1	D6.0A	6L	36.4	1.0061	103.97	10,874.80	Light oil/Condensate
2	D2.0X	6S	24.2	1.0028	207.43	43,144.53	Asphaltenic oil
3	D9.0A	7L	43.4	1.0268	28.42	829.14	Gas
4	D8.0A	7S	41.3	1.0143	29.83	902.37	Gas

Table 5. Contd.

5	D3.0X	19L1	24.7	1.0034	127.37	16,277.56	Waxy oil
6	D4.0X	19L2	34.2	1.0038	110.38	12,229.88	Light oil/Condensate
7	D9.0N	20L	41.6	1.0022	91.00	8,299.29	Waxy condensate
8	D2.0N	25S	24.9	1.0021	93.40	8,742.37	Waxy condensate
9	D6.0N	30S	36.4	1.0131	117.03	13,875.32	Waxy condensate
10	F2.7N	31L	40.3	1.0441	95.20	9,462.57	Light oil/Condensate
11	E3.8A	31S	39.6	1.0488	91.91	8,859.50	Waxy condensate

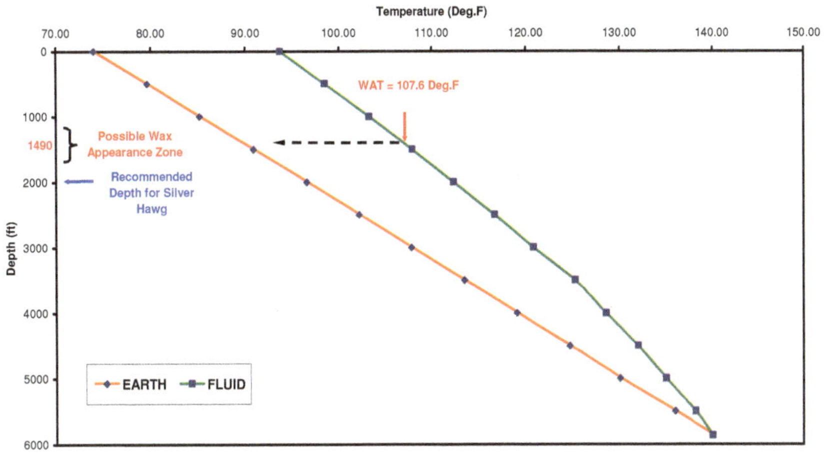

Figure 3. Temperature distribution profiles in production tubing for well 25S.

D3.000X reservoir fluid

There are a total number of 13 wells producing from D3.000X sand horizon but the analytical data for Well 19L1 were used as input data for analysis. The well was completed with 7" casing at 6,813 ft (2,076.6 m) b.d.f (below derrick floor) and interval 6,182 - 6,188 ft (1,884.3 - 1,886.1 m) open to production.

The predicted wax appearance temperature for this fluid is 91.4°F (33°C). This value does not correspond to any fluid temperature in the production tubing as shown in Figure 6. Thus, wax formation is unlikely inside the production tubing but precipitation may occur in the surface equipment (flowlines, separators, or storage tanks) when the fluid temperature drops below 91.4°F. Field observation has corroborated model prediction as no wax deposit has been found inside the tubing or surface flowline.

D3.000C reservoir fluid

The horizontal Well 59T is the only drainage well producing from D3.000C reservoir sand. The original reservoir static pressure was 2,675 psi while the initial bottom-hole temperature was 142°F. The well was completed in 1998 with 7" predrilled liner and 4½" tubing. Model predictions indicated that wax precipitation would start when the fluid temperature decreases from the initial bottom-hole temperature of 142°F to the predicted WAT, 114.8°F (46°C). This corresponds to a tubing depth of 835 ft (254.5 m) as illustrated on the temperature distributions plot (Figure 7). Allowing for a clearance or a depth interval of 500 ft (152.4 m), it suffices to install the wax-inhibiting tool at 1,300 ft (396.2 m). However, the closest profile for hanging the tool is the Baker sliding sleeve located at 6,008 ft (1,831.2 m). Therefore, it is recommended that the tool be set at this depth.

In 2002, the steel flowline was replaced with 'glass reinforced epoxy (GRE)' pipe to prevent wax formation in the pipeline and the WIT was installed inside the tubing at a depth of 6,008 ft (1,831.2 m). Production increased from 570 bbls/d in February 2002 to 3,155 bbls/d after the installation (Figure 8). A well entry with wireline in May 2003 indicated that the tubing was mostly wax-free but a3.74" gauge cutter had some wax cutting up to 700 ft (213.4 m). This is in line with the model prediction of 835 ft (254.5 m), which represents the depth at which the first wax crystal appears or separates from the crude. The

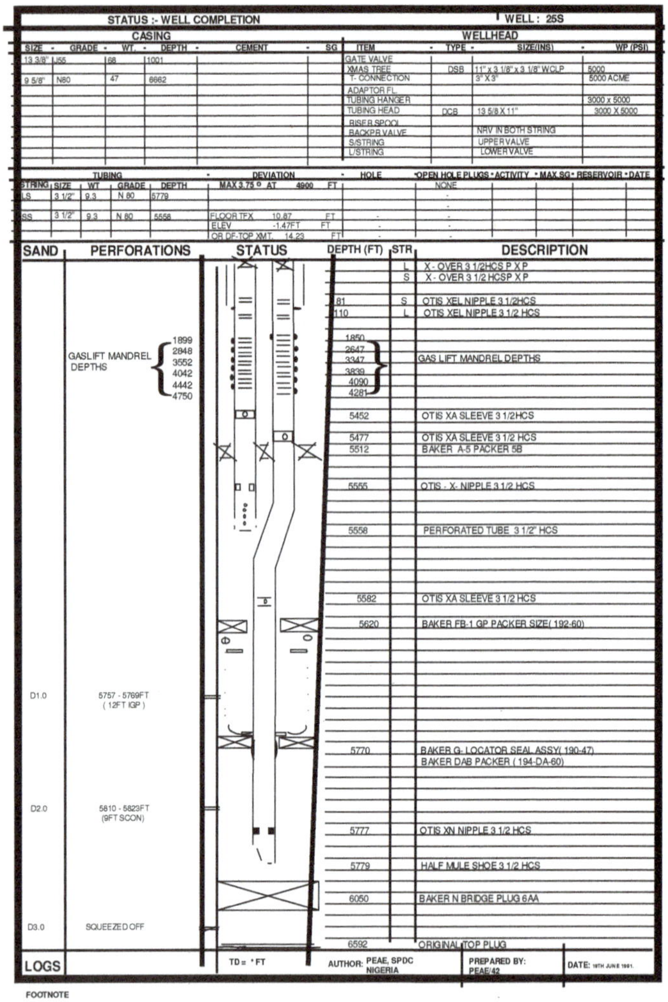

Figure 4. Completion diagram for well 25S.

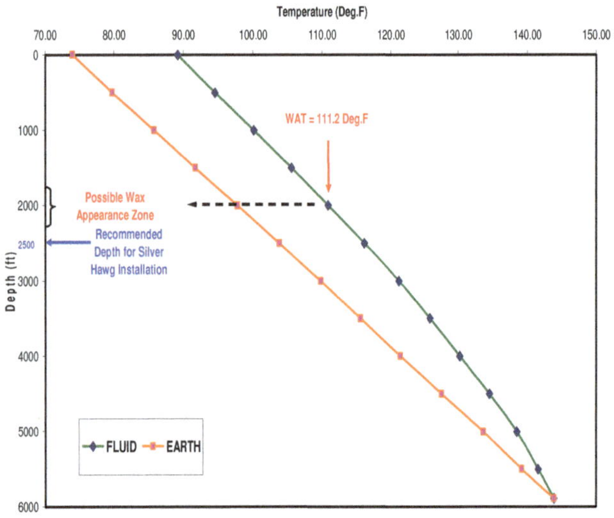

Figure 5. Temperature distribution profiles in production tubing for well 6S.

well was closed-in between August 2003 and October 2004 due to community problems.

Wax-free reservoir fluids

Model results have shown that some reservoir fluids in ND field are wax-free. The reservoirs containing wax-free crude oils are: D8.000A, D9.000A, E2.000A, F2.700N, F4.100N and F4.200N. Several model runs did not give any indication of possible wax formation at any pressure and temperature conditions. Field observation has shown that none of these reservoir fluids contain wax.

WAT variation with depth

Investigations have shown that shallow reservoir oils exhibit relatively higher WAT values than deeper reservoir fluids and thus, may experience more precipitations. This may be attributed to heat loss to the formation, loss of volatile components, or bacteria action (biodegradation). Although there is no direct relationship between 'Depth' and 'WAT', the trendline indicates that the WAT decreases with depth (Figure 9).

Conclusion

A proactive approach for dealing with wax deposition problems during petroleum production and processing operations is presented. The technique is based on the development of a robust computer algorithm based on the basic thermodynamic concepts to predict the onset conditions for wax precipitation, and application of a modified fluid temperature model to determine potential wax formation region within the production system.

The combined thermodynamic and fluid temperature distribution models can be used to reliably predict and locate potential wax precipitation point inside the production tubing. When appropriately installed within the predicted wax precipitation depth, the WIT can effectively prevent wax deposition. The ND field consists of more light/volatile oil and condensate reservoirs than black or heavy oil reservoirs. Moreover, the tendency of a reservoir fluid to precipitate wax decreases with depth.

RECOMMENDATIONS

Analysis to determine paraffin and asphaltene constituents in crude oil should be conducted during the early development of a reservoir in order to predict deposition problems and to develop methods of minimizing deposition. The WAT should be experimentally determined to validate and if necessary, fine-tune the thermodynamic model. The fluids from the wells producing from the identified waxy reservoir fluids should be monitored for

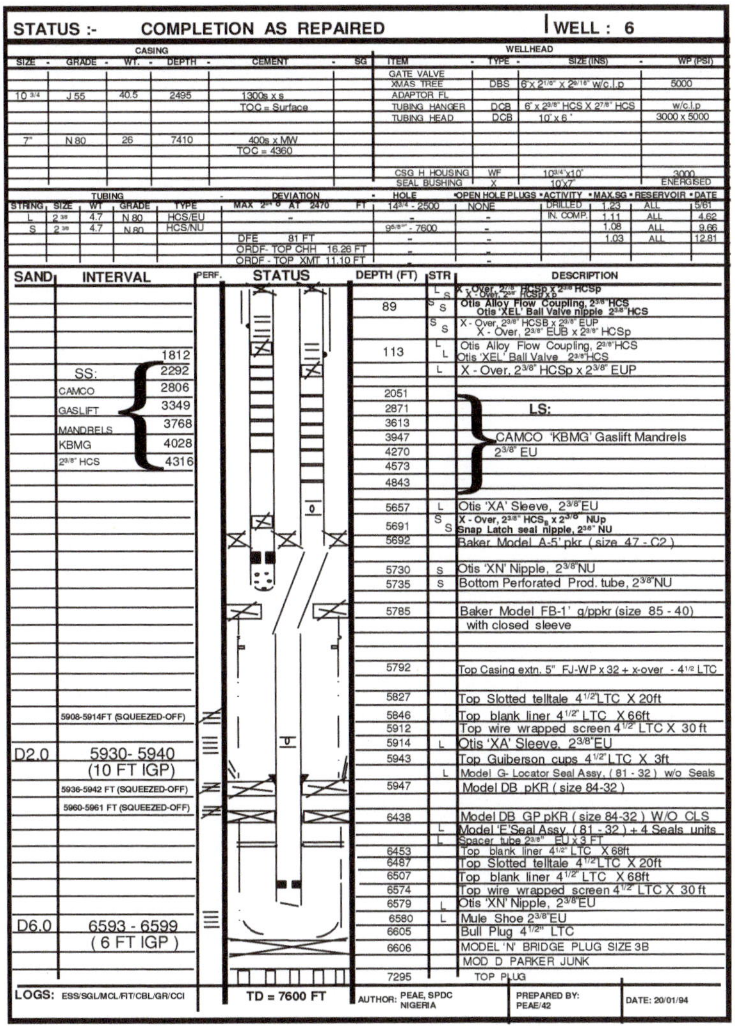

Figure 6. Completion diagram for well 6S.

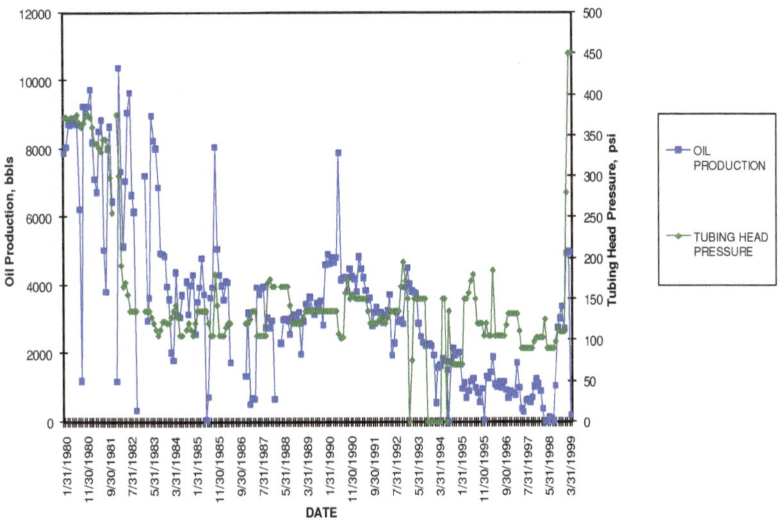

Figure 7. Monthly production and pressure profiles for well 6S.

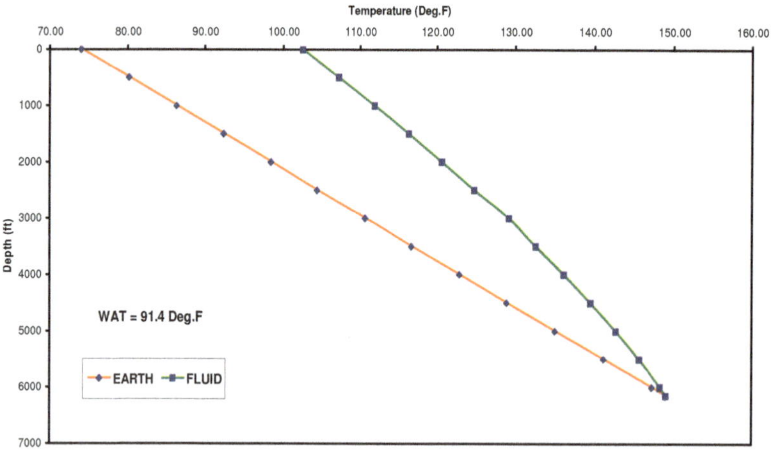

Figure 8. Temperature distribution profiles in production tubing for well 19L1.

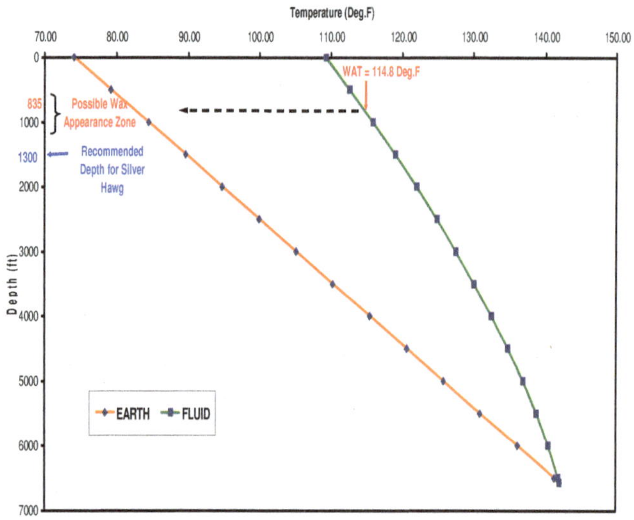

Figure 9. Temperature distribution profiles in production tubing for well 59T (D3.000C sand).

the following indicators of an imminent wax deposition:

1. Change in crude appearance.
2. Accumulation of paraffin in stock tanks.
3. Paraffin build-up in surface flowline and tubing.
4. Production loss.
5. Gradual increase in tubing or wellhead pressure.
6. Sharp increase in the Gas-Oil-Ratio (GOR).

ACKNOWLEDGEMENT

We thank the management of the Shell Petroleum Development Company (SPDC) Nigeria for permission to publish this work.

REFERENCES

Baker K (2003). Understanding Paraffin and Asphaltene Problems in Oil and Gas Wells. Based on a Workshop sponsored by PTTC's South Mid-continent Region held int Arkansas Natural Resources Museum, Smackover, July 16.

Dalirsefat R, Feyzi F (2007). A Thermodynamic Model for Wax Deposition Phenomena. Fuel, 86:1402-1408.

Erickson DD, Niesen VG, Brown TS (1993). Thermodynamic Measurement and Prediction of Paraffin Precipitation in crude Oil. Paper (SPE 26604) presented at the SPE 68th Annual Technical Conference and Exhibition held in Houston, Texas, 3-6 October. pp. 933-948.

Firoozabadi A (1999). Thermodynamics of Hydrocarbon Reservoirs. New York: McGraw-Hill., pp. 295-350.

Hammami A, Raines MA (1999). Paraffin Deposition from Crude Oils: Comparison of Laboratory Results with Field Data. SPE J., March. 4(1): 9-16.

Leontaritis KJ (2007). Wax Flow Assurance Issues in Gas Condensate Multiphase Flowlines. Paper (OTC 18790) Presented at the Offshore Technology Conference held in Houston Texas, U.S.A, 30 April - 3rd May. pp.1-10.

Ramey HJ (1962). Wellbore Heat Transmission. Paper (SPE 96) Presented at the 36th Annual Fall Meeting of the Society of Petroleum Engineers (SPE) held in Dallas, March 6, pp. 426-435.

Sulaimon AA (2009). Thermodynamic Modelling of Wax Precipitation in Oil Well Production Tubing, Ph.D. Thesis, University of Ibadan.

Sagar R, Doty DR, Schmidt Z (1991). Predicting Temperature Profiles in a Flowing Well. SPE Production Engineering, November, pp. 441-448.

WEAFRI (2005). "Catalytic Fluid Conditioning Treatment of Paraffin, Scale, Corrosion, Iron Sulfide, and Hydrogen Sulfide", Corrosion Inhibitor Systems Manual.

Weispfennig K (2006). Recent Advances in wax Deposition Modelling. A Baker Petrolite's PowerPoint Presentation Prepared for the Advances in Flow Assurance Technology Conference, Eni E&P Division and EniTechnologie, Milan, Italy. February 23.

Won KW (1985). Continuous Thermodynamics for Solid-Liquid Equilibria: Wax Formation from Heavy Hydrocarbon Mixtures. Paper 27A Presented at AIChE Spring National Meeting held in Houston TX, 26 March.

APPENDIX

Principle of operation of a WIT

Silver-Hawg is an alloy comprised of dissimilar metals. Its operation is based on scientific knowledge that physical characteristic of a flowing liquid changes or is modified by contact with a particular alloy (Figure 1). During oil production, crude oil is sucked up through the open lower end of the tool's inner tube by pressure into the inner tube. This causes multiple streams or jets of crude oil to emanate from the radically bored holes in the wall of the tube to bombard the copper nickel walls of the annular chamber and the center insert (WEAFRI, 2005). Electrons released from the copper in the walls of the chamber combine with molecules of the hydrocarbons and other minerals, thereby altering certain physical characteristics of the crude oil, produced water and the other entrained minerals which otherwise clog the tubing and impede the upward flow of the crude.

The crude oil and its entrained minerals treated in the Silver-Hawg pass through the string of tubing to the surface. The treated crude oil not only keeps paraffins and other waxes in suspension, but it also breaks up the long chain hydrocarbon molecules thereby making the oil "slicker". On high paraffin low gravity crudes, the treatment increases the API gravity of the resulting crude by at least two or three points, thus, increasing the marketability of the treated crude.

Design methodology for Silver-Hawg installation

When cloud point indicates that the precipitation is in the flowline, a flowline Silver-Hawg would be most appropriate and effective when placed before the nucleating point. If cloud point combined with fluid temperature predictions indicated wax/paraffin formation in the tubing, a specific Silver-Hawg design for tubing suspension would be appropriate. The tool should not only be installed a few feet below the cloud point but must also posses a flow area that will not reduce oil production. It is essential to locate the nearest profile such as "X", "XN", "Sleeve" that is closest to the nucleating point for hanging the tool. The closer the tool is to the nucleating point, the more effective is its performance.

Optimum site selection of natural gas vehicles station in Bangkok using geographic information system

Adsavakulchai S. * and Huntula C.

[1]School of Engineering, University of the Thai Chamber of Commerce 126/1 Vibphavadee Rangsit Road, Bangkok, Thailand.

This paper is to demonstrate the integration of spatial modeling and geographic information systems (GIS) can be used in the optimal site location for natural gas vehicles (NGV) station in Bangkok. Integration has been proposed as one of the characteristics which distinguish GIS from other information systems. Variables used as input in this process mainly address direct construction costs and station efficiency once the structure has been completed. Some of the variables examined include: proximity to population centers, distance from neighboring stations, the easements of using existing utility, car density and speed, land price, and the number of NGV registered car in Bangkok. Increasing levels of interest in the coupled systems, which are often referred to as spatial decision support systems (SDSS) has wide range of application areas. To do spatial analysis using ArcView GIS 3.2 as a tool. The results from this study can be concluded that six locations around Nongjok district were evaluated as the optimal sites for the NGV station in Bangkok.

Key words: NGV station, Bangkok, geographic information systems (GIS), spatial analysis, integration, spatial decision support systems (SDSS).

INTRODUCTION

Due to the increase of urban population and the growth of the number of cars, gas stations are among the places that residents refer to repetitiously. Considering the number of cars and urban population, distributing gas station sites can be effective in shortening urban trip paths. Thus, the number of natural gas vehicles (NGV) facilities and filling stations remained inadequate. Geographic information systems (GIS) provide the appropriate tools for analyzing the effective factors on spatial data and non-spatial data (Bowman, and Lewis, 2006). GIS are powerful computer-based tools for the capture, storage, management, retrieval, query, analysis and presentation of spatial data. GIS ability as spatial data processing and analyses tools available can be used to manage a wide range of Information (Boyles, 2002). GIS provides an integrated computing environment for social and data integration (ESRI, 2005). GIS facilitates the integration of disparate data sets, creation of new and derivative data sets, development and analysis of spatially explicit variables.

The integration has the potential to become a powerful analytical toolbox enabling regional and social scientists to gain fundamental insight into the nature of spatial structures of regional development (Dougherty, 2008). Increasing levels of interest in these coupled systems, which often referred to as spatial decision support systems (SDSS) which make up to wide application areas. Systems coupled with GIS, provide an efficient form for storing, retrieving spatial data and provide a mapping capability for data and final results (Batty and Longley, 1996). The main objective of this study is to find the optimum location of NGV station in Bangkok by integrating information with GIS.

Methodology

Study area

Bangkok is the capital city of Thailand as shown in Figure 1. The transportation and movements of residents is mainly done by private cars, although public buses and taxies are also functioning. The number of NGV stations in Bangkok consist of 76 stations and

*Corresponding author. E-mail: suwannee_ads@utcc.ac.th

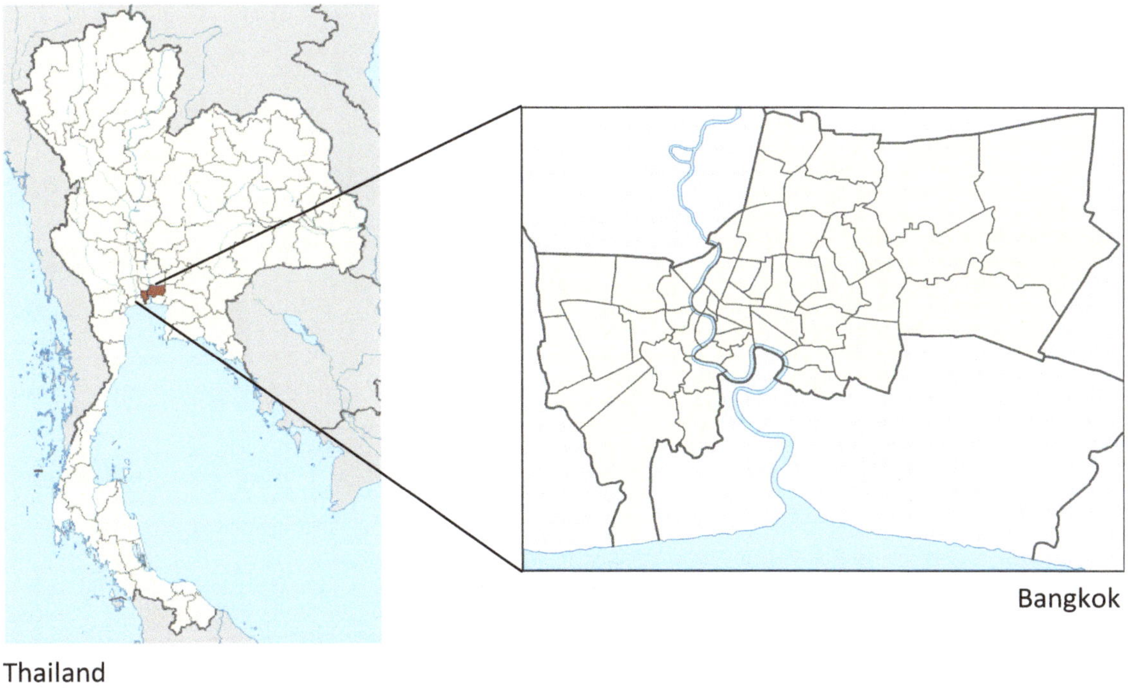

Figure 1. Study area in Bangkok.

the number of NGV car license in Bangkok totally is 122,576 vehicles in 2009.

Methodology used

There are three steps in this study as follows: in the first step, requirement analysis is done, in which the needed spatial and attribute data are identified. The information is gathered and entered to ArcView GIS software. In the second stage the input data is analyzed and the required information layers are made (Harder, 1997). In the last stage by examining the output information from the second stage conclusions and necessary suggestions are made. Several factors played a role in determining the proposed location of the NGV in this study are:

1. The digital topographic maps of Bangkok that contains 7 information layers.
2. The information related to transportation, proximity to population centers, distance from neighboring stations, the easements of using existing utility, car density and speed, land price, and the number of NGV registered car in Bangkok.

ArcView software is then used to generate:

1. The development of the base map for generation of GIS files.
2. The locations of gas stations are specified and their attributes is linked to each spatial feature.
3. Buildings information layer containing the polygons of urban regional divisions and their attributes containing the number of cars, gas stations, and populations.
4. The information layer containing the regions' centers.
5. The spatial decision support systems (SDSS) has wide range of application areas.

Spatial Analyst extension of ArcView is used to compute the distance of each region center to NGV stations. The suitability of each of these sites was assessed against technical, operational and economic criteria (Murphy, 2001). The technical considerations included the criteria for selection of the NGV stations which are either qualitative or quantitative types are as follows:

1. Site location (urban, rural and residential).
2. Site accessibility (Highway, principal or secondary roads).
3. Fuel and service capacities (No. of fuels tanks, No. of service pumps).
4. Presence of driving factors (supermarket, malls, and highway).
5. Other variables (ATM machines, catering, and area expansion).

RESULTS AND DISCUSSION

To do overlay technique from each parameter that is land price (Figure 2), NGV stations density (Figure 3), population density (Figure 4) and NGV car density (Figure 5), the results demonstrate that the optimal district site location for NGV station in Bangkok is Nongjok district (Figure 6). To do more criteria for specific location using the distance from UTURN is not more than 50 m, the entrance of NGV station nearby, the 12 to 16 meter-width road and also network analysis. Six locations around Nongjok district were evaluated as the optimal sites for the NGV station (Figure 7).

CONCLUSIONS AND RECOMMENDATIONS

The GIS knowledge and its practical software can be useful in analyzing and finding out how the optimal site locations for NGV stations in Bangkok. The results of the

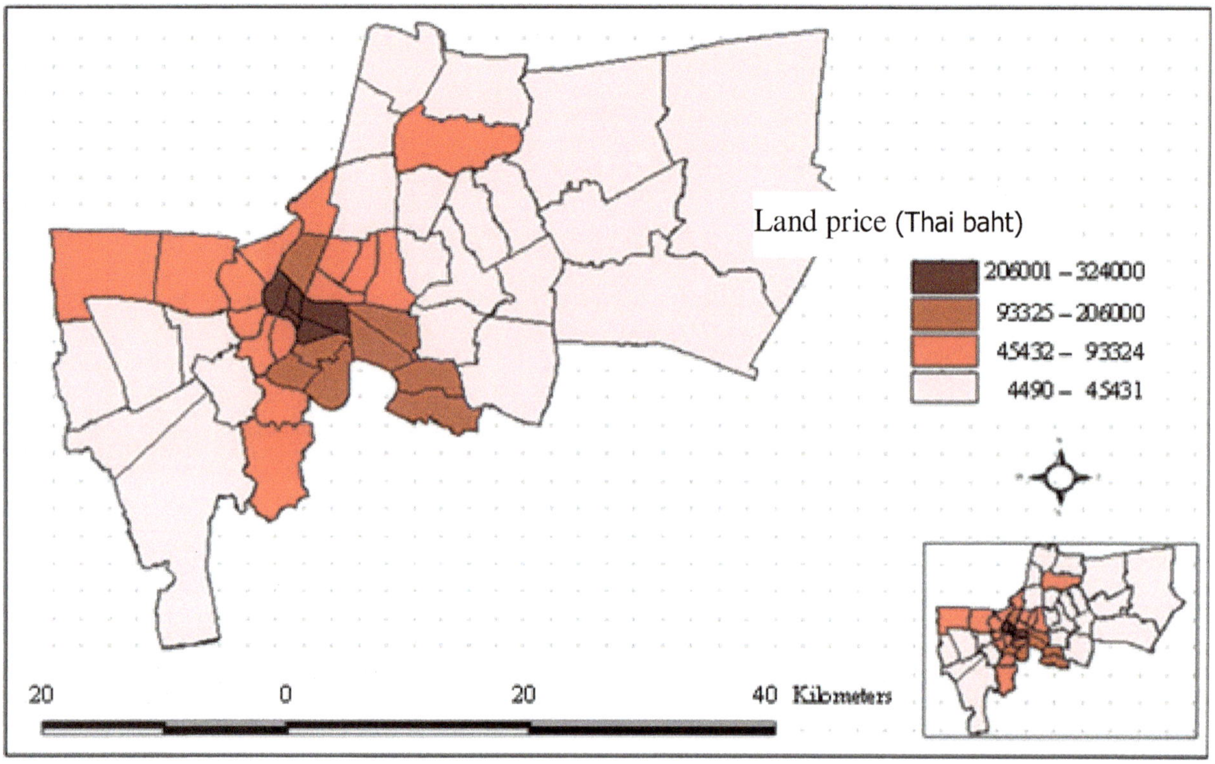

Figure 2. Land prince density.

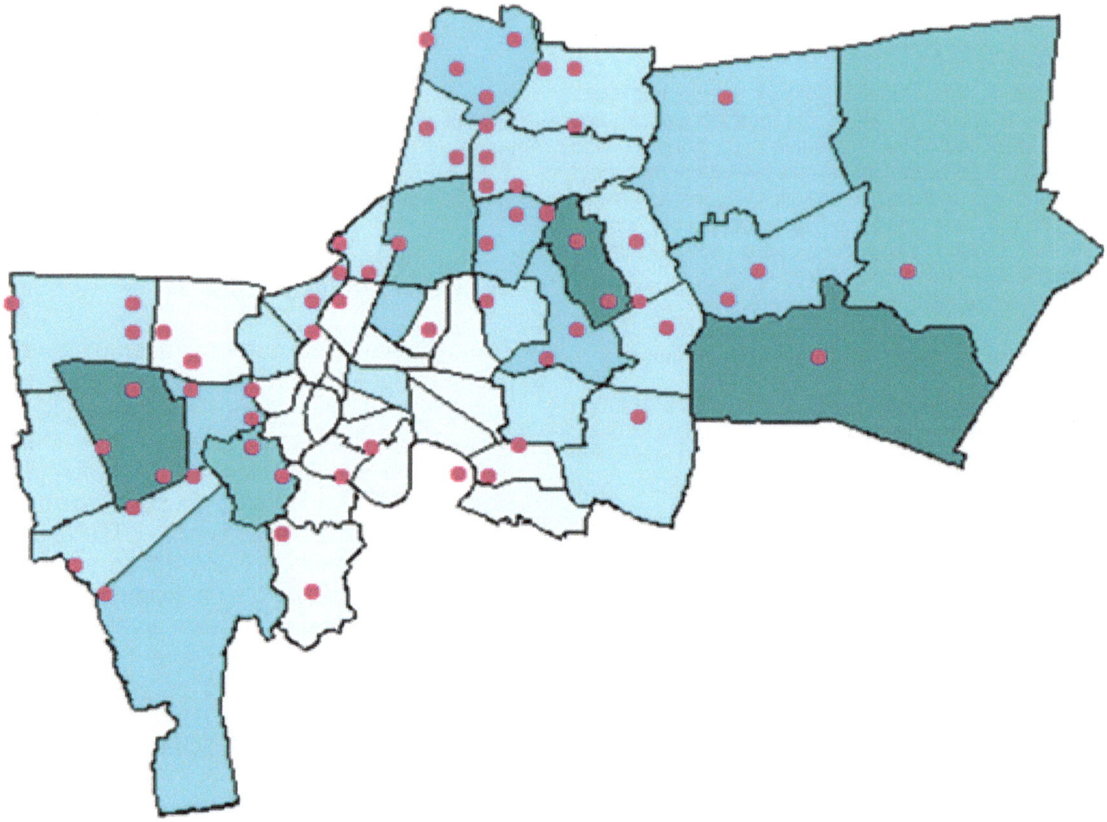

Figure 3. Current NGV stations position in Bangkok.

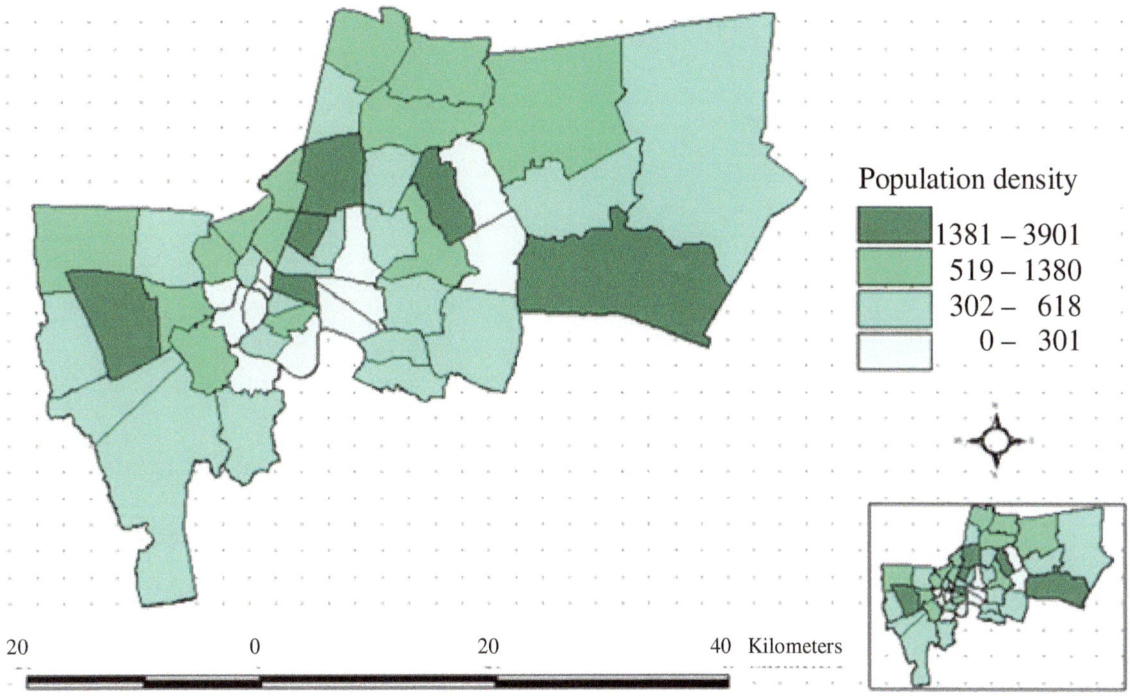

Figure 4. Population density in Bangkok.

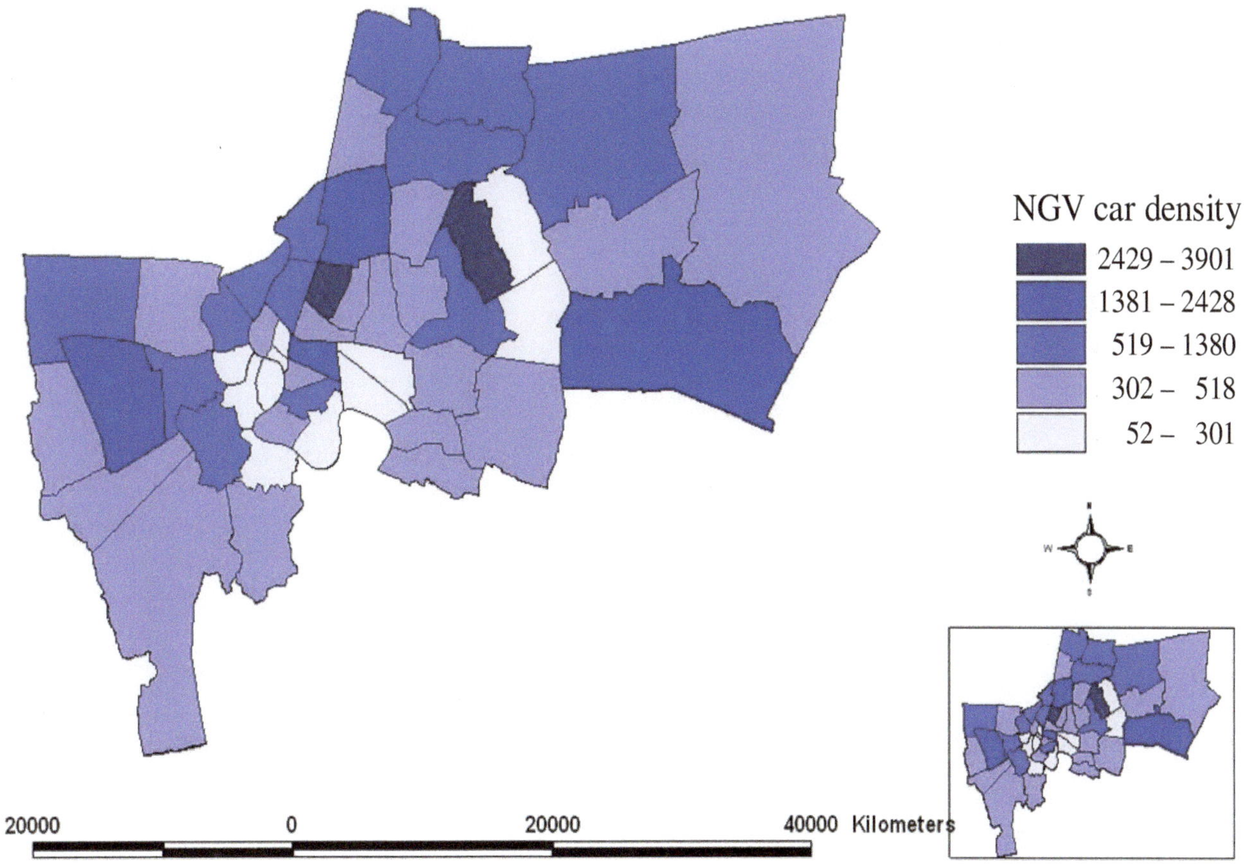

Figure 5. NGV car density in Bangkok.

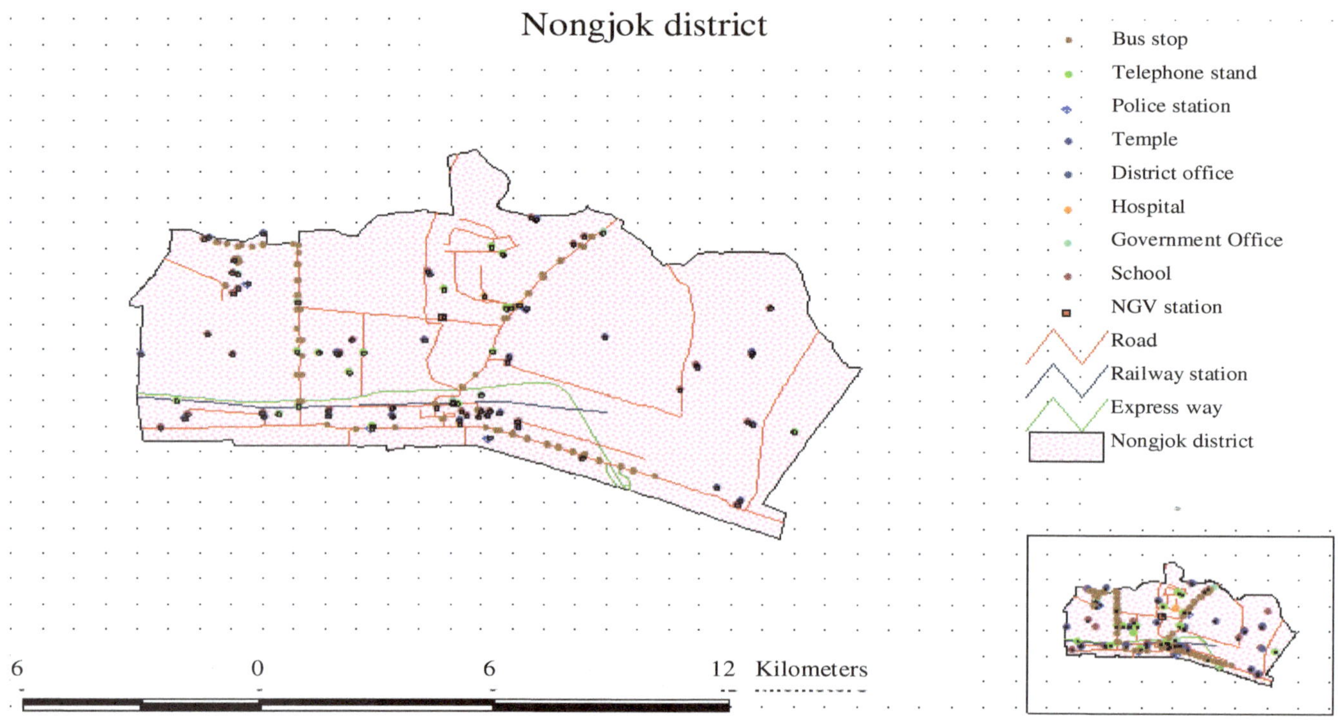

Figure 6. The optimal district site location for NGV station in Bangkok.

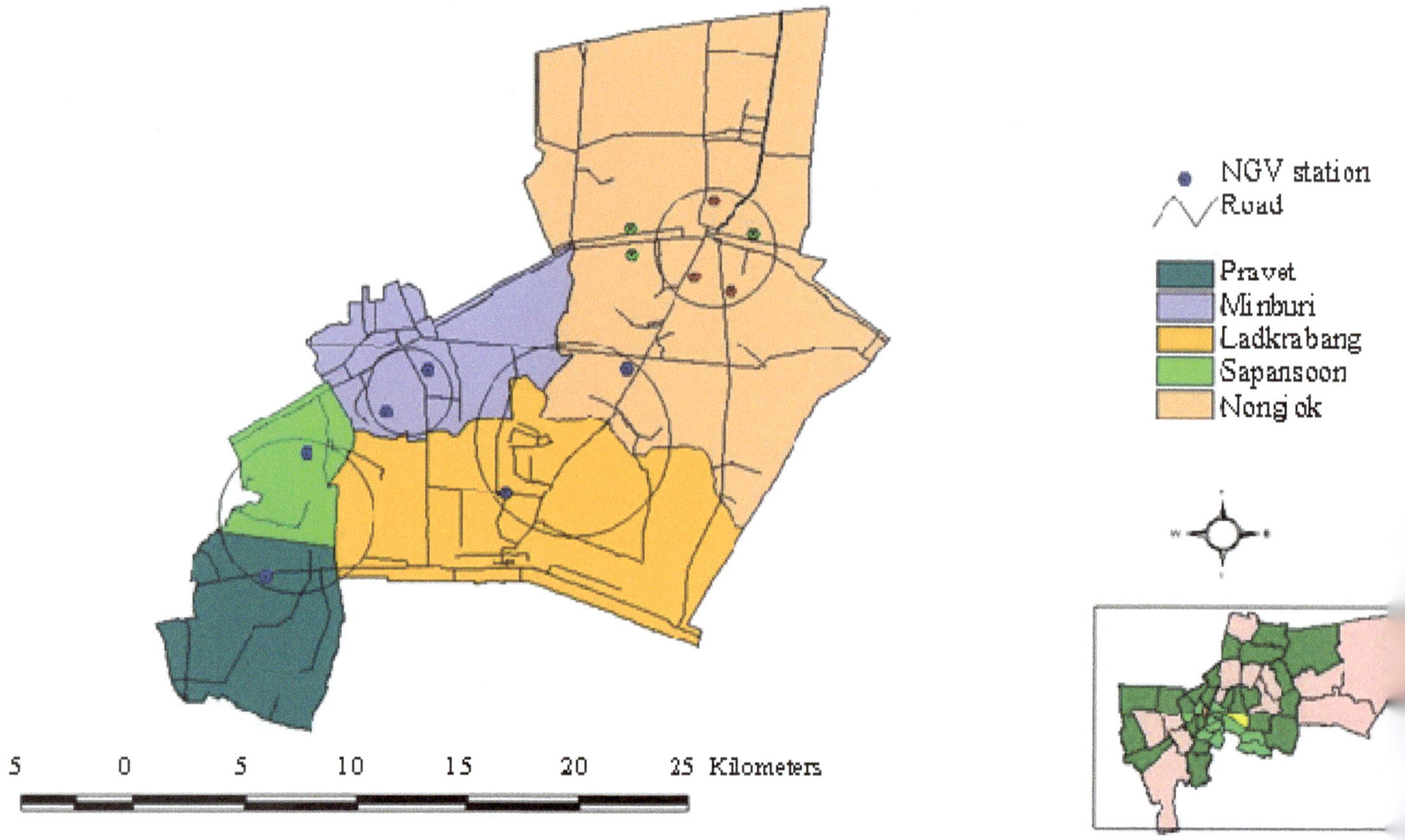

Figure 7. Six optimal locations around Nongjok district for NGV stations.

examinations show that we can specify the potential of the site locations for NGV stations for each region in comparison to other regions. In this examination by focusing on the site selection method of gas stations and high potential regions were specified (Swenson, 1996). The proposed sites at Nongjok district were selected for several other reasons that were close to the urban centre of Nongjok, but sufficiently far removed to minimize disturbance.

There are potential benefits related to the proximity to Industrial Real Estate in Chachengsoa Province which fit with accessing to local support services; not far from UTURN less ore than 50 m and also the entrance of NGV station nearby, the 12 to 16 meter-width road and also network analysis. The results showed that there are six optimal site locations for NGV stations in Nongjok district in Bangkok. It is suggested to study this subject by considering other factors such as accurate street data are critical for supporting geocoding, routing, architecture and road building methods in regions and compare the results with this research. Finally, GIS based decision support systems to decide where to set up a NGV station.

References

Batty M, Longley PA (1996). Fractal cities. London and San Diego: Academic press.

Bowman D, Lewis D (2006). Sears Holdings Corporation deploys GIS navigation and mapping system. ArcNews Online.

Boyles D (2002). GIS means business, volume two. Redlands, CA: Environmental Research Institute, Inc., Press.

Dougherty M (2008). Content management: Managing spatial data. IBM Database Magazine, 13(2): 224.

ESRI (2005). The Washington Times case study. http://www.esri.com/library/fliers/pdfs/cs-washingtontimes.pdf.

Harder C (1997). ArcView GIS means business. Redlands, CA: Environmental Research Institute, Inc., Press.

Murphy F (2001). Applying GIS and OR techniques to solve Sears's technician dispatching and home-delivery problems. Interfaces, 31(1): 130.

Swenson J (1996). Planning systems—GIS software goes corporate—government staple breaks out of niche. InformationWeek, 582: 103

4

Analytical critical drawdown (CDD) failure model for real time sanding potential prediction based on Hoek and Brown failure criterion

Gbenga Folorunso Oluyemi[1]* and M. Babs Oyeneyin[2]

[1]School of Engineering, Robert Gordon University, Schoolhill, Aberdeen, AB10 1FR, UK.
[2]School of Engineering, Robert Gordon University, Schoolhill, Aberdeen, AB10 1FR, UK.

The risks of failure in sand reservoirs and consequent sand production is now a stark reality in the upstream oil and gas industry. As a result, failure analysis of reservoir rocks for sanding potential prediction purposes has become a routine activity more than ever before. Owing to the huge economic, operational and safety implications of risks of sand failure, the efficient management of these risks for field operation optimization requires a reliable failure model, which can capture the failure processes adequately in real time. Mohr coulomb failure criterion has been more widely applied for rock mechanics problems relating to sand failure analysis and production in the oil industry and elsewhere, and has been used as the basis or platform for more than 80% of the failure models being used in the industry today for rock failure analysis and sanding potential prediction. The major reasons for this could be attributed to: (a) simplicity in understanding and use and (b) description by a simple mathematical expression. The mathematical expression of Mohr Coulomb criterion defines shear stress as a linear function of the normal stress, which is depictive of a linear failure envelope. In addition Mohr Coulomb is only applicable to intact rocks and cannot be applied to already failed rock. Failure envelope in petroleum formation rock has however been proved to be non-linear and as such Mohr Coulomb failure criterion and the models based on it cannot be trusted to capture the failure processes adequately and reliably. In this study, Hoek and Brown failure criterion has been used as a platform to develop a new time-coupled analytical failure model for the analysis of sanding potential prediction in real time. The basis for using the Hoek and Brown failure criterion lies in its ability to capture rock failure as a non-linear process and applicability to both intact and failed rocks. This model has been tested and validated on some field data; in addition, it has been compared with another Mohr Coulomb-based drawdown failure model. The results obtained from the testing and validation scheme are very encouraging and show that Hoek and Brown criterion can indeed help overcome the inherent problems in Mohr Coulomb criterion.

Key words: Rock failure, sand prediction, failure criteria, critical drawdown, failure envelope, uniaxial compressive strength (UCS).

INTRODUCTION

Analysis of risk of failure in reservoir rock and potential for sand production is now a routine activity in the petroleum industry. Sanding potential prediction in real time is crucial and important to be able to make timely reservoir management decisions on the sand control methods or techniques to be deployed in any field. Conventional sand prediction techniques used in the industry today are based on field observation and experience, laboratory sand production experiments and theoretical or numerical modelling (Veeken et al., 1991). Only recently, neural network based technique evolved through the work of Kanj and Abousleima (1999).

Techniques based on field observation and experience usually attempt to establish a correlation using multi-variable linear regression between the data collected

*Corresponding author. E-mail: g.f.oluyemi@rgu.ac.uk.

from a sand producing well and operational and field parameters relating to formation, completion and production e.g. strength, flow rate, drawdown etc. These correlations are usually established with a small selection from the vast assemblage of parameters that could possibly affect sand production. For example Stein and Hilchie (1972) and Stein et al. (1974) correlated sand production from the reservoir with production rate, neglecting the effects of other parameters that affect sand production. Operational parameters (for example bean-up pattern, flow rate and drawdown) affecting sand production are known to be many and to vary from field to field. Using just a small selection of these parameters and extrapolating the results from one field to another may give inaccurate sand production prediction.

Techniques based on laboratory sand production experiments involve observation and simulation of sand production in controlled laboratory environments. These experiments have suggested that sand production in unconsolidated sandstone is caused by the flow rate and capillary forces (Hall and Harrisberger, 1970) while in friable-consolidated sandstone, by boundary stress (Vriezen et al., 1975). Laboratory sand production experiments are usually performed on cores. The great setback of this technique is the fact that most wells are not cored, meaning that cores are not always available. Even when they are available, they may be affected adversely during retrieval, transportation and processing by a number of factors such as core damage, stress relief etc. All these add to the degree of uncertainty surrounding the results of laboratory sand production. Extrapolating these results to field conditions may therefore, constitute a source of error.

Techniques based on theoretical modelling suggest compressive failure, tensile failure and erosion as mechanisms responsible for sand production (Addis et al., 1998). Theoretical modelling also suggests that compressive failure can be triggered by both far-field stresses (depletion) and drawdown pressure; and tensile failure, exclusively by drawdown. Erosion is believed to occur when the drag forces exerted on a particle at the sand face exceed its apparent cohesion. However, theoretical modelling requires a mathematical approach to failure mechanisms (Veeken et al., 1991) and relies heavily on log-derived geomechanical parameters. However, the uncertainty in the formation strength obtained from log-derived parameter affects the reliability of theoretical modelling. Estimating formation strength from the mechanical properties log, may be inapplicable in some fields due to non-consolidation and high clay content (Cole and Ross, 1998). Often times these models are not validated with field sand production data, their results, in these circumstances, can best be described as qualitative.

Neural network based sand prediction was first reported by Kanj and Abousleiman (Kanj and Abousleiman, 1999). Parameters that were thought to affect production of sand in a gas well, were presented to a feed forward back propagation network (BPN) and a generalized regression neural network (GRNN) to predict important sanding indication parameters (SIP) for the gas wells of Northern Adriatic Basin. It was concluded that neural network proved capable of predicting sanding potentials with an unprecedented level of accuracy. However, the presentation of many input parameters to the network is capable of increasing the network complexity due to increased network size (Oluyemi et al., 2006). This may have a negative performance impact on the ability of the network to predict sanding potential accurately. In addition, one of the input data to the network - formation cohesive strength - can only be obtained using Mohr circles. Generation of Mohr circles is heavily dependent on core acquisition; this does not allow for real time sanding potential prediction.

PREDICTIVE SAND PREDICTION MODELS BASED ON MOHR COULOMB FAILURE CRITERIA MODEL

In recent times, predictive models and/or techniques based on the numerous failure criteria in use in the oil and gas industry have been more extensively used for sand prediction; most of these predictive models and techniques are based on Mohr Coulomb failure criterion. Essential features of some of the recently formulated and the most widely used of such models are discussed in the following paragraphs to underline their development framework and limitations.

McPhee and Enzendorfe (2004) and McPhee et al. (2000) developed a plastic (failed) extension model (PZE) based on Mohr Coulomb failure criterion. This is based on a yield zone approach that accounts for shear failure triggering sand failure, the existence of a plastic (failed) zone around perforations, and the effective stress state in near wellbore area. The calibration parameter used to define the critical conditions for sand production is the ratio of the plastic zone radius to the wellbore/perforation radius (r_p/r_w). The model is represented thus (equation 1):

$$\left[\frac{r_p}{r_w}\right]^{q-1} = \frac{q-1}{q+1}\frac{1}{C_o}\left\{2\sigma_h^{'}(t) - (2-\gamma)\left[P_i(t) - p(\infty,t)\right] + \frac{2C_o}{q-1}\right\} \tag{1}$$

Where:

r_p = radius of plastic zone, ft

r_w = radius of wellbore or perforation, ft

$$q = \tan^2\left(\frac{\Pi}{4} + \frac{\theta}{2}\right) \text{, in radian} \tag{2}$$

$\sigma_h^{'}$ = effective minimum horizontal stress (at time t), psi

$$\gamma = \frac{1 - 2\nu}{1 - \nu} \quad (3)$$

ν = Poisson ratio
θ = angle of internal friction (degree)
Pi (t) = Constant pore pressure around well (at time t), psi
p(∞, t) = far field (reservoir) pore pressure (at time t), psi
Co = uniaxial compressive strength, psi

Input parameters for the model include UCS (Co), frictional angle θ, Poisson ratio (ν), in-situ stress state – horizontal stresses, ($\sigma_h^{'}$ (t)), and well drawdown and planned depletion level – Pi (t) and p(∞, t).

Shear failure model used by BP in-house for sanding potential prediction is a stress-based model of shear failure around a perforation or an open hole wellbore (Wilson et al., 2002; Vaziri et al., 2002; Palmer et al., 2003) the mathematical representation is given in equation 4:

$$CBHFP \leq \frac{3\sigma_1 - \sigma_3 - \sigma_y}{2 - A} - P_r \frac{A}{2 - A} \quad (4)$$

CBHFP = critical bottom hole flowing pressure
P_r = current average reservoir pressure
σ_1 & σ_3 = the total principal major and minor stresses
A = poro-elastic constant (it is a function of Poisson ratio and formation compressibility)
σ_y = 3.1*TWC
The factor 3.1 includes the scale transformation from TWC laboratory sample (OD:ID = 3) to field (OD:ID = infinity).

Coates and Denoo (1981) model, based on Mohr coulomb theory, is also a shear failure model specifically developed for borehole stability analysis during drilling and sand prediction during production. The three principal stresses, x, y and z acting on a block of material deep down the earth bowel, are first written in terms of overburden stress, pore pressure and Poisson ratio as given in equations 5 - 7:

$$\sigma_x = \frac{\nu}{1-\nu}\sigma_{ob} + \alpha P_p\left(1 - \frac{\nu}{1-\nu}\right) \quad (5)$$

$$\sigma_y = \frac{\nu}{1-\nu}\sigma_{ob} + \alpha P_p\left(1 - \frac{\nu}{1-\nu}\right) \quad (6)$$

$$\sigma_z = \sigma_{ob} \quad (7)$$

ν is Poisson ratio; σ_{ob} is overburden pressure; α is Biot poroelastic constant; and Pp is pore pressure.

The three stresses are then written as radial coordinates for ease of analysis during drilling; and then transformed in a manner similar to the popular Kirsch's stress transformation to radial systems of overburden, tangential and radial stresses (equations 8 - 10).

$$\sigma_z = \sigma_{ob} + 2\nu(\sigma_1 - \sigma_2) \quad (8)$$

$$\sigma_\theta = 3\sigma_1 - \sigma_2 - P_{mud} \quad (9)$$

$$\sigma_r = P_{mud} \quad (10)$$

Analysis of the interplay between the radial and tangential stresses is then utilised for failure potential analysis in rocks and by extension sanding potential prediction.

Wang and Lu (2001) developed a model based on Equivalent Critical plastic strain. The model was implemented and developed using finite element numerical method by fully coupling a comprehensive geomechanic model to a multiphase reservoir model. The equivalent critical plastic strain level is considered as signifying the onset or initiation of hole collapse and sand production. The onset of plastic yielding, sand production and wellbore collapse, are defined based on a combined criterion in which stress concentration and strain are calculated and compared to critical strength and strain. Onset of wellbore instability or sand production is defined when the following criterion for effective or equivalent plastic strain is satisfied:

$$\varepsilon_e^p = a_o + a_1 J_1 = \sqrt{\frac{2}{3}\left[\left(\varepsilon_{11}^p\right)^2 + \left(\varepsilon_{22}^p\right)^2 + \left(\varepsilon_{33}^p\right)^2\right]} \quad (11)$$

a_o = 0.02 and a_1 = 0.008 have been suggested for sand production, provided the compression is taken to be positive (Wang and Lu, 2001).

$J_1 = \sigma_{11} + \sigma_{22} + \sigma_{33}$ (J_1 is the first stress invariant) (12)

Linear poro-elastic and brittle plasticity model developed by Wu and Tan (2002) is an improvement on the earlier one by Wang et al. (1991). It is based on linear poro-elasticity and brittle plasticity with a critical equivalent plastic strain on cavity surface as the sanding criterion. Inclusion of the effect of residual strength in the plastic zone surrounding the cavity is the major improvement in this model. The inclusion is based on the assumption that rock is linearly poro-elastic prior to the peak strength and become brittle plastic after the peak strength is exceeded, and with the stress exceeding the peak strength, the strength of the rock reduces to its residual strength.

For this model, onset of sanding initiation is defined by the equivalent critical strain equation given in equation 13:

$$\varepsilon_e^p = \sqrt{\frac{2}{3}\left(N_r^4 + 1\right)}\varepsilon_{\theta i}^p \tag{13}$$

Where N_r^4 is residual strength parameter related to the angle of internal friction and $\varepsilon_{\theta i}^p$ is the tangential plastic strain. The equations for calculating these parameters are given by Wu and Tan (2002).

One feature common to all the models discussed above is the Mohr Coulomb failure criterion platform for their formulation. However, as pointed out earlier, Mohr Coulomb failure criterion has two major shortcomings, which are: (a) Assumption of linearity of failure envelope and (b) Inapplicability in failed rock environment. The linear approximation of failure behaviour of rocks by Mohr-Coulomb is considered an oversimplification of the failure process whilst its inability to capture discontinuities in failed rocks constitutes a serious limitation of its capability to accurately capture failure process and mechanism as obtained in the field. The implication of this is that all the models discussed above cannot be applied for formation rocks with discontinuities. Besides, some of them are formulated using complex numerical solutions which may make their application technically demanding and unrealistic.

Hoek-Brown criterion was developed as a gap-bridging model to capture the non-linearity of failure envelope in rock and the influence of discontinuities in already failed rock (Hoek and Brown, 1980; Hoek and Brown, 1988). The criterion was developed based on field, laboratory and theoretical considerations as well as experience, which makes it applicable to both intact and failed rocks. The criterion has therefore been used as a platform for the development of a sanding potential prediction model in this work; the developed model is simple and easy to use.

CRITICAL DRAWDOWN FAILURE MODEL FORMULATION

Hoek-Brown failure criterion is given mathematically as (Hoek and Brown, 1980; Hoek and Brown, 19880):

$$\sigma_1 = \sigma_3 + \sqrt{m\sigma_{ucs}\sigma_3 + s\sigma_{ucs}^2} \tag{14}$$

Where σ_1 and σ_3 are effective principal stresses; m and s are constants; and σ_{ucs} is uniaxial compressive strength.

At the borehole wall the effective principal stresses σ_1 and σ_3 in equation (14) can be represented respectively by the effective tangential, σ_θ and radial, σ_r stresses. Equations for estimating both the tangential (for both maximum and minimum horizontal stresses) and radial stresses are given as (Abass et al., 2003):

$$\sigma_{\theta_a \equiv 90} = 3\sigma_H - \sigma_h - p_w \tag{15}$$

(where azimuth $\theta_a = 90$)

$$\sigma_{\theta_a \equiv 0} = 3\sigma_h - \sigma_H - p_w \tag{16}$$

where azimuth $\theta_a = 0$

$$\sigma_r = p_w \tag{17}$$

The Hoek-Brown failure criterion can therefore now be expressed in terms of the radial and tangential stresses, as given in equation (18):

$$\sigma_\theta = \sigma_r + \sqrt{m\sigma_{ucs}\sigma_r + s\sigma_{ucs}^2} \tag{18}$$

Substituting equations (15) and (17) (for estimating tangential and radial stresses respectively) in equation (18) and rearranging, we have:

$$3\sigma_H - \sigma_h + 2(\alpha P_r - P_w) = \sqrt{m\sigma_{ucs}(P_w - \alpha P_r) + s\sigma_{ucs}^2} \tag{19}$$

The parameter, α, introduced in equation (19), is a scaling factor called Biot's or poroelastic constant, which measures the effectiveness of the pore pressure response to the total applied stress (Brandt, 1955). Its value, which depends on the pore geometry and the physical properties of the constituents of the solid system, varies between 0 and 1 depending on the effectiveness of its response to the total applied stress. In this work, pore pressure response to the total applied stress is assumed to be 100% effective. However, this may not be true for formations at advanced stages of depletion; there may therefore be need for either model or data adjustment or recalibration to account for this in such formations.

Therefore let α = 1; and $y = CDD_l = (P_r - P_w)$. Replacing pressure drawdown term in equation (19) with y and taking the square of both sides, we have,

$$\left((3\sigma_H-\sigma_h)+(2y)\right)^2 = m\sigma_{ucs}(P_w-P_r)+s\sigma_{ucs}^2 \quad (20)$$

Expanding the left hand side of equation (20) and rearranging, we have:

$$4y^2+4(3\sigma_H-\sigma_h)y+(3\sigma_H-\sigma_h)^2=m\sigma_{ucs}(P_w-P_r)+s\sigma_{uc}^2 \quad (21)$$

Let A $=(3\sigma_H-\sigma_h)$; Substituting for A in equation (21) and rearranging, we have:

$$4y^2+4Ay+A^2+m\sigma_{ucs}(P_r-P_w)-s\sigma_{ucs}^2=0 \quad (22)$$

Recall that (P_r-P_w) = y; replacing the new pressure drawdown term in equation (22) with y and rearranging, we have:

$$4y^2+(4A+m\sigma_{ucs})y+(A^2-s\sigma_{ucs}^2)=0 \quad (23)$$

Equation (23) is a quadratic equation and can be solved using the solution of a quadratic equation. The equation can therefore be written in the form of a general quadratic equation given in equation (24):

$$ay^2+by+c=0 \quad (24)$$

The solution of the quadratic equation of this form can be written as:

$$y=\frac{-b\pm\sqrt{b^2-4ac}}{2a} \quad (25)$$

If equation (23) is matched with equation (24), then a = 4; b $=(4A+m\sigma_{ucs})$; and c $=(A^2-s\sigma_{ucs}^2)$

Substituting these parameters in equation (25) and recalling that y $=(P_r-P_w)=$ CDD$_i$, we have

$$CDD_i=\frac{(4A+m\sigma_{UCS})\pm\sqrt{(4A+m\sigma_{UCS})^2-16(A^2-s\sigma_{UCS}^2)}}{8} \quad (26)$$

The developed mechanistic model shown in equation (26) is coupled to time by using a term which describes reservoir formation depletion profile in a similar manner to the work of Hettema et al. (2006). The resulting time-coupled equation is given in equation (27).

$$CDD_c=\left[\frac{(4A+m\sigma_{UCS})\pm\sqrt{(4A+m\sigma_{UCS})^2-16(A^2-s\sigma_{UCS}^2)}}{8}\right]-\left[n\frac{(P_{ri}-P_{rc})}{1}\right] \quad (27)$$

P_{ri} is the initial reservoir pressure before the start of production; P_{rc} is the current reservoir pressure as a result of depletion; and n is a dimensionless parameter, which represents the ratio of change in critical drawdown pressure with reservoir depletion. Hettema et al. (2006) suggests that n = 1 when drawdown and depletion are equally important for sand production, which is the case in this work. Equation 27 is a mechanistic model for predicting the current critical drawdown (CDD$_c$) of reservoir formation rock (wrt time) which, when exceeded either during drilling or production, would result in failure of the formation and by extension, sanding. Input parameters into the model include field minimum and maximum horizontal stresses, uniaxial compressive strength (UCS), and Hoek and Brown constants which are readily available.

HOEK AND BROWN ESTIMATES OF MODEL CONSTANT PARAMETERS

Hoek and Brown (1988) developed, based on a series of experimental work, estimates of m and s for a wide range of rocks under a wide range of conditions that can be encountered in petroleum formation rocks.

The estimated values of these parameters reflected the level of disturbance undergone by the rock formations. The estimates as given by Hoek and Brown (1988) are given in Table 1. These estimates can be used for the constant terms m and s in the CDD model developed in this work. Caution must however be exercised in their use especially in situations where it is difficult to assign formation rock to the correct failure state.

ANALYSIS, TESTING AND VALIDATION OF MODEL

Two sets of North Sea field data, obtained from Field A and B were used to analyse, test and compare the CDD model (Equations 26 and 27) with another onset of sand prediction model developed by Abass et al. (2003). Abass et al. (2003) model was used for the comparison because of its similarity to the model developed in this work in terms of input parameters and availability of the input data. The results are shown in Figures 1 - 6 and Table 2.

Figure 1 shows the results obtained when data from Field A was utilised for comparison of the performance of the current model with Abass et al. (2003) model in terms of accurate and reliable predictions of Critical Drawdown for Field A. The results, as shown in the Figure, indicate

Table 1. Hoek and Brown (11) estimates of m and s.

Rocks	Carbonates rocks E.g limestone	Lithified argillaceous rocks e.g. shale	Arenaceous rock e.g. sandstone	Fine grained igneous rocks e.g. rhyolite	Coarse grained igneous rocks e.g. granite
Intact rocks	M=7 S=1	M=10 S=1	M=15 S=1	M=17 S=1	M=25 S=1
Undisturbed rocks	M=4.10 S=0.189	M=5.85 S=0.189	M=8.78 S=0.189	M=9.95 S=0.189	M=14.63 S=0.189
Moderatly weathered rocks	M=9.2-2.0 S=0.00198-0.0205	M=1.35-2.86 S=0.00198-0.0205	M=2.03-4.298 S=0.00198-0.0205	M=2.301-4.871 S=0.00198-0.0205	M=3.383-7.163 S=0.00198-0.0205
Heavily weathered rocks	M=0.219 S=0.00002	M=0.313 S=0.00002	M=0.469 S=0.00002	M=0.532 S=0.00002	M=0.782 S=0.00002

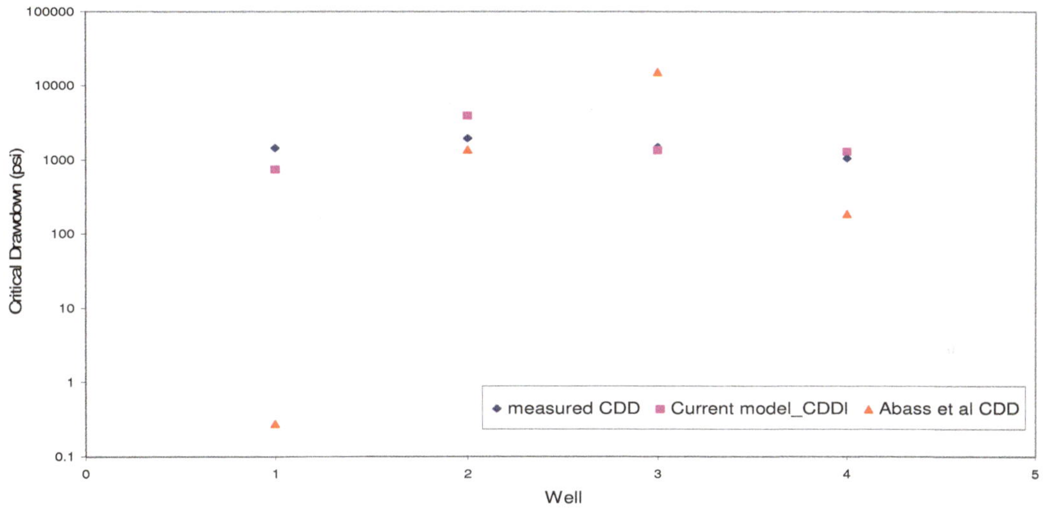

Figure 1. Comparison of CDD model with Abass et al. CDD model using a North sea field data (field A).

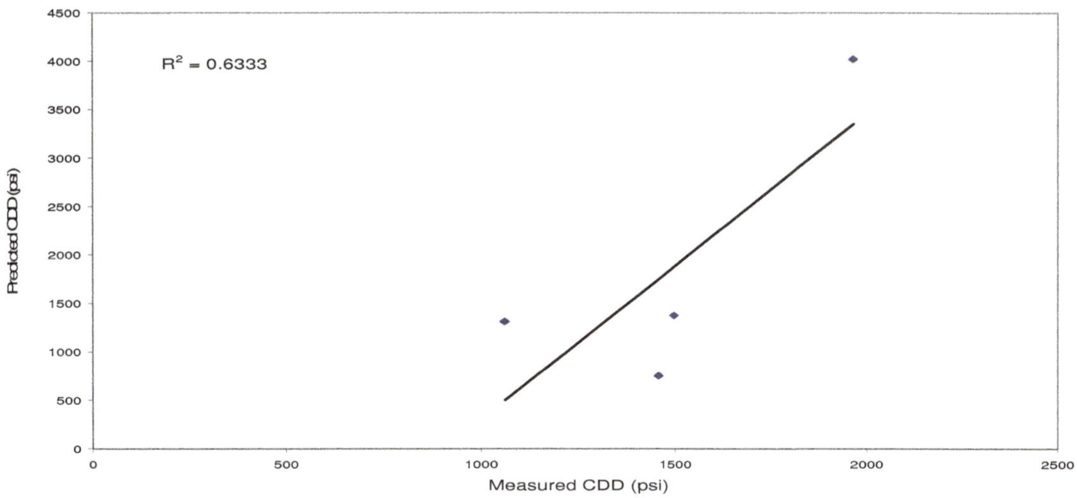

Figure 2. Correlation coefficients of predicted versus measured CDD for the current model (field A).

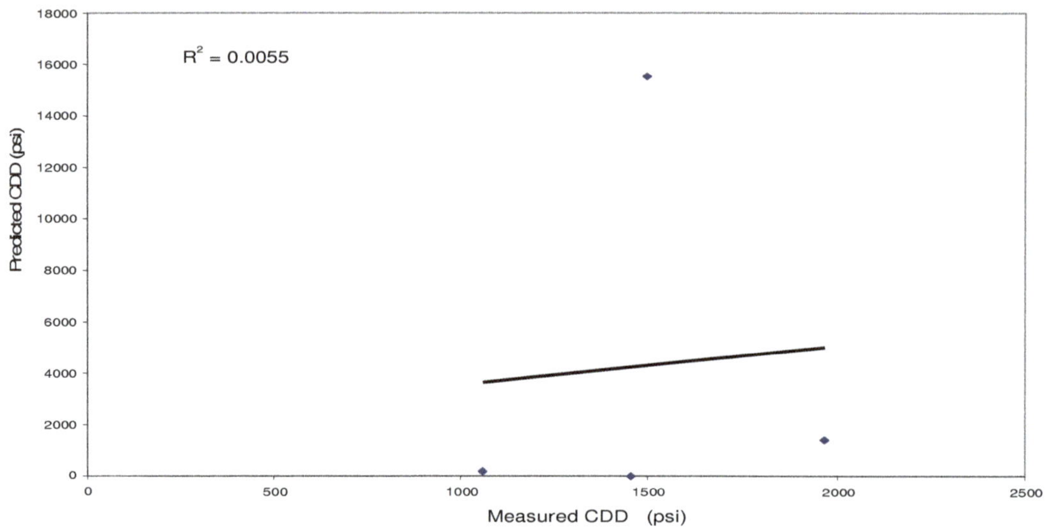

Figure 3. Correlation coefficients of predicted versus measured CDD for Abass et al model (Field A).

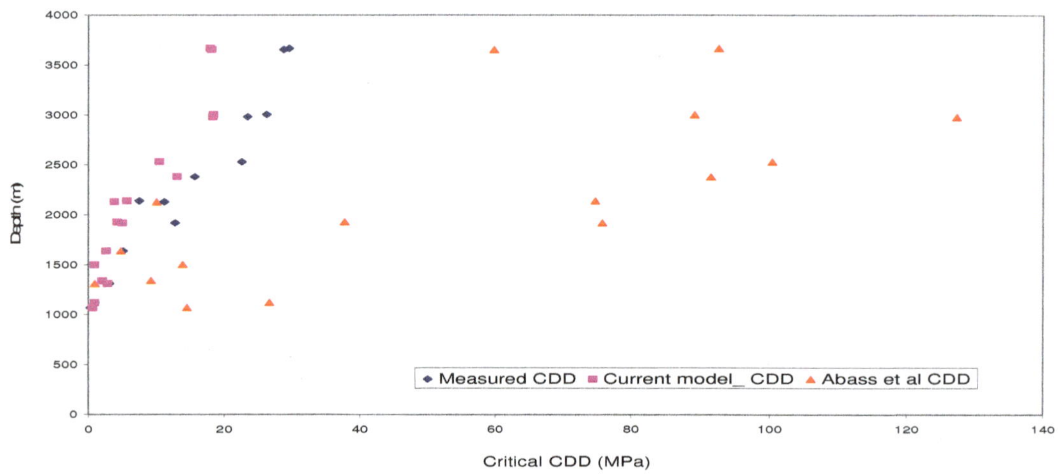

Figure 4. Comparison of CDD models with Abass et al. CDD models using a North Sea field data (Field B).

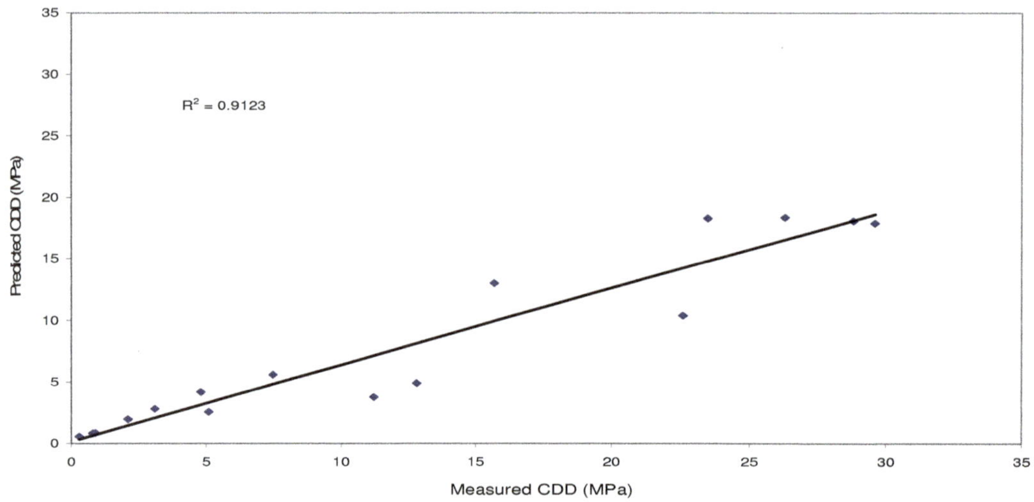

Figure 5. Correlation coefficients of predicted versus measured CDD for the current model (Field B).

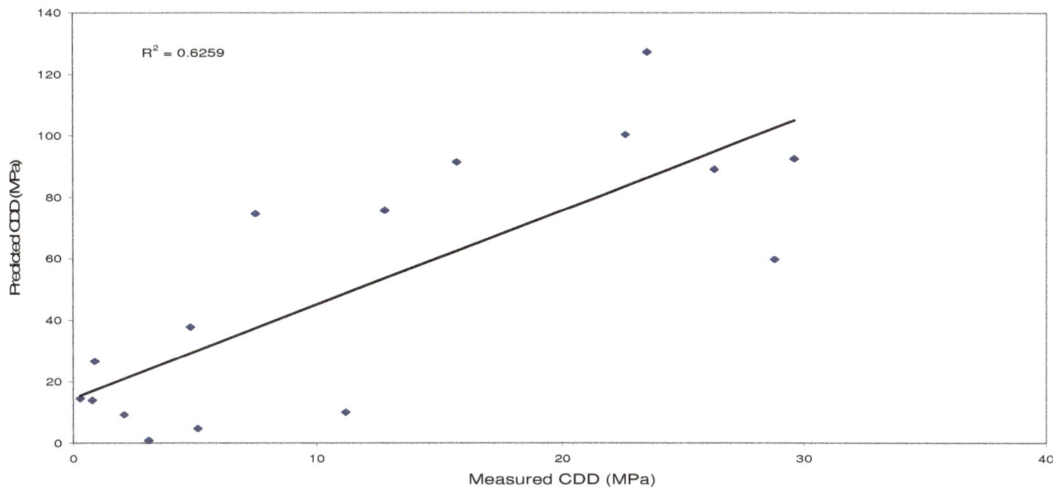

Figure 6. Correlation coefficients of predicted versus measured CDD for Abass et al. model (Field B).

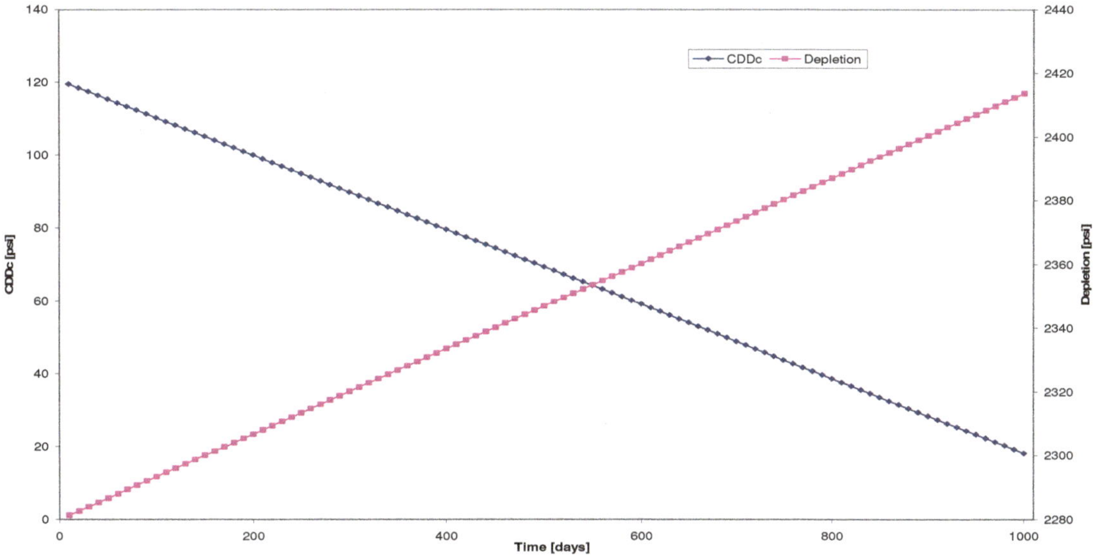

Figure 7. CDD predictions in real time for a North Sea well at 17000ft.

Table 2. Correlation coefficients.

Model	Field A	Field B
Current model	0.6533	0.9123
Abass et al. model	0.0053	0.6259

that the current model's predictions agree better with the measured CDD than Abass et al. (2003) model's predictions. The correlation coefficients between the predicted CDD by the two models and measured data are shown in Figures 2 and 3. The current model's predictions have a correlation coefficient of 0.6333; whilst for Abass et al. (2003) model's predictions, the correlation coefficient is 0.0055, which is undoubtedly poor.

Figure 4 shows the results obtained from the use of Field B data; the results also indicate that the current model's predictions are closer to the measured CDD much more than Abass et al. (2003) model' predictions. Determination of the correlation coefficients for the models' predictions for Field B data shown in Figure 5 and 6 shows that the current model's predictions have a correlation coefficient of 0.9123 whilst Abass et al. (2003) model predictions have a correlation coefficient of 0.6259, again showing that the current model has a better predictive capacity and exhibits better reliability. Shown in Figures 7 and 8 are the examples of CDD

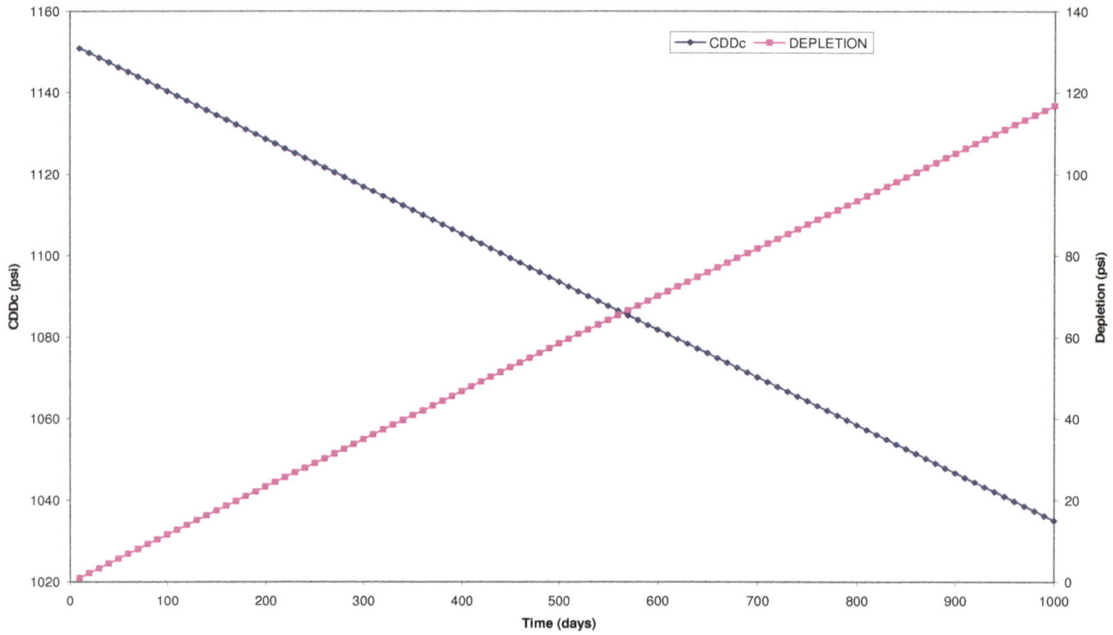

Figure 8. CDD predictions in real time for a North Sea well at 6412ft.

profiles obtained from real time prediction of critical drawdown (CDD) by the time coupled variant of the CDD model (equation 27) for a North Sea well at 6412 ft and 17000 ft.

In summary, the results show that the CDD model developed in this work has a better predictive capability than the corresponding model developed by Abass et al. (2003).

Conclusion

Many of the most widely used predictive models for predicting sanding potential and sanding initiation in the oil and gas industry are developed using Mohr Coulomb failure criterion as a platform for their development. Mohr Coulomb criterion is however, not able to capture the failure process adequately and accurately due to the assumption of non linearity of failure envelope. To address this problem, a new model for predicting sanding potential has been developed using Hoek and Brown failure criterion as a platform. The model has been tested, analysed and validated against a Mohr Coulomb based sand initiation model and found to perform better.

Nomenclature

r_p = Radius of plastic zone

r_w = Radius of well/perforation

θ = Angle of internal friction

υ = Poisson ratio

$\sigma_h^{'}$, $\sigma_H^{'}$ = Effective minimum and maximum horizontal stresses.

$p(t)$ = Constant pore pressure around well at time t

$p(\infty,t)$ = Far field reservoir pressure at time t

σ_{ucs}, C_o = Uniaxial compressive strength

CBHFP = Critical bottom hole flowing pressure

p_r = Current average reservoir pressure

σ_1, σ_2, σ_3 = Total principal stresses

A = Poroelastic constant

TWC = Thick Walled Cylinder strength

σ_{ob} = Overburden stress

σ_x, σ_y, σ_z = Principal stresses acting in x, y and z directions

α = Biot poroelastic constant

p_{mud} = Mud column pressure

CDD = Critical drawdown pressure

CDD_i = Initial Critical Drawdown

CDD_c = Current Critical Drawdown

σ_r = Radial stress

σ_θ = Tangential stress

ε_e^p = Effective or equivalent plastic strain

ε_{11}^p, ε_{22}^p, ε_{33}^p = Plastic strain in reference directions

J_1 = First stress invariant

p_w = Wellbore pressure

P_{ri} = Initial reservoir pressure

P_{rc} = Current reservoir pressure

N = Ratio of change in critical drawdown pressure with reservoir depletion.

REFERENCES

Abass HH, Habbtar AH, Shebatalhamd A (2003). Sand Control During Drilling, Perforation, Completion and Production. SPE 81492. In: proceedings of Middle east oil Show, Bahrain, 9-12, June 2003.

Addis MA, Choi X, Gunning J (1998). The Influence of the Reservoir Stress-Depletion Response on the Lifetime Considerations of Well Completion Design. SPE/ISRM 47289, SPE/ISRM Eurock '98, Trondheim, Norway, 8-10 July 1998.

Brandt H (1955). A Study of the Speed of Sound in Porous Granular Media. Trans. AIME, 1955, 77: 479.

Coates DR, Denoo SA (1981). Mechanical Properties Program Using Borehole Analysis and Mohr's Circle. In: proceedings of SPWLA 22nd Annual Logging Symp., June 23-26, 1981.

Cole RC, Ross C (1998). Sand Stabilization and Exclusion; In: Petroleum Well Construction, ed. Economides, M. J., Larry, T. W. and Dun-Norman, S; John Wiley and Sons Ltd., West Sussex, England, 1998.

Hall CD, Harrisberger WH (1975). Stability of Sand Arches: A key to Sand Control. JPT (July, 1970) pp. 821-829.

Hettema MH, Andrews JS, Blaasmo M, Papamichos E (2006). The Relative Importance of Drawdown and Depletion in Sanding Wells: Predictive Model Compared with Data from the Statfjord Field. SPE 97794. In: proceedings of Int. Sym. Exh. On Formation Damage Control, 15-17 Feb. 2006; Lafayette, Louisiana, USA.

Hoek E, Brown ET (1980). Underground Excavations in Rock. London: Institution of Mining and Metallurgy; 1980; p. 527.

Hoek E, Brown ET (1988). The Hoek-Brown Failure Criterion – A 1988 Update. In: proceedings of the 15th Canadian Rock Mechanics Symposium, 1988; pp. 31-38.

Kanj MY, Abousleiman Y (1999). Realistic Sanding Predictions: A Neural Approach. SPE 56631, SPE Ann. Techn. Conf. Exhibition, Houston, Texas, 3-6 Oct. 1999.

McPhee CA, Enzendorfer CK (2004). Sand Management Solutions for High-Rate Gas Wells, Sawan Field, Pakistan, SPE 86535. In: proceedings of SPE Int. Symp. Exh. on Formation Damage, Lafayette, Louisiana, USA 18-20 Feb, 2004

McPhee CA, Lemanczyk ZR, Helderle P, Thatchaichawalit D, Gongsakdi N (2000). Sand Management in Bongkot Field, Gulf of Thailand: an Integrated Approach. SPE 64467. In: proceedings of SPE Int. Symp. Exh. on Formation Damage, Lafayette, Louisiana, USA, 18-20 Feb, 2004

Oluyemi G, Oyeneyin MB, Macleod C (2006). Prediction of Directional Grain Size Distribution: An Integrated Approach. SPE 105989, resented at the 30th SPE Annual International Conf. Exh., Abuja, Nigeria, July 31 – Aug. 2, 2006.

Palmer I, Vaziri H, Wilson S, Moschovidis Z, Cameron J, Ispas I (2003). Predicting and Managing Sand Production: a New Strategy. SPE 84499. In: proceedings of SPE Annual Tech. Conf. Exh., Denver, Colorado, USA, 5-8 2003.

Stein N, Hilchie DW (1972). Estimating the Maximum Production Rate Possible from Friable Sandstones Without Using Sand Control. JPT (Sept. 1972) pp. 1157-1160.

Stein N, Odeh AS, Jones LG (1974). Estimating the Maximum Sand-Free Production Rates from Friable Sand for Different Well Completion Geometries. JPT (Oct. 1974) pp. 1156-1158.

Vaziri H, Xiao Y, Palmer I (2002). Assessment of Several Sand Prediction Models with Particular Reference to HPHT Wells. SPE 78235. In: proceedings of the SPE/ISRM Rock Mechanics Conference, Irving, Texas, 20-23 Oct. 2002.

Veeken CAM, Davies DR, Kenter CJ, Kooijman AP (1991). Sand Production Prediction Review: Developing an Integrated Approach. SPE 22792; 1991. In: proceedings of the SPE Annual Technical Conference and Exhibition, Dallas, Texas, USA. 6 – 9 October.

Vriezen PB, Spijker A, Van der Vlis AC (1975). Erosion of Perforation Tunnels in Gas Wells. SPE 5661; 1975.

Wang Y, Lu BA (2001). Coupled Reservoir-Geomechanics Model and Applications to Wellbore Stability and Sand Prediction., SPE 69718. In: proceedings of the 2001 SPE Int. Thermal Operations & Heavy Oil Sym., Margarita, Venezuela, March 12-14, 2001

Wang Z, Peden JM, Damasena ESH (1991). The Prediction of Operating Conditions to Constrain Sand Production from Gas Well. SPE 21681. In: proceedings of Production Operation Symposium, Oklahoma, USA, April 7-9, 1991.

Willson SM, Moschovidid ZA, Cameron JR, Palmer ID (2002). New Model for Predicting the Rate of Sand Production. SPE/ISRM 78168. In: proceedings of the SPE/ISRM Rock Mechanics Conference, Irving, Texas, 20-23 Oct. 2002.

Wu B, Tan CP (2002). Sand Production Prediction of Gas Field – Methodology and Field Application. SPE/ISRM 78234. In: proceedings of the SPE/ISRM Rock Mechanics Conf., Irving, Texas, 20-23 Oct. 2002.

5

Thermal analysis of evacuated solar tube collectors

Siddharth Arora[1], Shobhit Chitkara[1], R. Udayakumar[1]* and Muhammad Ali[2]

[1]Department of Mechanical Engineering, Bits Pilani Dubai.
[2]ADC Energy Systems. Dubai.

A detailed design of the evacuated tube collectors is enumerated followed by the thermal network analysis of this collector. The temperature of each component is determined empirically. Numerical analysis is applied to find the coefficient of heat transfer across the air gap from absorber plate to the copper tubes. The performance characteristics of these collectors are analyzed and compared with commercially available brands to test their capability to power the generator of an absorption chiller. The generator of a typical 1 KW capacity refrigeration system is designed based on the hot water made available from outlet of evacuated tube collectors.

Key words: Evacuated tubes, heat transfer applications and absorption chiller.

INTRODUCTION

Various researches in the field of solar thermal conversion have deemed the prospect of using solar energy for space heating and cooling. It is a known fact that during the day, when the solar radiation is at its maximum and the ambient temperatures are soaring, the cooling requirement reaches its peak. Research is concentrated on the conventional compression cycle or the vapor absorption refrigeration cycle. With regard to the use of alternative source of energy, the vapor absorption system is a more viable option since it uses thermal heat to power the cycle. This energy could be sourced from the sun. As the cooling requirements are concurrent with the amount of solar radiation depending upon the time of the day, scientists and researchers find the idea of using solar thermal collectors viable. Although flat plate collectors are commonly in use but evacuated tube collectors are more appropriate for the space heating and cooling applications.

These collectors perform well in both direct and diffuse solar radiation and offer the advantage that they work efficiently with high absorber temperatures (Salah, 2009).

Evacuated tube collectors

Conventional flat plate collectors are suited for warmer climates and for the times when the intensity of the solar radiation is substantially high. Their benefits are however reduced when they are exposed to cold, cloudy and windy days. Furthermore, when exposed to weathering conditions, the tubes and the insulation has a tendency to deteriorate thereby causing loss of performance. Each evacuated tube consists of two glass tubes made from extremely strong borosilicate glass. The outer tube has very low reflectivity and very high transmisivity that radiation can pass through. The inner tube has a layer of selective coating that maximizes absorption of solar energy and minimizes the reflection, thereby locking the heat. The ends of the tubes connected to the copper header are fused together and a vacuum is created between them. This process is called as evacuation, as by definition, it means that the air is pumped out from the cavity. The vacuum is created to recreate the thermos flask effect as vacuum acts as an insulator and does not allow short wave radiation to escape through the glass tube. This traps the solar radiation much more effectively and hence higher temperatures can be achieved. This process is shown in Figure 1. It means that the evacuated tube collectors have a capacity to perform better than flat plate collectors in cold and cloudy climates. On the internal surface of the inner borosilicate glass tube there is a absorber plate which collects the radiation that passes through the glass layer. This absorber plate is mostly of aluminum or copper as both of these metals have a high heat reflectivity and transmitivity quotient. It is also painted black so as to allow it to absorb maximum amount of solar radiation.

*Corresponding author. E-mail: udaya@bitsdubai.com.

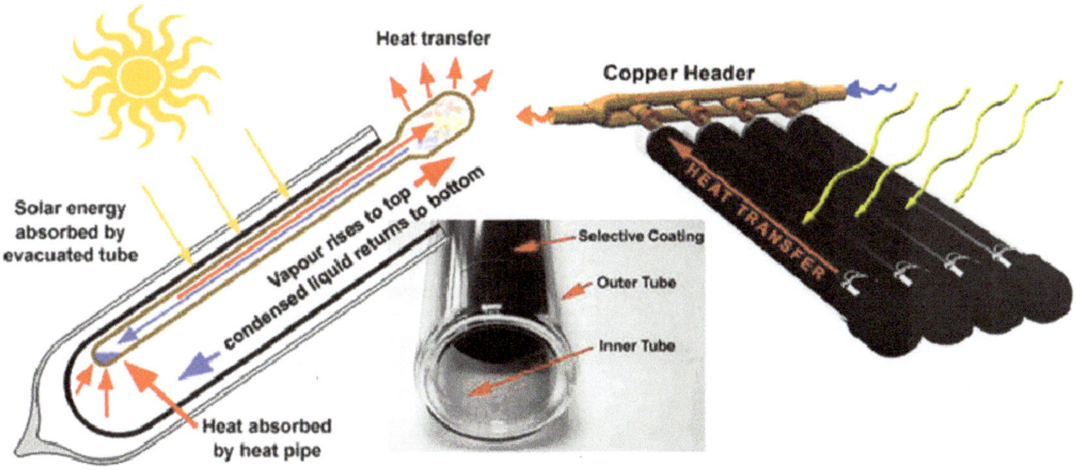

Figure 1. Design and working of evacuated (or vacuum) tubes pane.

Table 1. General parameters for design of evacuated tube collectors (Holman, 2005).

Parameter	Value
Glass tube diameter	65 mm
Glass thickness	1.6 mm
Collector length	1965 mm
Absorber plate material	Copper
Coating	Selective
Absorber area	0.1 m^2

Table 1 shows the typical dimensions of the vacuum tube collectors with type of used material. The glass tube diameter should be just enough to allow maximum amount of solar radiation to transmit through the collector system. The absorber tube is a blackened surface so it acts as a black body and ideally has $\alpha = \varepsilon \sim 1$ (Holman, 2009). Because the absorber area has to be maximized, the absorber plate is placed in direct contact with the inner glass tube. Evacuated tube collectors are used for a variety of applications such as: space heating and cooling, industrial and domestic hot water requirements. The design of evacuated tube collectors is complicated and is based on the thermos flask effect that is. heat loses from the low intensity radiation emitting from the interior tube are minimized. Various brands are available with similar designs and most manufacturers are based in Germany and in United States. Since Germany is a big market for solar energy conversion based products, their products are much more accomplished and have a better quality.

To enable the heat output of the evacuated tube collectors to power the generator of an absorption chiller, a pool type heat exchanger needs to be retrofitted to the generator section. This is done to facilitate optimum heat exchange and to maintain regulated flow.

Viability of use of evacuated tube collectors

A lot of research has been performed to find the viability of the use of evacuated tube collectors for applying them in single direct heat exchanger type absorption chiller. Various factors must be considered before designing a solar powered system.

Intensity of solar radiation

Measurements of intensity of solar radiation are important because of the increasing number of solar heating and cooling applications. Because the rate of receipt of solar radiation on a given surface on the ground depends on the orientation of the surface with reference to the sun it is important that we design the system is based on solar variables (NASA, 2007).

1. Performing the study of various products available in the market and finding which one is feasible for the particular application.
2. Finding out the heat transfer characteristics and solution of the thermal network of various components (Jean, 2005).
3. In order to model the evacuated tube collector, a number of simplifying assumptions have to be made (Jean, 2005).
4. These assumptions must be in accordance with actual values used in industry.

As most of the products are certified by various test labs in Europe and America, they conform to a certain test standard. The ANSI/ASHRAE test standard 93-2003 specifies the design of solar collectors and stipulated the

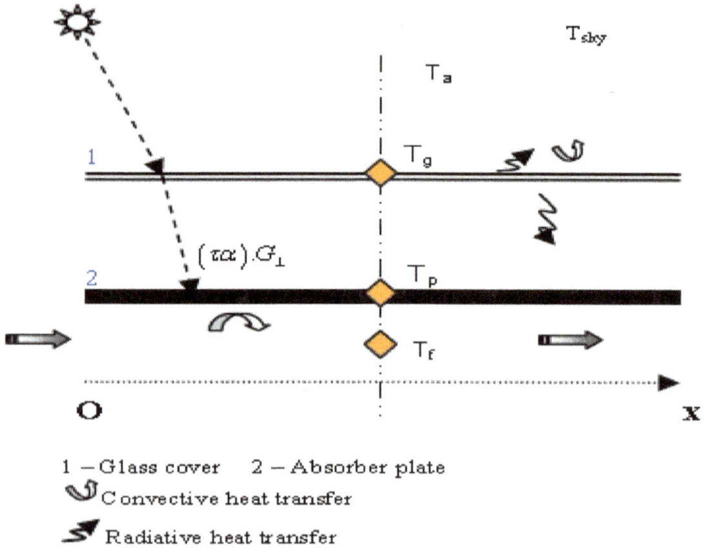

Figure 2. General description of the collector model (Jean, 2005).

Table 2. Properties of the outer borosilicate tube.

Property	Value
Transmitivity (τ)	0.92 (given)
Refractive index, n	1.474
Mean diameter	47 mm
Length	1500 mm

minimum criteria for a product to be made commercially available.

THERMAL ANALYSIS OF EVACUATED TUBE COLLECTORS

To create a thermal circuit so that heat transfer from the surface of the collector tube to the copper tubes can be explained. The amount of solar radiation incident on the face of the tube can be determined from the empirical relations. In effect this analysis is only suitable for a similar value of solar radiation.

General description and thermal network of collector tubes

The evacuated tube collectors have an outer and an inner casing of borosilicate glass 3.3. These tubes are sealed at each end and between them a vacuum is maintained to allow minimal heat loss due to radiation from the absorber plate. As solar radiation is incident on the outer layer of the borosilicate glass tube, a part of it is absorbed, part reflected back into the atmosphere and most of the radiation is transmitted to the next layer of glass tube. The properties of the outer borosilicate tube are given in Table 2. This arrangement is shown in Figure 2. Since,

$\tau = 0.92$
$\alpha = 1 - 0.92 - 0.04 = 0.04$

From the empirical relations, the insolation at Dubai's latitude and longitude incident on a horizontal collectors is measured to be approximately = 8.35 KWh/m^2 day. Figure 3 depicts a typical solar radiation curve over a clear day. The readings on the graph are been plotted every 15 min to obtain accuracy and correctness of the readings. The graph clearly explains the gradual rise and fall of the intensity levels of radiation. The area under the curve of this graph gives the value in KWh/m^2 day as radiation is only available from 7 AM to nearly 7 PM depending upon the season. The data calculated indicates a value of 8.35 KWh/m^2 day incident in Dubai, it represents the entire area under the curve of the Figure 3. To obtain the peak value of insolation, it is required that this value is divided with an approximate number of bright sunshine hours.

Assuming that May 15 is a clear day with no clouds and 8 h of bright sunshine, the radiation received/m^2 at every incident would be = 8.35/8 ~ 1050 KW /m^2. The further calculations are based on this value of incidence solar radiation flux.

$$G\,\alpha_{sun} = \alpha_{low\ temp}\,\sigma\,(T^4 - T^4_{surr})$$

Assuming the value of α_{sun} = 0.12 as given in Rai (2009), and the T_{surr}= 40°.

The temperature of the outer glass plate is calculated to

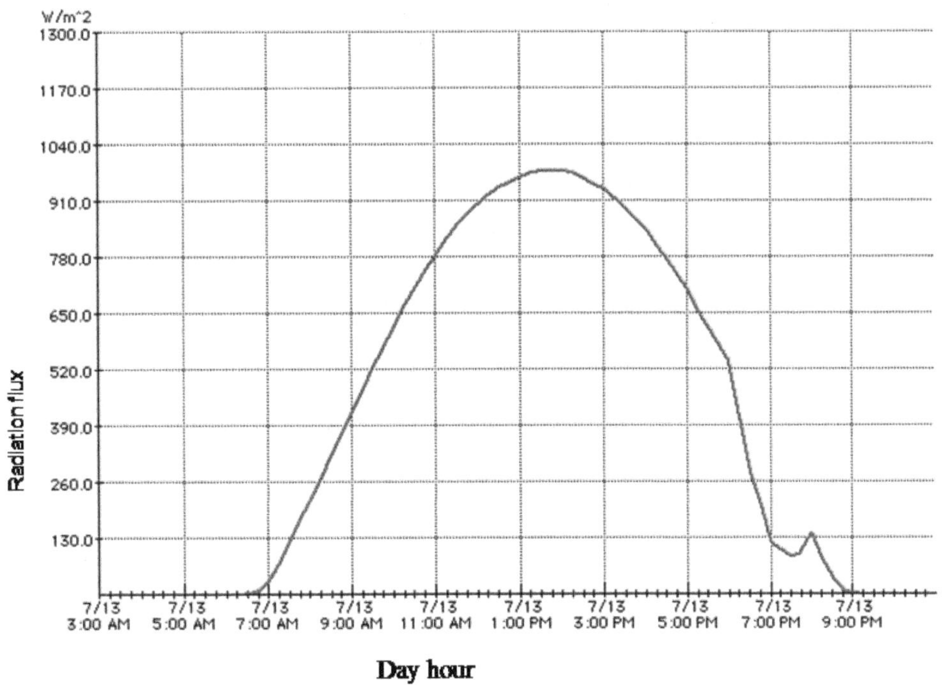

Figure 3. Daily solar radiation map.

be 232°.

It is further assume that the total incident energy available on the surface is;

G x Aperture Area.

Total incident energy = 1050 x 0.110685
= 116.219 W.

The temperature of the outer tube is determined and it is therefore required to calculate the temperature of the inner glass tube. As the two tubes are separated by a vacuum and we have assumed that no convection and only transmission takes place from the top layer.

$\text{Energy}_{trans} = \tau \times \text{Energy Incident}$.
$\text{Energy}_{trans} = 0.92 \times 116.219$ W
= 106.92171 W

Using the aforementioned equation:

$106.92171 = 3.14 \times (0.037) \times 1.5 \times 0.04 \times 5.669 \times 10^{-8} \times (T^4 - 313^4{}_{surr})t = 180°$

At steady state, the inner borosilicate glass tube is maintained at a temperature of 180°C. This is shown in Figure 4. The absorber plate is in contact with the glass tube and at steady state, the temperature of the plate approximately equals the temperature of the inner glass tube. Since the aluminum absorber plate is a black body, its absorptivity and emmisivity are assumed to be 1 that is it absorbs the entire solar radiation incident on it and is capable of reflecting all radiation onto the Cu tube.

Heat transfer coefficient across the air gap

The heat available at the absorber tube is transferred to the copper tubes that carry the transfer fluid. This thermal exchange is both convection and a radiation problem. The amount of heat transfer through convection cannot be calculated unless the heat transfer coefficient of the medium is known. Figure 5 is a schematic diagram of a single copper tube running at the centre of the glass-tube absorber assembly, to arrive at an approximate value in estimation of the total transfer of heat, the area is adjusted in the final heat transfer equation.

Heat transmitted by the absorber place = 106.9271 W
Temperature of the plate = 180°

The diameter of the tube is assumed to be 10 mm. And the nature of the air gap is assumed to be vertical.

To calculate the heat transfer coefficient of the air gap, it is desired to perform a numerical analysis with the 1st iteration assuming the temperature of the copper tube to be 150°C and under ideal conditions when maximum heat is transferred from the absorber plate to the Copper tubes, the amount of heat transferred is the same as the amount of heat absorbed by the absorber tube.

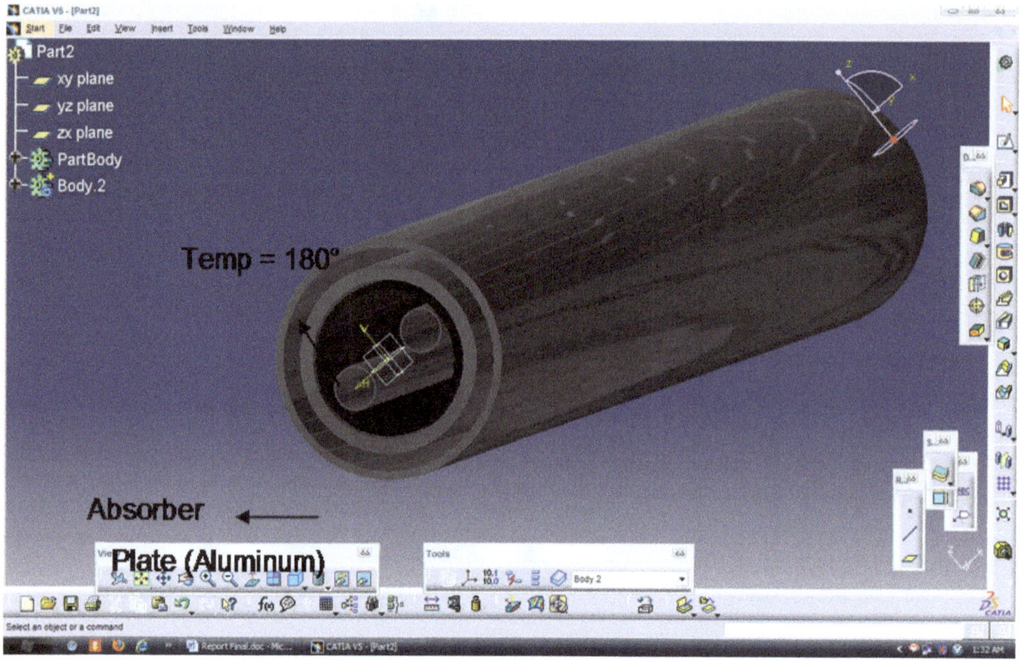

Figure 4. Model of the evacuated tube collector depicting the temperature of the absorber plate.

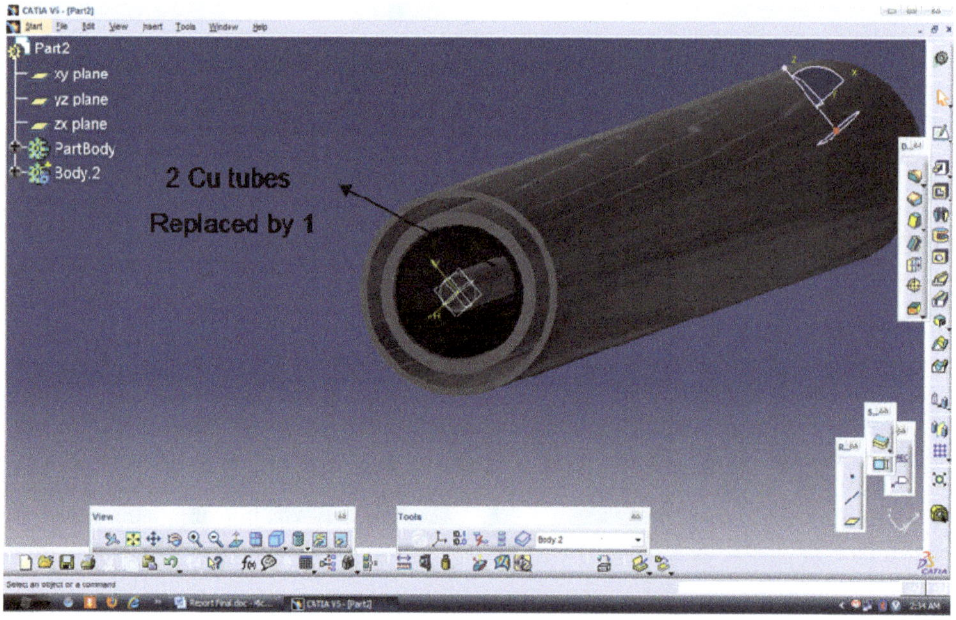

Figure 5. Depiction of the model of the evacuated tube collector with stated assumptions.

It can be noticed from Figure 6 that the heat transfer coefficient of the vertical air gap has refined itself to a particular value. At the start of the iteration the slope of the graph indicates a massive deviation from the actual value. As the number of iterations increases, the heat transfer coefficient holds to a constant value thereby indicating that the observed value of the variable is approaching its real value.

The refining of the value further suggests that the indicated value of heat transfer coefficient can be utilized to calculate the convective heat transfer between the absorber plate and the Copper tube. It can be noticed from graph in Figure 7 that the temperature of the central copper tube has refined itself to a particular value. At the

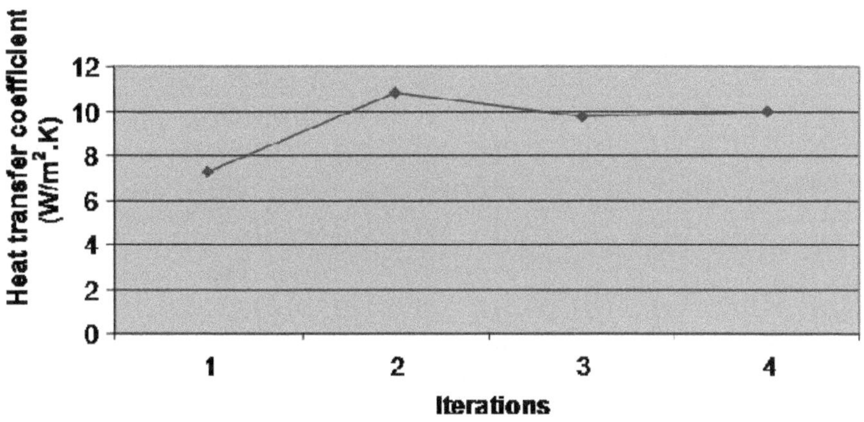

Figure 6. Numerical approximation of heat transfer coefficient with iteration

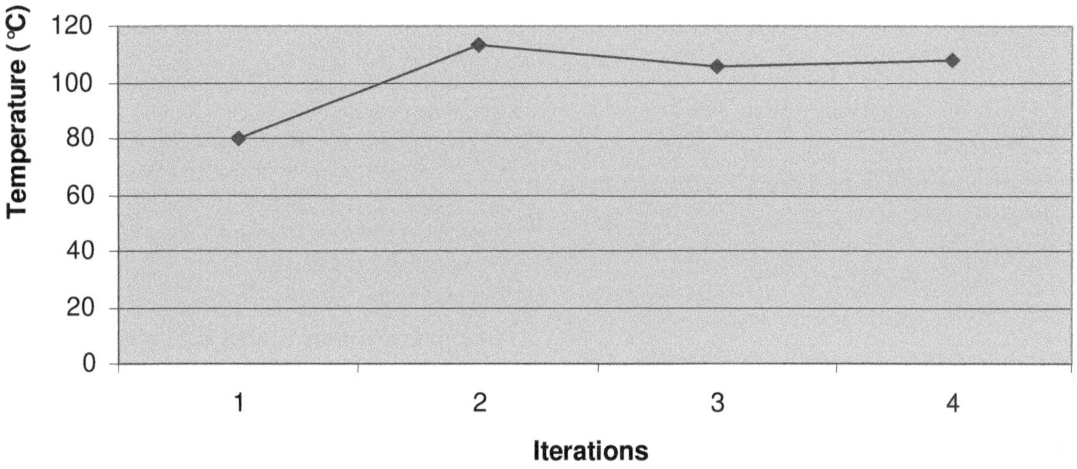

Figure 7. Numerical approximation of Cu tube temperature with iterations.

beginning of the iteration the slope of the graph illustrates a massive deviation from the actual value. With the increment in number of iterations, the slope of the graph decreases thus suggesting that the observed value of the variable is nearing its real value. The refining of the value further suggests that the indicated value of copper tube temperature can be utilized to calculate the convective heat transfer between the absorber plate and the Cu tube.

Calculation of the total heat transfer

The design of the evacuated tube collectors is such that a vertical gap exists between the absorber and the Cu tube; this facilitates convective heat transfer from the absorber plate and because they are maintained at different temperatures, heat transfer also takes place through radiation (NASA, 2007).

Total heat transfer:

$$Q_{total} = Q_{rad} + Q_{conv}$$
$$= 54.71 + 24.60 \text{ W}$$
$$= 79.31 \text{ W}$$

The mass flow rate of water into the collector array is approximately in the range of 15 to 30 LPH.

Assuming the output of the collector is fixed at 20 LPH:

This corresponds to a flow rate $= 5.55 \times 10^{-6}$ m^3/s
The cross sectional area of the Cu Tube $= 3.14 \times 10^{-4}$ m^2
Density of water $= 980$ kg/ m^3
Mass flow rate $= 5.439 \times 10^{-3}$ kg/s

Assuming there is no other heat loss and all the energy transferred from the absorber to the Cu tubes, we can calculate the outlet temperature of water.

$Q_{total} = m\ Cp\ (\Delta\ T)$.

At steady state we assume the inlet temperature of water

Table 3. Performance characteristics of commercially available solar collectors.

Description	Consolar (Tubo 12)	Viessman (SD2 20)	Ritter (CPC 12 OEM)
Number of tubes	12	20	12
Optical efficiency	0.62	0.82 #	0.66
1st order heat loss	0.395	1.62 #	0.82
2nd order heat loss	0.02	0.0068 #	0.0064
Aperture area	2.06	2.11	2.0
Angle correction K @ perpendicular	1.04	1.04	1.04
Other parameters			
Operating Pressure (max)	10 Bar	6 Bar	10 Bar
Stagnation Temperature	250°	282°	292°

Efficiency with respect to absorber area = 2.05 m².

to be 70°C. Most commercially available evacuated tube collectors allow 6 tubes to be connected in series.

$79.31 \times 6 = 5.439 \times 10^{-3} \times 4220 \times (T-70)$

$T \quad = 90.732°C$

Hence with a series connection of 6 tubes, a temperature difference of 20.72°C can be targeted and if the water inlet is considered to be 70°C at steady state, an outlet temperature of 90.732°C can be achieved.

Study of commercially available evacuated tube collectors

For the purpose of the study 3 German suppliers were identified for evacuated tube solar collectors. As efficiency of these collectors is a function of output temperature, desired output temperature was fixed at 120°C. The insolation value was obtained assuming optimum tilt of the module.

Test products:

1. Consolar Solare Energiesysteme GmbH- (Tubo 12)
2. VIESSMAN-Vitosol Range- (SD2 20)
3. Ritter Solar- (CPC 12 OEM)

Table 3 discusses the performance characteristics of commercially available solar collectors mentioned above. Efficiency is expressed as a function of incident radiation and temperature difference. To power the generator of an absorption air conditioner temperature required is in the range 90 to 120° at certain flow rate. at 10 bar = 179.88° and at 6 bar = 158.84°.

Therefore, when the maximum attainable temperature difference is taking average water inlet was 25°C and Ambient temperature is assumed to be at 30° (160 + 25)/2 = 92.5. Temperature Difference, $\Delta T = 92.5 - 30 = 62.5°$. Temperature of about 110 - 120° can be attained in evacuated tube collectors. Temperature difference corresponding to $120^{\circ} = 42.5°C$

From graph in Figure 8 we observe efficiency corresponding to 42.5°C temperature difference to be around 65%. Table 4 summarizes the overall efficiency and heat output of various collectors.

The available heat from the collector array is used to heat the refrigerant-absorbent solution in the generator of an absorption chiller. The design of the generator and the selection of the tubes play an important part in facilitating optimum heat transfer.

DESIGN OF THE GENERATOR OF AN ABSORPTION CHILLER

Single effect water-lithium bromide systems are ideal for air conditioning operations where the refrigerant flow normally does not exceed 10°C under normal conditions. The generator of these systems can be powered easily with a low grade energy source such as hot water from evacuated tube collectors. The hot water heats up the refrigerant-absorbent mix in a pool boiling type heat exchanger. Water-Lithium Bromide solution enters the generator at a particular concentration and when heat is supplied, fraction of the water in the solution vaporizes leaving behind a strong solution of water and lithium bromide in the heat exchanger. The change in concentration of the solution affects its heat transfer coefficient. The heat transfer coefficient is found to increase as the heat flux increase. Average heat transfer coefficients were found to vary between 1600 and 7500 W/m2°C (Kalogirou, 2001).

Design of the generator

The generator of a Water-Lithium Bromide absorption chiller is a pool type heat exchanger that is capable of extracting heat from low grade energy source such as hot water available from a solar water heater. The design of the generator is specific to the use of evacuated tube

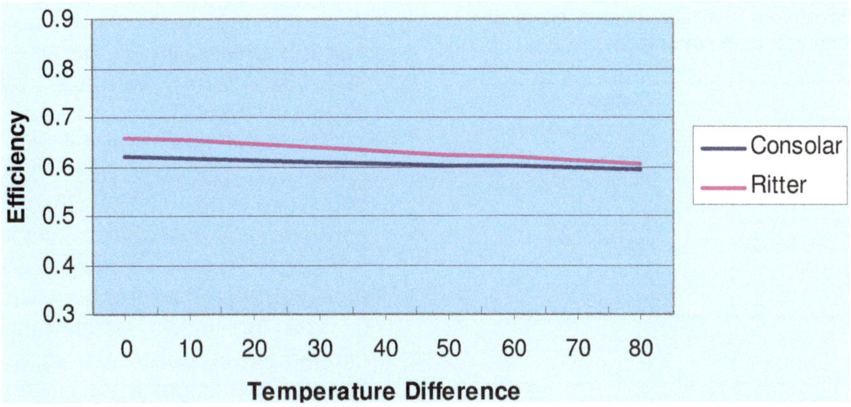

Figure 8. Efficiency as a function of the mean temperature difference at W/m^2.

Table 4. Overall efficiency and heat output of various collectors.

Description	Consolar (Tubo 12)	Viessman (SD2 20)	Ritter (CPC 12 OEM)
Overall efficiency @ dT= 42.5	0.60601	0.7626 #	0.630958
Heat output, Q_{out}	1498.05 W	1876.05 W	1514.30 W

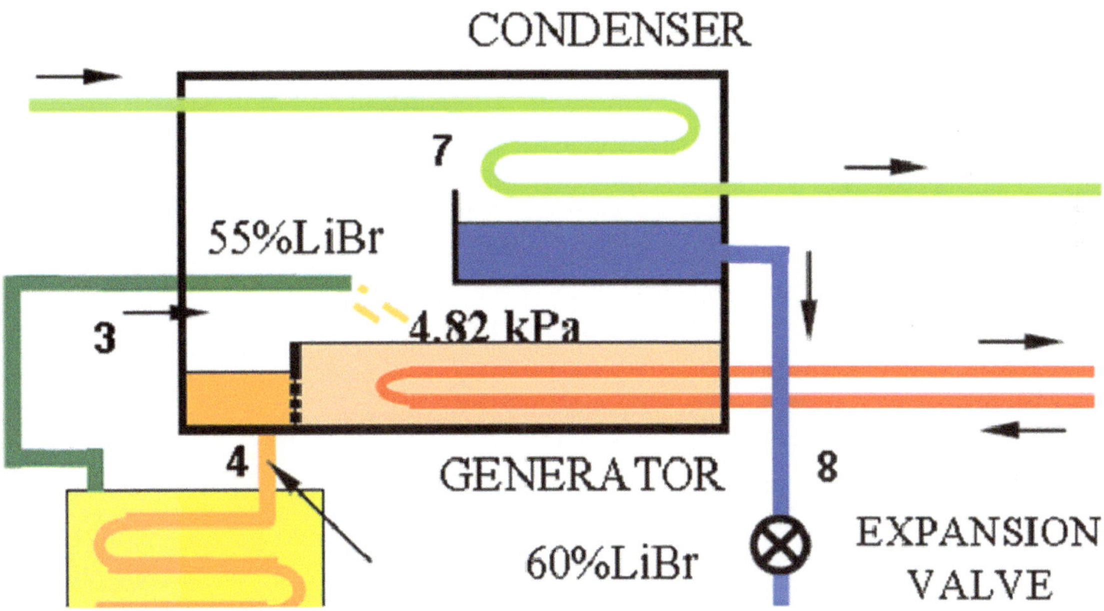

Figure 9. Typical conditions for an absorption chiller with 1KW Capacity (Duffie and Beckman, 1980).

collectors. Typical conditions for an absorption chiller with 1KW capacity are shown in Figure 9 and listed as follows (Kalogirou, 2001):

Generator pressure: 4.82 KPa
Generator solution Entering: 55% LiBr
Leaving: 60% LiBr
Generator water vapor mass flow rate: 0.000431 kg/s

With these typical values, the amount of heat transferred in the generator can be determined. As the generator is a pool type heater and the heat is utilized to raise the sensible heat of the water-Lithium Bromide Solution and to vaporize a part of the fluid. Lithium Bromide from the solution does not exert any vapor pressure on the liquid therefore; analyzer and dephlagmator are not required as ancillary units. Using the dimensions aforementioned, a

Table 5. Specifications of heat exchange tube in a pool type heat exchanger in the generator.

Particulars	Value
Material	Copper
Flow rate	1.85×10^{-5} m^3/s
Length	2.1673 m
Inner diameter	10 mm
Outer diameter	11 mm
Total capacity	1575 W

generator heat exchanger for an absorption refrigerator having the refrigerating effect of 1 KW can be designed. The specifications of this design are mentioned in Table 5.

CONCLUSION

The evacuated tube solar collectors provide a substantial difference in the temperature of the working fluid. The amount of heat transferred to the working fluid depends on the incident flux of solar radiation and on the aperture of the tubes. The heat transfer coefficient of the air gap in the tubes plays a major role in facilitating the heat transfer. Moreover, the report uses typical data for the design of a 1 KW absorption unit and analyse the design of the generator section using the heat available from evacuated tube collectors. The following conclusions could be drawn from the analysis.

1. The average influx of solar radiation in Dubai (Latitude 25°18'N, 55°20'E) was calculated and verified at 8.35 KWh/m^2.day. With 8 h of estimated sunlight 1050 W/m^2 is available for thermal conversion.
2. The temperature of the absorber plate is theoretically determined to be 180°C and the heat transfer to the Copper tube is suggested to be due to both convection and radiation.
3. Numerical analysis is used to calculate the heat transfer coefficient on the air gap: the value is calculated as 9.992 W/ m^2.K with the temperature of the copper plate as 107.62°C.
4. The total heat transfer from the array of 6 collector tubes in parallel was determined as 475.86 W commensurate with the performance of most commercially available collectors.
5. Approx 20 tubes are required to power the generator of an Absorption machine with 1 KW (0.285 TR) capacity.
6. The log mean temperature difference in the heat exchanger is calculated to be 17.65°C with the heat transfer coefficient 1192 W/ m^2.K. The generator must be a single pass pool type heat exchanger with 2.2 m long copper tube.

REFERENCES

Arora CP (2009). Refrigeraion and Air Conditioning, 3rd Edition, Tata McGraw Hill, India, pp. 402-431

Duffie JA, Beckman W (1980). Solar Engineering of thermal Processes, 1st Edition, John Wiley & Sons Inc., pp. 3-39, 147-178, 216-239, 250-301.

Holman JP (2005).Heat Transfer, 8th Edition, Mc Graw Hill, Singapore, pp-336-360, 394-484, 553-587.

Jean PP, François G, Franck L (2005). Dynamic modeling and elements of Validation of Solar Evacuated Tube Collectors, Ninth International IBPSA Conference, Montreal, Canada.

Kalogirou S (2001). Design and Construction of a Lithium Bromide Water Absorption Refrigerator, Clima (2000) /Napoli World Congress, Napoli, Cyprus.

Kalogirou S (2009). Solar Energy Engineering: Processes and Systems, 1st edition, Elsevier Publications, London, UK, pp- 53-99, 123-160.

NASA (2007). surface Meteorology and Solar Energy-<http://eosweb.larc.nasa.gov>, Software for Measurement of Solar Radiation.

Rai GD (2009). Non Conventional Energy Sources, Khanna Publishers, 4th edition, New Delhi, pp. 47-122. http://pvcdrom.pveducation.org/SUNLIGHT/RADDATA.HTM

Salah AA (2009). Masheiti , Modelling of lithium-bromide (LiBr) chillers and organic Rankin cycle (ORC) powered by low-temperature geothermal heat source, Research, New Castle University.

6

Methods of water saturation estimation: Historical perspective

Andisheh Alimoradi[1]*, Ali Moradzadeh[1] and Mohammad Reza Bakhtiari[2,3]

[1]Department of Mining, Petroleum and Geophysics Engineering, Shahrood University of Technology, Shahrood, Iran.
[2]Department of Petroleum Engineering, Amir Kabir University of Technology, Tehran, Iran.
[3]National Iranian Oil Company, Exploration Directorate, Tehran, Iran.

One of the most important tasks in reservoir engineering is characterizing different parameters of the reservoir. Water saturation is a parameter which helps evaluating the volume of hydrocarbon in reservoirs. Determination of this parameter started from 1942 by integrating some well logs in clean sandstones. After that, many scientists introduced some equations to validate this procedure in shaly sands and carbonates. To treat the problem of dependency of water saturation estimation on core analysis in previous works, other scientists proposed using rock physics and arrived at improved models of water saturation estimation. More recently, interpreters have used seismic attributes to evaluate water saturation values directly or estimating proper rock physical properties such as shale volume which are useful in water saturation estimation process.

Key words: Characterizing, well log, water saturation, rock physics, seismic attribute.

INTRODUCTION

Hydrocarbon reservoirs are the main properties of exploration and production companies. Most of the reservoirs atleast consist of two different phases of liquids. These phases are gas-water or oil-water. Some of the reservoirs have all of the three phases of gas, oil and water (Dandekar, 2006). Sandstone, as the most famous reservoir rock, has many spaces to reserve hydrocarbon. Carbonate rocks are also important kips to reserve considerable quantities of hydrocarbon.

One of the most important tasks in reservoir engineering is determining the characteristics of reservoirs to estimate their performance in future. Without a proper and accurate characterization, Many prominent errors can enter the reservoir performance prediction. These errors can finaly lead to lose the noticable values of hydrocarbone in extracting process.

Reservoir characterization is a process of describing various reservoir properties using all the available data to provide reliable reservoir models for accurate reservoir performance prediction (Jong, 2005). This process can be either qualitative or quantitative. In qualitative characterization the quality of rock is evaluated in order to see if it can be a reservoir rock. In recent years a new topic called improved reservoir characterization or quantitative reservoir characterization, has been propounded between scientists. This characterization has an important role in new reservoirs management. Quantitative reservoir characterization is the process of numeral statement of some characteristics in reservoir such as permeability, porosity, saturation, pressure and pores size. Suitable data in this characterization are cores, logs, production and seismic data.

One of the most important parameters in improved reservoir characterization is hydrocarbon saturation. Kamel and Mabrouk (2002) assumed that all void spaces in a reservoir consist of water and hydrocarbon, therefore:

$$S_h = 1 - S_w \qquad (1)$$

where S_h is hydrocarbon saturation and S_W is water saturation. Since estimation of hydrocarbon saturation is

*Corresponding author. E-mail: andisheh@mine.tus.ac.ir

not generally an easy task, it is recommended to determine the water saturation and predict the saturation of hydrocarbon using Equation 1.

History of water saturation estimation

First time, in 1942 Archie proposed that it is possible to determine the value of water saturation for a clean (non shaly) formation using porosity and electrical resistivity values of the formation (Archie, 1942):

$$S_W = \left[\frac{\alpha . R_W}{\varphi^m . R_t}\right]^{\frac{1}{n}} \tag{2}$$

In this equation, S_W is the water saturation of the clean formation, R_t is the corrected total electrical resistivity of formation obtained from resistivity logs, φ is porosity of the rock obtained by porosity logs such as acoustic or density, R_W is the water resistivity obtained from self potential (SP) logs or production tests, α is a constant which depends on rock type and tortuosity of the fluid path, and m and n are cementation and saturation exponents which are constant. The value of α is 0.6 for unconsolidated sand stones, 0.8 for consolidated sand stones and 1 for carbonates (Kamel and Mabrouk, 2002). Cementation exponent can get values between 1.6 to 2.4, but generally it is equal to 2 and the value of saturation exponent is diferent from 1.7 to 2.2 (Moradzadeh and Ghavami, 2001).

It should be mentioned that there are other well logs which can give the values of water saturation. As an example by combining the density log with hydrogen index obtained from NMR log, it is possible to determine the water saturation value. This procedure is only sensitive to the gas saturations and can be applied in the regions near the wells (Ellis and Singer, 2007; Holstein, 2007). Therefore the most important well logs in water saturation measurements are resistivity logs.

The main problem in application of Archie's method is determination of R_W when there is no enough production tests or SP log data (as is common in many oil fields). The other problem is the precision of the estimated φ when the rock matrix type is unknown. In addition, there is an inherent uncertainty in estimations of m and n values. To solve these problems, Hingle (1959) and Pickett (1963) presented a graphical method using two different plots that were essentially obtained from the Archie formula (Hingle, 1959; Pickett, 1963). Although these plots better predict the water saturation, their of some simplistic assumptions in Archie's formula and consequently in Hingle and Pickett's plots. More specifically, it is not possible to use these methods in shaly (unclean) and heterogeneous formations. In shaly sands, the presence of clay adds an additional conductivity.

This additional conductivity will cause an error in water saturation estimation (Dandekar, 2006). In carbonate formations, wide range and irregular distribution of pore sizes change the rock conductivity and adversely affects the precision of Archie's formula (Van Golf-Rocht, 1982).

To alleviate this problem in shaly sands, works have been ongoing towards better determination of the effects of shale in formations (Patnode and Wyllie, 1950; Winsauer and McCardell, 1953; Wyllie and Southwick, 1954; Waxman and Smits, 1968; Poupon and Leveaux, 1971; Clavier et al., 1984; Worthington, 1985; Herron, 1986; Sen et al., 1988; Schwartz and Sen, 1988). Worthington introduced four general kinds of equation which consider the effect of shale in sands to determine the values of water saturation as bellow:

$$C_t = \frac{C_W}{F} S_w^n + X \tag{3}$$

$$C_t = \frac{C_W}{F} S_w^n + X S_w^s \tag{4}$$

$$\sqrt{C_t} = \sqrt{\frac{C_W}{F}} S_w^{n/2} + \sqrt{X} \tag{5}$$

$$\sqrt{C_t} = \sqrt{\frac{C_W}{F}} S_w^{n/2} + \sqrt{X} S_w^{S/2} \tag{6}$$

In these equations, C_t is total conductivity of fully saturated sand, C_W is water conductivity, S_w^n is water saturation of non-shaly sand, S_w^s is water saturation of shaly sand, *n* and *s* are saturation exponents for non-shaly and shaly sands, *X* is the additional conductivity added by clay and *F* is the formation factor and equals to:

$$F = \frac{R_0}{R_W} = \frac{\alpha}{\varphi^m} \tag{7}$$

application is limited in clean formations. This is because in the first kind of the Worthigton equations it is assumed that shale and sand are independent from each other in conducting the electrical current and shale has not been affected by hydrocarbon. In situation of

Table 1. The values of X according to different values of Δt_{ma} (Kamel and Mabrouk, 2002)

Matrix	$\Delta t_{ma} (\mu s / ft)$	X
Silica	55.5	1.60
Calcite	47.6	1.76
Dolomite	43.5	2.00

scattered shale, hydrocarbon will influence the shale. According to interaction between shale and sand (or shale and saturation), a third parameter will appear in the model (Models 4, 5 and 6). A classic example of the fourth kind of Worthington model is Indonesian equation introduced by Poupon and Leveaux (Equation 8). This equation is empirical and is valid for most shaly reservoirs in Indonesia (Poupon and Leveaux, 1971).

$$C_t = \frac{C_W S_W^2}{F} + 2\sqrt{\frac{C_W V_{sh}^{(2-V_{sh})} C_{sh}}{F}} S_W^2 + V_{sh}^{(2-V_{sh})} C_{sh} S_W^2 \tag{8}$$

Although all these methods could potentially detect the effect of shale and yield more realistic values of the formation conductivity and water saturation, a major obstacle remains in these methods due to their strong dependency on cores and logs analyses which are costly and time consuming. Opposite to shaly sands, there are not any highlighted works on the estimation of water saturation in carbonate formations. A few have worked on the effect of pore size and distribution in the evaluation of water saturation in these kind of rocks (Alger et al., 1989; Obeida et al., 2005; Lucia, 2007) perhaps the most significant contribution is the equation by Lucia.

$$S_W = a \times H^b \times \varphi^c \tag{9}$$

In this equation, H is the reservoir height (vertical thickness of the reservoir zone), *a, b,* and *c* are constant coefficients which are the functions of rock type and grain size. Unfortunately, Lucia formula does also depend on the analyses of cores and logs. To treat the problem of dependency of water saturation estimation on core analysis, Kamel and Mabrouk (2002, 2003) proposed using rock physics and arrived at an improved model of water saturation estimation. Their method is based on two equations as follows:

i) A first equation to determine the water saturation using a combination of Archie and Raiga formulas with two logs of acoustic and electrical resistivity (Kamel and Mabrouk, 2002) as follows:

$$S_W = \sqrt[m]{\frac{\alpha R_W}{\left[1 - \left(\frac{V_P \Delta t_{ma}}{10^6}\right)^{\frac{1}{X}}\right] R_t}} \tag{10}$$

Where V_P is P-wave velocity in rock obtained from acoustic log, Δt_{ma} is the acoustic wave transition time in rock matrix, and X is a parameter obtained from Δt_{ma} using Table 1.

ii) A second equation to determine shale volume in shaly sands using a combination of three porosity logs of neutron, density, and acoustic (Kamel and Mabrouk, 2003).

$$\left(\frac{\rho_{sh}-\rho_{ma}}{\rho_f-\rho_{ma}}\right)V_{sh}^2 - \left(\varphi_N + \frac{\rho_b-\rho_{ma}}{\rho_f-\rho_{ma}} + \frac{\rho_{sh}-\rho_{ma}}{\rho_f-\rho_{ma}} - 2\frac{\Delta_{tsh}-\Delta_{tma}}{\Delta_{tf}-\Delta_{tma}}\right)V_{sh} + \left(\varphi_N + \frac{\rho_b-\rho_{ma}}{\rho_f-\rho_{ma}} - 2\frac{\Delta_t-\Delta_{tma}}{\Delta_{tf}-\Delta_{tma}} \times \frac{100}{\Delta_{tsh}}\right) = 0 \tag{11}$$

In this equation, ρ_{sh} is density of shale, ρ_{ma} is density of rock matrix, ρ_f is density of fluid, φ_N is Notron log porosity, ρ_b is total density, Δ_{tsh} is acoustic wave time in rock matrix, Δ_{tf} is acoustic wave transition time in transition time in shale, Δ_{tma} is acoustic wave transition time in rock matrix, Δ_{tf} is acoustic wave transition time in rock fluid, and Δ_t is total transition time of acoustic wave. The roots of Equation 11 are the values of shale volume.

Kamel and Mabrouk's (2002 to 2003) procedures give reasonably good results of water saturation estimation especially in shaly sands.

A review of literature, as documented here, shows that most valuable contributions to date have been focused on determination of water saturation using well logs and cores data. Graphically illustrated, the process of reservoir characterization will much benefit from more detailed studies in the first part of the process which involves seismic data, as is shown in Figure 1.

There seems to be a lack of coherent methodologies to

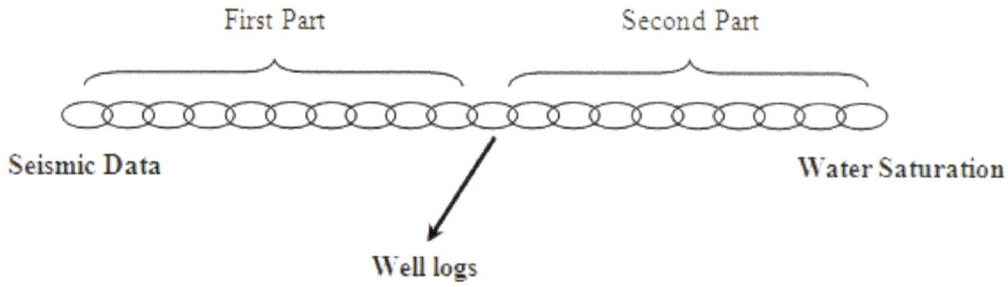

Figure 1. Reservoir characterization chain.

Table 2. Two kinds of error in seismic attributes selection (Kalkomey, 1997).

Decision	Property and attribute are uncorrelated	Property and attribute are correlated
Keep seismic attribute as a predictor	Type I Error	Correct decisio (no error)
Reject seismic attribute as a predictor	Correct decision (no error)	Type II Error

incorporate seismic data into water saturation evaluation instantly; alternatively such procedures could be used to determine only the necessary well logs that could help estimation of water saturation level to complete the first part of the chain in Figure 1.

Seismic attributes in reservoir characterization

Generally, the three most important aspects of seismic reservoir characterization are considered to be quality of seismic data, determination of proper seismic attributes, and existence of a physical relationship between attributes and the reservoir property of interest. Kalkomey (1997) introduced two kinds of errors which may occur during selection of seismic attributes. Table 2 illustrates these errors.

A Type I Error will occur if no relationship exists between the seismic attribute of choice and the reservoir property of interest, yet we opt to use the seismic attribute as a predictor. On the other hand, a Type II Error occurs when a physical relationship exists between the seismic attribute and the reservoir property of interest, but we fail to consider the seismic attribute as a predictor. The cost of a Type I Error is inaccurate prediction biased by the attribute and the cost of a Type II Error is less accurate prediction than if we would have used the seismic attribute.

After selecting the proper attributes, the next step is integration of the attributes to predict reservoir property of interest.

Pan and Ma (1997) introduced three kinds of techniques for integrating seismic attributes:

1. Techniques based on statistical relationships, such as correlation and cross plotting.
2. Techniques based on expert experience and information.
3. Techniques which use artificial neural networks, fuzzy logic, and genetic algorithms to find the relationship between seismic attributes and the reservoir property of interest.

Seismic attributes in water saturation estimation

A review of previous studies on correlation between seismic attributes and water saturation is presented next. Seismic attributes such as amplitude, instantaneous amplitude, and impedance are shown generally well-correlated with water saturation. There also seem to be a consensus amongst researchers on applicability of prestack seismic data and AVO to provide reliable information about liquids and their identification in reservoirs (Van, 2000; Varela, 2003; Li et al., 2007; Zhou et al., 2009). Some of the findings are as follows:

Investigation of oil saturation in a sand stone formation in China using 3D seismic data (Pan and Ma, 1997). Their result is presented in Figure 2. Pan and Ma use real values of oil saturation from three wells, with interpolation throughout the reservoir using 3D seismic data that makes their final results highly dependent on the output of limited number of wells.

Balch et al. (1999) predicted the water saturation in a sandstone reservoir in Mexico using artificial intelligence and seismic attributes. The reservoir in their study had two zones of hydrocarbon (L and K). Three dimensional seismic data and values of water saturation at 19 wells were used first step in a fuzzy logic algorithm to detect five attributes (reflection coefficient, frequency, instantaneous phase, amplitude and energy) that were strongly correlated with water saturation. Then, a back-propagating artificial neural network was used to find the relationship between these attributes and the value of water saturation. The results of training the network is

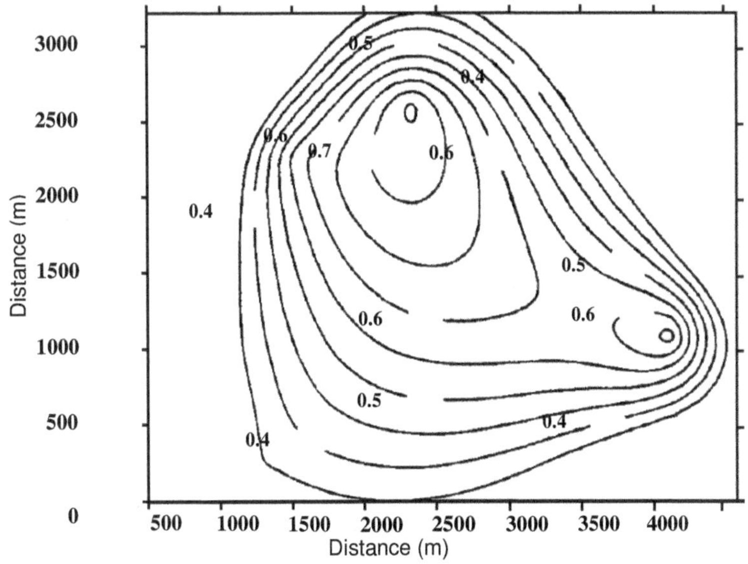

Figure 2. Oil saturation values (Pan and Ma, 1997).

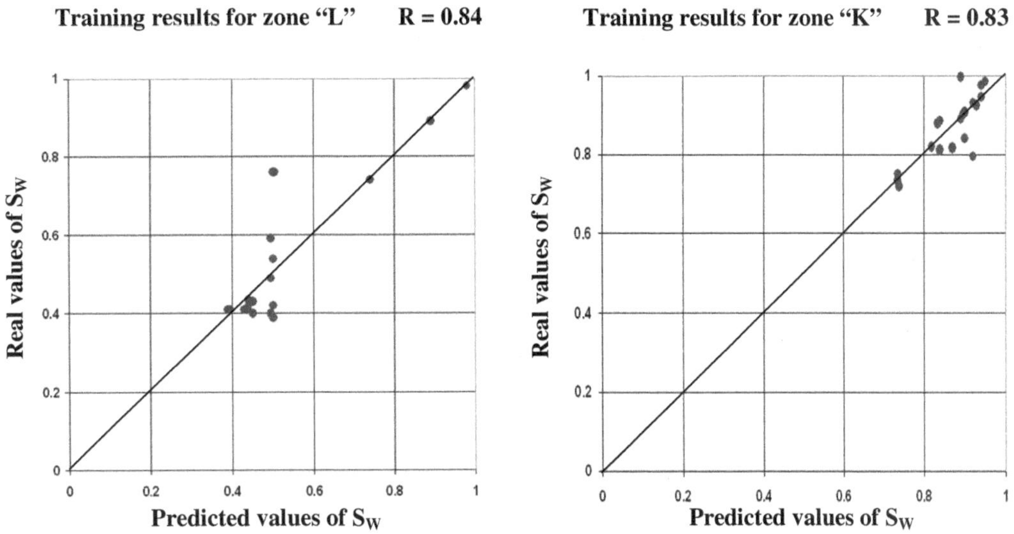

Figure 3. Correlation coefficient between real and predicted values of water saturation in zone "L" (left panel) and "K" (right panel), from Balch et al. (1999).

shown in Figures 3. The main contribution in the study by Balch et al. (1999) is using seismic attributes for water saturation estimation.

Boadu (2001) studied the effect of change in oil saturation level on seismic wave velocities and their ratio in a laboratory experiment. He applied changes to the temperature and values of oil saturation and observed a relationship between these variables and the values of P and S wave velocities and their ratio. He used an artificial neural network to express this relationship.

One of the most interesting studies in this field is the work done by Mu and Cao. They created a physical model of a sandstone reservoir in laboratory scales (Mu and Cao, 2004). They isolated the model and drilled two holes (injecting and discharging) in it. Saturating the sandstone layer with water, oil, CO_2 and CH_4 from 10 to 100 percents respectively, they succeeded in simulating seismic surveying by application of ultrasonic data acquisition; therefore, creating an environment to study the effect of change in the fluids type and saturation value on P-wave amplitude and absorption coefficient (Figures 4 and 5). The outcome of their study was an expression for determining absorption coefficient profile using Biot theory and reflection amplitude spectrum. They applied this formula for a sandstone reservoir in China (Figure 6) and reliably detected the gas zones (Figure 7) that could

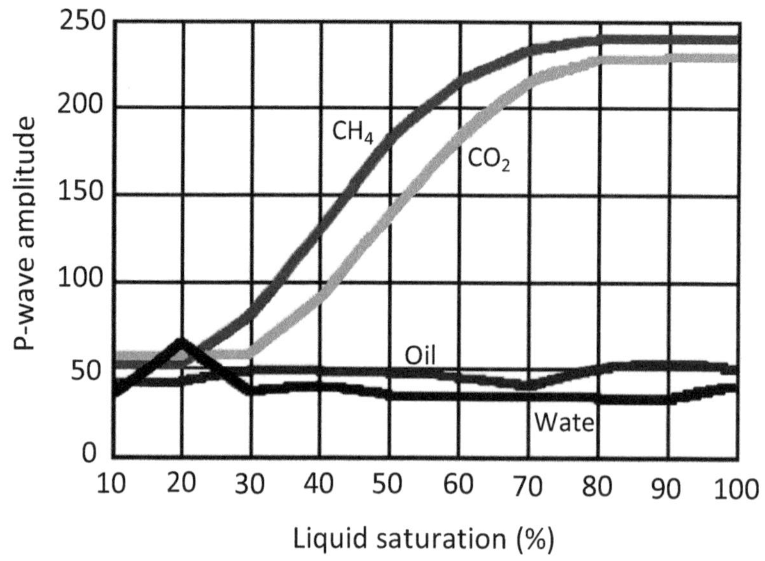

Figure 4. The relationship between amplitude of P-wave reflected from top of the sandstone layer and saturation values of different liquids (Mu and Cao, 2004).

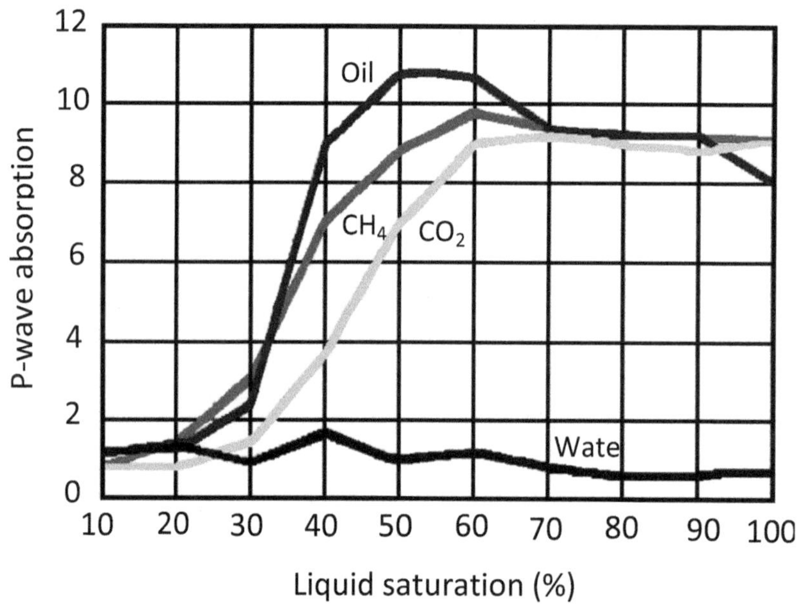

Figure 5. The relationship between P-wave absorption coefficient in sandstone layer and saturation values of different liquids (Mu and Cao, 2004).

not be seen on the initial seismic section in Figure 6. It is clear from Figure 7 that Mu and Cao's formula had detected other zones as gas; further drillings indicated that these zones in fact belonged to the coal seems that lied underneath the gas layer. This potentially presents a problem in the discrete gas and coal detection methodology of Mu and Cao.

Kitamura et al. (2006) studied the effect of water and gas saturation on P and S wave velocity values in sandstone samples. They changed the temperature at restricted pressure value to 130 MPa for each saturation degree and determined the P and S wave velocities for each temperature. Their results are shown in Figure 8.

CONCLUSIONS

Water saturation is one of the most important parameters in reservoir characterization procedure. This parameter can be either predicted from core data, well logs, or seismic attributes directly or can be estimated from an intermediate parameter such as shale volume in sand

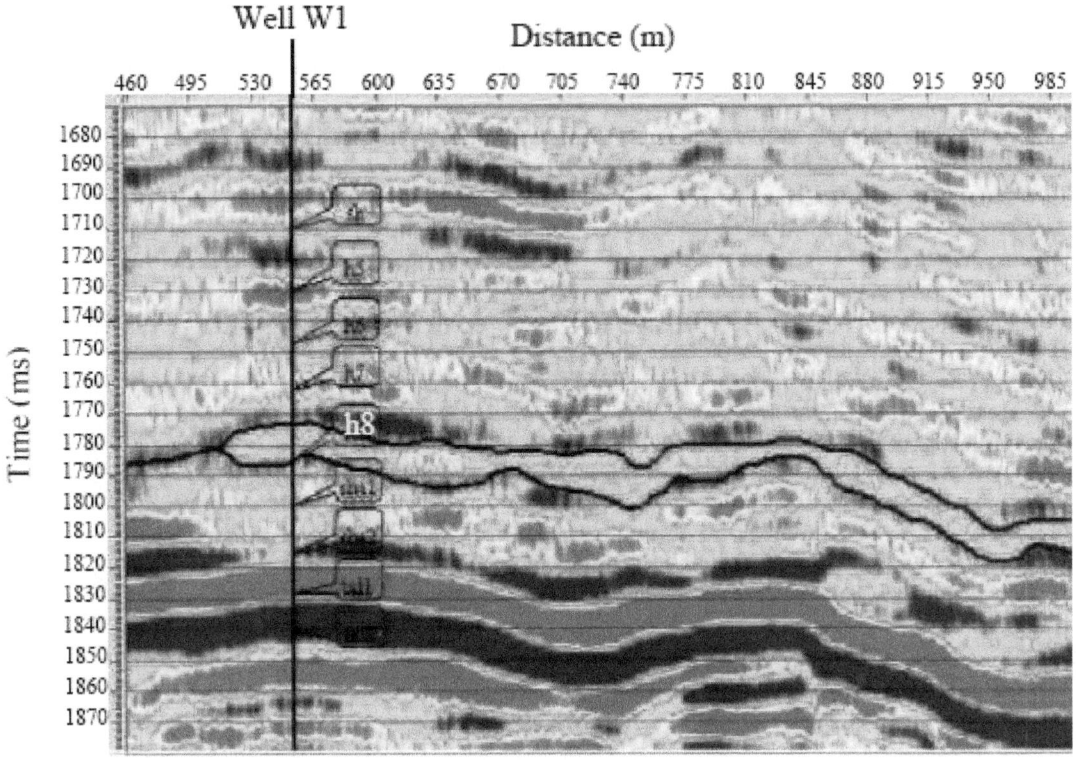

Figure 6. Seismic profile of the reservoir – h8 is a zone of gas (Mu and Cao, 2004).

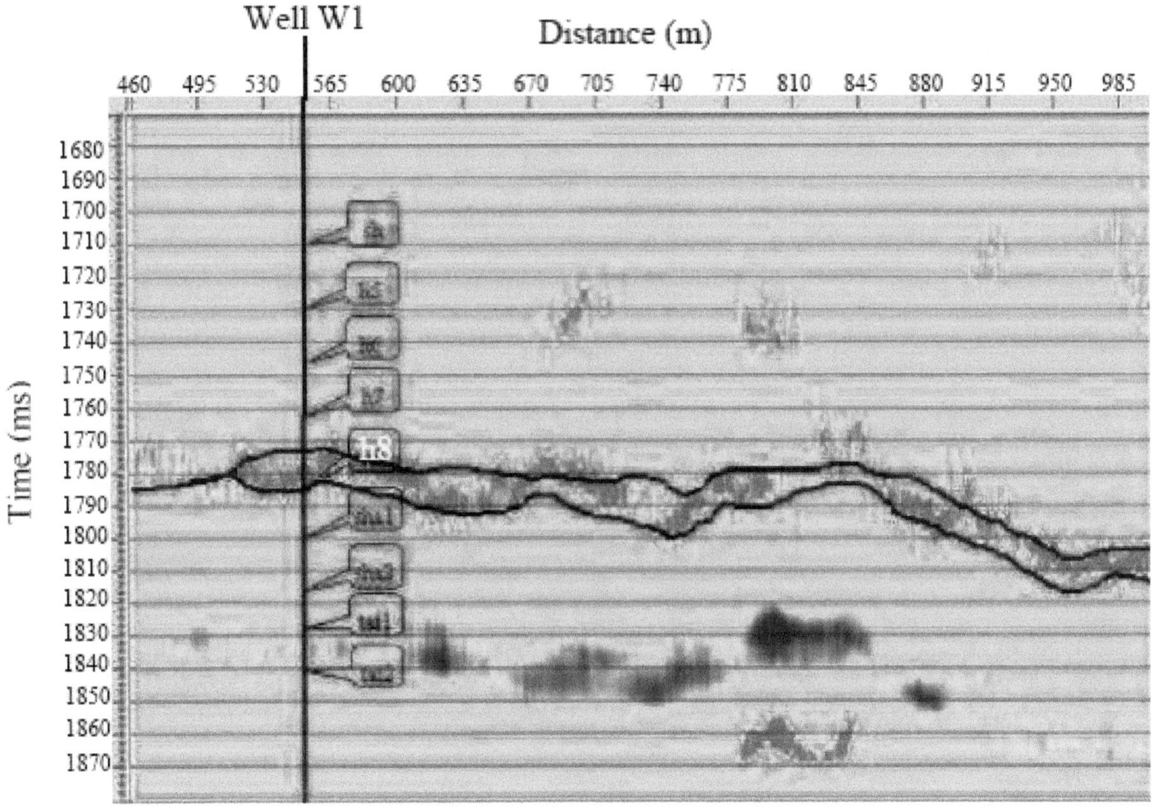

Figure 7. Black zones are zones of high absorption coefficient and indicate the existence of gas in the formation (Mu and Cao, 2004).

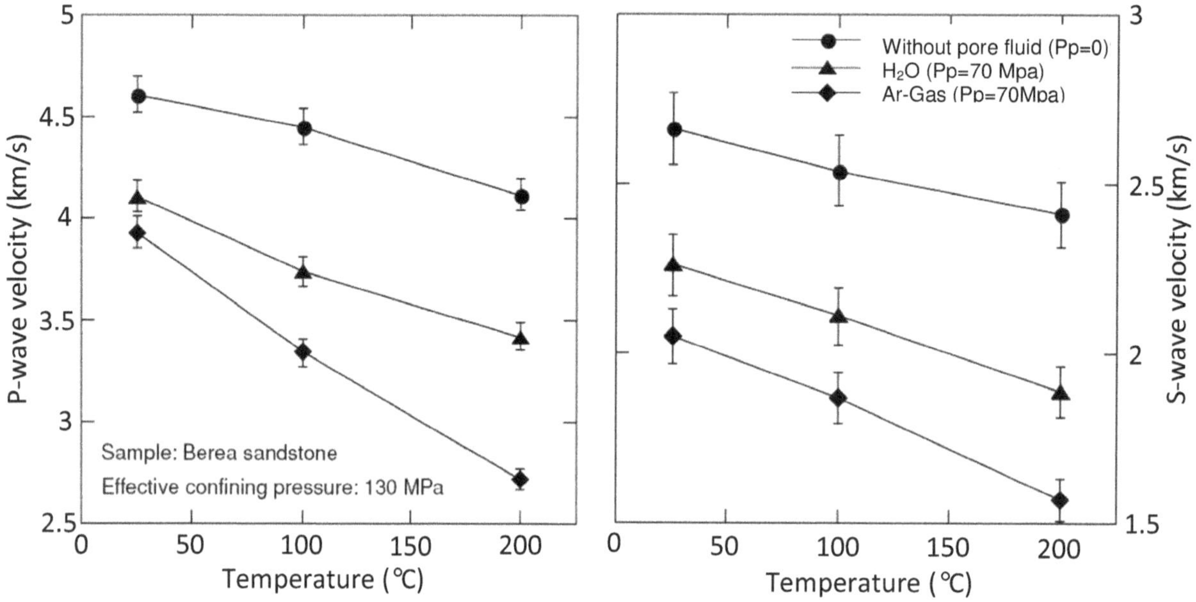

Figure 8. The P and S wave velocity values for dry, saturated with water, and saturated with gas samples (Kitamura et al., 2006).

stone reservoirs. Over the past decades, we have witnessed developments in using well log data to estimate water saturation. This process was started from Archie formula in 1942 and progressed by the works of Patnode and Wyllie (1950), Winsauer and McCardell (1953), Wyllie and Southwick (1954), Waxman and Smith (1968), Poupon and Leveaux (1971), Clavier et al. (1984), Worthington (1985).

To treat the problem of dependency of water saturation estimation on core analysis in previous works, Kamel and Mabrouk (2002, 2003), proposed using rock physics and arrived at an improved model of water saturation estimation. All of the proposed procedures till that time reveal the superior progress in second part of the reservoir characterization chain (from well logs to water saturation estimation).

More recently, interpreters have used seismic attributes to evaluate water saturation values directly or estimating proper rock physical properties such as shale volume which are useful in water saturation estimation process (Pan and Ma, 1997; Balch et al., 1999; Boadu, 2001; Mu and Cao, 2004; Kitamura et al., 2006; Lucia, 2007). These methods apply artificial intelligence computational agents such as "fuzzy logic", "genetic algorithms" and "neural networks" or "statistical approaches" to detect unknown non-linear relationships between different seismic attributes and the reservoir property of interest which is here the water saturation value.

ACKNOWLEDGMENTS

A review paper spanning more than six decades cannot be a one- or two-man job. The authors therefore acknowledge the advice, support, and suggestions given by many prominent individuals, some of whom were actually involved in the development of the methods of reservoir characterization. The many constructive comments and advice received from Professor Majid Nabi Bidhendi of Geophysics Institute of University of Tehran, Dr. Iraj Abdollahi Fard and Mr. Torabi of National Iranian Oil Company, Dr. Misaghi from Research Institute of Petroleum Industry, Dr. Iraj Pirouz and Dr. Abolghasem Kamkar Rouhani from the department of mining, petroleum and geophysics engineering of Shahrood University of Technology is appreciated. The authors also would like to gratefully acknowledge their sponsor: NIOC, Exploration Directorate for their support.

Nomenclature

Sw, water saturation; **Sh,** hydrocarbon saturation; tortuosity factor; **Rw**, water resistivity; φ, porosity; **R_t,** corrected total electrical resistivity of formation; **m,** cementation exponent; **n,** saturation exponent; **C_t,** total conductivity of fully saturated sand; **C_w,** water conductivity; S_w^n, water saturation of non-shaly sand; S_w^s, water saturation of shaly sand; **s,** saturation exponent for shaly sand; **X,** additional conductivity added by clay; **F,** formation factor; **V_{sh},** shale volume; **C_{sh},** shale conductivity; **H,** reservoir height; **V_p,** P-wave velocity; Δ_{tma}, Acoustic wave transition time in rock matrix Δ_{tsh}, Acoustic wave transition time in shale; Δ_{tf}, Acoustic wave transition time in rock fluid; Δ_t, Total transition time of acoustic wave; ρ_{sh}, shale density; ρ_{ma}, rock matrix density; ρ_f, fluid density; ρ_b, total density; φ_N, notron log porosity.

REFERENCES

Alger RP, Luffel DL, Truman RB (1989). New Unified Method of Integrating Core Capillary Pressure Data with Well Logs, SPE, 16793.

Archie GE (1942). The Electrical Resistivity Log as an Aid in Determining Some Reservoir Characteristics, Trans., AIME, 146: 54–62.

Balch RS, Stubbs BS, Weiss WW, Wo S (1999). Using Artificial Intelligence to Correlate Multiple Attributes to Reservoir Properties, SPE, 56733.

Boadu FK (2001). Predicting Oil Saturation from Velocities Using Petrophysical Models and Artificial Neural Networks, J. Petroleum Sci. Eng., 30: 143–154.

Clavier C, Coates G, Dumanoir J (1984). Theoretical and Experimental Basis for the Dual Water Model for Interpretation of Shaly Sands, SPE, 6859.

Dandekar AY (2006). Petroleum Reservoir Rock and Fluid Properties, Taylor and Frencis.

Ellis DV, Singer JM (2007). Well Logging for Earth Scientists, Chapters 4 and 23, Springer.

Herron MM (1986). Mineralogy from Geochemical Well Logging, Clay and Clay Minerals 34: 204-213.

Hingle AT (1959). The Use of Logs in Exploration Problems, Presented at the 29th SEG Annual International Meeting, Los Angeles.

Holstein ED (2007). Petroleum Engineering Handbook, Volume 5: Reservoir Engineering and Petrophysics, Society of Petroleum Engrs.

Jong SL (2005). Reservoir Properties Determination Using Fuzzy Logic and Neural Networks from Well Data in Offshore Korea, J. Petroleum Sci. Eng., 49: 182-192.

Kalkomey CT (1997). Potential Risks When Using Seismic Attributes as Predictors of Reservoir Properties, The Leading Edge, March, pp. 247–251.

Kamel MH, Mabrouk WM (2002). An Equation for Estimating Water Saturation in Clean Formations Utilizing Resistivity and Sonic Logs: Theory and Application, J. Petroleum Sci. Eng., 36: 159– 168.

Kamel MH, Mabrouk WM (2003). Estimation of Shale Volume Using a Combination of the Three Porosity Logs, J. Petroleum Sci. Eng., 40: 145–157.

Kitamura K, Masuda K, Takahashi M, Nishizawa O (2006). The Influence of Pore Fluids on Seismic Wave Velocities Under High Temperature and High Pressure Conditions: Development of a New Technique with Gas Apparatus at AIST, Japan, Earth Planets Space, 58: 1515–1518.

Li Y, Feng J, Jiao M (2007). Prestack Seismic Data Analysis with 3D Visualization – A Case Study, CSPG, CSEG Convention.

Lucia FJ (2007). Carbonate Reservoir Characterization, an Integrated Approach, Second Edition, Springer.

Moradzadeh A, Ghavami RR (2001). Well Logging for Engineers, Shahrood University of Technology Press, First Edition, Shahrood, (In Persian).

Mu YG, Cao SY (2004). Seismic Physical Modeling and Sandstone Reservoir Detection Using Absorption Coefficients of Seismic Reflections, J. Petroleum Sci. Eng., 41, 159–167.

Obeida TA, Al-Mehairi YS, Suryanarayana K (2005). Calculations of Fluid Saturations from Log-Derived J-Functions in Giant Complex Middle-East Carbonate Reservoir, E-J. Petrophysics, 1(1): 1-9.

Pan R, Ma X (1997). An Approach to Reserve Estimation Enhanced with 3-D Seismic Data, Nonrenewable Resources, .6(4): 251-255

Patnode HW, Wyllie MRJ (1950). The Presence of Conductive Solids in Reservoir Rock as a Factor in Electric Log Interpretation, Pet Trans, AIME, 189: 47–52.

Pickett GR (1963). Acoustic Character Logs and Their Application, J Pet Tech June, 659–667.

Poupon A, Leveaux J (1971). Evaluation of Water Saturations in Shaly Formations, SPWLA 12th Annual Logging Symposium, paper O.

Sen PN, Goode PA, Sibbit AM (1988). Electrical Conduction in Clay Bearing Sandstones at Low and High Salinities, J. Applied Physics, 63: 4832-4840.

Schwartz LM, Sen PN (1988). Electrical Conduction in Partially Saturated Shaly Formations, 63rd Annual Conference of SPE, Paper 18131.

Van Golf-Racht TD (1982), Fundamentals of fractured Reservoir Engineering, Elsevier.

Van RP (2000). The Past, Present, and Future of Quantitative Reservoir Characterization, Jason Geosystems, Rotterdam, The Netherlands.

Varela OJ (2003). Stochastic Inversion of Pre-stack Seismic Data to Improve Forecasts of Reservoir Production, Ph.D. Thesis, The University of Texas at Austin.

Waxman MH, Smits LJM (1968). Electrical Conductivities in Oil-Bearing Shaly Sands, Paper 1863-A, SPE J June, 107–122.

Winsauer WO, McCardell WM (1953). Ionic Double-Layer Conductivity in Reservoir Rock, Pet Trans, AIME, 198: 129–134.

Worthington PF (1985). The Evolution of Shaly-Sand Concepts in Reservoir Evaluation, The Log Analyst, SPWLA, 26: 23–40.

Wyllie MRJ, Southwick PF (1954). An Experimental Investigation of the S.P. and Resistivity Phenomena in Dirty Sands, Pet Trans, AIME, 201: 43–56.

Zhou YJ, Tao JQ, Guo YB, Zhang XH, Qiang M (2009). Rock Physics Based Prestack Seismic Reservoir Characterization - Application to Thin Bed, 71th EAGE Conference & Exhibition – Amsterdam, The Netherlands.

Application of nonlinear reduction techniques in chemical process modeling

Ahmmed S. Ibrehem* and Hikmat S. Al-Salim

Department of Chemical and Petroleum Engineering, UCSI University, Kuala Lumpur, 56000, Malaysia.

Model reduction techniques have been used widely in all engineering fields especially in electrical, mechanical as well as chemical engineering. The basic idea of reduction technique is to replace the original system with much smaller state space dimension. A reduced order model is more beneficial to process and industrial field in terms of control purposes by using proper orthogonal decomposition method (POD), which finds applications in computationally processing large amounts of high-dimensional data with the aim of obtaining low-dimensional descriptions that capture a large amount of the phenomena of interest. The discrete version of the POD, which is the singular value decomposition (SVD) of matrices, is described in some detail.

Key words: Model reduction technique, nonlinear model, proper orthogonal decomposition.

INTRODUCTION

The developments of mathematical models are based on: (i) theoretical models using the principles of chemistry and physics, (ii) empirical models, obtained from the statistical analysis of the operating data's process and (iii) semi empirical hybrid models. The issue of development of the nonlinear process model is very important and is more challenging in an industrial setting. The fundamental models have several advantages over the other models, for example, it requires less process data, the model parameters can be estimated from laboratory experiments and routine operating data instead of time consuming plant test, and it can be extrapolated to operate on regions which are not represented in the data set used for model development (Henson, 1998).

The major drawback of the fundamental dynamic model is the highly nonlinear state dimension of the system and its complexity too. Thus, the models are not well suited for incorporation into the control scheme. Moreover, the derivation of dynamic models for large scale process is difficult.

Modeling is a common theme within the simulation, control and optimization of processes. However, model reduction techniques have become more popular. The aim of almost any reduction technique is to provide a reduced order model that can be used for controller design.

There are two approaches to reduce the model (Johan et al., 2005):

1. Reducing the system with respect to the physics of the system.
2. Project the system onto smaller subspace.

The concept of the first approach will be better if we have good knowledge regarding the system. However, the drawback is that this approach can not be used to black box the model because they have different techniques for each system. The second approach has a disadvantage which is, the reduced system is not as good as expected for a physical meaning anymore. The limits of applicability of these techniques are represented by a big difference between the real and theoretical systems. Also, reducing the effects of inputs parameters inside the system does not lead to a clearer view about the real system.

In this paper, we will focus on a proper orthogonal decomposition (POD) method that reduced the nonlinearity of the system.

*Corresponding author. E-mail: ahmadsaadi1@yahoo.com

Lall et al. (1999) have developed a method of model reduction for the nonlinear system. The method requires only standard matrix computations for the used balanced truncation in the linear system. For the nonlinear system, the method is used for the POD.

Hahn and Edger (2002) have improved the technique by Lall et al. (1999) which reduced the nonlinear system especially for the ordinary differential equation system. The proposed technique is a balancing technique used for POD and Galerkin projection.

Hedengren and Edger (2005) have developed the adaptive order reduction of large scale nonlinear differential algebraic. This method includes three steps: adaptive with POD, algebraic state and *in situ* adaptive tabulation which is used to transform the implicit sets into linear explicate approximation.

Sun and Hahn (2005) have implemented a reduction technique on stable nonlinear differential algebraic equation systems. This method (POD) reduces the other differential equations for a model of distillation column as a case study.

The POD has been used to obtain approximately, low-dimensional descriptions of turbulent fluid flows (Holmes et al., 1996), structural vibrations (Cusumano et al., 1994) and insect gait, and has been used for damage detection to name a few applications in dynamic systems. Also, it has been extensively used in image processing, signal analysis and data compression.

METHODOLOGY

Suppose we wish to approximate a function of z(x, t) over some domain of interest as a finite sum in the variables separated form:

$$z(x,t) = \sum_{k=1}^{M} a_k(t)k(x) \tag{1}$$

It is reasonably expected that the approximation becomes exact in the limit as M approaches infinity, except possibly on a set of measure zero (readers unfamiliar with the measure theory may ignore it if they deal with finite-dimensional calculations; and consult, for example, Rudin 9 otherwise).

While in Equation (1) there is no fundamental difference between t and x, we usually think of x as a spatial coordinate (possibly vector-valued) and t as a temporal coordinate. Thus, the representation of Equation (1) is not unique. For example, if the domain of x is a bounded interval of x on the real line, then the functions of f k(x) can be chosen as either a Fourier series, Legendre polynomials, Chebyshev polynomials, and so on. For each, such choice of a sequence of f k(x) that forms a basis for some suitable class of functions z(x, t) and the sequence of time-functions ak(t) is different. That is, for sines and cosines we get one sequence of functions for ak(t), while for Legendre polynomials we get another, and so on. The POD is concerned with one possible choice of the functions of f k(x). However, if we have chosen the orthonormal basis functions, we will have:

$$\int_x k_1(x)k_2(x)dx = \begin{cases} 1 \text{ } if \text{ } k_1 = k_2 \\ 0 \text{ } otherwise \end{cases} \tag{2}$$

$$u_k(t) - \int_x z(x,t)k_x d(x)$$

For the orthonormal basis functions, the determination of the coefficient function of ak(t) depends only on f k(x) and not on the other f 's. What criteria should we use for selecting the functions of f k?

Orthonormality would be useful; moreover, while an approximation to any desired accuracy in Equation (1) can always be obtained if M can be chosen adequately, we may likely choose the f k(x) in such a way that the approximation for each M is as good as possible in a least squares sense. That is, we would try to find, permanently, a sequence of orthonormal functions f k(x) such that the first two of these functions give the best possible two-term approximation, the first seven give the best possible seven term approximation, and so on. These special, ordered, orthonormal functions are called the proper orthogonal nodes for the function of z(x, t). Nonetheless, orthonormality would be the expression that is called the POD of z(x, t) in Equation (1).

Mathematical expression

Consider a system where measurements of m state variables are taken (these could be from m strain gauges on a structure, or m velocity probes in a fluid, or a mixture of two kinds of probes in a system with flow-induced vibrations, etc.), and assume that at N instants of time, measurements of N sets of m are taken simultaneously at these m locations. We arrange the data in an N × m matrix A, such that element Aij is the measurement from the jth probe taken at the ith time instant. The m state variables are not assumed to be measured by transducers that are arranged in some straight line in the physical space. We merely assume that the transducers have been numbered for identification, and that their outputs have been placed side by side in matrix A. In the actual physical system, these measurements might represent one spatial dimension (for example, accelerometers on a beam), or more than one spatial dimension (for example, pressure probes in a three dimensional fluid flow experiment). Each physical transducer may itself measure more than one scalar quantity (for example, triaxial accelerometers). In such cases, the different scalar time series from the same physical transducer are arranged in different columns of A. Here, the final result of the data collection is assumed to be the N × m matrix A.

It is common to subtract, from each column of A, the mean value of that column. Whether or not this is done does not affect the basic calculation, though it affects the interpretation of the results.

Notwithstanding the previous warnings, through a combination of engineering judgement and luck, the POD continues to be fruitfully applied in a variety of engineering and scientific fields. Judging from the study, it is a useful tool at least for people who regularly deal with moderate to high dimensional data.

The singular value decomposition

We now compute the singular value decomposition (SVD) of matrix A, which is of the form (Joan et al., 2006):

$$A = U\Sigma V^T \quad (3)$$

where U is an N × N orthogonal matrix, V is an m × m orthogonal matrix, the superscript T indicates matrix transpose, and Σ is an N × m matrix with all its elements equal to zero, except along the diagonal. The diagonal elements of Sii consist of r = min (N, m) and nonnegative numbers of si, which are arranged in decreasing order, that is, r ≥ 0. The s's are called the singular values of A (and also of A^T) and are unique. The rank of A equals the number of nonzero singular values. In the presence of noise, the number of singular values which is larger than some suitably small fraction of the largest singular value might be taken as the 'numerical rank'. Since the singular values are arranged in a specific order, the index k of the kth singular value will be called the singular value number as shown in Figure 1.

Correspondence with Equations (1) and (2): In Equation (3), let US = Q. Then the matrix Q is N × m, and $A = QV^T$. Letting qk be the kth column of Q and v_k the kth column of V, we write out the matrix product as:

$$A = QV^T = \sum_{k=1}^{M} q_k \quad v_k^T \quad (4)$$

Equation (4) is the discrete form of Equation (1). The function of z(x, t) is represented here by matrix A. The function of ak(t) is represented by the column matrix qk, while the function of f k(x) is represented by the row matrix v_k^T. The approximation of Equation (1) is now exact because the dimension is finite. Due to the orthonormality of the columns of V, Equation (2) corresponds to the multiplication of Equation (4) by one of the V's on the right.

Relationship between temperature, distance and time represents a suitable example of the high nonlinearity system as shown in Equation (5):

$$T(x,t) = \exp^{-(x-0.5)(t-1)} + \sin(x,t), \quad 0 \le x \le 1,\ 0 \le t \le 2 \quad (5)$$

Imagine that we 'measure' this function at 25 equally spaced x points, and 50 equally spaced instants of t. The temperature of z(x, t) is shown in Figure 1a. Arranging the data in matrix Z, we compute the SVD of Z, and then compute rank 1, rank 2 and rank 3 approximations to Z, as shown in Figure 1. MATLAB software is used for Equation (5) to reduce the order by calculating the SVD and POD as shown in Figure 1. The rank 3 approximation (Figure 1) looks indistinguishable from the actual temperature as shown in Figures 1 and 2. This is explained by Figure 1 which shows the singular values of Z. Note how the singular values decrease rapidly in magnitude, with the fourth one significantly smaller than the third (The numerical values are 47.5653, 2.0633, 2.0256, 0.0413, 0.0106 . . .).

Note that in this example without noise, the computed singular values beyond number 14 flatten out at the numerical roundoff floor around 10 to 15. The actual singular values beyond number 14 should be smaller, and an identical computation with more digests of precision should show the computed singular values flattening out at a smaller magnitude. Conversely, perturbing the data matrix by zero mean random numbers of typical magnitudes, 10 to 8 causes the graph of singular values to develop an obvious elbow at about that value. For experimental data with noise, the SVD of the data matrix can sometimes provide an empirical estimate of where the noise floor is. So far in this example, we have merely computed the lower rank approximations to the data, and the use of the SVD in the calculation may be considered incidental. Now, suppose we wish to interpret the results in terms of mode shapes, that is, in the context of the POD, the first 3 columns of V provide the 3 dominant x-direction mode shapes, and on projecting the data onto these mode shapes we can obtain the time histories of the corresponding modal 'coordinates'. The calculation of the modal coordinates is straightforward. Using Equation (4), t he kth modal

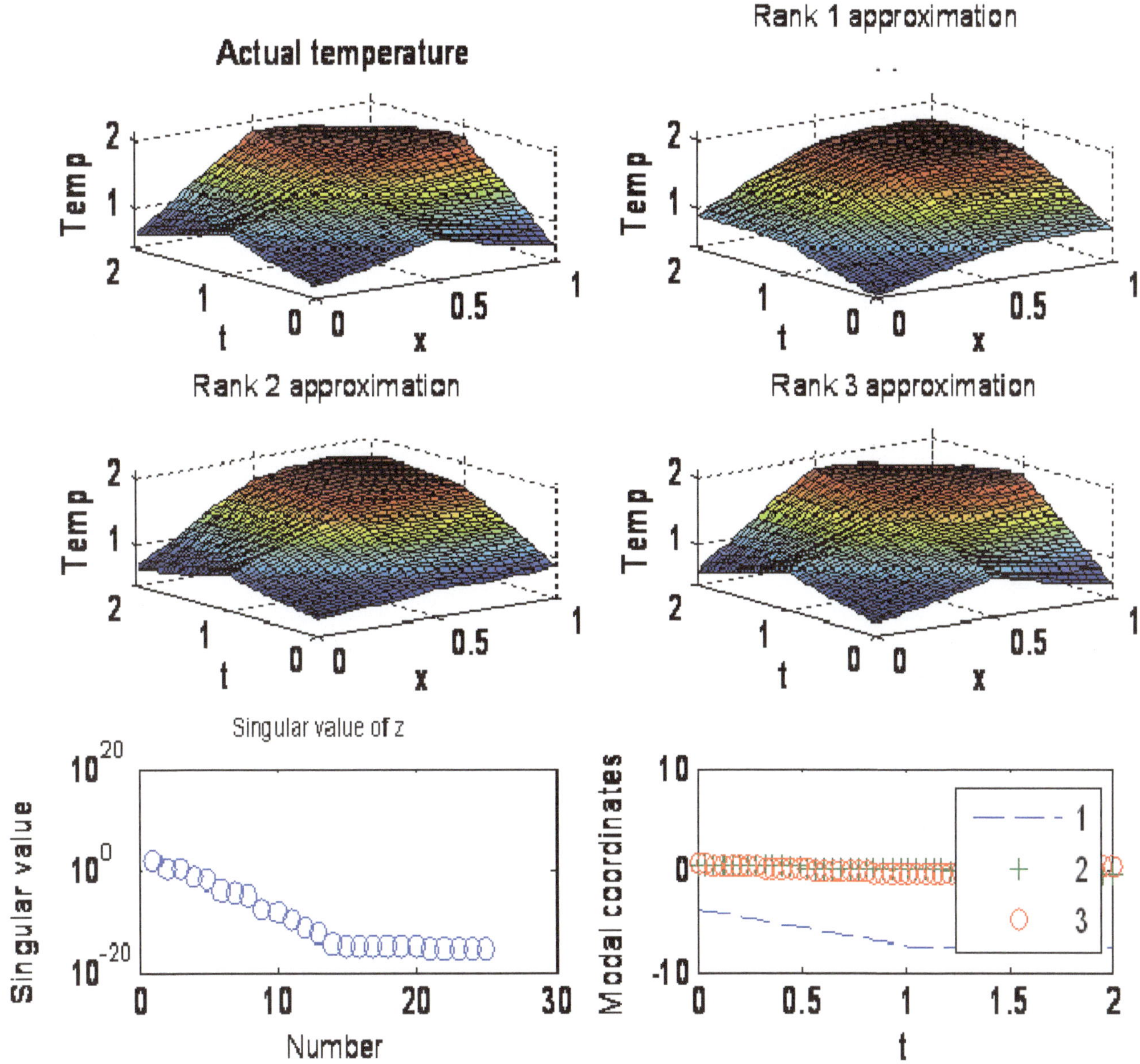

Figure 1. Approximation of temperature.

coordinate of q_k is simply uksk, where uk is the kth column of U (assuming U is available from the SVD). Alternatively, if only the proper orthogonal modes of V are available, then the projection calculation is simply $q_k = Av_k$, where v_k is the kth column of V. The modal coordinates for the temperature given by Equation (5) are plotted in Figure 1. The first coordinate is obviously dominant (the first singular value is dominant), while the second and third have comparable magnitude (singular values 2 and 3 are approximately equal) as shown in Figure 2. Rank 3 is the best because the difference between the actual value and approximation valuesfor Ranks 1, 2 and 3 gave high approximation for rank 3 which reaches 97.34% as shown in Figure 2.

Consider the differences between the SVD and eigenvalue decomposition. The SVD can be computed for non-square matrices, while the eigenvalue decomposition is only defined for square matrices. The SVD remains within the real arithmetic whenever A is real, while eigenvalues and eigenvectors of unsymmetric real matrices can be complex. Each of the left and right singular vectors (columns of U and V, respectively) is orthogonal, while eigenvectors of unsymmetric matrices need not be orthogonal even when a full set exists. Finally, while an eigenvector (say) ψ and its image A ψ are in the same direction, a right-singular vector v_k (kth

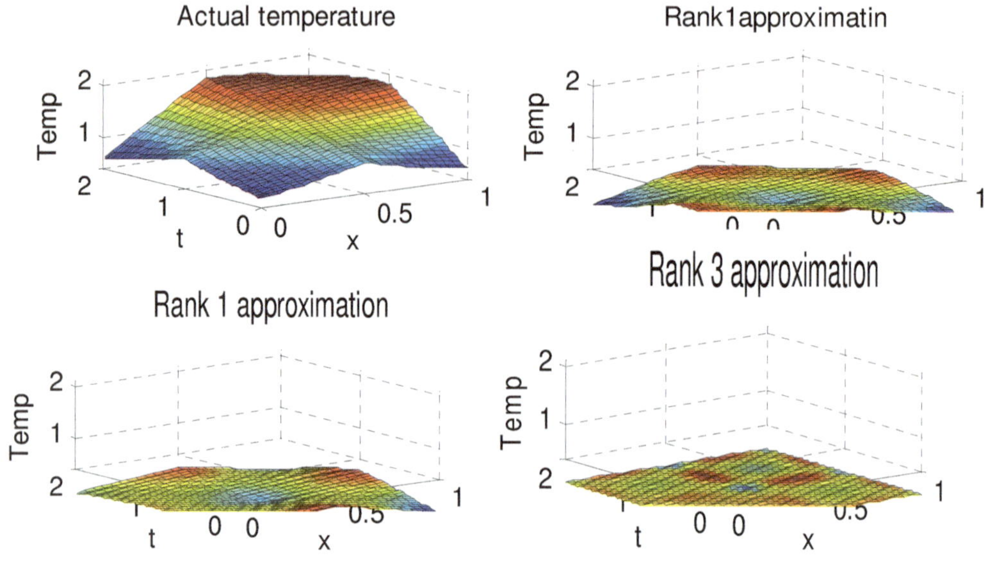

Figure 2. Difference between actual and rank approximation methods.

column of V) and its image Av_k need not be in the same direction or even in spaces of the same dimension. However, the SVD does have strong connections with the eigenvalue decomposition. On premultiplying Equation (3) with its transpose and noting that $V^{-1} = V^T$, we see that V is the matrix of eigenvectors of the symmetric $m \times m$ $A^T A$ matrix $A^T A$, while the squares of the singular values are the $r = \min(N, m)$ largest eigenvalues of $A^T A$.

CONCLUSIONS

A few reduction techniques applied in the chemical process have been reviewed. It was found that the number of reduced techniques application in the chemical processes was still low. However, POD represented one of the important methods that can be used to reduce the nonlinearity of the system without having high effects on the behavior of the system.

REFERENCES

Cusumano JP, Sharkady MT, Kimble BW (1994). Philos.Trans. R. Soc. London, Ser., A347: 421-438.

Hahn J, Edgar TF (2002). "An improved method for nonlinear model reduction using balancing of empirical gramians". Comput. Chem. Eng., 26(10): 1379-1397.

Hedengren JD, Edgar TF (2005). "Order reduction of large scale DAE models". Comput. Chem. Eng., 29: 2069-2077.

Henson MA (1998). "Nonlinear model predictive control: current and future directions". Comput. Chem. Eng., 23: 187-202.

Holmes P, Lumley JL, Berkooz G (1996).Turbulence, Coherent Structures, Dynamical Systems and Symmetry, CambridgeMonogr. Mech., Cambridge University Press.

Lall S, Marsden JE, Glavaski S (1999). "Emprical model reduction of controller nonlinear systems". 14th IFAC World Congress, 473-478, Beijing, China.

Sun C, Hahn J (2005). "Reduction of stable differential-algebraic equation systems via projections and system identification". J. Process Control, 15(6): 639-650.

Johan S, Kenji F, Torkel G (2005). "Model reduction of nonlinear differential-algebraic equation ".M.Sc Thesis, University of Wuppertal, Ge, pp. 77-79

Rudin W (1976).Principles of Mathematical Analysis, International Series in Pure and Applied Mathematics, McGraw-Hill, , 3rd edn., p.144-153.

8

Recovery energy from the separated and gravity type of heat pipe exchanger in China

Huang Wei[1] and You Hongjun[2]*

[1]Liaoning Shihua University, Fushun, Liaoning, P.R. China
[2]SAIT Polytechnic, Calgary AB, Canada.

The national and international circumstance and the application of heat pipeline exchanger were simply introduced in this study, while an application of the separated and gravity types of heat pipeline exchanger was discussed at the recovery energy of the electrical factory, the desulphurisation unit, the construction industries and the air-conditioning system, respectively. Their heat pipeline exchanger's utilization not only effectively decreased energy loss and reduced equipment corrosion and environmental pollution, but also improved the factory's competition, with great benefits reached by the industry.

Key words: Heat pipe exchanger, separated type, gravity type, China.

INTRODUCTION

Heat pipe as a patent technology was named as a "heat element" by USA Gaugler Company in 1944. Grover, firstly pointed out heat pipe as a patent technology in 1963 and he publicly published their experimental results, while Cotter described the theory of heat pipe in detail in 1965 (Shen et al., 2000; Zhang, 2010). Heat pipe is a very high heat transfer element, which can continually transfer energy internally. Its thermal conduction arrives at 10^3 and 10^4 of the metal. Heat pipe exchanger has much more advantage than other exchangers, such as having high heat transfer, having even temperature at the heat pipe surface, finishing heat transfer of solid-solid or solid-liquid, low pressure loss, no requirement for heat offset, safe operation, compaction construction, no mixture of heat flow with cold flow, controlling corrosion, etc (Guo et al., 1998; Luo, 2010).

It can show special advantage for solving the energy problem, for example, it can be obtained by good economic income and social benefits via the use of heat energy, recovery residual energy, saving feedstock and decreasing capital (Le Mingchong et al., 1996; Jiang, 2010). Heat pipe exchangers, having special exchanger , were gradually being developed since 1970, and it could become one of the important saving energy. However, this was studied at Nanjing Industry University in 1973, where the No.1 heat pipe exchanger was researched with the Nanjing refinery plant. The different heat pipe exchangers were widely used, for example, the separated heat pipe exchanger had high heat transfer, long distance heat transfer and can be easily set up, etc., and its advantages are separately heating and condensation sectors (Yang et al., 2001). Moreover, a lot of researches were done on the heat pipe exchanger, but in this paper, the economic advantage of the different separated heat pipe exchanger and the gravity of the heat pipe exchanger will be discussed.

ECONOMIC ADVANTAGE OF SPECIAL HEAT PIPE EXCHANGER

Appliance of the separated heat pipe exchanger at the recovery energy of the electrical factory

There are a lot of flue gases given off (100000 m^3/h) at the large electrical factory, and its heat pipe is between 10 and 20 m, while the single pipe transfer power is 20 kW. There is difference in the making and transportation of this heat element and in the installation and

*Corresponding author. E-mail: youhongjun@hotmail.com.

Table 1. The separated heat pipe residual recovery unit.

Type	The separated heat pipe residual heat recovery unit
Heat source	Burning residual heat
Cold fluid	Water
Cold wind temperature	200 - 600°C
Steam pressure	5×105 - 1.4×106 Pa
Recovery heat	4650 kW (4000000 Kcal/h)

Table 2. The separated heat pipe exchanger at the desulphurisation unit.

Type	The separated heat pipe exchanger
Heat fluid	Flue gas
Cold fluid	Air
Heat fluid's temperature	250 - 350°C
Cold fluid's temperature	120 - 220°C
Recovery heat	5800 kW (5000000 Kcal/h)

maintenance of it. The pipe diameter has to increase and reduce the compaction of the heat pipe because of transferring high power and transforming limit and stiff requirement. The single heat pipe exchanger cannot reach the technology requirement because the hot fluid is not allowed to be discharged from the heat pipe exchanger. The separated heat pipe exchanger gradually takes the place of the single heat pipe exchanger at the recovery energy of the electrical factory. The work mechanism of the separated heat pipe exchanger shows that, the heat fluid is firstly heated to go up at the evaporation sectors and then condensate at the condensation sectors; the condensation liquid goes back to the evaporation sectors due to gravity; no condensation gas separated pipe is installed at the condensation sectors; there is a discharge valve above the condensation sectors; no condensation gas can be discharged outside at anytime and the condensation liquid can go back to the evaporation sectors because of the potential energy.

Its main advantages are: (1) good and suitable for the large heat pipe exchanger unit; (2) it can furnish heat transfer between the cold and heated fluid; (3) it can fully separate the cold and heated fluid; (4) it can transfer heat between one fluid and many fluids; (5) it can be easily set up; and (6) it can adjust the heating surface ratio of the heating and condensation sectors. Table 1 shows the separated heat pipe residual recovery unit, which can recover energy (about 4650 kW or 4000000 Kcal/h) in the operation of Laioning electrical plant. Consequently, good financial benefit and potential favourable social impact could be obtained.

Utilization of the separated heat pipe exchanger at the desulphurisation unit (Wang, 2001, 2002)

In order to protect the environment and decrease pollution (such as SO2, ash, etc.), the flue gas from the boiler must go through the washing unit and then it will get rid of the ash and SO2 content. As such, it has to satisfy the national standard. Therefore, the flue gas' temperature was kept at 60°C after wash. The water content in the flue gas has reached a saturation state and the flue gas still contains a little bit of sour gas. When the water begins to condense, it can meet the sour gas to corrode the internal surface at the pipe and chimney. So the flue gas after wash must be heated, in order to avoid corrosion of the pipe and chimney.

Currently, a lot of factories use to heat the flue gas for a second time. It not only increases the fuel investment capital, but also easily causes environment pollution. Using the separated heat pipe exchanger, to heat gas after wash, can save money from burning fuel the second time and further prevent pollution in the second operation. It can be economic and provide protection for the environment. Table 2 indicates the separated heat pipe exchanger at the desulphurization unit. It can save 5800 KW of energy by using the separated heat pipe exchanger and it can be beneficial to the factory and society as well (Table 2).

Appliance of a gravity type of heat pipe exchanger at the construction industries

On flowing through the concrete structure, the flue gas can discharge a lot of energy to increase the internal temperature of the concrete more than the outside temperature. It makes the internal and external surfaces of the concrete to increase the pulling stress. When the pulling stress exceeds the anti-cracking ability, the construction of the concrete will crack to have serious effects on the whole construction and cause weariness (Xie, 1994). Xu Jianfeng (Xu et al., 2003) obtained the different temperature and position temperature distribution

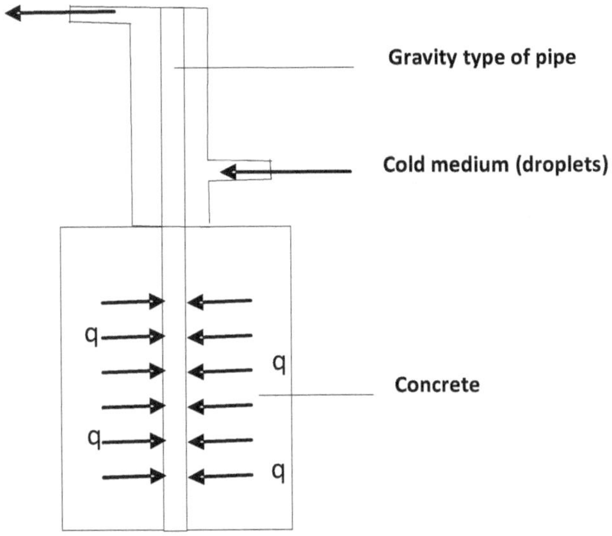

Figure 1. Heat exchange from the gravity type of heat pipe in concrete.

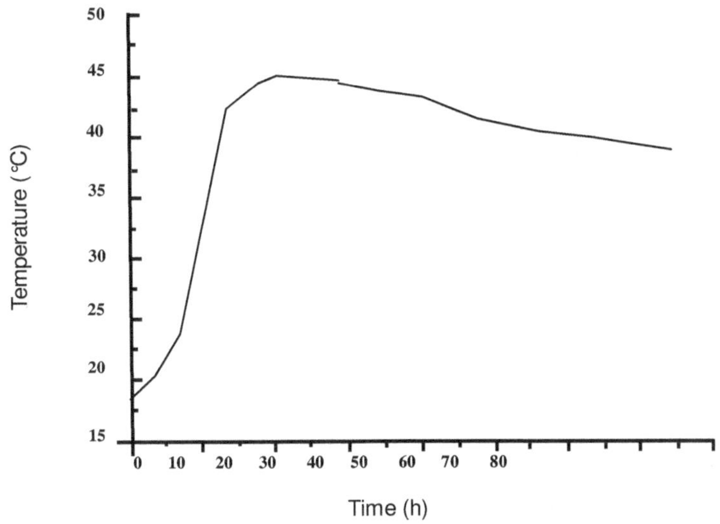

Figure 2. The curve of the concrete construction's temperature with time.

by theoretical analysis and mathematical calculations, and proved that his theoretical model was appropriate. The theory and practice proved that a gravity type of heat pipe exchanger could control the temperature degree of the concrete's internal and surface temperature at the allowable stage.

Figure 1 shows how to exchange heat for the gravity type of heat pipe exchanger. The gravity type of heat pipe is set up at the large concrete construction. The heat energy given off from the concrete passes to the evaporation sectors by conduction. It makes the working medium of the heat pipe to evaporate and exchange the heat energy with the cold medium (droplets) and then, it condensates to liquid and goes back to the heating sectors due to the gravity function. As such, the concrete can transfer the heat energy to the cold medium by this circulation.

The discrete equation will be set up based on the thermal conduction differential equation, while the temperature fields of the concrete during the appliance are calculated, using the finite difference calculus. Figure 2 represents the different initial and terminal conditions, the heating field and the temperature distribution of the concrete construction that are obtained at different times. After the concrete sets, the heat energy given off from the concrete is higher than the heat that is transferred from the heat pipe. So, the heat pipe's temperature quickly increases and the concrete's temperature increases with

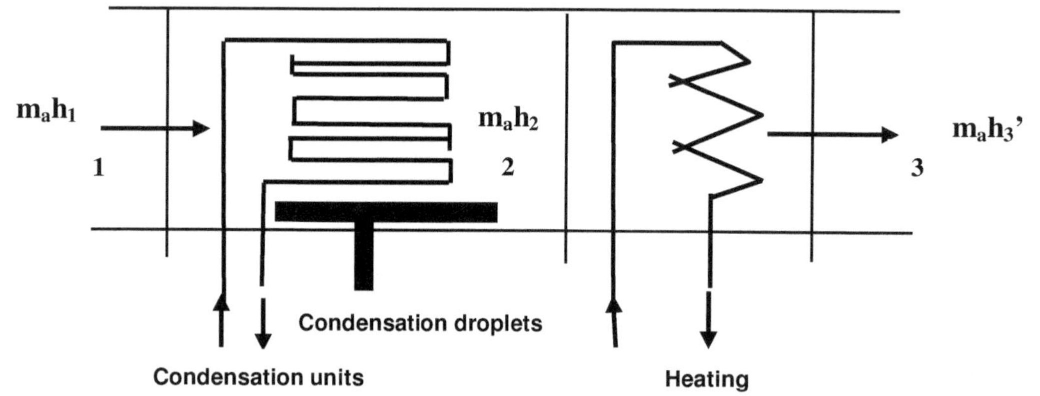

Figure 3. The de-humidity method of traditional air conditioning.

the increase of the heat pipe's energy. When the time increases, the heat energy given off from the concrete will decrease, and when the heat energy given off from the concrete equals the heat energy of the heat pipe and reaches the maximum temperature, the concrete's temperature and the heat energy of the heat pipe gradually decreases. So, the gravity type of the heat pipe can be used at the large concrete construction. However, it can cause a decrease in the concrete's temperature, reduce the temperature degree and improve the stability of the concrete. At the same time, it can decrease the heat pulling stress, reduce the crack and improve the concrete quality (Figure 2).

Utilization of the gravity type of heat pipe exchanger at the air conditioning system

There is hot climate and high relative moisture in air in the southern area of China, and as such, air conditioners are used in this area. The bedrooms are not good for people because the latent energy of the total heat loads of the air conditioner is immensely great in moist areas. So, the air conditioners have no good effect on the bedroom (Yu et al., 2004; Cai, 2003). The common temperature scope and the relative moisture for people are between 22 and 25°C, respectively. So, it is very important for people to control the air conditioner's temperature and moisture. Figure 3 indicates the de-humidity method. When the cold air goes through the condensation unit, its temperature will decrease below the dew point, to condensate droplets at the surface of the pipe. The air after discharging water will be heated to reach the required temperature.

This method not only wastes a lot of the heat energy, but it also increases the factory investment. If the gravity type of the heat pipe exchanger is used in the system, it can increase to dehumidify and decrease the total heat energy and provide suitable temperature and moisture for people (Wu et al., 1997). Figure 4 shows the application principle of the gravity type of heat pipe exchanger in the air conditioning system. Before the cold air goes into the bed room, the cold air's temperature is very low and it has a high relative moisture content. The temperature of the cold air must be decreased by going through the evaporation sectors (from 1 point), and a lot of condensation droplets are discharged outside. The residual air can exchange the heat energy with the gravity type of the heat pipe exchanger. Finally, the ideal temperature and moisture can be sent to the bedroom (Figure 4). Two methods can be calculated as follows:

(1) Common method:

Condensation sectors energy Qc = ma*(h1-h2) (1)
Evaporation sectors energy Qh = ma*(h3-h2) (2)

(2) Heat pipe exchanger method:

Condensation sectors energy Qc' = ma*(h1-h2') (3)
Evaporation sectors energy Qh = ma*(h3-h2) (4)

EH = (Qc - Qc') / Qc = (h2'- h2)/ (h1 - h2) (5)

Figure 5 represents the two methods that are contrasted for the heating load and it clearly shows that (h2'-h2) and (h3-h2) are different. This is the saving energy from the gravity type of the heat pipe exchanger. 1, p, 2 and 3 in Figure 5 represent 4 positions for the air flow (before dry, after dry, before condensation and after condensation, respectively). The gravity type of the heat pipe exchanger can improve the refrigerant capacity of the air conditioning system and decrease the operational cost to reach the saving energy.

Conclusions

Based on the aforementioned review, the heat pipe exchanger is one of the highly efficient and widely used elements. Researchers conducted different studies on the theory and practical of the heat pipe exchanger in order

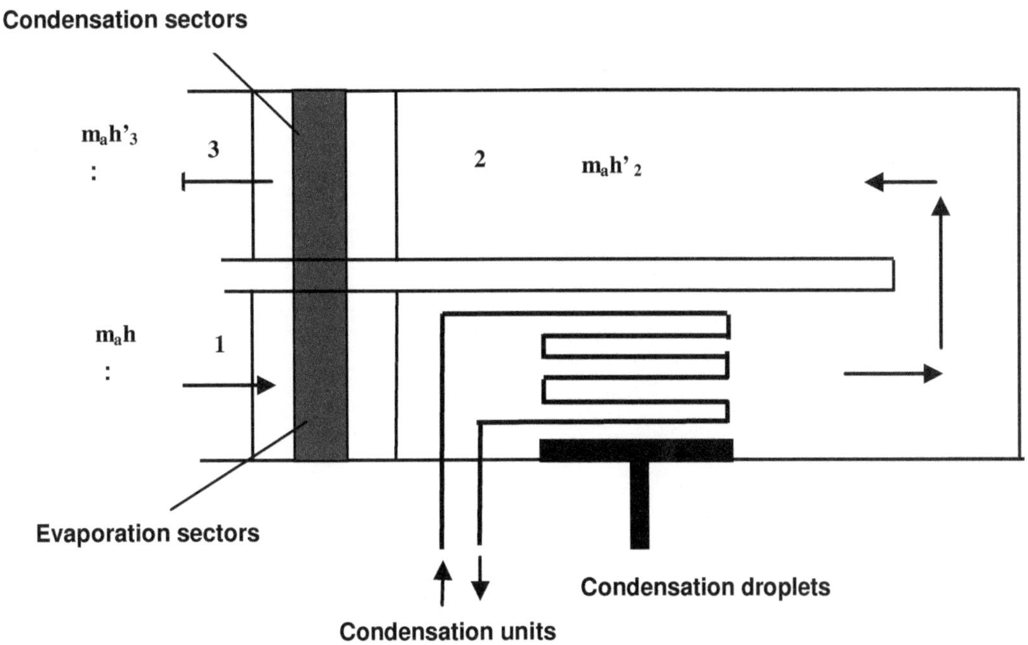

Figure 4. The application principle of the gravity type of heat pipe exchanger in the air conditioning system.

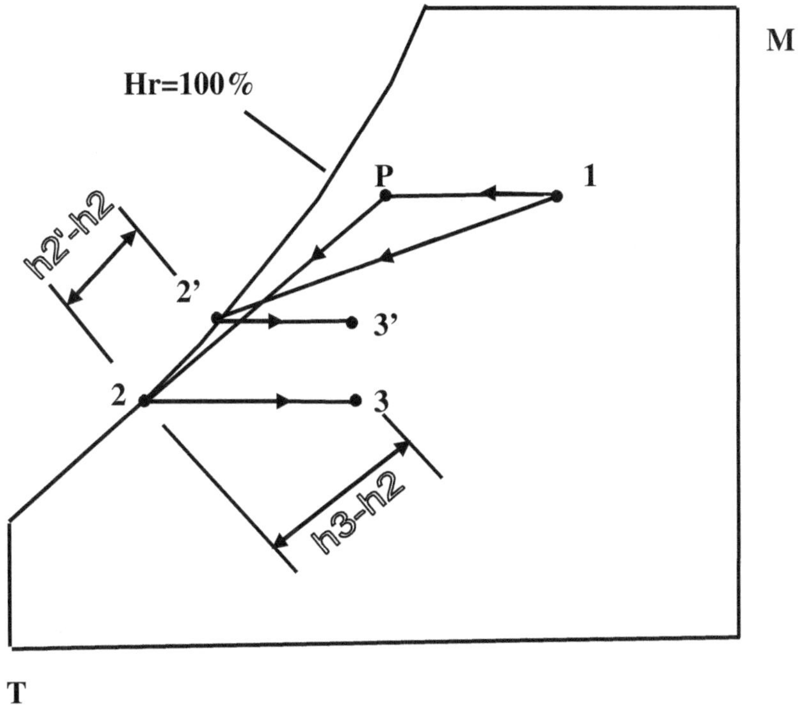

Figure 5. The heating load contrast.

to improve the factory income and reduce environment pollution. An application of the separated type of heat pipeline exchanger was discussed in detail at the recovery energy of the electrical factory and the desulphurisation unit.

However, it can separate both the heat and cold resources and transfer the heat energy from the long distance without an extra power. At the same time, the gravity type of heat pipeline exchanger was introduced at the construction industries and the air conditioning system, and can increase the refrigerant capacity of the air conditioning system and reduce the operational cost to reach the saving energy. Nonetheless, both the separated and gravity types of heat pipeline exchanger will be widely used at the different industries in the future.

REFERENCES

Cai WG (2003). Study on the heat pipe technology and use in the air conditioner industry. Refrigeration and air conditioner, 3: 31-36.

Guo YC, Liu JC (1998). Appliance and circumstance of heat pipe exchanger, 25(2): 18-20.

Jiang ZH (2010). Appliance of heat pipe exchanger in refinery. Chem. Equipment, p. 3.

Le MC, Chen YG (1996). Heat pipe and heat pipe exchanger. Chongqi Press.

Shen YY, Chen BD, Fang GW (2002). Design and optimize gas-gas exchanger. Fushun Petroleum Institute J.I, 20(2): 52-54,62.

Luo Ying (2010). Appliance of the heat pipe exchanger in the practical. Technol. Commun., p. 15.

Wang L (2001). Use the separated heat pipe exchanger at the desulphurisation unit. Appliance Energy Technol., (5): 24-25.

Wang L (2002). Utilize the separated heat pipe exchanger at the desulphurisation unit. LPG and heat energy, 22(3):274-275.

Xie M (1994). New method for reinforcing frozen earth basement-appliance of heat pipe. Constr. Technol., 6: 51-53.

Wang XP, Peter J (1997). Appliance of heat pipe exchanger to humidity control in air-air conditioning systems. Appl. Ther. Eng., 17(6): 561-568.

Xu J, Hong RH, Sun ZJ (2003). Study on the heat pipe to control the internal temperature of the concrete construction. 37(5):591-595.

Yang J, Xu TM (2001). Design and apply separated gas-gas heat pipe exchanger. Stud. Apply Energy, (2): 35-36.

Yu X, Wang W, Wang WZ (2004). Appliance of the heat pipe in the air conditioner, 17(3): 39-41.

Zhang DY. 2010. Study on the different heat pipe exchanger. Mech. Eng., p. 6.

Water based mud lifting capacity improvement by multiwall carbon nanotubes additive

Ariffin Samsuri* and Amir Hamzah

Department of Petroleum Engineering, Faculty of Petroleum and Renewable Energy Engineering, Universiti Teknologi, Malaysia.

Poor hole cleaning remains one of the major concerns in oil and gas industry since it will induce expensive drilling problems such as stuck pipe, slow drilling rate, high torque and drag, lost control of density and poor cement jobs. Various drilling fluids have been widely used in the oil and gas industry to improve lifting capacity of drilled cuttings. In this study, the experiment was conducted using water based mud with Multiwall Carbon Nanotubes (MWNTs) additive to study its lifting capacity. The study focused on the amount of MWNTs used, cutting size and mud annular velocity effect on the mud lifting capacity. The results show that lifted cuttings increase as the amount of MWNTs added increases. MWNTs associated with water based mud displays the stability against base mud since surface forces easily balance the gravity force and attached to drilled cuttings, resulting in increase of drag force acts to drilled cuttings and easily lifted cuttings to the surface. The MWNTs also will improve viscosity which will significantly increase carrying capacity of the mud. For small and medium cuttings, the improvement relatively simplified compare to the big cuttings. The impact will significantly increase as the annular velocity increase.

Key words: Water based mud, lifting capacity, mud additives, drilled cuttings, mud viscosity.

INTRODUCTION

One of the most important functions of mud in drilling operations is to transport the drilled cuttings to the surface through the well bore annulus. This ability of a mud to transport cuttings from bottom to the surface is called lifting capacity or carrying capacity. Mud lifting capacity depend on mud rheological properties and flow rate, particles settling velocities, particle size and size distribution (geometry, orientation and concentration), drill bit penetration, rotary speed, mud density, annulus inclination, drill pipe position in the well bore (eccentricity) and axially varying flow geometry. In vertical and inclined well, cuttings transport remains one of the major problems effecting drilling operation. One of the major factors of this problem is, the cuttings settle at the bottom side of the borehole due to gravitational force.

When the hole cleaning is adequate, there is a tendency for expensive drilling operation due to the problem of stuck pipe, lost circulation, slow drilling, high torque and drag, poor cementing jobs, wellbore enlargement, accumulation of cuttings at lower side of annulus, formation and accumulation of mud cake at porous formation. To overcome these problems, various expensive operations such as wiper trips or pumping out of the hole, washing and back reaming must be carried out. Specialized petroleum laboratory at China's Shandong University has developed an advanced fluid mixed with nanosized particles and superfine powder that significantly improve drilling speed (Saeid et al., 2006).

The nanoparticles improve the fluid rheological, mechanical, optical and thermal properties. Suspension of nanosized particles may also enhance stability against sedimentation since surface forces easily balance the gravity force. Recent experiments have demonstrated that, nanofluids have attractive properties for application where heat transfer, drag reduction, binding ability or sand consolidation, gel formation, wettability alteration, and corrosive control is of interest (Phuoc et al., 2007). Abouzar et al. (2008) shows that, carbon black nano particles in drilling mud produced a more continues and

*Corresponding author. E-mail: ariffin@petroleum.utm.my

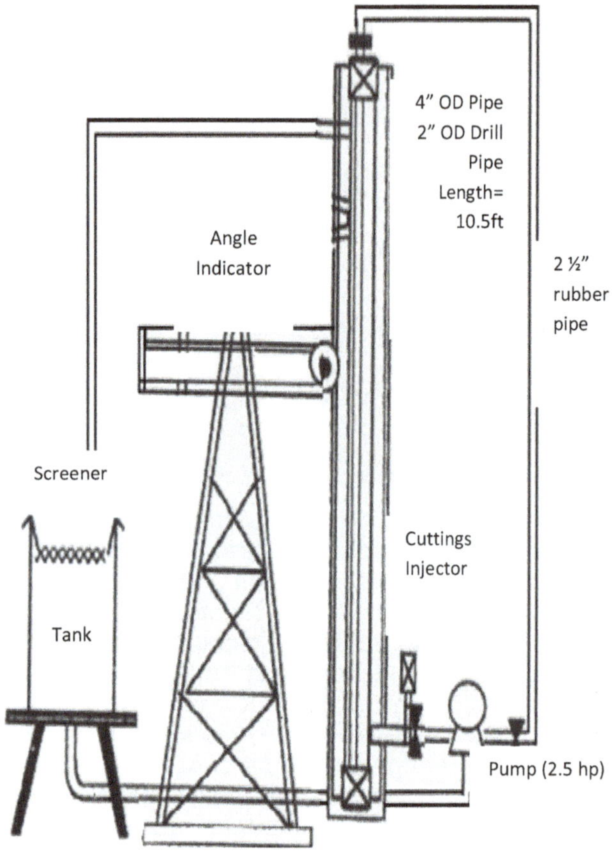

Figure 1. Schematic of rig model.

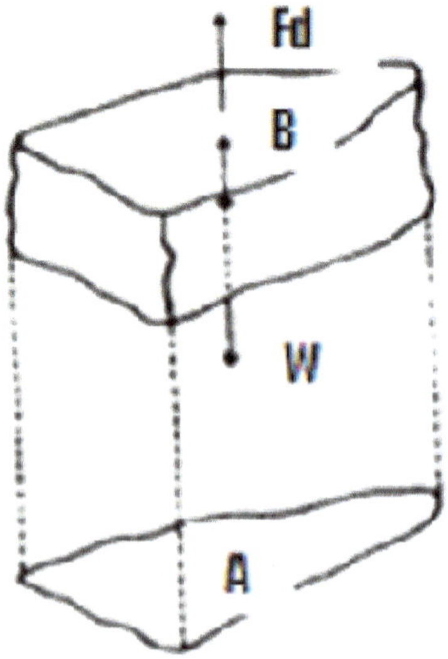

Figure 2. Scheme of force in a particle settling. Fd = drag force due to slipflow of the particle; B = boyant force; W = gravitational force; A = cutting's projected area.

integrated mud cake, therefore less filtrate and mud cake thickness.

This research is aimed to use Multiwall Carbon Nanotubes (MWNTs) as an additive for water based mud lifting capacity improvement. This will also improve thewater based mud's rheological, mechanical, optical and thermal properties. The MWNTs can also enhance stability against sedimentation since surface force easily balances the gravity force. A rig model as shown in Figure 1 has been used in the experimental works with cutting sizes of 2, 2.8 and 4.8 mm. A 10 lb/gal water based mud has been added with various amounts of MWNTs and annular velocity of 29.2 to 102.1 ft/min.

Literature review

Lifting capacity

The lifting capacity of the mud system must consider four forces acting on the drilled cuttings (Figure 2; Machado and Aragao, 1990): Firstly, a downward gravitational force due to the drilled cuttings being settled down, then, an upward buoyant force due to the cutting being immersed in the mud, followed by a drag force, parallel to the direction of the mud flow due to the mud flowing around the cutting particle and lastly, a lifting force perpendicular to the direction of the mud flow. Study conducted by Tianping and James (2007) have shown that nanofluids have tendency to increase drag acting on thedrilled cuttings, gel formation, binding ability for sand consolidation and heat transfer.

Sifferman et al. (1973) used several fluids and particles to study the cutting transportation in a full scale vertical annulus of a clear plastic pipe. In their experimental work, they conclude that annular flow velocity and fluid rheological properties have a major influence on cutting transportation, whereas, other variables have only moderate or little effects.

Multiwall carbon nanotubes (MWNTs)

A multiwall carbon nanotube is a rolled up stack of graphene sheet into concentric cylinders (Meyappan, 2004). The distance between the atomic graphitic sheets of the wall in the MWNTs is about 0.34 nm, which similar to the interplanar spacing in graphite (Dojin, 2004). The diameter of MWNTs is in the range of a few to ten nanometers and its length can be from micrometer to millimeters with the density of 2.6 g/cu.cm.

Nanofluids are the fluids containing at least, a very small quantity of nanoparticles with size of 1 to 100 nanometer (Amanullah and Al-Tahnini, 2009) that are uniformly and stabily suspended in a liquid.

In this study, the Multiwall Carbon Nanotubes (MWNTs) which is produced using custom built Catalytic Chemical

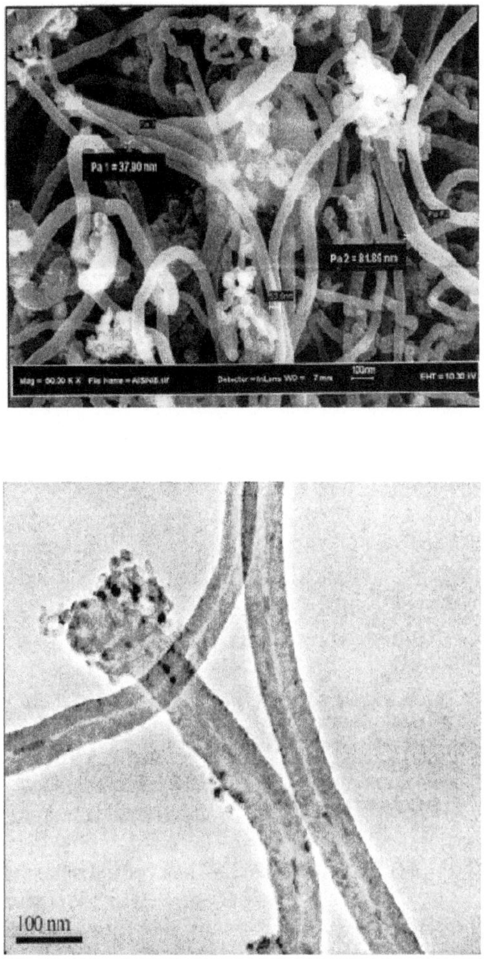

Figure 3. Structure of grown MWNTs under FE-SEM and TEM micrograph.

Vapor Deposition (CCVD) and DC Plasma Enhanced Chemical Vapour Deposition systems (PECVD) has been used as an additive. In the CCVD method, catalyst material, such as Ni, Co and Fe is heated to 500 to 1000°C (usually 700°C) and exposed to hydrocarbon gas which flows through the reaction zone. The dissociation of the gas will occur at the hot catalyst surface. The precipitation of carbon from the saturated metal particle leads to formation of tubular carbon structure in the form of nanotubes (Yusran et al., 2004). Plasma Enhanced CVD method allows an alternative at substantially lower wafer temperature to growth CNTs. The low temperature operation is possible since precursor dissociation is enabled by the high energy electrons in otherwise cold plasma (Meyappan, 2004).

The production of MWNTs is carried out by using carbon source gaseous fed over supported catalysts and is synthesize using oxygen, aluminum, silica and nickel. Figure 3 shows the structure of grown MWNts under FE-SEM and TEM micrograph (Tee and Goh, 2008). MWNTs have the diameter of 8 to 40 nm and length greater than 200 µm. When nanoparticles are dispersed in a base fluid, they will change fluid's thermal conductivity and viscosity conditions respectively.

METHODOLOGY

Experimental rig

As illustrated schematically in Figure 1, the rig model consists of simulated wellbore made by transparent pipe with a nominal length of 14.5 ft, an OD of 4 in. (10.16 cm), an ID of 3.57 in. (9.07 cm). The inner drill pipe consisted of 2 in. (5.08 cm) OD.

The prepared mud was injected into the annulus by a 2.5 hp duplex pump. The flow rate was control by the valve in front of the duplex pump. Cuttings must be weighted and separated based on sizes before injected into the annulus manually. A stop watch was used to measure the time for cuttings lifting from bottom to surface.

Preparation procedure

1. Prepared water based mud was circulated through the rig model at room temperature to prevent any occurrence of leakage.
2. Slowly adjust the rig model to vertical position.
3. Prepare the cuttings and weigh. Before each run, the cuttings were screened, washed and dried.
4. Prepare the water based mud with various amount of MWNTs added as an additive.

Experimental procedure

1. From mud tank, pump the mud into the annulus.
2. Control the flow rate by using the first pump valve to allow stable flow into the annulus and make sure the mud level in the tank is constant.
3. Set the flow rate at 10 gpm and when the flow rate is stabilized, open the second valve to allow cutting to be injected into the annulus through cutting injector.
4. After all cuttings have been injected, start the watch and after 1 min of flow, cuttings recovered at screener are collected.
5. Then, the collected cuttings are washed, dried and weighted.
6. Remove all cuttings from the annulus.
7. Repeat steps 3 to 6 with difference flow rate of 25 and 35 gpm, and then change with other sizes of cuttings.
8. For difference of mud composition, repeat steps 1 to 7.

RESULTS AND DISCUSSIONS

Water based mud properties

The mud properties had been tested in accordance of API Spec. 13B-1, 2009 (American Petroleum Institute, 2009) and Table 1 shows the experimental results. As can be seen, viscosity of the water based mud increases as the amount of MWNTs added increased, particularly after 0.005% of volume MWNTs was used.

Cutting slip velocity

Sand disc with various diameter and density of 20.6 lb/cuft had been used as a cutting in the experiment.

Table 1. Properties of mud used with various amount of MWNTs.

Mud	Viscosity at 600 rpm (cP)	Apparent viscosity (cP)	Gel strength (gms)
Water Based Mud (WBM)	30	19.5	30 (51)
WBM+ 0.001% MWNTs	40	20	28 (36)
WBM + 0.003% MWNTs	40	20	30 (30)
WBM + 0.005% MWNTs	41	20.5	32 (45)
WBM + 0.01% MWNTs	44	22	28 (45)

Table 2. Properties and slip velocities of cuttings.

Cutting size	Diameter (mm)	Slip velocity at				
		0.00% (ft/min)	0.001% (ft/min)	0.003% (ft/min)	0.005% (ft/min)	0.01% (ft/min)
Small (SAS)	2.0	9.12	9.04	9.04	8.97	8.76
Medium (MAS)	2.8	12.77	12.67	12.67	12.56	12.27
Big (BAS)	4.8	21.91	21.72	21.72	21.54	21.04

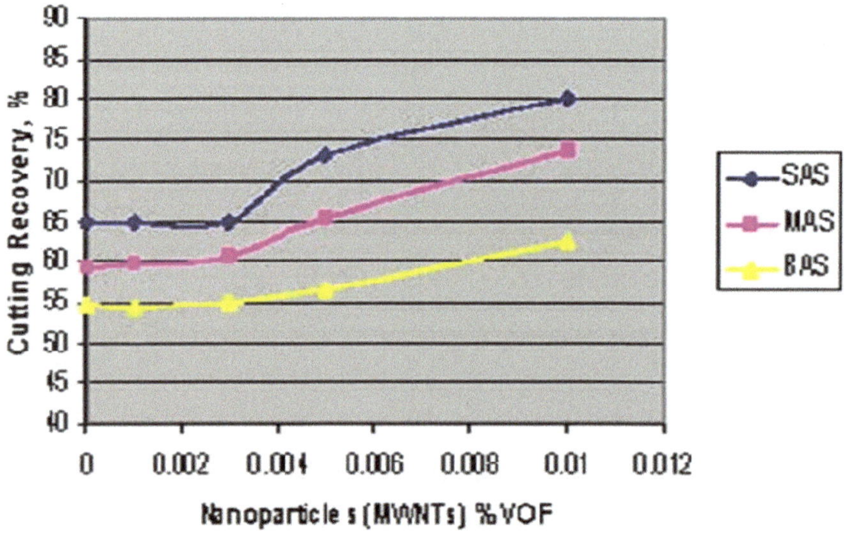

Figure 4. Effect of nanoparticles (MWNTs) on cuttings recovery for 10 gpm flow rate.

Table 2 shows the results of cutting slip velocity in various composition of water based mud. From the table, it's clearly seen that slip velocity increases as the size increases but as the amount of MWNTs added increases, the cutting slip velocity slightly decreases. This is due to the mud system becoming more viscous as more MWNTs were added to the mud system.

Effect of MWNTs on lifting capacity

Figures 4 to 6 show the effect of adding MWNTs to the water based mud lifting capacity for various flow rates. The percentage of the cutting recovery is the indication of the mud ability to lift cuttings from the bottom hole to the surface. These graphs show that by increasing the MWNTs added to the water based mud, the cuttings recovery or cutting lifting capacity also increased.

For lower amount of MWNTs used (0.001 to 0.003% of volume), the impact was not very significant, but as more MWNTs was added, the impact significantly increased. For example, the cutting recovery increased about 5 to 15% when 0.005% of volume MWNTs was added to the water based mud, depending on the cutting size and annular velocity. For 0.01% of volume MWNTs added, the cutting recovery increased in the range of 5 to 21%. The improvement of cutting lifting capacity is very significant for the small cutting size as compared to the

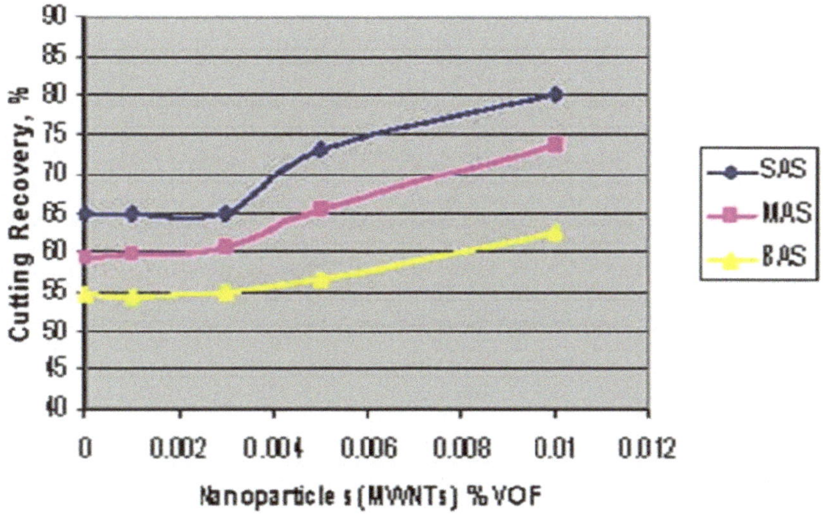

Figure 5. Effect of nanoparticles (MWNTs) on cuttings recovery for 25 gpm flow rate.

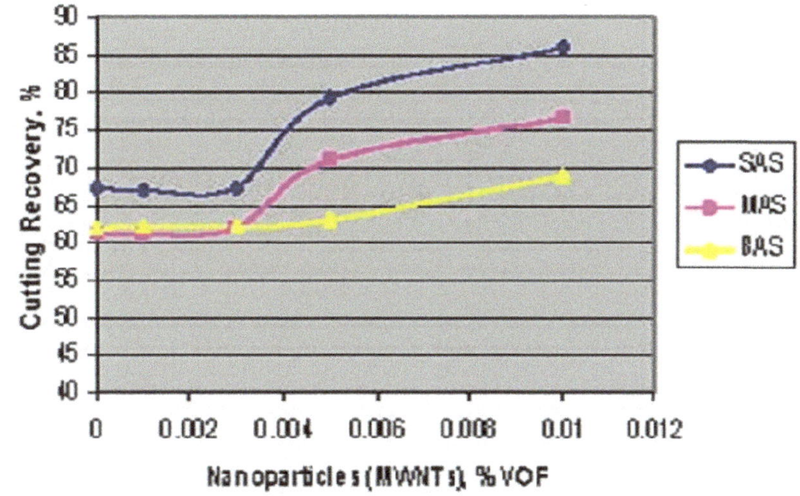

Figure 6. Effect of nanoparticles (MWNTs) on cuttings recovery for 35 gpm flow rate.

bigger cutting size.

The improvement in mud cutting lifting capacity is due to addition of MWNTs to the mud, since the MWNTs will improve the mud rheological properties. The nanoparticles material can enhance the stability against based mud since surface forces can easily balance the gravity force. When the force acts downward decreasing, the cuttings have less potential to settle down to the bottom of borehole. Therefore, under these conditions, the cuttings can easily be transported to the surface.

Effect of mud viscosity and gel strength on lifting capacity

In general, high gel strength of mud will have high viscosity. But, this is not always true because, the high viscosity can be achieved when the dispersion of clays in the mud is high. For this study, adding MWNTs into the water based mud will influence mud viscosity and gel strengths (Table 1). These unique nanoparticles can function as fluids viscosity stabilizer to significantly improve the carrying capacity.

Figure 7 shows the relationship between the slip velocity and percentage of nanoparticles, MWNTs used. Theoretically, slip velocity is inversely proportional to viscosity of fluids in the laminar flow. The viscosity of fluids increases and the slip velocity will also decrease. So, as a result, the transportation ratio for particles recovered at the screener will be increased, as clearly seen from Figures 8 to 10.

Figure 7shows that the slip velocity of three different

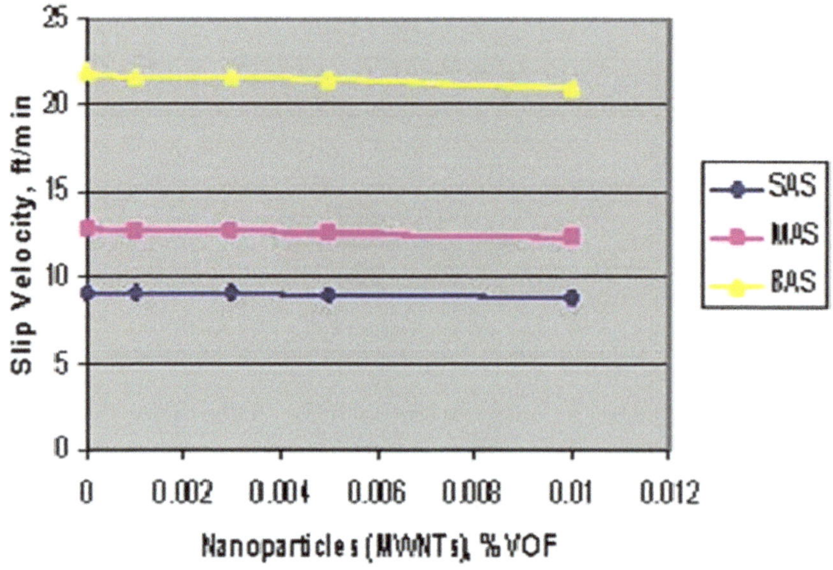

Figure 7. Effect of nanoparticles (MWNTs) on slip velocity.

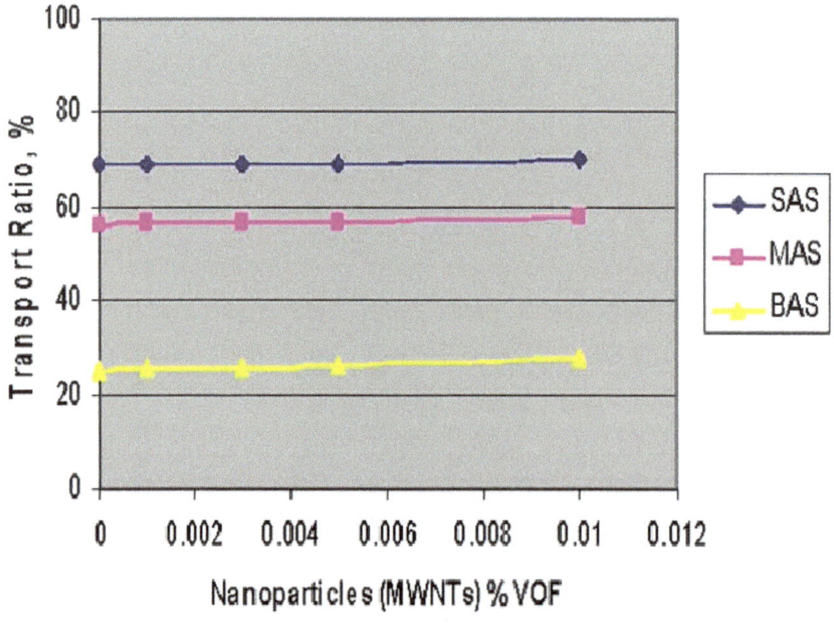

Figure 8. Effect of nanoparticles (MWNTs) on transportation ratio for 10 gpm flow rate.

sizes of cutting, slightly decreased when the percentage of MWNTs increases. This phenomena was influenced by nanoparticles (MWNTs) presence which (will) increased the viscosity of the water based mud. For bigger cutting, the slip velocity is higher, compared to medium and followed by the smaller cuttings. With 0.001 and 0.003% volume of MWNTs used, the viscosity of water based mud is 40 cp. But, when 0.005 and 0.01% of volume MWNTs was used, the viscosity of water based mud increased to 41 and 44 cp, respectively.

Effect of cutting size on lifting capacity

Figure 11 shows that the effect of cutting sizes on cuttings recovery with presence of MWNTs in the mud. Results in Table 2 indicates that by increasing percenttage of MWNTs used, lifting capacity of a small cutting has improved, followed by the medium and bigger cutting size. For big cuttings, at three different annular velocities, it shows minor effect on cuttings recovery. It is because the slip velocity of big cuttings is relatively high,

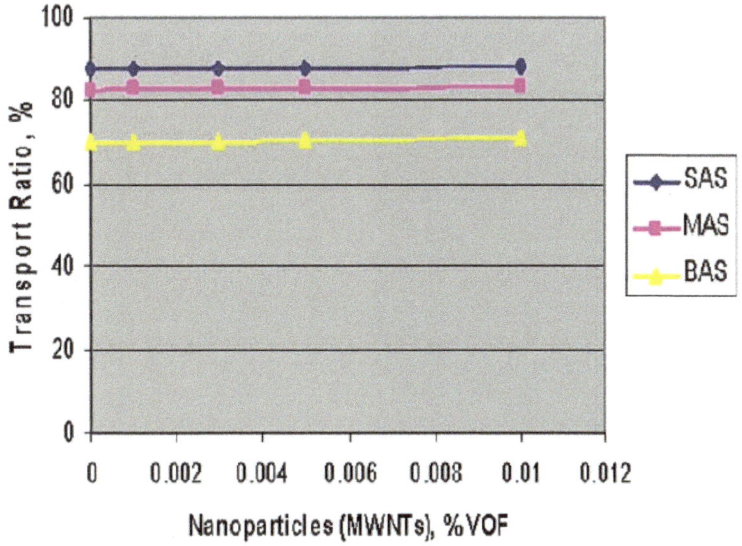

Figure 9. Effect of nanoparticles (MWNTs) on transportation ratio for 25 gpm flow rate.

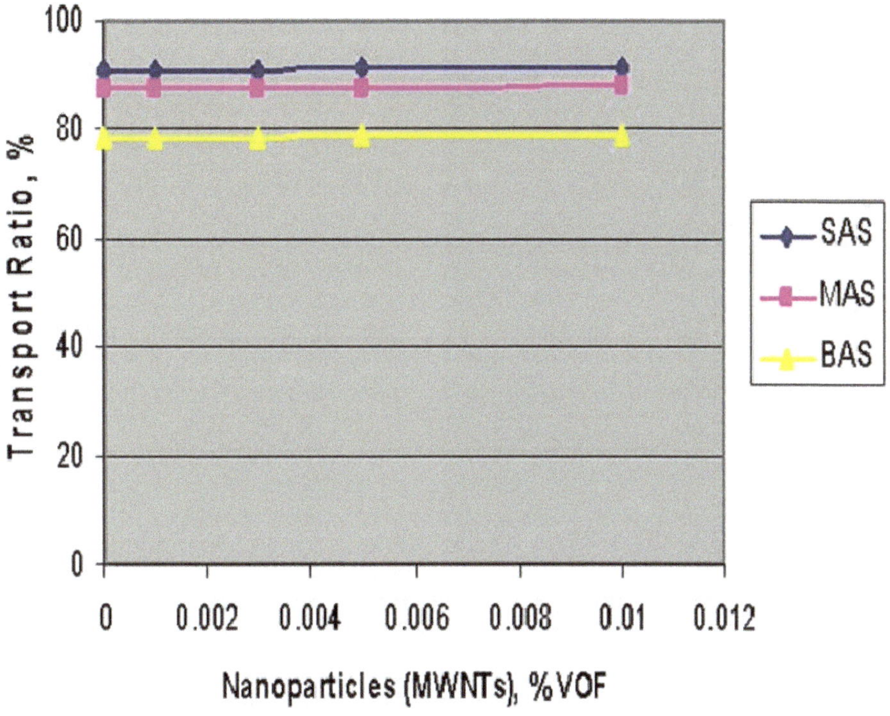

Figure 10. Effect of nanoparticles (MWNTs) on transportation ratio for 35 gpm flow rate.

compared to both small and medium cuttings. Therefore, it has tendency to settle at the bottom of annulus. Otherwise, for small and medium cuttings at lower percentage of MWNTs used (0.001 and 0.003% of volume), the recovery is not significant, but at 0.005 and0.01% of volume MWNTs used, the improvement are relatively significant.

For small cuttings, cuttings recovery increased 8.6 from 61.6 at 0.005 % of volume MWNTs used and 5.2 from 70.2% at 0.01% of volume MWNTs used. And for medium cuttings, cuttings recovery increased 6.5 from 55.8% at 0.005 % and 9 from 61.3% at 0.01% of volume

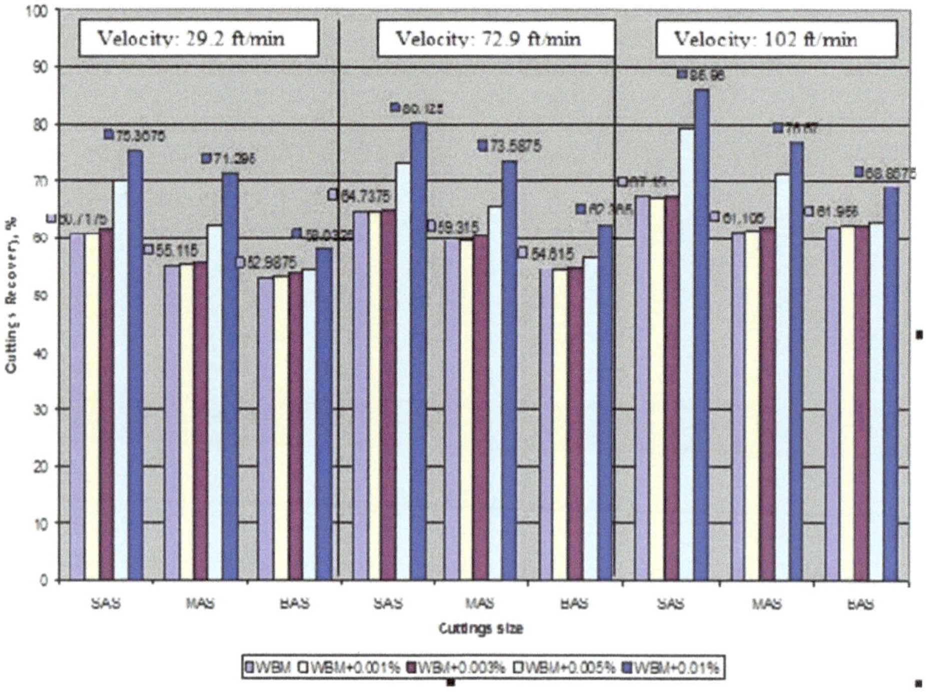

Figure 11. Effect of cutting sizes with nanoparticles, MWNTs in the mud.

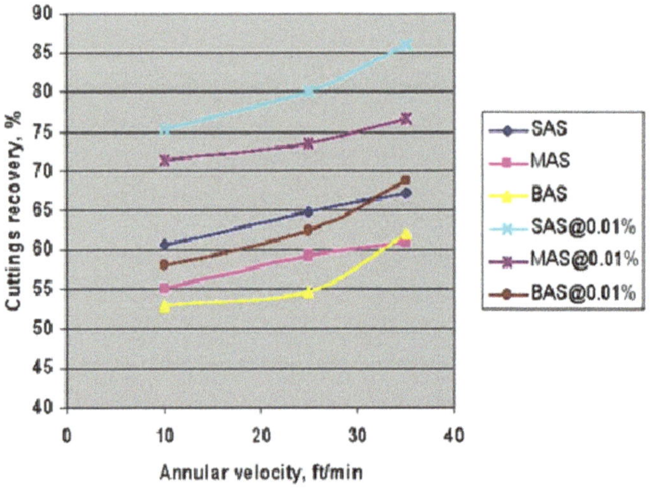

Figure 12. Effect of annular velocity on present of nanoparticles (MWNTs).

MWNTs used.

Effect of annular velocity on lifting capacity

Figures 4 to 6 show the cutting recovery for various annular velocities. Generally, cuttings recovered increases when annular velocity increases under the laminar flow condition.

Results show that the effect of annular velocity is not much pronounced with MWNTs in the mud. Figure 12 indicates that for 0.01% of volume MWNTs used, increase in the annular velocity will slightly increase the cuttings recovery. When the annular velocity increases from 10 to 25 gpm, percentage of cutting recovery increases about 5% only, from 75.4 to 80% and 6% at 35 gpm. Without MWNTs in the mud, small cuttings recovery at, 10, 25 and 35 gpm are 60.7, 64.7and 67.2% respecttively. The same trend occurred for medium and bigger cutting sizes. For bigger cuttings, with MWNTs in the mud, the percentage of cuttings recovered increased from 5 to 6% when the flow rate changed from 10 to 25 gpm and 25 to 35 gpm, respectively. Without MWNTs in the mud, the difference is 2 and 5%, respectively.

Conclusions

From the study, some conclusion can be drawn:

1. MWNTs in the water based mud, can improve lifting capacity of drilled cuttings depending on the cuttings sizes, annular velocity and flow rate. Therefore, the easy drilled cutting is raised to the surface for disposal.
2. When percentage of MWNTs in the mud increases, the viscosity of water base mud also increased. Therefore, the lifting capacity also increased. The MWNTs are dispersed in water based mud because, water absorbs

into it and becomes agglomerated. These phenomena will increase the viscosity of mud.

3. Lifting capacity of the mud will increase when the annular velocity increases, with or without the MWNTs in the mud.

REFERENCES

Abouzar MP, Bandar DA (2008). Using Nanoparticles to Decrease Differential Pipe Sticking and Its Feasibility in Iranian Oil Fields. Oil and Gas Business.

Amanullah Md, Al-Tahini AM (2009). Nano-Technology-Its Significance in Smart Fluid Development for Oil & Gas Field Application. SPE 126102 paper presented at the 2009 SPE Symposium and Exhibition, Alkhobar, Saudi Arabia, 11 May.

American Petroleum Institute (2009). Recommended Practice for Field Testing Water-based Drilling Fluids. 4th. American Petroleum Institute

Dojin K (2004). Multiwall Carbon Nanotubes. In H.S. Nalwa ed. Encyclopedia of Nanoscience & Nanotechnology California USA: American Scientific Publishers, pp. 879-894.

Machado JCV, Aragao AFL (1990). Gel Strength as Related to Carrying Capacity of Drilling Fluids. SPE Latin American Petroleum Engineering Conference. October 14-19. Rio d Jeneiro: SPE Pap., 21106: 1-7

Meyappan M (2004). Carbon Nanotube Growth by Chemical Vapor Deposition. In H.S. Nalwa ed. Encyclopedia of Nanoscience & Nanotechnology California USA: American Scientific Publishers, pp. 581-589.

Phuoc X, Tran R, Gupta LW (2007). Nanofluids for Use as Ultra Deep Drilling Fluids. R&D Facts National Energy Technology Laboratory.

Saeid M, Mariela AF, Rafiqul IM (2006). Applications of Nanotechnology in Oil & Gas E&P. Journal of Petroleum Technology, 58 (JPT online)

Sifferman TR, George NM, Elera DH, Harry AW (1973). Drill Cutting Transport in Full Scale Vertical Annuli. 48th Annual Full Meeting of the Society of Petroleum Engineers of AIME. Sept. 30 – Oct. 3. Las Vegas: SPE Pap., 4514: 1295-1302.

Tianping H, James BC (2007). Nanotechnology Applications in Viscoelastic-Surfactant Stimulation Fluids. European Formation Damage Conference. 30 May-1. Scheveningen, Netherdelands: SPE paper 107728, 1-7.

Tee JC, Goh PS (2008). As Grown Multi-Walled Carbon Nanotubes. Advanced Membrane Technology & Research Centre. Universiti Teknologi Malaysia.

Yusran S, Buang NA, Sanip S, Ismail AF (2004) Production of Nano-Carbon Using Arc Discharge in Liquid Nitrogen. In H. Saidi, M. M.Nasef & Inayati eds. Advanced in Fuel Cell Research & Development in Malaysia, pp.195-201

10

Dehyrogenation catalysts of higher normal paraffins on a nanocrystalline γ-Al_2O_3 support: Different impregnation sequences

Mandana Akia[1,2,3]*, Seyed Mahdi Alavi[1], and Zi-Feng Yan[3]

[1]Chemical Engineering Department, Iran University of Science and Technology, Tehran, Iran.
[2]Chemical Engineering Department, Kermanshah University of Technology, Kermanshah, Iran.
[3]State Key Laboratory for Heavy Oil Processing, Key Laboratory of Catalysis, CNPC, China University of Petroleum, Dongying 257061, China.

In this paper, high surface area nanocrystalline gamma alumina was synthesized by the sol-gel method using cationic surfactant. The Pt-based dehydrogenation catalysts consisting of promoters and support modifiers (Sn, In, Li and Fe) were prepared using the synthesized gamma alumina and employed in the dehydrogenation reaction of n-dodecane. The effects of different successive impregnation methods for Pt, Sn and In elements were investigated in dehydrogenation of normal dodecane. The prepared samples were characterized by X-ray diffraction (XRD), N_2 adsorption (BET), Scanning electron microscopy (SEM), Transmission electron microscopy (TEM) and Temperature programmed reduction (TPR) techniques. The catalytic performances of the prepared samples showed the best catalytic results for the catalyst which was prepared by the co-impregnation of In, Sn and Fe in the first step followed by Pt and Li impregnation in the second and third step, respectively. The synthesized support also performed well as an efficient carrier for the dehydrogenation catalyst of higher normal paraffins.

Key words: Nanomaterials, sol–gel preparation, γ-Al_2O_3, dehydrogenation catalyst, catalyst modifiers, successive impregnation.

INTRODUCTION

Dehydrogenation of hydrocarbons is an important commercial process because of the great and expanding demand for dehydrogenated hydrocarbons for use in the manufacture of various chemical products (Bell, 1992). All the dehydrogenation catalysts of long chain paraffins are designed to produce linear olefins for the manufacture of biodegradable detergents from raw materials such as C_{10} to C_{13}, C_{11} to C_{14}, C_{11} to C_{15}, and mixtures of the like (Padmavathi et al., 2005). Platinum and platinum-containing bimetallic catalysts supported on alumina are widely used for naphtha reforming and for heavy linear alkanes (C_{10}-C_{15}) dehydrogenation in the petrochemical industries. However, recently these kinds of catalysts have also been used for dehydrogenation of light paraffins. The difference between dehydrogenating and reforming catalysts relies mainly on the acidity of the support. Reactions catalyzed by the acidic sites of the support (such as isomerization, cracking and polymerization) must be inhibited in order to increase the yield to olefins in the dehydrogenation processes (Miguel et al., 1995; Pieck et al., 2005).

The key role of dehydrogenation catalysts is to accelerate the main reaction while controlling the other side reactions (Bhasin et al., 2001). In the case of dehydrogenation of long chain alkanes, a multimetallic alumina supported platinum catalyst is implemented, which contains In and Sn promoters as platinum modifiers, and alkaline and alkaline-earth metals as support modifiers. Promoters improve the activity, selectivity and stability of platinum–alumina catalysts (Gaidai et al., 2001;

*Correspondingauthor. E-mail: akia.mandana@gmail.com.

Gokak et al., 1996). Multicomponent oxides usually show higher activity, longer life time, and better resistance to poisons than the single-component oxide. The majority of multicomponent catalysts for the dehydrogenation of higher normal paraffins have been reported in patents (Dongara et al., 1997; Wilhelm, 1993).

The addition of Sn to Pt/Al_2O_3 catalysts improves the activity and selectivity to olefins in the dehydrogenation reaction, while increasing the catalyst stability (Passes et al., 2000; Castro and Catal, 1993; Sanfilippo et al., 2006). Indium as a modifier decreases the hydrogenating capacity of alumina and suppresses olefin cracking and isomerization, probably as a result of a decrease in the acidic function of the support. The addition of Sn to Pt/Al_2O_3 catalysts has been investigated in several studies, while the effect of In received less attention (Passos et al., 1998). Incorporation of iron in this catalyst increases both the activity and the stability of the catalyst (Dongara et al., 1997). Studies the effect of impregnation sequences in dehydrogenation catalysts are mostly reported for bimetallic catalysts (Baronetti et al., 1985; Kappenstein et al., 1995).

Alumina is the most widely used support material for the dehydrogenation catalysts because of its superior capability to maintain a high degree of platinum dispersion which is essential for achieving high dehydrogenation activity.

But its strong acidity causes side reactions and coke formation (Bhasin et al., 2001). The addition of alkaline and alkaline-earth ions (for example, Li) to γ-Al_2O_3 selectively poisons the isomerization active sites, without affecting the dehydrogenation capacity of the catalytic systems (Siri et al., 2005). The sol-gel techniques have been widely applied for the synthesis of highly porous alumina which originally was developed by Yoldas (1975). The advantages of the sol-gel method includes the ability of maintaining high purity, changing the physical characteristics such as pore size distribution, pore volume and preparing samples at low temperatures (Farias et al., 2003; Zhang et al., 2002).

In this study high surface area nanocrystalline gamma alumina support was synthesized by the sol-gel method and employed as a catalyst carrier for n-dodecane dehydrogenation catalyst (Pt-based catalyst). The effects of different successive impregnation methods for Pt, Sn and In elements were investigated in normal dodecane dehydrogenation reaction.

EXPERIMENTAL

Materials

Aluminium isopropoxide, (AIP, 99 wt%), hexadecyl trimethylammonium bromide (C_{16}TMABr, 99 wt%), and nitric acid were used for the synthesis of gamma alumina. Hexachloroplatinic acid (H_2PtCl_6), tin chloride ($SnCl_2$), indium chloride ($InCl_3$), hydrochloric acid (HCl), lithium nitrate ($LiNO_3$) and iron nitrate ($Fe(NO_3)_3$) were used as metal precursors for catalyst preparation. Normal dodecane and high purity hydrogen (99.99%) were used for the catalytic reactions.

Support preparation

Preparation of the nanocrystalline gamma alumina support based on the method which was described by Akia et al. (2009, 2010). In short, the aluminium isopropoxide and hexadecyltrimethyl ammonium bromide were first dissolved in water. The molar ratios of water to AIP and hexadecyltrimethyl ammonium bromide to AIP were chosen as 90 and 0.8, respectively. The hydrolysis step was carried out at a temperature of 80°C for a time period of 30 min, under vigorous stirring. Subsequently the mixture was peptized using nitric acid (10 wt%) under vigorous stirring by careful pH adjustment to 6.5. The mixture was aged at ambient temperature for 5 h. The condensation of the mixture resulted from the evaporation of the solvent by heating the reaction mixture and subsequent drying in an oven at 110°C for 15 h.

Finally, the dried sample was calcined at 550°C for 5h in order to remove the surfactant and obtain the gamma crystallite phase.

Catalyst preparation

The dehydrogenation catalyst consists of a synthesized or conventional gamma alumina support containing 0.5 wt% platinum, 0.5 wt% of tin, 0.3 wt% of indium, 0.2 wt% Fe, 0.6 wt% of lithium and less than 0.1 wt% of chloride. The catalyst samples were prepared by the successive incipient-wetness impregnation technique. In all the catalysts Fe was impregnated in the first step and Li in the final step. In a typical preparation method, Sn, Pt and Fe were first co-impregnated by the incipient-wetness technique on the gamma alumina support in a solution containing $SnCl_2$, $H2PtCl_6$, $Fe(NO_3)_3$ and 10 wt% HCl. This mixture was aged at room temperature for 4 h under stirring conditions. It was subsequently dried at 80°C for 18 h in a vacuum oven and calcined at 540°C for 2 h in air atmosphere.

In the second step, the calcined sample was impregnated with an aqueous solution containing a mixture of In $(Cl)_3$ and 10 wt% HCl to obtain the desired content of indium. The aging and drying step were done in the same way as the first stage. In order to reduce the chloride content (that is, less than 0.1 wt%), which is responsible for the side reactions, wet calcination was performed at a specific atmosphere which was a mixture of air and water (50/50 molar ratio). The resultant sample was further calcined in air atmosphere at 500°C for 2 h. In the final step, the dehalogenated sample was impregnated with an aqueous solution of lithium $LiNO_3/HNO_3$, followed by drying and calcination under the same conditions as mentioned earlier.

The other catalysts were prepared with the same procedure with the changes in the impregnation sequences. The actual concentrations determined by ICP–AES analyses are in good agreement with the theoretical values.

Characterization

To examine the crystallinity of the prepared samples, powder x-ray diffraction (XRD) analysis was carried out with a MAC Science Co. M18XHF diffractometer using CuKα X-ray radiation (λ= 0.1540 nm). The surface area (BET), pore size distribution and pore volume were determined by nitrogen adsorption at -196°C using an automated gas adsorption analyzer (Tristar 3000, Micromeritics). Scanning electron microscopy (SEM) was performed with HITACHI S-4800 FE-SEM operated at 5 Kv. Transmission electron microscopy (TEM) investigation was performed with JEOL JEM-2100UHR operated at 200 Kv. Temperature programmed reduction

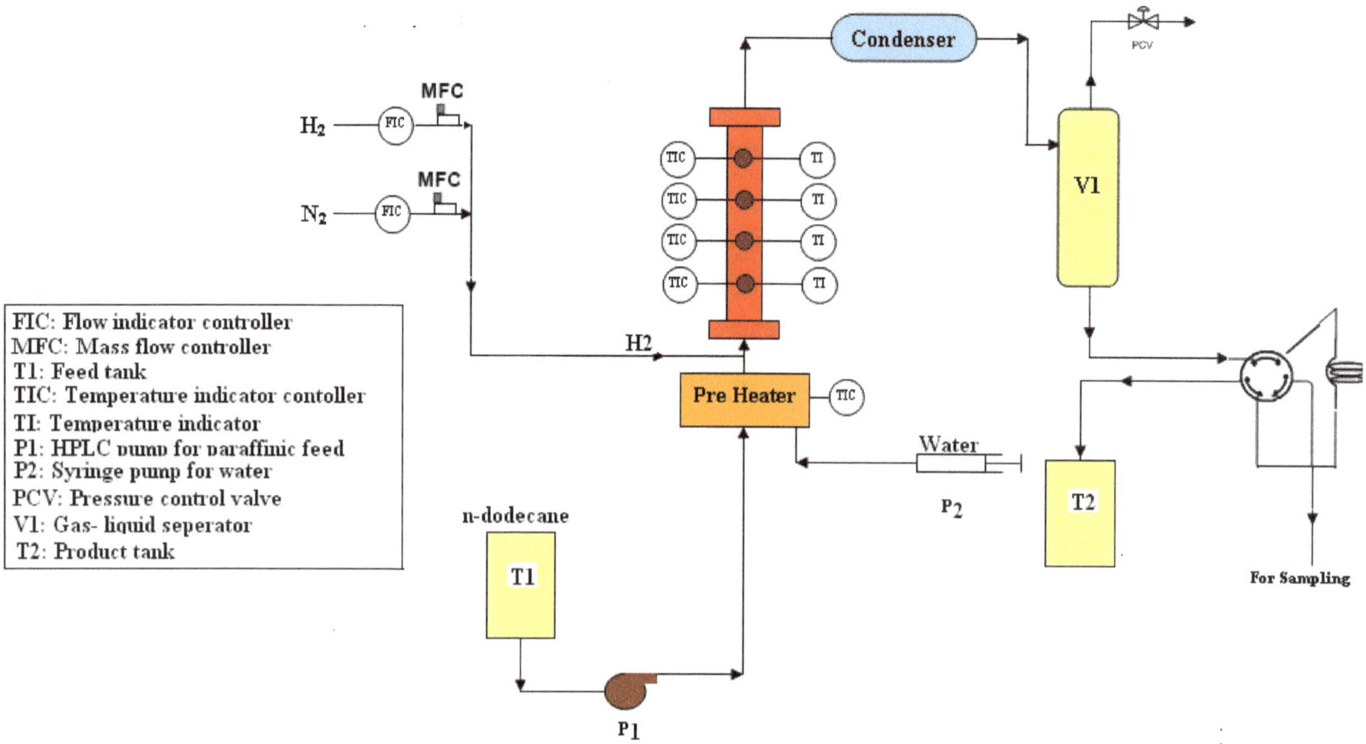

Figure 1. The schematic of the set- up for the dehydrogenation of normal dodecane paraffin.

Table 1. The operating parameters in the dehydrogenation of C_{12} paraffin.

Feed	Operating parameters	Reactor type
Normal paraffin C_{12} Hydrogen (99.99%)	Temperature: 460-490˚C. Pressure: 1-1.7 bar Liquid hourly space velocity (LHSV) of feed: 20 h−1, Hydrogen/hydrocarbon molar ratio: 6 Water: 2000 ppm wt%.	Up-flow fixed bed reactor.

(TPR) was carried out using an automatic apparatus (ChemBET-3000 Quantachrome) equipped with a thermal conductivity detector.

The fresh catalyst (500 mg) was subjected to a heat treatment (10°C/min) in a gas flow (50 ml/min) containing a mixture of H2:Ar (10:90). Before conducting the TPR experiment, the sample was heat treated under an inert atmosphere at 400 ˚C for 3 h.

Experimental set-up for catalyst evaluation

The schematic diagram of the experimental set-up used for the catalyst evaluation is shown in Figure 1. Major sections in the unit are: (1) Hydrogen flow control module, (2) Liquid pumping system, (3) Syringe pump for water, (4) Preheater and mixer for hydrogen, feed and water, (5) Reactor, (6) Condenser and gas–liquid separator, (7) Adequate flow meters to account material balance. The reactor is a tubular fixed-bed type with an inner diameter of 1.9 cm which is operated in isothermal and up-flow mode. Reactor internal temperature is measured with a central temperature probe with 4 thermocouples. Reactor inlet pressure is continuously monitored with a pressure transmitter.

Prior to the experiments all the catalysts, previously 20 to 40 mesh sieved were reduced under flowing H_2 at 430°C for 4 h with a heating rate of 5°C/min. These catalysts were tested for 36 h on stream. The dehydrogenation reactions were carried out at three temperatures; 460, 475 and 490°C. The reactor effluent is cooled down in a water cooler and then enters the separator. Liquid analysis is performed on the product for analysis of the following components: Paraffins, olefins and diolefins, iso-paraffins, alkyl aromatics and alkyl naphthenes and cracked products. This is achieved with FID detector and HP PONA column.

Reaction conditions, which were the same for all the tests, are summarized in Table 1. As it can be seen in Table 1, hydrogen is utilized in amounts sufficient to ensure hydrogen to hydrocarbon mole ratio of about 6:1. Excess hydrogen is necessary in the dehydrogenation of lower and higher paraffins. Hydrogen serves a dual-function in both diluting the paraffin and suppressing the formation of hydrogen deficient, carbonaceous deposits on the catalyst composite.

When hydrogen is used as the diluent, a preferred practice is to add water or a water-producing compound to the dehydrogenation zone (Gaidai et al., 2001; Wilhelm, 1993).

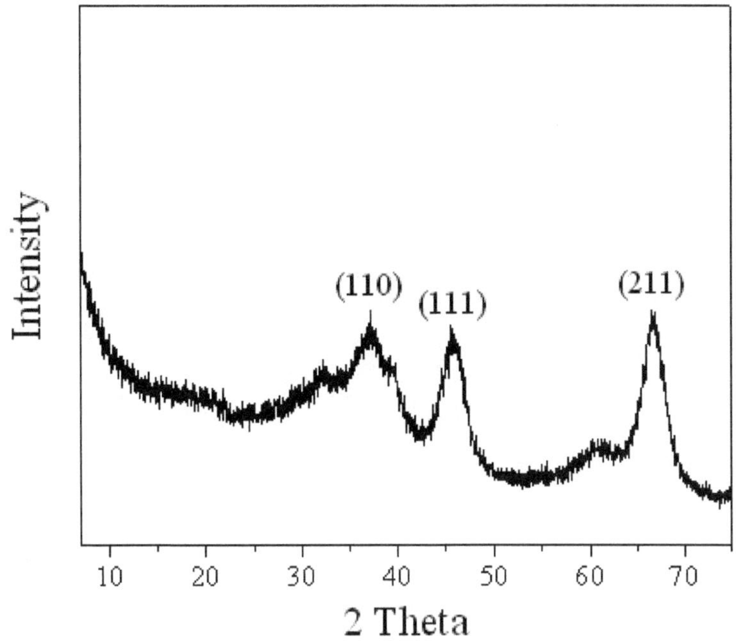

Figure 2. XRD pattern of the prepared Al_2O_3 calcined at 550°C.

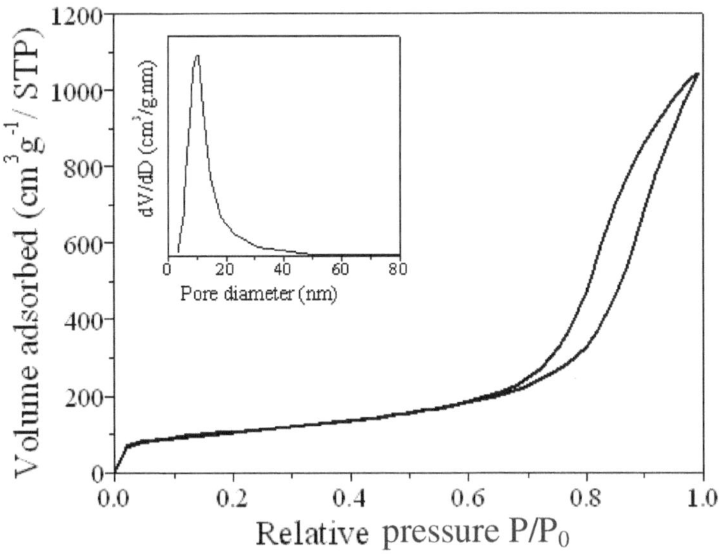

Figure 3. Nitrogen adsorption/desorption isotherm and pore size distribution (upper inset) of the Al_2O_3 calcined at 550°C.

RESULTS AND DISCUSSION

Texture and surface properties of γ-Al_2O_3

Aluminum alkoxide is hydrolyzed with water, producing aluminum mono- or trihydroxides. In both cases X-ray diffraction patterns of the initial products are the same. Poorly crystallized boehmite (pseudoboehmite) develops after a few hours while aging the precipitate. Boehmite is the only phase occurring if the temperature exceeds 77°C during the hydrolysis reaction or during aging of an initially amorphous precipitate (Dilsiz et al., 2002). The XRD pattern of the alumina sample calcined at 550°C is shown in Figure 2. As it can be seen, the sample is in the gamma crystallite phase.

The nitrogen adsorption/desorption isotherm and pore size distribution of the Al_2O_3 calcined at 550°C are shown in Figure 3. It can be concluded that the sample exhibits the classical shape of a type IV isotherm according to the IUPAC classification, typical for mesoporous solids

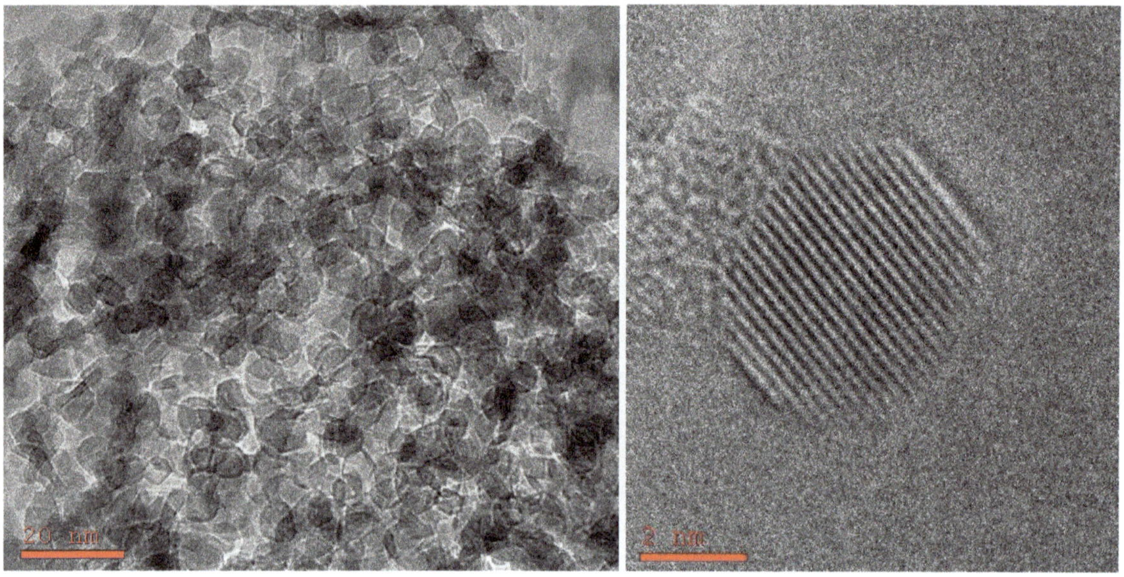

Figure 4. TEM images of the prepared γ -Al_2O_3 sample calcined at 550℃.

(Leofantia et al., 1998). However, for this sample, a hysteresis loop (type H1) occurs at a higher relative pressure range (p/p_0= 0.7 to 0.9) suggesting a broad pore size distribution with uniform size and shape. The pore size distribution (Figure 3, upper inset) indicates a mesoporous structure and also confirms a relatively broad pore size distribution. The BET surface area, average pore diameter and pore volume for the synthesized support were 328 m^2g^{-1}, 16.21 nm and 1.61 cm^3g^{-1}, respectively.

Figure 4 shows the TEM images of the synthesized gamma alumina. As it can be seen the prepared alumina highly crystallized with almost uniform size and/or shape particles. The TEM images indicate the presence of hexagonal crystal structure. The results also confirm the nanocrystallinity of the synthesized gamma alumina support (less than 4 nm).

Catalyst characterization and evaluation of catalytic performance

The results listed in Table 2 represent the different impregnation steps and the structural properties of the prepared catalysts. The BET results reported in Table 2 indicate that the catalysts for which platinum was impregnated in the first step had smaller pore volumes compared to other catalysts for which platinum was impregnated in the second step. The first step platinum impregnated catalysts also showed narrower pore size distributions in comparison to the other catalysts. The results indicate that the largest pore diameters (more than 6 nm) were obtained in catalysts for which platinum was impregnated in the second step. All the catalysts for which platinum was first impregnated had smaller pore diameters (less than 6 nm). Figure 5 shows the nitrogen adsorption isotherms at -196℃ and the BJH pore size distributions (upper inset) of the two catalyst samples (2 and 7 as specified in Table 2) prepared with different impregnation sequences. The isotherms show the classical shape of the type IV according to the IUPAC classification, typical for mesoporous solids (Leofantia et al., 1998). However, for these samples, a hysteresis loop (type H3) occurs at lower relative pressures, suggesting a narrow pore size distribution. The pore size distributions (Figure 5, upper inset) confirm this assertion. Furthermore the pore size distribution for Cat. #7 shows a wider range in comparison to Cat. #2. It has been reported that alumina support with surface area higher than 150 $m^2\ g^{-1}$ shows good dispersion of metals.

Based on the research which investigated the effect of pore size on the platinum dispersion, it was revealed that dehydrogenation catalysts supported on alumina with small pores (more than 30% in the range of 2 to 10 nm) have dispersion more than 70% (Sharma et al., 2002). These results are also confirmed in (Merlen et al., 1996; Gomez et al., 1996; Bocanegra et al., 2006). The pore size distribution of all the prepared samples are mainly in the range from 2 to 10 nm, which means there is a good dispersion of platinum on the gamma alumina support with high surface area (328 m^2g^{-1}). Figure 6 represents the SEM image of Cat. #2 and Cat. #7. The surface image of Cat. #7 shows better distribution of the metals, but Cat. #2 indicates larger aggromerization of particles on the support.

Micrographs of transmission electron microscopy of Cat. #7 are shown in Figure 7. The dark spots in Figure 7(a) are believed to be metal clusters, while the lighter areas correspond to the gamma alumina support. In high resolution micrograph (Figure 7b), the Pt nanoparticles can be observed. Different catalysts prepared were

Table 2. The Textural properties of the dehydrogenation catalysts prepared with various impregnation sequences.

Catalyst #	Steps of impregnation	BET (m^2 g-1)	Pore diameter (nm)	Pore volume (cm^3g-1)
1	Pt+Sn+Fe – In – Li	179.03	4.35	0.25
2	Pt+Sn+In+Fe– Li	199.92	5.16	0.33
3	In+Fe– Pt+Sn – Li	187.87	6.03	0.47
4	Pt+Fe – In+Sn – Li	187.90	4.95	0.27
5	Sn +Fe – Pt+In– Li	217.69	6.76	0.51
6	Pt+Fe+In – Sn – Li	197.42	5.51	0.35
7	In+Sn+Fe – Pt – Li	227.74	6.41	0.51
8	In+Sn+Fe – Pt – Li	116.5	6.25	0.41

evaluated in the n-dodecane dehydrogenation reaction. Table 3 shows the catalytic performances of the samples at three different temperatures. The total conversion, olefin percentage and the selectivity for olefinic products are presented in Table 3. It is evident for all the samples, with an increase in temperature, the total conversion increased but the selectivity to olefinic products decreased because of the increase in the side reactions. The olefinic products which presented in Table 3 are mostly alfa monoolefins which is the desired product in higher normal paraffins dehydrogenation. In the samples 1, 2 and 3, platinum and tin co-impregnated. In sample 1, indium impregnated after them, in sample 2 indium co-impregnated and in sample 3, indium impregnated before them. As it can be observed, Cat. #3 showed better catalytic performance in comparison to the Cat. #1 and Cat. #2. In the samples 2, 4 and 7, tin and indium co-impregnated. In sample 2, platinum also co-impregnated, in sample 4 platinum impregnated before them and in sample 7, platinum impregnated after them. The catalytic results showed better performance for Cat. #7 in comparison to the Cat. # 2 and Cat. #4. In samples 2, 5 and 6, platinum and indium co-impregnated. In sample 2, tin co-impregnated with the other mentioned elements. In sample 5, tin impregnated before them and in sample 6, tin impregnated after them. Cat. #6 showed better catalytic results in comparison to the Cat. #2 and Cat. #5.

The results presented in Tables 2 and 3 indicate that the catalysts with smaller pore sizes and pore volumes also show good catalytic performances. The critical dimension for spherical molecules, computed by using the Lennard–Jones potential, was 1.24 (nm) for n-dodecane (Cruz et al., 2004). The results obtained in Table 2 showed that the pore volumes and pore diameters for the prepared catalysts were high enough for this reaction. But it must be noted that the large pores can minimize diffusion resistance. Sufficiently large pores are required for diffusion as the catalyst ages and fouls.

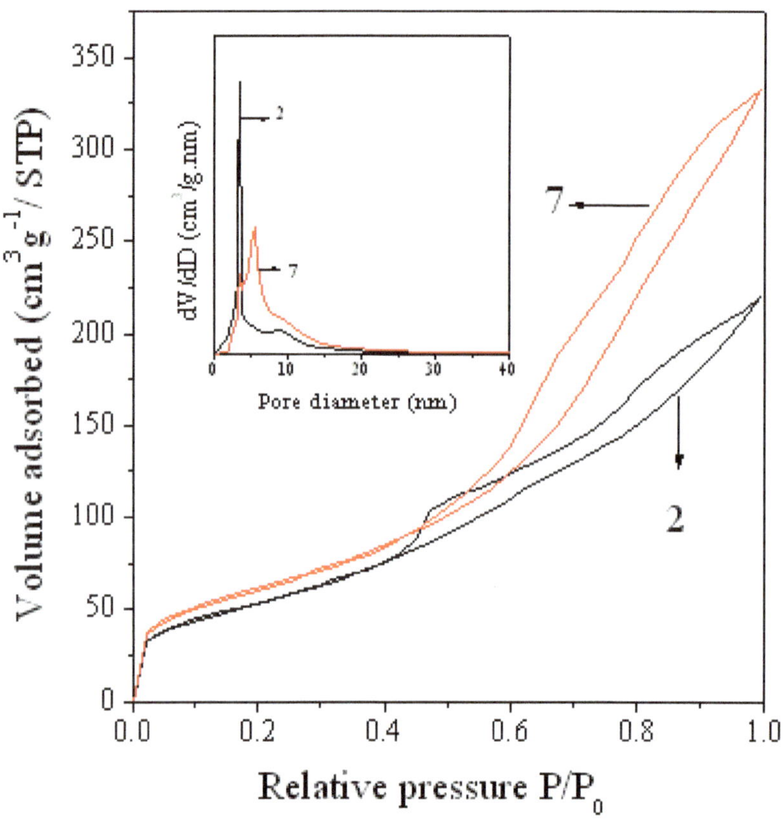

Figure 5. Nitrogen adsorption/desorption isotherm and pore size distribution (upper inset) of Cat. #2 and Cat. #7 prepared with different impregnation sequences.

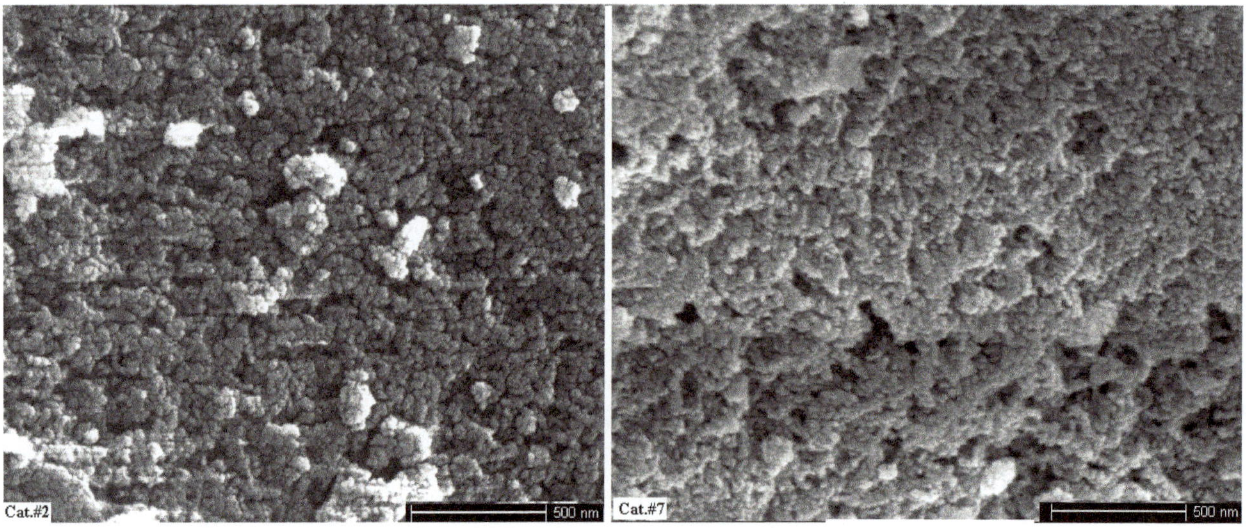

Figure 6. Micrographs of scanning electron microscopy of Cat. #2 and Cat. #7.

Our experimental results indicate that the catalyst which was prepared by co-impregnation of indium, tin and iron in the first step, impregnation of platinum in the second step, followed by lithium impregnation in the third step (Cat. #7), showed the best catalytic performance. The pore volume and the pore diameter for this sample were almost the highest among the other catalysts and the BET surface area was also the highest among the samples prepared. The results clearly showed the lower catalytic activity for the sample, which was prepared by

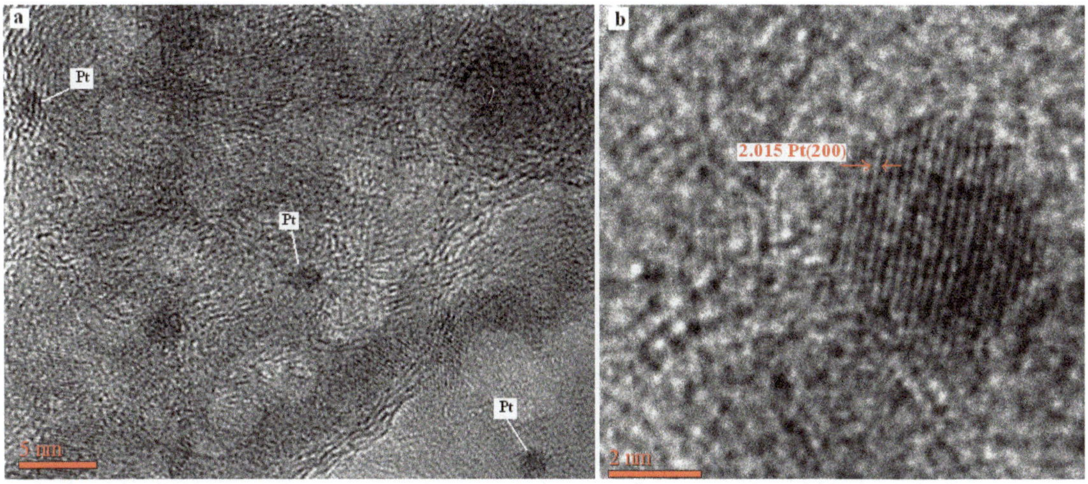

Figure 7. (a) Transmission electron micrograph and (b) high-resolution image of a Pt particle in Cat. #7.

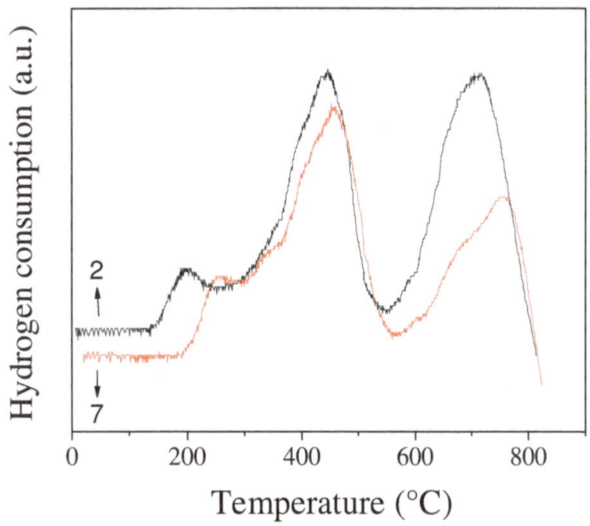

Figure 8. TPR profiles of two dehydrogenation catalysts (Cat. #2 and Cat. #7) prepared with the changes in the impregnation sequences.

co-impregnation of all the metals except lithium in the first step (Cat. #2).

The Cat. #8 was prepared with the conventional gamma alumina support to compare the effect of support on the performance of the dehydrogenation catalyst. The specific surface area, pore diameter and pore volume of the conventional gamma alumina support were 190 m^2 g^{-1}, 10 nm and 0.5 cm^3 g^{-1}, respectively. This catalyst was prepared with the impregnation sequences used for the Cat. #7. As it can be seen in Table 3, the catalytic results revealed that the synthesized support performed well as a good support to be used in dehydrogenation catalysts.

The TPR results for two catalyst samples (Cat. #2 and 7) are presented in Figure 8. For both catalysts, three main reduction zones appeared in the 150 to 320°C, 320 to 550°C and 550 to 800°C. The first small peak is related to the reduction of large Pt oxide or oxychloride crystals weakly interacted with the support. The second sharp peak indicates the co-reduction of Pt with the other metals (Sn and In).

The results of catalytic testing indicate that the catalysts with the first peak at relatively higher temperatures showed better catalytic activity. The shift of the first peak to the higher temperatures indicates the co-reduction of Pt with the other metals (Sn and In), which suggests a strong interaction between the metals with probable alloy formation. The alloys formation may break larger Pt ensembles into more finely distributed Pt species and thus improve the dispersion indirectly. As it was seen in Table 3, the highest and lowest catalytic

Table 3. The catalytic testing results of the prepared samples in the dehydrogenation reaction of n-dodecane at three different temperatures.

Catalyst #	Temp. (°C)	Total conversion (%)	Monoolefins (%)	Selectivity (%)
	460	9.06	7.08	78.20
1	475	11.75	8.03	68.37
	490	14.65	8.10	55.35
	460	6.70	5.00	74.76
2	475	8.88	4.66	52.49
	490	11.66	5.26	45.12
	460	10.08	7.58	75.2
3	475	13.2	8.59	65.07
	490	18.04	9.38	52.01
	460	8.64	6.92	80.13
4	475	12.27	8.84	72.08
	490	17.47	9.96	57.02
	460	11.14	8.00	71.8
5	475	12.93	8.47	65.46
	490	16.73	9.92	59.3
	460	10.23	8.04	78.572
6	475	12.75	9.45	74.15
	490	17.40	11.29	64.91
	460	11.39	8.80	77.31
7	475	14.03	10.18	72.55
	490	18.21	11.58	63.58
	460	10.18	7.23	71.04
8	475	12.86	8.38	65.2
	490	15.42	8.94	58.60

activity obtained for Cat. #7 and Cat. #2, respectively. As it is shown in Figure 8, the first reduction temperature for sample 7 was shifted to the higher temperatures in comparison to sample 2. This result also indicates the good interaction between active phase and its modifiers and consequently fine distribution of the platinum in Cat. #7. The TPR results for all the catalysts showed a reduction peak centered around 700°C, which is related to some portions of the oxides might be strongly bonded among themselves or with the support.

CONCLUSIONS

In this study high surface area nanocrystalline gamma alumina was prepared by the sol-gel method. The synthesized support was used as a carrier for dehydrogenation catalysts of higher normal paraffins. Different dehydrogenation catalysts were prepared with changes in the impregnation sequences of active phase (Pt) and its modifiers (Sn and In). It was observed that different impregnation methods lead to the different catalytic behaviors. The BET results revealed that smaller pore volumes were obtained for the catalysts which were impregnated with platinum in the first step. Based on the computed molecular size of n-dodecane, the results indicated that the pore sizes and pore volumes of all the catalysts were suitable for this reaction. But a higher pore catalyst improves accessibility of the hydrocarbons and reduces the possibility of the pores being plugged due to coke or metals deposition.

The best performance was obtained by Cat. #7, for which Fe, Sn and In were first impregnated, with Pt impregnation in the next step, followed by Li impregnation in the final step. The surface area, pore size and pore volume of Cat. #7 was almost the highest among the catalysts prepared. The catalytic results also confirmed the ability of the synthesized nanocrystalline gamma alumina as a promising support for dehydrogenation catalysts of higher normal paraffins.

ACKNOWLEDGEMENTS

The authors acknowledge financial assistance of Research and Technology center of the petrochemical company in Iran.

REFERENCES

Akia M, Alavi SM, Rezaei M, Zi- Feng Y (2009). Optimizing the sol–gel parameters on the synthesis of mesostructure nanocrystalline γ-Al_2O_3. Microporous Mesoporous Mater., 122: 72

Akia M, Alavi S M, Rezaei M, Zi- Feng Y (2010 Synthesis of high surface area γ- Al2O3 as an efficient catalyst support for dehydrogenation of n-dodecane. J. Porous Mater., 17: 85

Baronetti GT, Miguel SR, Scelza OA, Fritzler A (1985). Pt-Sn/Al_2O_3 catalysts: studies of the impregnation step. Castro AA Appl C, 19: 77

Bell AT (1992). Catalysis Looks to the Future: National Academy Press, Washington D.C, p. 1992. 23.

Bhasin MM, McCain JH, Vora B V, Imai T, Pujado PR (2001). Dehydrogenation and oxydehydrogenation of paraffins to olefins. Appl Catal. A., 221: 397

Bocanegra SA, Castro AA, Ruız AG, Scelza OA, Miguel SRD (2006). Characteristics of the metallic phase of Pt/Al_2O_3 and Na-doped Pt/Al_2O_3 catalysts for light paraffins dehydrogenation. Chem. Eng. J., 118: 161

Castro AA (1993). Catalysts for the selective dehydrogenation of high molecular weight paraffins. Catal. Lett., 22: 123.

Cruz FJ, Laredo GC, (2004). Molecular size evaluation of linear and branched paraffins from the gasoline pool by DFT quantum chemical calculations. Fuel, 83: 2183.

Dilsiz N, Akoval G (2002). Study of sol–gel processing for fabrication of low density alumina microspheres. Mater. Sci. Eng., 332: 91

Dongara R, Basrur AG, Gokak DT, Gokak K V, Rao V, Krishnamurthy K R, Bhardwaj IS (1997 Catalyst composite for dehydrogenation of paraffins to mono-olefins and method for the preparation thereof. Bhardwaj I S US 5 677 260. 1997

Farias RF, Arnold U, Martinez L, Schuchardt U, Jannini MJDM, Airoldi C (2003). Synthesis, Characterization and Catalytic Properties of Sol-Gel Derived Mixed Oxides. J. Phys. Chem. Solids, 64: 2385

Gaidai N A, Kiperman S L, Kinet C (2001). Kinetic Models of Catalyst Deactivation in Paraffin Dehydrogenation Kinet Catal., 42: 527

Gokak DT, Basrur AG, Rajeswar D, Rao GS, Krishnamurthy KR (1996). Lithium promoted Pt/Sn/Al_2O_3 catalysts for dehydrogenation of n-decane: Influence of lithium metal precursor. React. Kinet. Catal. L, , 59: 315

Gomez R, Bertin V, Lopez T, Schifter I, Ferrat G (1996). Pt-Sn/Al_2O_3 sol-gel catalysts: catalytic properties. J. Mol. Catal. A. Chem., pp. 109: 55

Kappenstein C, Saouabe M, Guerin M, Marecot P (1995). Characterization and activity of Pt-Sn/Al_2O_3 catalysts of different preparation: coimpregnation and new Pt-Sn precursor. Catal J., 31: 9

Leofantia G, Padovan M, Tozzola G, Venturelli B, (1998). Surface area and pore texture of catalysts. Catal Today, 1998, 41: 207

Merlen E, Beccat P, Bertolini JC, Delichere P, Zanier N, Didillon B (1996). Characterization of Bimetallic Pt–Sn/Al_2O_3 Catalysts: Relationship between Particle Size and Structure. J Catal., 159: 178

Miguel S R, Castro A A, Scelza O A (1995). Catal Lett. Effect of the addition of alkali metals on the metallic phase of Pt/Al_2O_3 catalysts.Catal Lett., 32: 281

Padmavathi G, Chaudhuri K K, Rajeshwer D, Rao GS, Krishnamurthy KR, Trivedi PC, Hathi K K, Subramanyam N, Chem Eng Sci, (2005). Kinetics of n-dodecane dehydrogenation on promoted platinum catalyst. Chem. Eng. Sci., 60: 4119.

Passes FB, Andera DAG, Schmal M (2000 The state of Tin on Pt–Sn/Nb2O5 catalysts. Catal. Today, 57: 283

Passos FB, Andera DAG, Schmal M (1998). Characterization and Catalytic Activity of Bimetallic Pt-In/Al_2O_3 and Pt-Sn/Al_2O_3 Catalysts Catal J., 178: 478

Pieck CL, Vera CR, Querini CA, Parera JM (2005). Differences in coke burning-off from Pt–Sn/Al_2O_3 catalyst with oxygen or ozone. Appl. Catal. A., 278: 173

Sanfilippo D, Miracca I (2006). Dehydrogenation of paraffins: synergies between catalyst design and reactor engineering. Catal Today, 111: 133

Sharma LD, Kumar M, Saxena AK, Chand M, Gupta JK (2002). Influence of pore size distribution on Pt dispersion in Pt-Sn/Al_2O_3 reforming catalyst. J. Mol. Catal A. Chem., 185: 135

Siri GJ, Ramallo-Lopez JM, Casella ML, Fierro JLG, Requejo FG, Ferretti OA (2005). XPS and EXAFS study of supported PtSn catalysts obtained by surface organometallic chemistry on metals Application to the isobutane dehydrogenation. Appl. Catal. A, 278: 239

Wilhelm FC (1993). Dehydrogenation of hydrocarbons with a multimetallic catalytic composite. US 3 998 900. 1993

Yoldas BE (1975). Alumina sol-preparation from alkoxides. Am. Ceram. Soc. Bull., 54: 289.

Zhang Z, Hicks RW, Pauly TR, Pinnavaia TJ (2002). Mesostructured forms of gamma- Al2O3. J. Am. Chem. Soc., 124: 1592

11

Optimization of plug utilized in lost circulation treatment while drilling

Abdelkader Khiari[1], Djemoi Merzoug[1], Noureddine Gherraf[1]*, Mourad Kourichi[2] and Noureddine Maza[1]

[1]Laboratoire des Ressources Naturelles et Aménagement des Milieux Sensibles, Larbi Ben M'hidi University, Oum el Bouaghi, 04000 Algeria.
[2]Laboratoire de génie de procédés, UKMO Ouargla University 30000 Algeria.

The major problem of oil wells drilling is circulation lost that can occur near weak formations and before well cementing. Conventional methods such as lost circulation material pills, cement plugs and dual injection were used to combat different types of massive mud loss in spite of their small success rate and high cost and risk. A different circulation treatment has been introduced to optimize plugging of loss zones by minimizing cost and time. It consists of combining InstanSeal (seal formation instantly) and cement to improve the mechanical performance of the system. InstanSeal-cement plugging fluid is a loose invert emulsion (water-in-oil) that gels rapidly in normal drilling operations after passing through the drill bit nozzles resulting in a rigid direct emulsion (oil - in- water) which will be squeezed and penetrated into weak formations to seal it. It is the only gelling system that uses mechanical means as a trigger mechanism. The present work aims to optimize the emulsion mechanism, the composition effect and the InstanSeal-cement as follows: (i) Introduce the InstanSeal-cement as an approach to cure massive losses where conventional methods fail. (ii) Analyze the evolution of InstanSeal-cement fluid by changing each product concentration. (iii) Find the proper emulsion of each kind of losses. (iv) Optimize the system according to the equipment taking into account the mixing energy. (v) Find the most effective formulation using mathematic models. (vi) The laboratory pre-checks will be taken as helping data to minimize the failure risk. After the success results obtained from the yard test, the InstanSeal-cement proved that it is competent to cure massive losses by improving the success rate (80%). This economic success can be matched by reducing risks, achieving excellent results and decreasing drilling time through a critical section.

Key words: InstanSeal, drilling fluid, circulation lost, optimization.

INTRODUCTION

Drilling fluids, such as mud, are fluids used to drill a hole through the earth crust. They circulate down the rotating drill pipe through the bit and up the annular space between the pipe and the formation or steel casing; to the surface (Henry et al., 1983).

In many parts of the world specific techniques and methods are being implemented to combat massive mud loss and save the well in spite of their cost, time consumption and risk (Tailleur, 1963). Instanced-cement is a "patented circulation treatment" which was developed as a solution to cure lost circulation problems while drilling when conventional methods fail. It is defined as a single fluid able to cure massive mud losses instantly by being pumped through the drill bit nozzles (www.connect.slb.com. *Mark of Schlumberger, 1999).

This shear-sensitive plugging fluid was designed to gel rapidly after passing through the bottom hole assembly in normal operations, thus forming a solid mass. It is a gelling system which uses mechanical means as a trigger mechanism to develop its initial compressive strength but not to consolidate the formation. Before developing some gel strength, the plugging fluid maintains the rheology of the drilling fluid allowing its penetration into the loss zone

*Corresponding author. E-mail: ngherraf@yahoo.com.

Table 1. Some test samples.

Test number	1	2	3	4	5	6
Oil-water ratio (%)	30	25	20	30	25	20
Gas oil (ml)	120	100	80	120	100	80
Emulsifier (ml)	5	5	5	8	8	8
Bentonite (g)	12.5	12.5	12.5	12.5	12.5	12.5
Cross linker (g)	4	4	4	4	4	4
Viscosifier (g)	7.5	7.5	7.5	7.5	7.5	7.5
Cement (g)	250	250	250	250	250	250
Water (ml)	280	300	320	280	300	320

where it actually stops and provides a permanent sealing.

STANSEAL-CEMENT REACTION MECHANISM

This plugging fluid is activated by mechanical means to give rigid gel that allows the lost circulation zone to be closed off quickly and continue drilling (David et al., 1999).

Definition

InstanSeal-cement is a single reactive fluid which is based upon what is termed a " loose" invert emulsion that is (water - in - oil) or " shear-sensitive" invert emulsion. This is due of the degree of instability towards high shear forces. It consists of: oil emulsifier, bentonite, cross-linker, viscosifier, class G cement and water.

Functioning principal

The shear activated gel consists of two phases -water and oil- emulsion. The internal phase of the loose plugging fluid (water) consists of a high polymer concentration (water soluble polymer) and the continuous phase (oil) contains a cross linker and cement. Upon exposing the fluid to high shear forces (pressure drop across small orifice) the loose invert emulsion undergoes very rapid deformation and flips to more stable direct emulsion (oil in water). Enough energy must be imparted to the system to invert the emulsion. This is accomplished by shearing the fluid with about 500 psi differential pressure at the drill bit.

Reaction mechanism

To provoke and initiate the emulsion inversion (the mechanical reaction between the reactive species) a minimum shear threshold must be achieved. When the loose invert emulsion is subjected to high shear forces, the emulsion flips and results in a rupture that releases both the encapsulated cross linker and the cement from the oil phase to the new water continuous phase, allowing the cross linker to contact the polymer that triggers the gel, forms a solid mass and initiates the cross-linking reaction almost instantaneously. The final product is a rigid, robust, very hard, but plastic gel created in few minutes after passage through the bit.

The loose emulsion is maintained by a low concentration of an emulsifier. The surface of the cross linker is oil wet and remains in the oil phase. The high shear triggers a rupture in the interfacial membrane of the emulsion by making the large water droplets ruptured to give an emulsion containing more stable numerous and smaller diameter oil droplets. The interfacial tension between the two phases is maintained by the emulsifier (surfactant) that reduces the surface tension between the water droplets and the oil (Parvazinia and Nassehi, 2006).

Chemistry of cross linked gel

The viscosifier is a water soluble polymer (polysaccharide) which has hydroxyl groups along its chain. In a specific pH region an oligomerisation is occurred by the creation of cross linked bond between hydroxyl groups and metallic polyvalent ions or other organic molecules.

The transfer of the cross linker into the high concentration polymer slurry is rapid. This is why this system (gelling time) is temperature independent http://www.slb.com/services/cementing/lost_circulation/instant_seal.aspx. The instability of the loose emulsion is exploited to create the new technology.

EXPERIMENTAL

Check ratio and interpretation

Table 1 discusses the test sample and Table 2 discusses the manipulation observation. However, the other tests have been performed in a similar way.

Table 2. Manipulation observation.

Test number	Observation
1	Good mixability and very viscous fluid, pH: 11.62, Gelling time: 7 s
2	Good mixability and less viscous fluid, pH: 11.65, Gelling time: 6 s
3	Difficult mixability, pH: 11.66, Gelling time: 5 s
4	Good mix ability, pH: 11.37, Gelling time: 65 min
5	Good mixability, pH: 11.67, Gelling time: 55 min
6	Acceptable mixability, pH: 11.69, Gelling time: 55 min.

Table 3. OWR effect.

OWR	pH	Gelling time (min)	Gel strength (psi)	Plastic viscosity (cp)
20	11.69	55	5.829	42
25	11.67	55	6.119	12
30	11.37	65	6.003	03

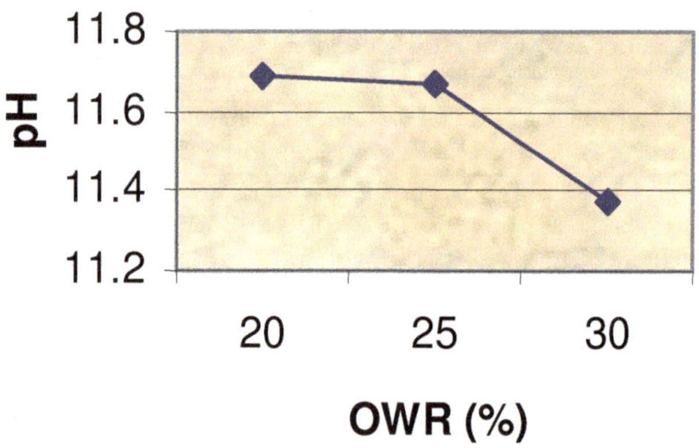

Figure 1. OWR effect on pH.

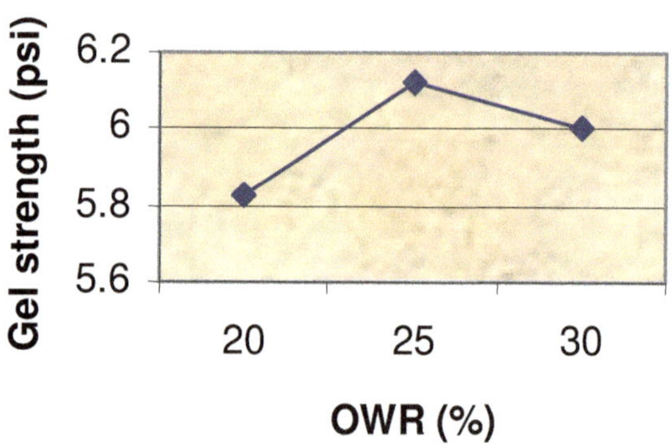

Figure 3. OWR effect on gel strength.

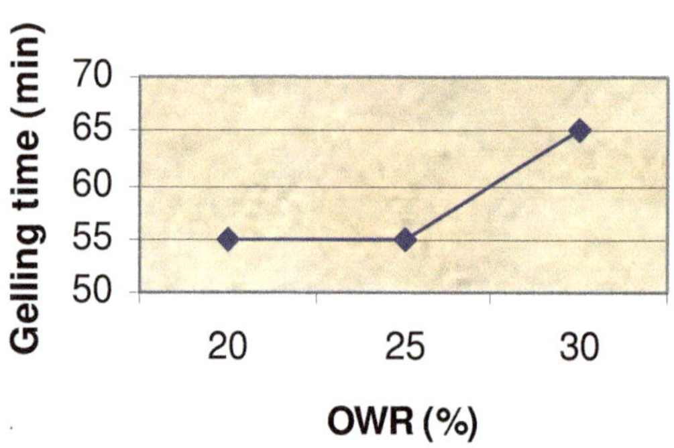

Figure 2. OWR effect on gelling time.

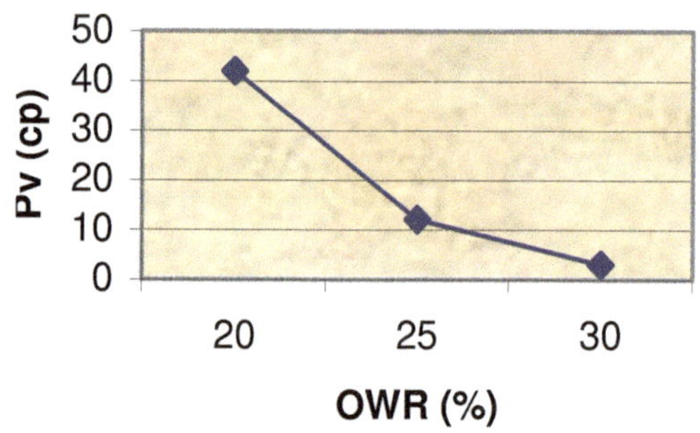

Figure 4. OWR effect on plastic viscosity.

RESULTS AND DISCUSSION

OWR (oil-water-ratio) effect

Table 3 and Figures 1 – 4 discuss the OWR effect. The impact of the oil-water-ratio on the other parameters is less important except on the viscosity.

Emulsifier effect

Table 4 and Figures 5 to 7 discusses the emulsifier effect.

Table 4. Emusifier effect.

Emulsifier (ml)	Plastic viscosity (cp)	Gelling time (min)	Gel strength (psi)	
			4 h	24 h
1.5	48.00	1.50	5.539	6.011
2.0	37.50	1.66	7.012	8.528
2.5	33.02	2.00	7.111	8.873
3.0	30.02	2.16	8.211	8.886
3.5	21.00	2.50	9.302	9.418
4.0	18.00	3.00	9.302	10.711
4.5	15.01	4.00	9.302	11.153
5.0	15.00	4.50	9.302	11.850
5.5	15.00	12.00	9.201	11.502
6.0	15.00	14.00	0	11.414
6.5	15.00	25.00	0	11.072
7.0	15.00	26.00	0	10.794
7.5	15.00	28.00	0	10.223
8.0	14.80	180.00	0	10.105

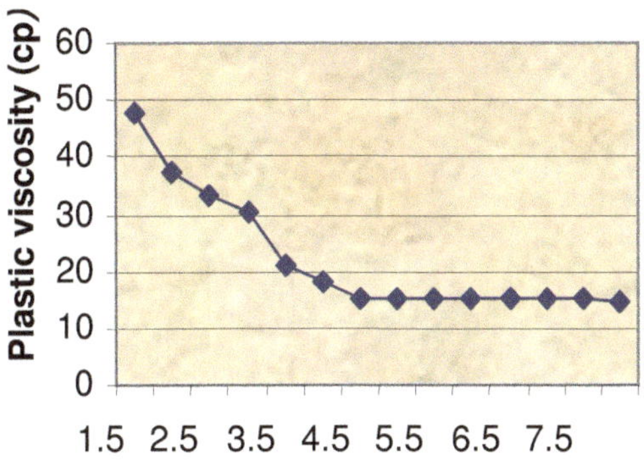

Figure 5. Emulsifier effect on rheology.

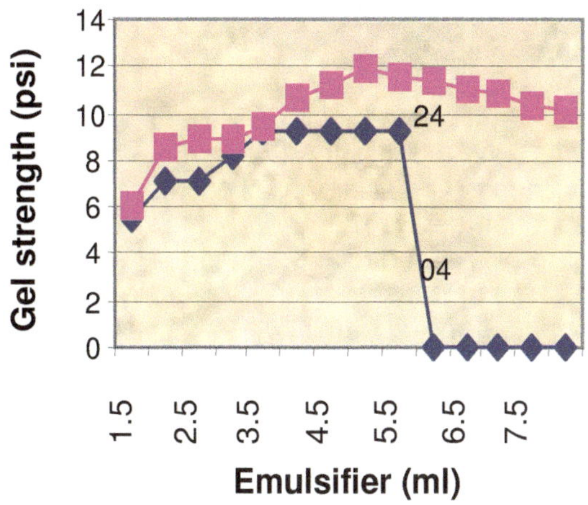

Figure 7. Emulsifier effect on gel strength.

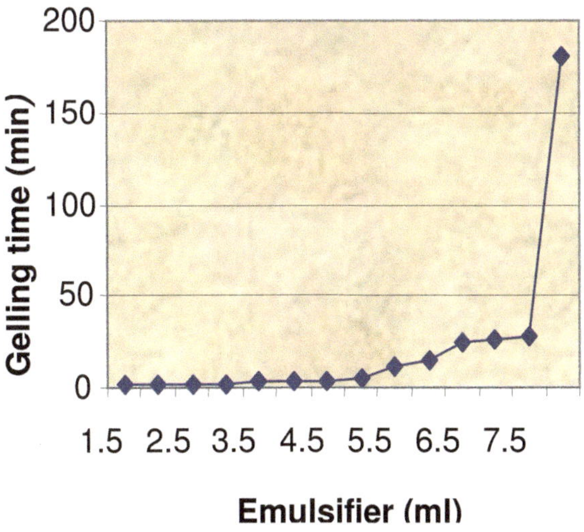

Figure 6. Emulsifier effect on gelling time.

Formulations with emulsifier = 5 ml

5 ml as emulsifier concentration is enough to facilitate the mixing and create the emulsion. However; the gelling time obtained with this concentration is not adequate to cure losses. It is very short and it can prevent the InstanSeal-cement from the penetration into the loss zone to seal it. Therefore, to continue the study a high concentration of emulsifier is added with maintaining the same reference design.

Formulations with emulsifier = 8 ml

If the OWR is low the mix ability can not be easy because the water volume is high. High OWR leads to improve the gel strength which means that the InstanSeal-cement is hard. For a high OWR, the obtained pH value is not high in comparison to that for low OWR. This means that the cross linker and cement are well protected in the oil phase (cross linking reaction is slow). The loose emulsion is more stable.

The slow release of the cement and the cross linker give a delayed gelling time. This is observed where OWR is high. High rheology property indicates the best transfer of the cement and the cross linker. This can be observed in the formulation where OWR is low. A reduced gelling time and an acceptable plastic viscosity with a better gel strength were observed with OWR: 25.0% and OWR: 75%. Therefore this ratio is chosen as the base ratio.

Observations while checking

For OWR 25% and OWR 75% the InstanSeal-cement cubic samples treated for 24 h in the curing chamber are deformable by applying a high finger pressure. This means that this formulation can be used as a plug to cure losses but it stays soft after 24 h from injecting.

Interpretation

For the formulation without emulsifier it was not observed any emulsion (bad mixing). But a high emulsifier concentration leads to an easier mixing. Increasing the emulsifier concentration delays the gelling time owing to the good curvature of this surfactant around the water droplets by minimizing the transfer of cross linker and cement from the oil phase to the water phase. This slow diffusion is the reason for getting a very long gelling time. The minimal transfer gives reduced values of rheology and pH. An exceeded concentration of emulsifier may prevent the InstanSeal-cement from setting (as it is shown in the last test where emulsifier = 8 ml which gives more than three hours as a gelling time and makes the formulation useless.

The formulations leading to a short gelling time should not be prepared to cure losses (InstanSeal-cement can not penetrate into the loss zone) whereas a long one can cure minor losses.

The high concentration of emulsifier reduces the gel strength. It is also observed that the evolution of the gel strength between 4 and 24 h is not well developed. Therefore the operator can continue drilling before 24 h from pumping. This proves that the emulsifier has a slight effect on the gel strength.

Emulsifier has a strong effect on the gelling time of the InstanSeal-cement

Bentonite effect

Table 5 and Figures 8 and 9 discuss the effect of bentonite.

Table 5. Bentonite effect.

Bentonite (g)	Gelling time (min)	Gel Strength (psi) 4 h	Gel Strength (psi) 24 h
0	125	10.994	11.082
5	6	11.512	11.862
10	5	11.563	12.636
11.5	5	12.056	12.697
12	4.5	12.205	12.679
12.5	4.5	12.218	12.884
13	3.5	12.220	13.039
15	3	12.242	13.114
20	3	0	15.328
25	3	0	0

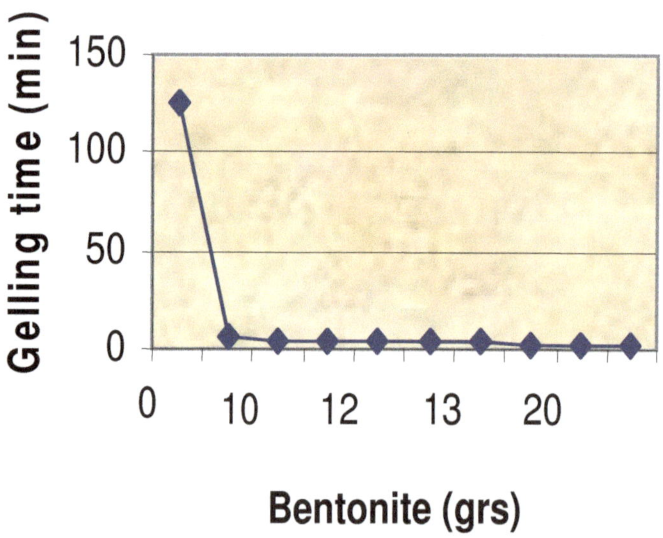

Figure 8. Bentonite effect on gelling time.

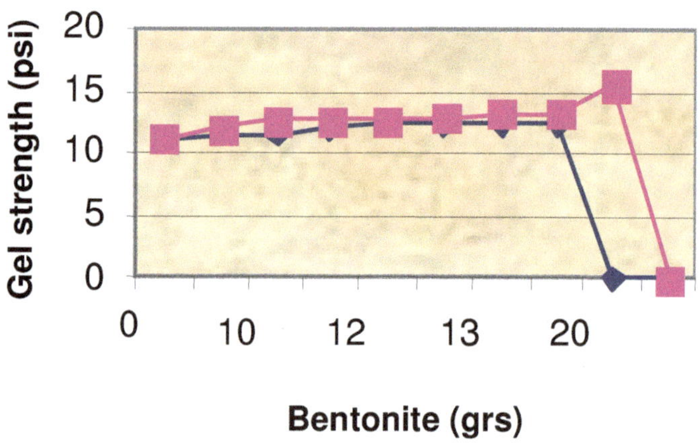

Figure 9. Bentonite effect on gel strength.

Table 6. Cross linker effect.

Cross linker (g)	Gelling time (min)	Gel Strength (psi) 4 h	Gel Strength (psi) 24 h
0	120	0	11.011
2	20	13.353	14.516
2.5	12	14.209	14.827
3	8	14.209	14.843
3.5	6	14.210	14.853
4	4.5	14.312	14.950
4.5	3	14.820	15.054
5	2	15.101	15.679
5.5	1.5	15.516	16.037
6	1.33	15.269	17.329

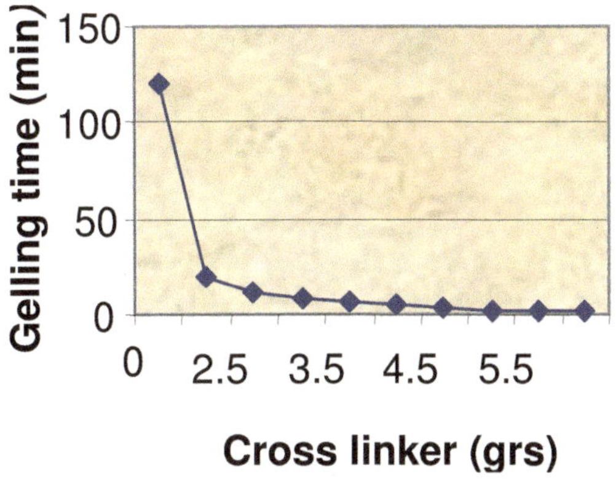

Figure 10. Cross linker effect on gelling time.

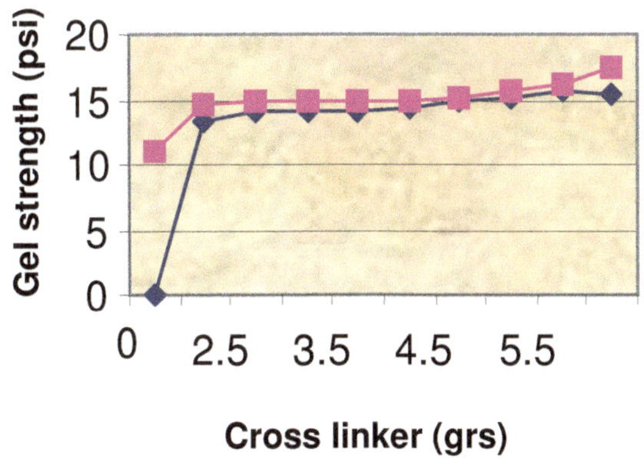

Figure 11. Cross linker effect on gel strength.

Observations while checking and results interpretation

Preparing formulation without bentonite gives an inconsistent emulsion. Thus the solids were sedimented at the bottom of the container and a very high quantity of water was not absorbed. This proves that the bentonite is the water absorption responsible. The bentonite is an accelerator (gelling time is reduced by adding more bentonite). Thus the presence of solid particles provides another shear forces to the system. In the range of 15 - 25 g the bentonite could not reduce the gelling time less than 3 min. This means that it has a slight impact on the gelling time of the InstanSeal-cement. According to the charts built, it is noted that the bentonite has a slight impact on the gel strength since its increase results in a slight increase of the gel strength.

The impact of the bentonite on the InstanSeal-cement behavior is too slight

Cross linker effect

Table 6 and Figures 10 and 11 discuss the cross linker effect.

Observations while checking and interpretation

Preparing the InstanSeal-cement without cross linker gives a much delayed gelling time (several days). This means that the cross linking reaction (which results in the gelling) did not occur. The buffer resists the change in its hydrogen ion concentration; therefore the pH is maintained in the following range [1.45 -1.58].

The cross linker is a solid that gives additional shear energy to the system and reduces the gelling time. Adding more cross linker allows to more transfer of cross linker and cement. But using a very high amount may provoke an early gelling that prevents the InstanSeal-cement from the penetration into the loss zone.

Fast release of cross linker results in a reduced gelling time and a high gel strength

Viscosifier effect

Table 7 and Figures 12 -14 discuss the viscosifier effect.

Observations while checking and interpretation

Preparing the InstanSeal-cement without viscosifier gives a fluid with solids at the bottom of the container and a partial phase separation. It was also observed that this formulation did not accept any shearing (stays loose emulsion). The presence of the viscosifier that is cross linked after shearing induces of the cross linking reaction which means improving the gel strength. The gelling time is reduced and the viscosity increased by adding more viscosifier.

Table 7. Viscosifier effect.

Viscosifier (ml)	Plastic viscosity (cp)	Gelling time (min)	Gel Strength (psi) 4 h	24 h
3	17.087	30	4.666	5.628
6	18.012	7	9.9972	10.139
6.5	18.012	5.5	13.377	13.884
7	21.014	4.5	13.651	14.276
7.5	24.017	4.5	13.879	14.386
8	27.019	3.5	14.545	15.997
9	36.025	3.5	15.029	17.304
10	40.001	3	15.279	18.464

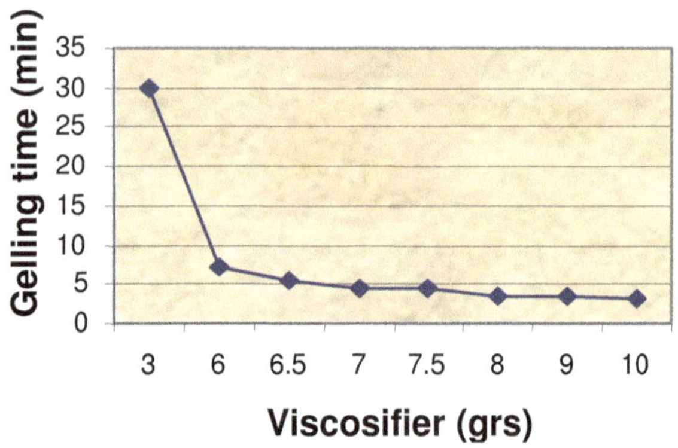

Figure 12. Viscosifier effect on gelling time.

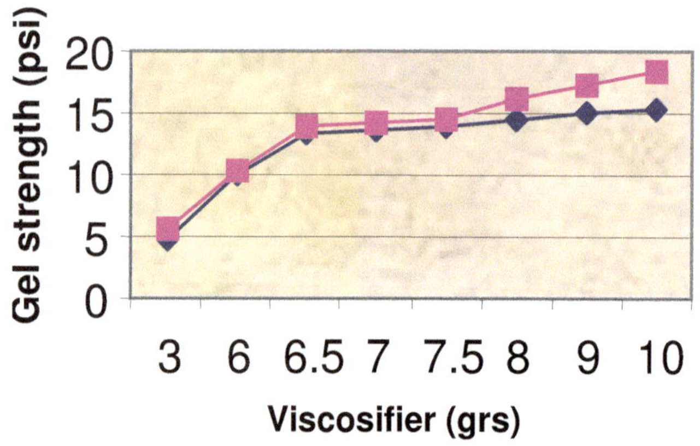

Figure 14. Viscosifier effect on gel strength.

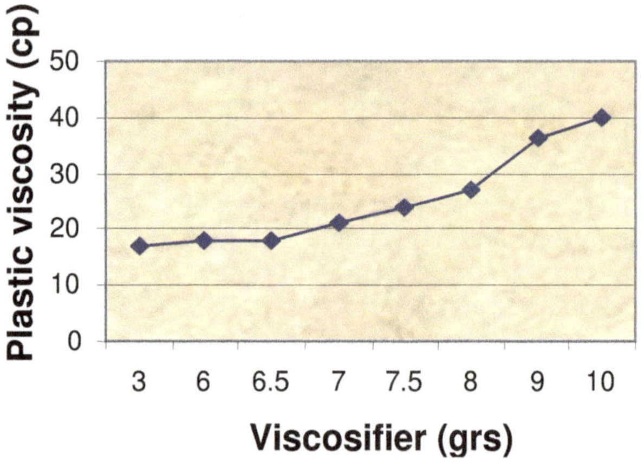

Figure 13. Viscosifier effect on rheology.

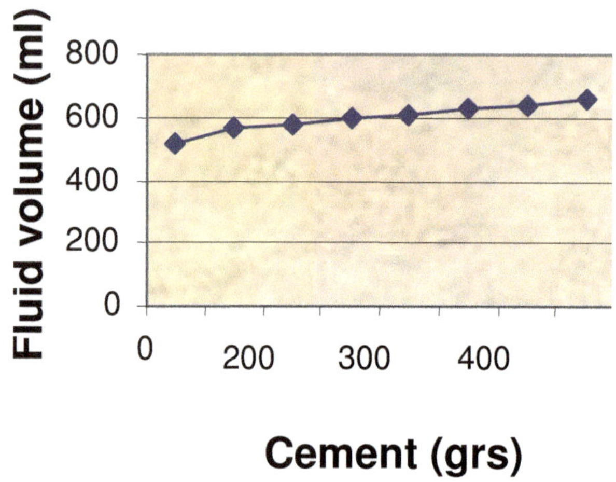

Figure 15. Cement effect on fluid volume.

An increase of the viscosifier content has a little effect on the gelling time but improves the gel strength

Cement effect

Table 8 and Figures 15 -20 discuss the cement effect.

Observations while checking and interpretation

Preparing the InstanSeal without cement gives a very soft touch to the sample therefore introducing the cement to the classic InstanSeal-cement leads to improved gel

Table 8. Cement effect.

Cement (g)	Fluid volume (ml)	Fluid density	Shear stress (lbf/100 ft)	pH	Gelling time (min)	Gel	Strength (psi)
0	516.64	0.97	142.37	9.41	100	0	0
150	563.81	1.17	146.14	10.70	20	0	11.584
200	579.53	1.22	157.37	11.42	17	11.703	13.731
250	595.26	1.25	174.37	11.48	4.5	13.650	14.380
300	610.98	1.33	182.38	11.85	1.5	14.021	16.031
350	626.70	1.35	188.04	12.02	0.2	14.304	16.873
400	642.43	1.42	194.04	12.07	0	14.537	17.996
450	658.15	1.43	194.38	12.08	0	15.203	18.838

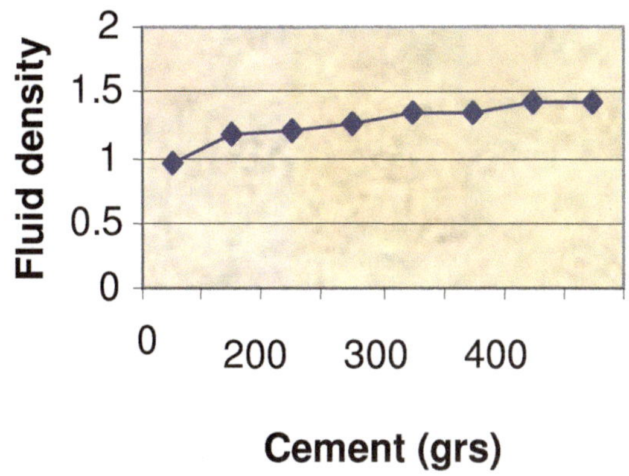

Figure 16. Cement effect on fluid density.

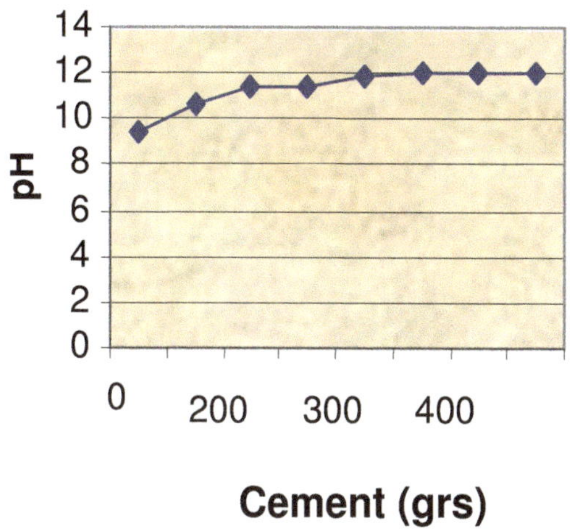

Figure 18. Cement effect on pH.

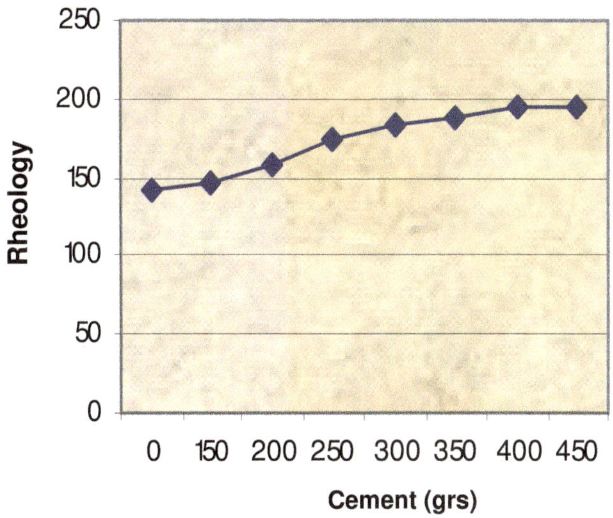

Figure 17. Cement effect on rheology.

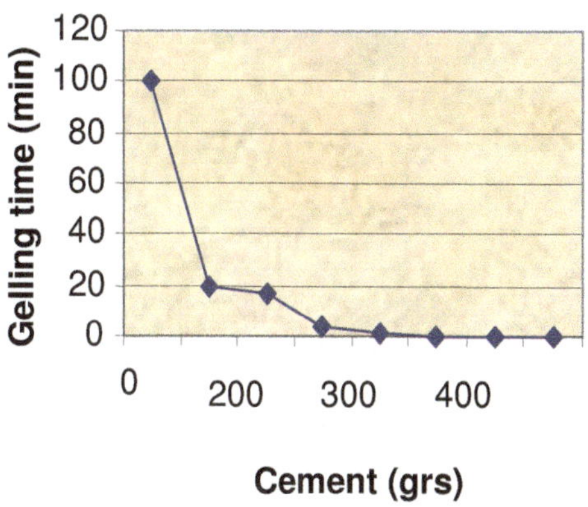

Figure 19. Cement effect on gelling time.

strength. Using a very high amount of cement (more [redacted] n 400 g) results in a difficult mixing and shearing. A [redacted] h amount of cement reduces the gelling time becaus [redacted] e cement hydration contributes to the premature ge [redacted] g through the release of divalent calcium ions. It leads also to high rheological properties (TY) that necessitates applying an initial pressure to facilitate the plugging fluid movement while pumping. Increasing the cement content leads to the increase of the gel strength and pH value (more than 12). This indicates that the reaction starts at

Table 9. Barite effect.

Barite (g)	Density	pH	Gelling time (min)	Gel	Strength (psi)
0	1.27	11.68	5	13.120	14.702
100	1.36	11.61	5	13.567	14.958
150	1.41	11.60	5	14.822	15.321
200	1.46	11.58	5	15.321	16.442
250	1.54	11.52	2	15.802	16.824
300	1.56	11.47	1.5	16.049	17.210
350	1.60	11.41	1	16.520	18.401
400	1.62	11.37	0.75	17.201	19.450

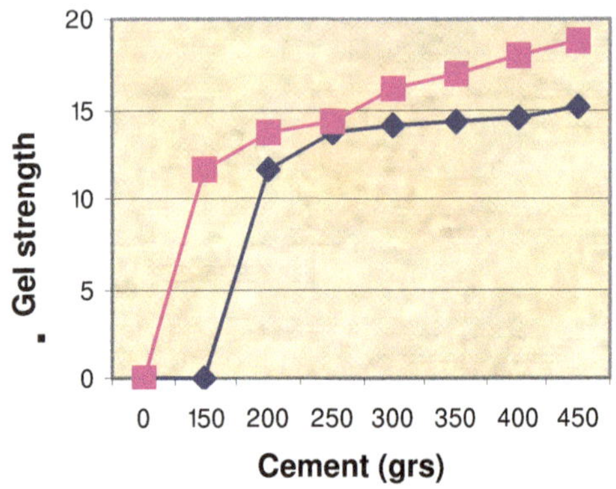

Figure 20. Cement effect on gel strength.

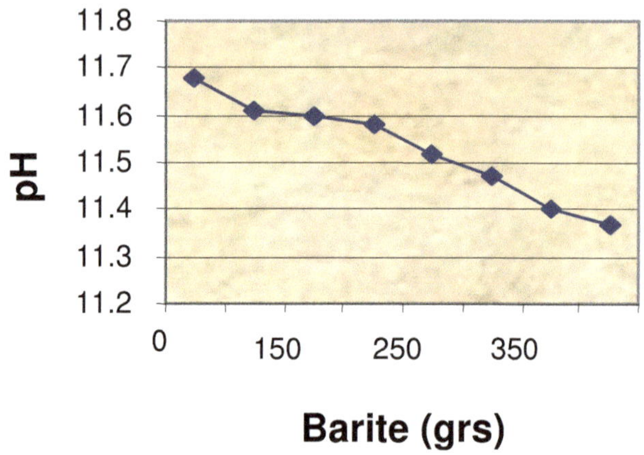

Figure 22. Barite effect on pH.

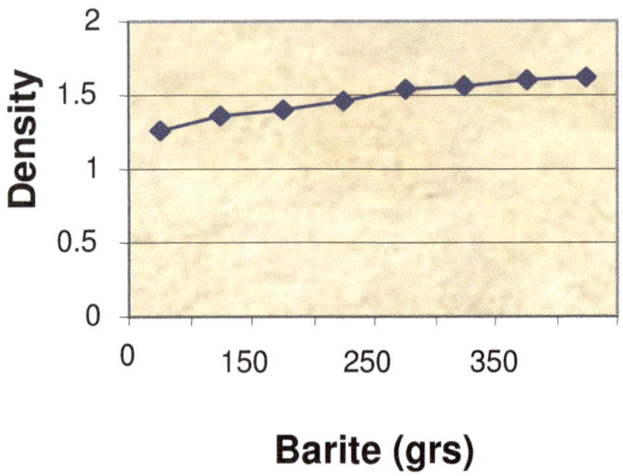

Figure 21. Barite effect on density.

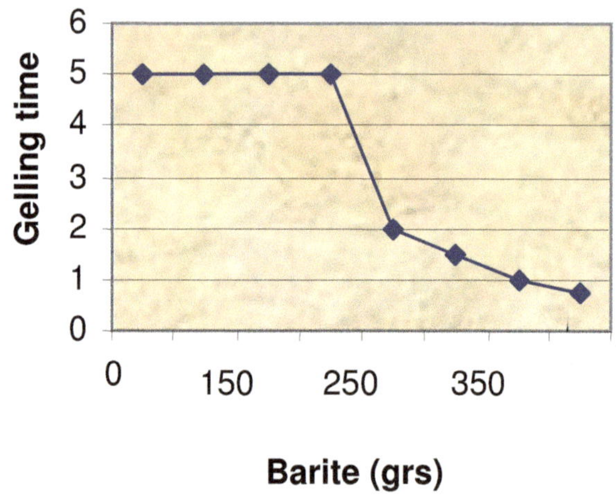

Figure 23. Barite effect on gelling time.

the surface (before shearing) because the cement hydration provides a second source of cross linker. That is why the cement should be added just before the job. Adding cement increases the InstanSeal-cement volume. This must be taken into consideration because the job application is limited by the equipment.

The cement leads to the better final mechanical properties and gains the compressive strength

Barite effect

Table 9 and Figures 21 -24 discuss the barite effect.

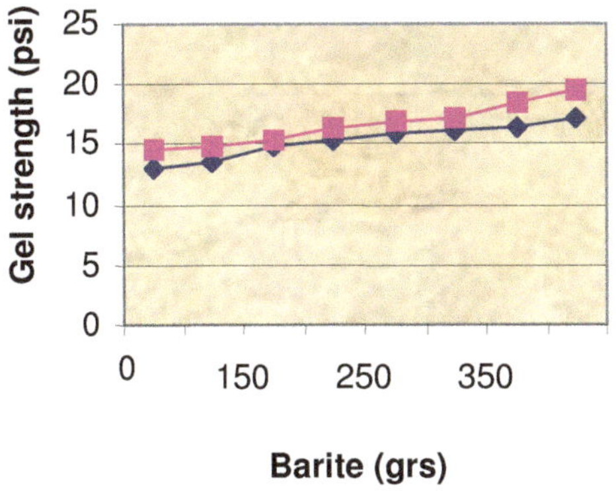

Figure 24. Barite effect on gel strength.

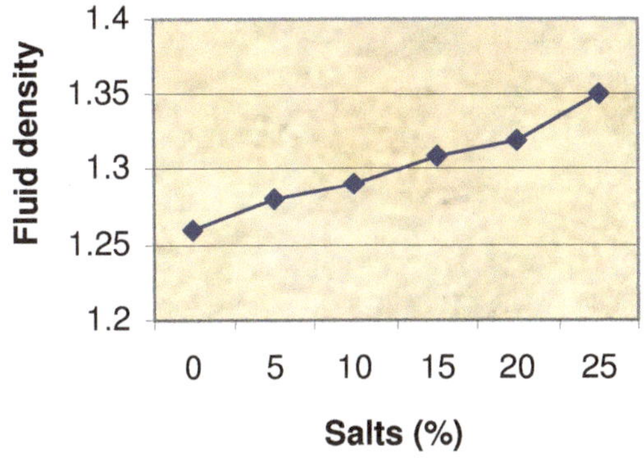

Figure 25. Salts effect on fluid density.

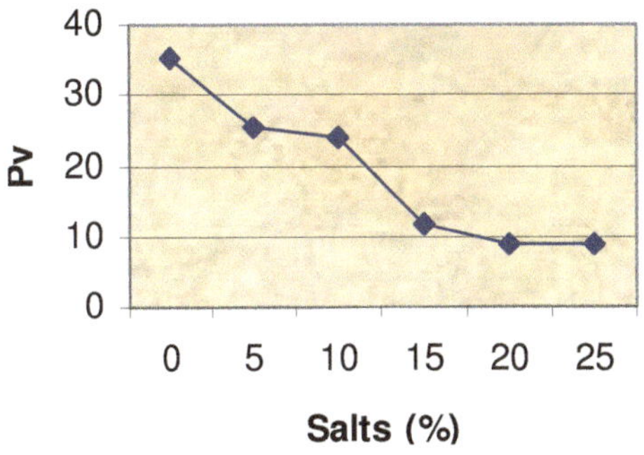

Figure 26. Salts effect on rheology.

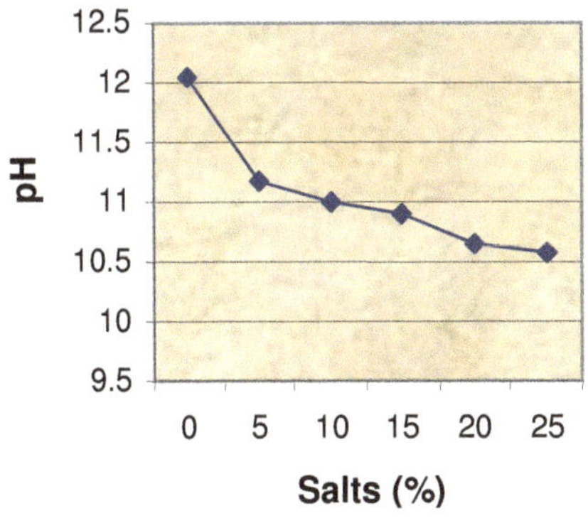

Figure 27. Salts effect on pH.

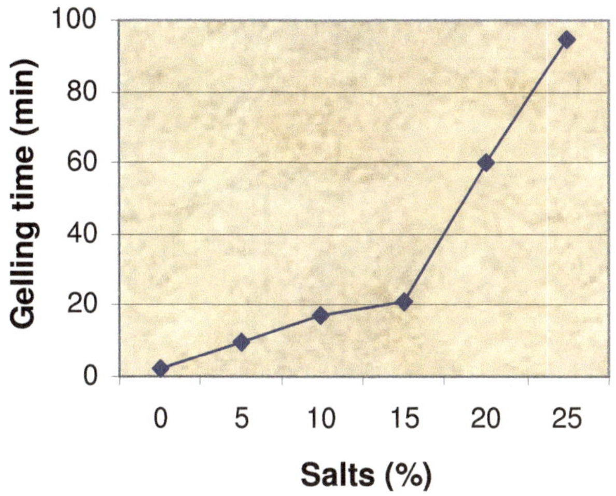

Figure 28. Salts effect on gelling time.

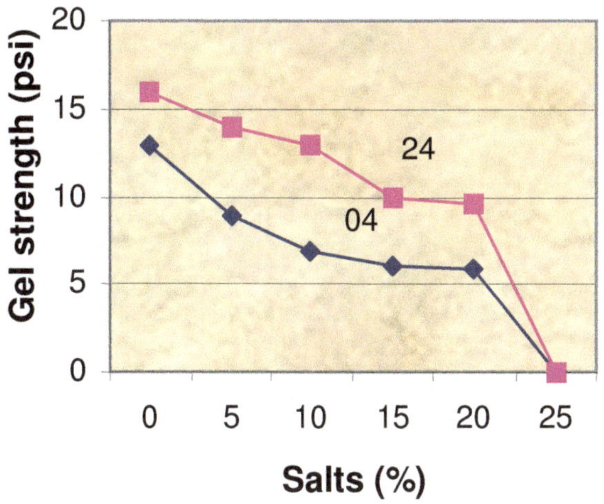

Figure 29. Salts effect on gel strength.

Observations while checking and interpretation

Adding weighting agents to the InstanSeal-cement leads to higher rheological properties and improves the gel strength. This weighting agent (barite) can accelerate the gelling time because the solid particles provide additional shear energy to the system and slightly impair the release of the cross linker and cement. This is why the pH value is reduced while increasing the weighting agent amount. Adding more barite gives a high value of density that reduces the contamination between the InstanSeal-cement and the well bore drilling fluid. The barite improves the stability of the gel.

Salts effect

Figures 25 -29 and Table 10 discuss the salts effect.

Table 10. Salt effect.

Salt (%)	Fluid density	Plastic viscosity (cp)	pH	Gelling time (min)	Gel	Strength (psi)
0	1.26	35.385	12.05	2	12.891	16.031
5	1.28	25.517	11.17	10	8.983	13.962
10	1.29	24.017	11.01	17	6.821	12.863
15	1.31	12.012	10.91	21	6.033	9.905
20	1.32	9.006	10.64	60	5.951	9.565
25	1.35	9.011	10.57	95	0	0

Observations while checking and interpretation

The salts increase the viscosity of the water. High amount of salt added to the water used to prepare the InstanSeal-cement can impair the cross linking reaction. This gives a reduced pH value, a delayed gelling time and weak gel strength.

Conclusions

(i) The resulting gel is robust in the initial phase and in the following days as the cement gains the compressive strength.
(ii) The cross linker must be released at a delayed rate to generate a suitable gelling time.
(iii) The gelling time can be increased or decreased by making the emulsion more or less shear sensitive.
(iv) The gelling time depends upon pressure and emulsifier concentration.
(v) The gelling time should be sufficient for fluids to enter the formation.
(vi) The InstanSeal cement requires follow-up with trying to mix with other additives to improve its feature and application range.
(vii) The salt is detrimental to the gel strength.

REFERENCES

Henry CH, Darley George Robert Gray (1983). Composition and properties of drilling and completion fluids; 5th edition golf publishing company, Houston TX; pp. 2-19.

Tailleur RJ (1963). 6th World Petroleum Congress. June pp. 19 - 26. Frankfurt am Main, Germany www.connect.slb.com. *Mark of Schlumberger. 1999.

David Quinn SPE, Schlumberger Dowell, Egil Sunde SPE (1999). Statoil A/S; and J. F. Baret. SPE. Schlumberger Dowell, Mechanism of a Novel Shear-Sensitive Plugging Fluid to Cure Lost Circulation. SPE International Symposium on Oilfield Chemistry. pp. 16-19, February. Houston, Texas.

Parvazinia M, Nassehi V (2006). Study of shear thinning fluid flow through highly permeable porous media, International Communications in Heat and Mass Transfer. 33(4):401-410

http://www.slb.com/services/cementing/lost_circulation/instant_seal.aspx.

12

Corrosion-erosion on waste heat recovery boiler system via blowdown optimization

Zainal Zakaria* and Nor Ismail Hashim

Department of Gas Engineering, Faculty of Petroleum and Renewable Energy Engineering, Universiti Teknologi Malaysia, 81310 Johor Baru, Johor, Malaysia.

It is estimated that problems due to corrosion-erosion of boiler system cost industries billions of dollars per year. Method of corrosion-erosion control varies depending upon the type of problem encountered. This study is focused on the optimization of the boiler blowdown frequency and its effect on the chemical values of the water-steam cycle. Parameters such as conductivity, iron and silica concentrations were closely monitored via on-line monitoring devices. The conductivity in the boiler water and steam cycle was between 0.3 to 0.6 uS/cm and 0.1 to 0.15 uS/cm, respectively. The average iron concentration in the water cycle was less than 0.01 ppm and in the steam cycle the average reading was less than 0.005 ppm and the average of silica concentration in the boiler fluctuated between 0.02 to 0.08 ppm.

Key words: Boiler, conductivity, corrosion, heat recovery, iron.

INTRODUCTION

In a waste heat recovery boiler system, heat of the combustion gases from the gas turbine exhaust is used to produce steam to drive a steam turbine and to heat up the main condensate. The combustion gases from the gas turbine exhaust first flows horizontally through an exhaust gas diffuser, the diverter damper and the exhaust gas duct before deflect at the boiler inlet. Then, the exhaust gas flows through the vertical part of the boiler which houses the heating surfaces which is fabricated from the finned tubes. Effective corrosion control monitoring is essential in ensuring the efficiency and trouble-free operation of boiler (Jiamin and Zhansong, 2009; Klien and Rice, 2000; Otsuka, 2002; Sandberg et al., 2011; Smiyan et al., 2002; Srikanth et al., 2003).

Appropriate corrosion-erosion monitoring techniques vary with different system and the testing should be performed at least once per shift. When the control system is difficult or during periods of more variables operating conditions, increasing the testing frequencies should be considered.

All monitoring data either spot sampling or continuous, should be recorded. Boiler feed water hardness (magnesium), oxygen, iron and pH should be measured as well (Sandberg et al., 2011). Iron and oxygen can be measured on a daily basis by installing continuous oxygen meter in a feed water system to detect oxygen intrusions. Iron should be measured with care to avoid sample contamination.

However, the corrosion process is a complex electro chemical reaction. It takes many forms and may produce general attack over a large metal surface or by pinpoint penetration of metal. Corrosion is a significant problem caused by water in boilers. The whole boiler system is generally vulnerable to corrosion caused by oxygen, which mainly occurs in the form of pitting. However, other factors such as acid conditions, stresses and specific chemical corrodents also play an important influence and caused different forms of attack. Thus, it is necessary to control the allowable quantity of the various harmful substances in the boiler water without jeopardizing the boiler (Otsuka, 2002). Corrosion may occur in the feed water system as a result of low pH water and the water impurities.

Corrosion can cause serious problems in feedwater and boiler system. The corrosive action will waste the metal away and reduced its overall strength of metal thus leading to eventual failure (Jiamin and Zhansong, 2009).

When the corrosion occurs, metals obtained from the

*Corresponding author. E-mail: zainalz@petroleum.utm.my.

corresponding metallic oxide tend to revert to their original state. Some of these metallic oxides will then form a stable coating which adheres firmly to the surface of the parent metal protecting it from further attack. However, other oxides form coatings which are either porous or unstable and can break down easily. For example, a special form of iron oxide known as magnetite can form in boilers. This provides a protective film on the metal but in the presence of oxygen breaks down to form ferric oxide which is porous and did not provide protection. In the boiler, the presence of water in an acidic condition provides the electrolyte required for corrosive action. The water can become acidic due to the presence of various dissolved gases, such as oxygen and carbon dioxide, or various dissolved metallic salts, such as magnesium chloride. Direct chemical attack can occur when metals at high temperature come into contact with the air or other gases resulting in oxidation or sulphidation of the metal.

Electrochemical corrosion usually takes place in the presence of moisture and a formation of corrosion cell with cathodic and anodic areas in the presence of an electrolyte, resulting in corrosion of the anode. This form of corrosion covers most of the corrosion that takes place in boiler system. In this group, there are two main forms of corrosion, the general wastage and the pitting (Jiamin and Zhansong, 2009). Other form of electrolyte corrosion which would be a combination of electrolytic and mechanical or chemical attacks are corrosion fatigue, acid corrosion and caustic corrosion which is sometimes known as caustic gouging, caustic attack or ductile gouging (Khajavi et al., 2007; Srikanth et al., 2003). The ASME Consensus for Industrial Boilers specifies that feedwater oxygen, iron and copper content should be very low (for example, < 7 ppb oxygen, 20 ppb iron and 15 ppb copper for a 900 psig boiler) and the pH should be maintained between 8.5 and 9.5 for system corrosion protection (Klien and Rice, 2000).

In order to minimize boiler system corrosion, an understanding of the operational requirements for all critical system components is necessary (Smiyan et al., 2002). Without proper mechanical and chemical deaeration, oxygen in the feedwater will enter the boiler. Much is flashed off with the steam but the remainders can attack the boiler metal. The point of attack varies with boiler design and feedwater distribution. Pitting corrosion is a localized form of corrosion which takes place in the passive state of the metal and is frequently visible in the feedwater distributor, at the steam drum waterline and in the downcomers tubes (Kritzer, 2004). Oxygen is highly corrosive when present in hot water and significantly more corrosive than carbon dioxide (Klien and Rice, 2000; Ahmad, 2006) as dissolved oxygen is six times more corrosive than a molar equivalent concentration of carbon dioxide in pure water at 65°C (Ahmad, 2006).

Thus, even small concentrations can cause serious problems. Due to pits that can penetrate deep into metal, oxygen corrosion can result in rapid failure of feedwater lines, boiler tubes, economizers and condensate lines. Additionally, iron oxide generated by the corrosion can produce iron deposits in the boiler. The acceptable dissolved oxygen level for any system depends on many factors, such as flow rate, feedwater temperature, pH, dissolved solids contents and the metallurgy and physical condition of the system (Smiyan et al., 2002). Based on experience, 3 to 10 ppb of feedwater oxygen is not significantly damaging the economizers.

Corrosion-erosion is defined as the acceleration or increase in rate of deterioration or attack on a metal because of relative movement between a corrosive fluid and the metal surface. Generally, the movement is quite rapid and mechanical ware effects or abrasions are involved. Metals removed forms at the surface as dissolved ions, or forms solid corrosion products that are mechanically swept form the metal surface. In many cases, failures caused by erosion occur in a relatively short time and they are unexpected large because evaluation corrosion test were run under static condition (Sidhu and Prakash, 2006; Uusitalo et al., 2004). Velocity often strongly influences the mechanisms of the corrosion reactions. It exhibits mechanical ware effect at high values and particularly when the solution contains solids in suspension. Increase in velocity generally results in increased attack, particularly if substantial rate of flow is involved (Sandberg et al., 2011). Erosion-corrosion failures also can occur due to turbulence flow condition. Turbulence results in greater agitation of the liquid at metal surface than in a case of laminar flow and in more intimate contact between the environment and the metal. The type of flow obtained depends on the rate and quantity of fluid handled and also the design of the equipment. In addition to high velocities, crevices, deposit, ledges, sharp changes in cross sections and other obstructions that disturb the laminar flow pattern may result in erosion-corrosion (Sidhu and Prakash, 2006). The objective of this study is to monitor the effect of the chemical values such as the conductivity, iron and silica concentration in the heat recovery boiler circuit.

METHODOLOGY

The study on the optimization of the boiler blowdown frequency was conducted on 2 boilers. The test was carried out in 2 different stages, that is, Stage 1, which did not involve any blowdown and Stage 2, which involved blowdown activities.

Stage 1

In order to achieve a stable chemical condition, a 3 h of boiler blowdown was carried out for each boiler prior to the test. The boilers were operated at base load condition and did not involve in any start-up or shutdown during the test period of 20 days. This will minimize the variation in the operating parameters such as steam pressure, temperature and flow rates. Since no shutdown was planned for the boiler, the blowdown valves were kept closed throughout the test period. Important chemical parameters such as conductivity, Iron and Silica concentration in the water and steam

Figure 1. On-line sampling station.

cycle were closely monitored and daily recorded in the monitoring sheet.

Stage 2

The boilers were operated at base load operation and involved in startup and shutdown. Boiler blowdown was carried out for 3 h for every startup. Important chemical parameters conductivity, Iron and Silica concentration in the water and steam cycle were closely monitored and daily recorded in the monitoring sheet. In addition, laboratory analysis on dissolved solids such as Calcium, Magnesium, Iron, Silica, Sodium, Sulphates and Chlorides were also recorded. The monitoring of the important values for water-steam cycle is done by continuous on-line measurement and analysis in the sampling station as shown in Figure 1.

RESULTS AND DISCUSSION

Method of corrosion-erosion control varies depending upon the type of problem encountered. The high flow rate during blowdown caused frequent erosion on the control valves as well as, the boiler internal tubes. The readings were recorded and compared with the limits and specification proposed by the manufacturer, Babcock and Wilcox.

Stage 1 test (Without blowdown)

The following parameters were recorded and analyzed during the test period of 20 days. During this period, the boilers were running at base load operation without any blowdown operation. Theoretically, evaporation in a boiler caused impurities such as dissolved and suspended solids to concentrate. Corrosion beneath certain types of boiler deposits can weaken the metal and tube failure may occur after certain period of time.

Conductivity analysis

The specific conductance of boiler water provides an indirect measure of dissolved solids. Figures 2 and 3 show the conductivity trend for water and steam cycle respectively. The limit for the boiler water conductivity was less than 3 uS/cm. From the trend, it shows that the boiler water conductivity for both boilers has gradually reduced by 0.2 to 0.4 uS/cm and stabilized around 0.35 uS/cm despite not doing any blowdown. The conductivity for the steam side also shows a reduction as well. The conductivity values of the steam for both boilers have gradually reduced and stabilized at 0.1 uS/cm from 0.15 uS/cm. The limit for the steam conductivity was less than 0.2 uS/cm.

Total Iron analysis

The limit of iron concentration in main condensate and feedwater was less than 20 ppb and for the boiler water, the limit was less than 100 ppb. For the steam system, the limit was below 20 ppb as well. Figures 4 and 5 show the total iron in the water cycle and steam cycle, respectively. The total iron content in the water cycle went slightly up to a maximum of 40 ppb for the boiler water. However, all the values were still below the permissible concentration limits of 100 ppb. From the trends, the iron remained non-detectable in the main condensate and feedwater.

Silica analysis

A high Silica deposit is very hard resembling porcelain. The crystals of Silica were extremely small foaming a very dense, impervious scale. The limit of Silica

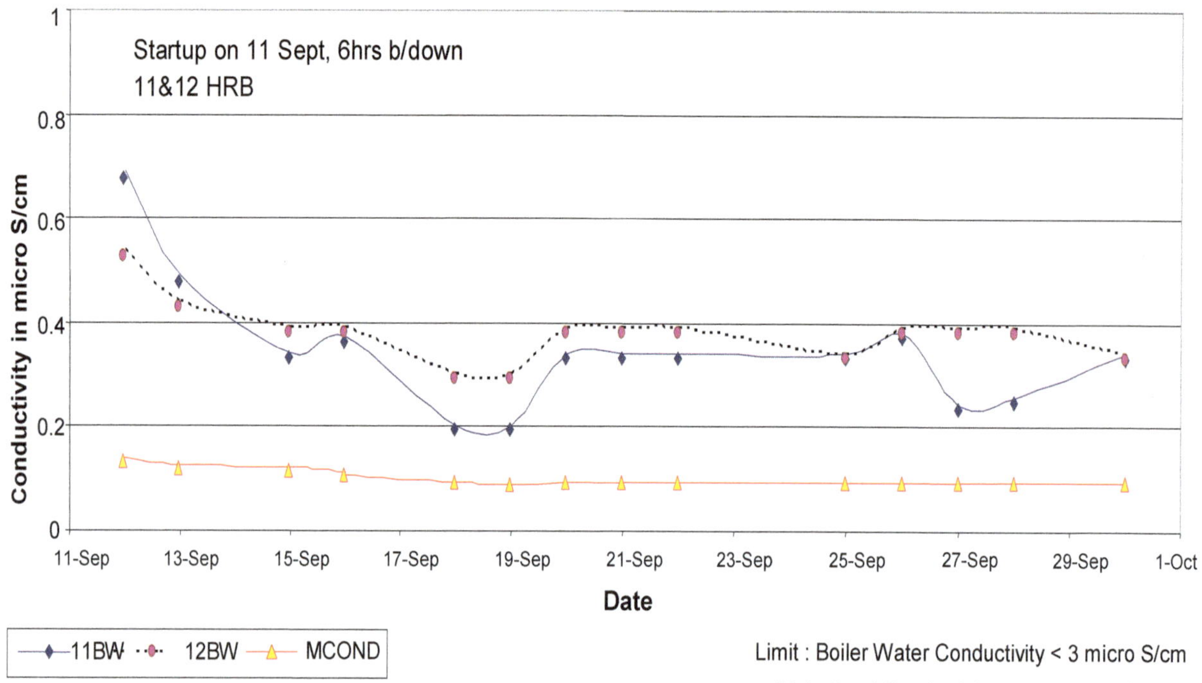

Figure 2. Conductivity profile for water cycle without boiler blowdown.

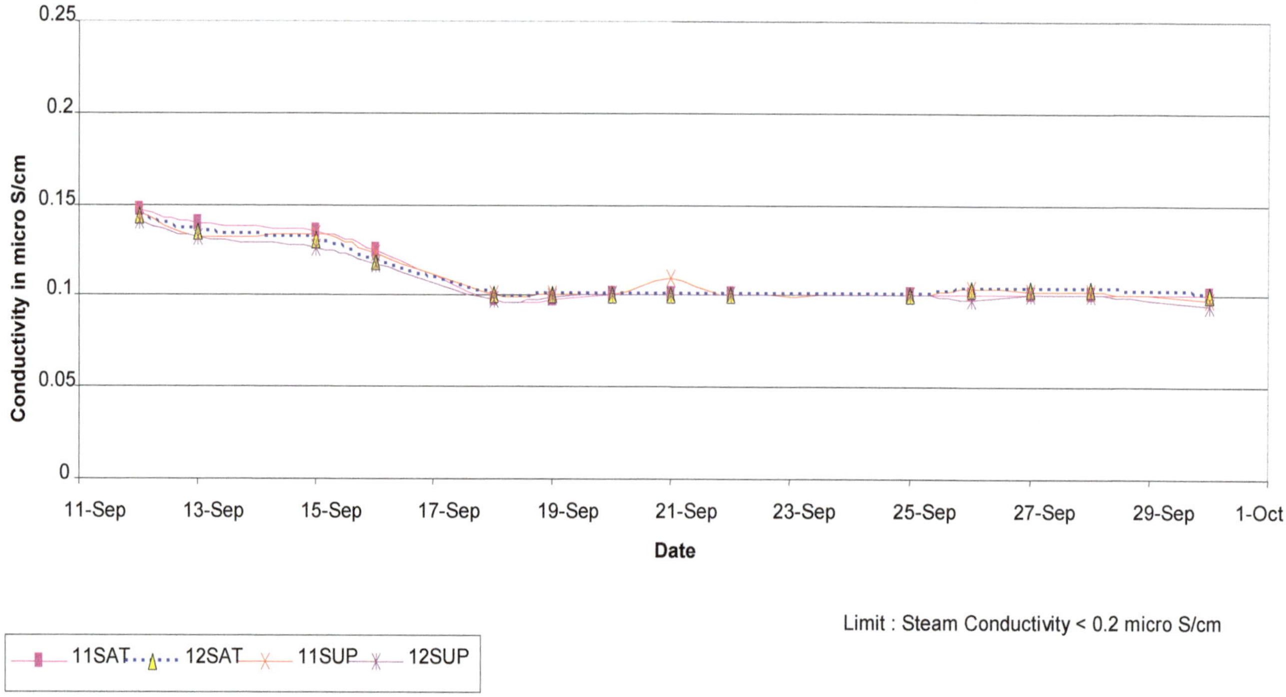

Figure 3. Conductivity profile for steam cycle without boiler blowdown.

concentration in main condensate, feed water and steam system was less than 0.02 ppm. For the boiler water, the limit was less than 0.5 ppm. Figures 6 and 7 show the Silica concentration in the water and steam cycle respectively. From the trend, it was observed that the Silica concentration in the boiler water fluctuated between

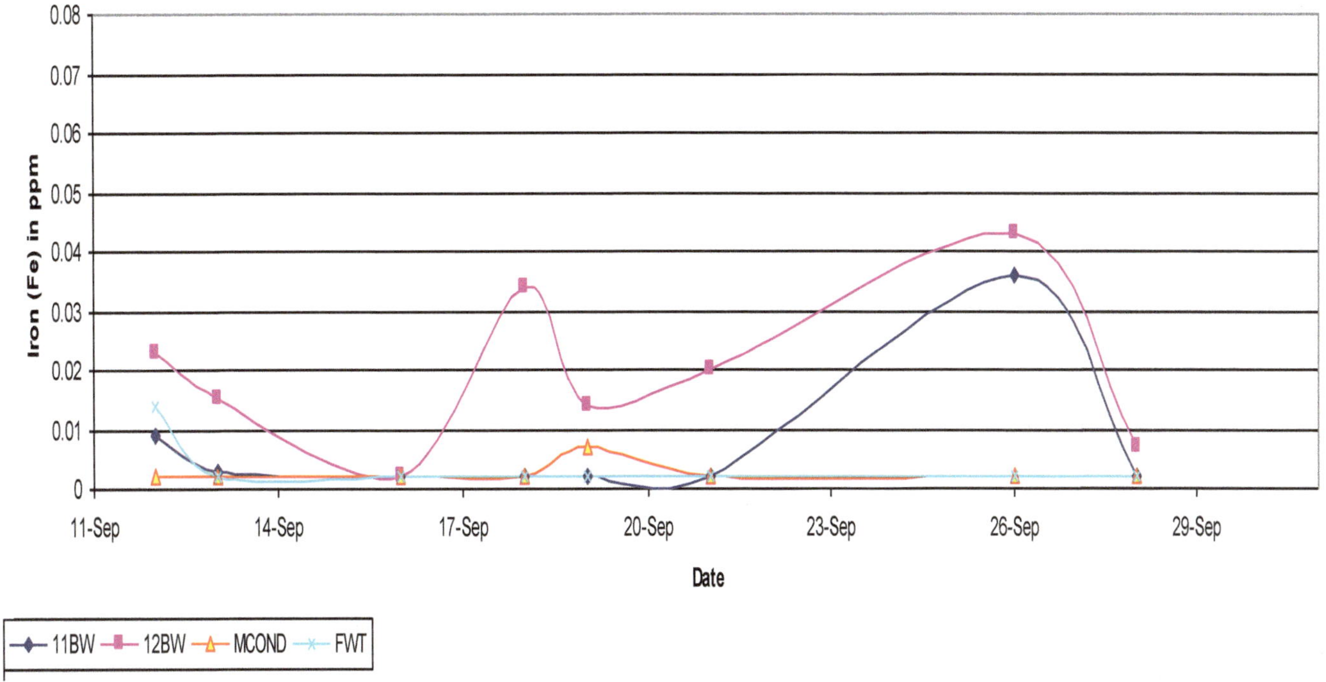

Figure 4. Total Iron (Fe) concentrations in water cycle without boiler blow down.

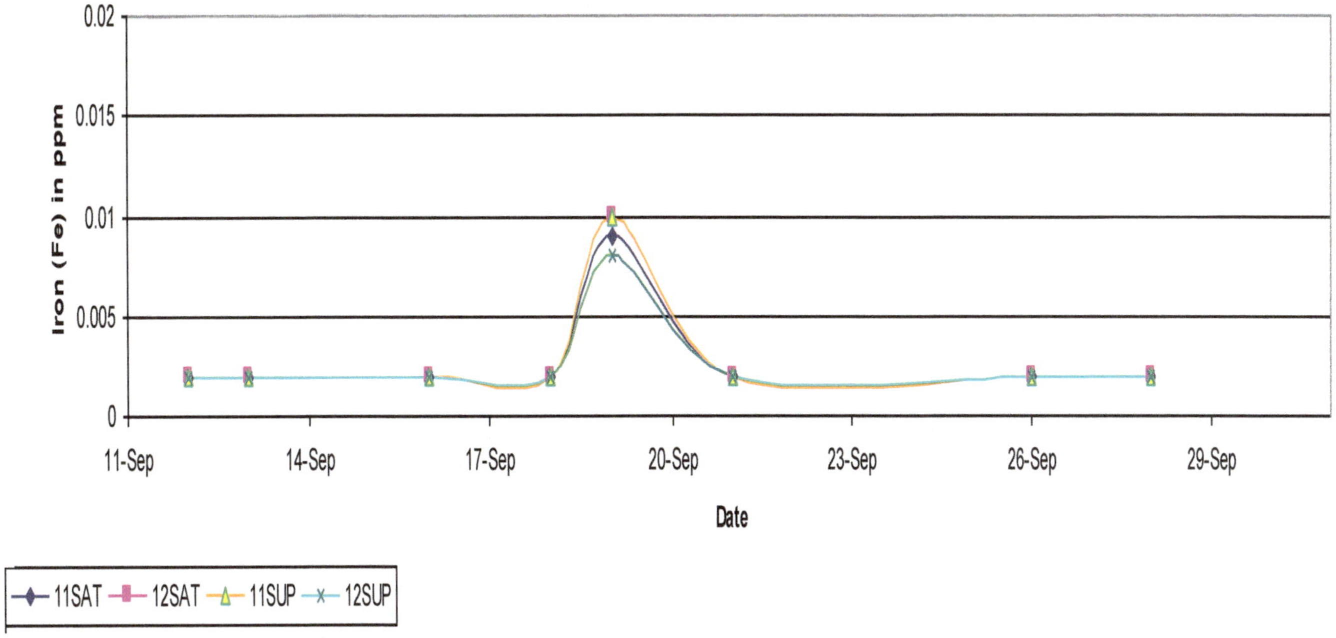

Figure 5. Total Iron (Fe) concentrations in steam cycle without boiler blow down.

0.02 to 0.05 ppm. Silica concentration in boiler water shows a slight increase. However, this value is still far below the limit. The silica concentration in the steam cycle remained constant at 0.01 ppm throughout the test.

The observation on the conductivity, iron and silica concentrations has proven that the chemical concentration in the boiler can still be maintained below the allowable limit even without being blowdown for 20 days. In general, based on the trends and monitoring sheets, it can be clearly monitored that all chemical parameters remain below their specified limits throughout the test.

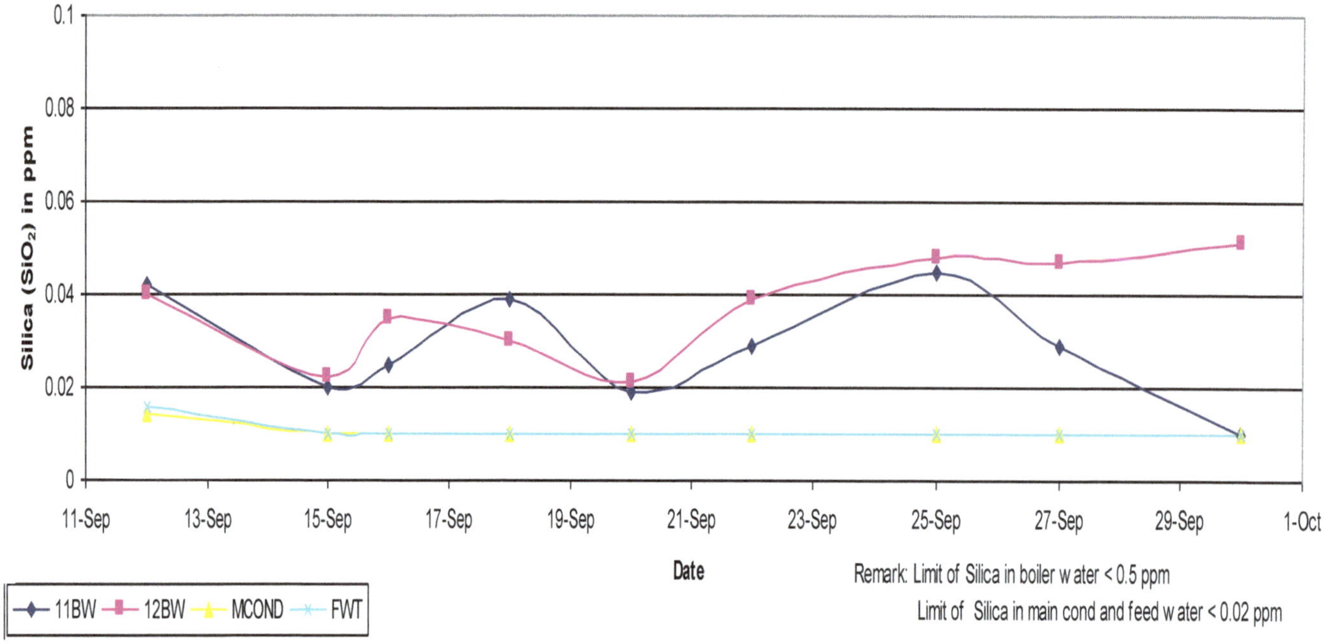

Figure 6. Silica concentrations in water cycle without boiler blowdown.

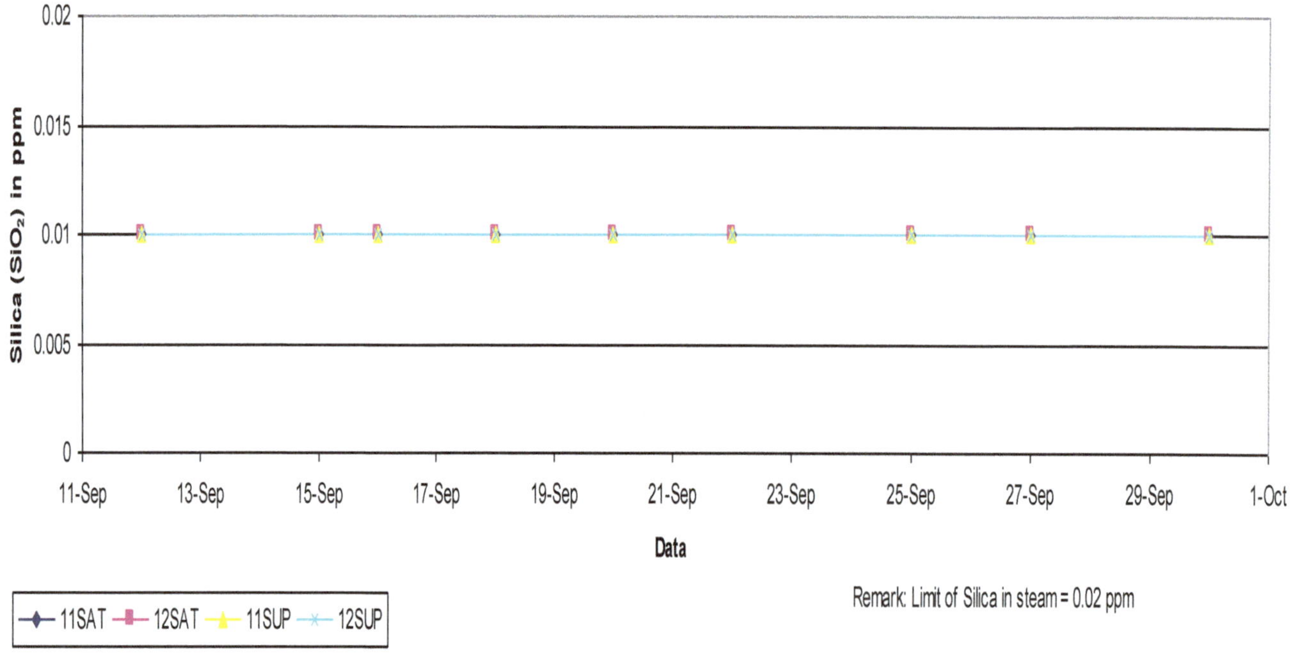

Figure 7. Silica concentrations in steam cycle without boiler blowdown.

Stage 2 test (Blowdown optimization)

Blowdown, either a periodic or continuous water removal process is essential in addition to chemical treatments to maintain the concentration of impurities within certain limit and to control the build-up of dissolved solid levels in the boiler. The required blowdown rate is determined according to the total dissolved solids, the alkalinity, silica or suspended solid levels. The amount of solids in the boiler water can be monitored and controlled by regulating the amount of blowdown. However, the amount of blowdown must properly established to avoid

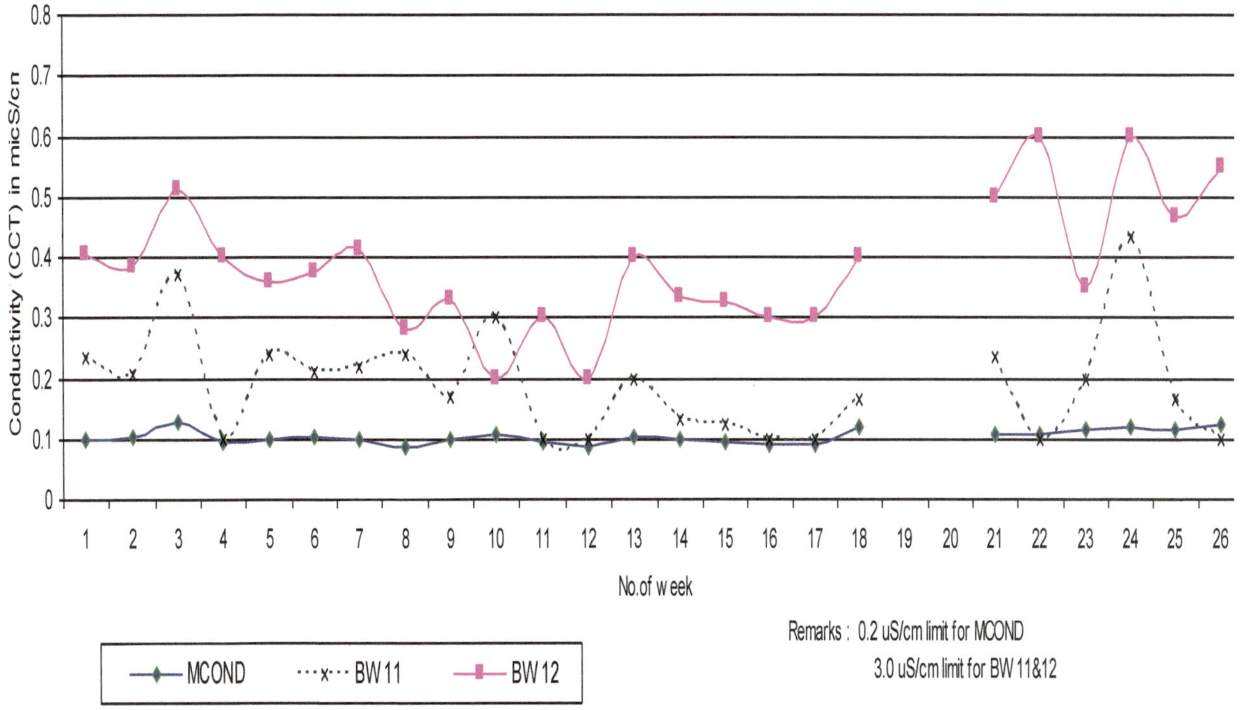

Figure 8. Weekly averages conductivity of water cycle with blowdown.

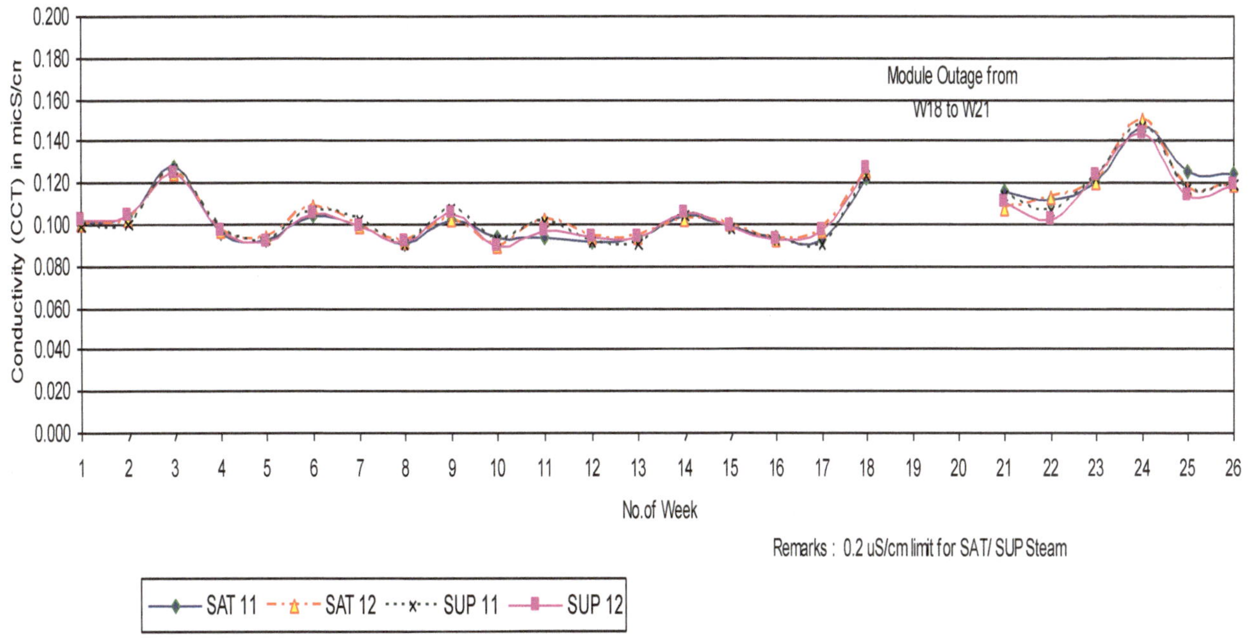

Figure 9. Weekly averages conductivity of steam cycle with blowdown.

sludge deposits which results in carryover if too little blowdown occur. On the other hand, too much blowdown will increase boiler fuel, boiler feedwater and boiler chemical requirements due to removal of excessive hot water (Bahadori and Vuthaluru, 2010).

Conductivity analysis

Figures 8 and 9 show the trend of conductivity for the water cycle and the steam cycle respectively. From the trends, we can see that the conductivity trends for the

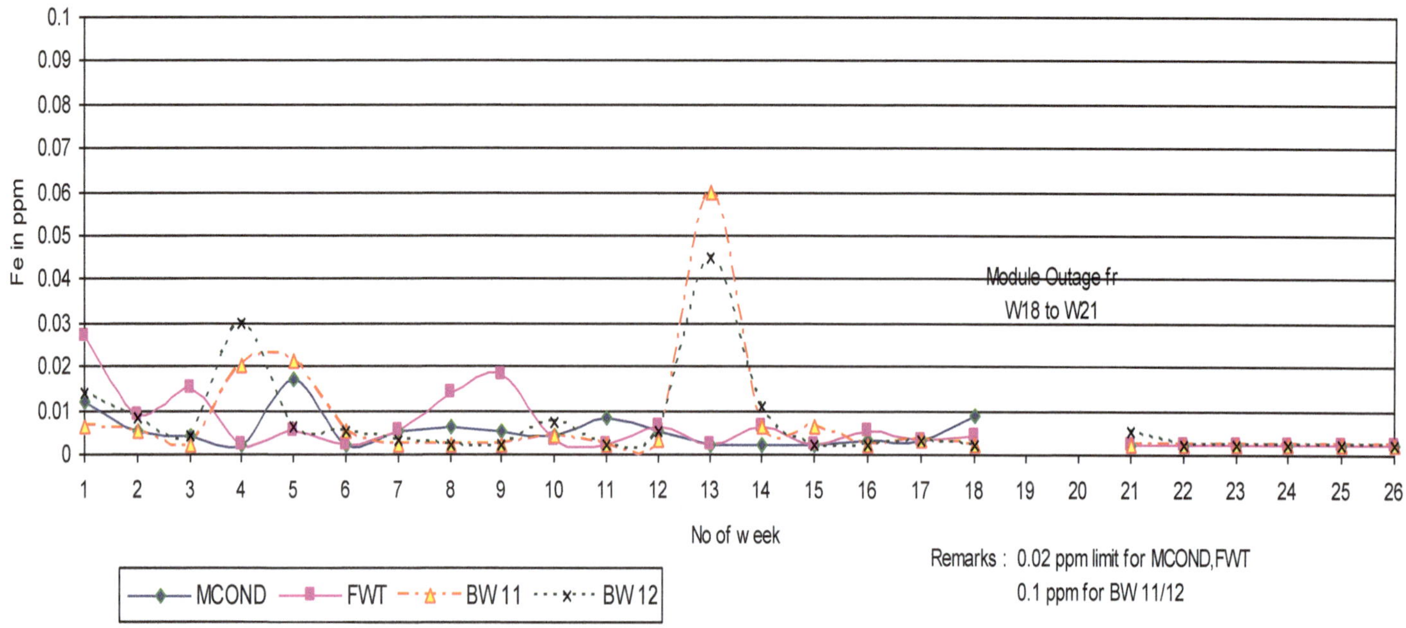

Figure 10. Weekly averages of iron concentrations in water cycle with blowdown.

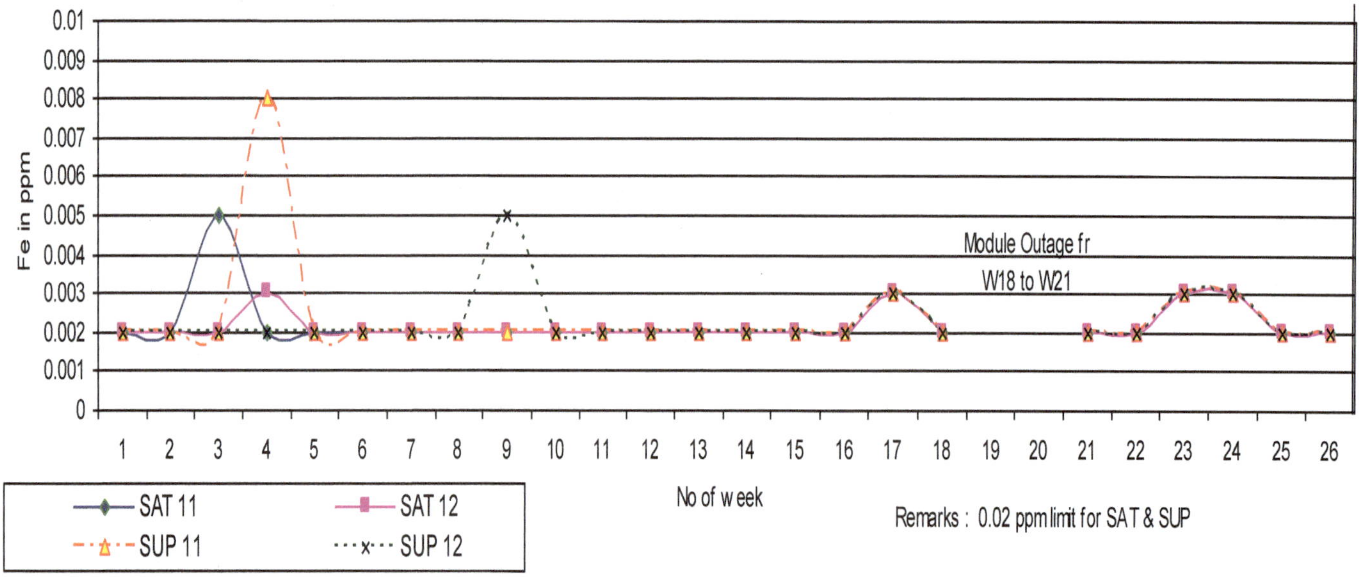

Figure 11. Weekly averages of iron concentrations in steam cycle with blowdown.

water cycle were observed below the limits. The boiler water conductivity in HRB11 and HRB12 fluctuated within the range of 0.1 to 0.3 uS/cm and 0.2 to 0.6 uS/cm. respectively. As shown in the trend, the temporary increase in the boiler water conductivity was expected during start-up before it reduced and stabilized back after a few days of the boilers in operation. The conductivity for the steam cycle for both boilers fluctuated between the range of 0.1 to 0.15 uS/cm which is still below the limit of 0.2 uS/cm, during normal operation. The trends also show that the conductivity values were still stable within this range after start-up of the boilers. This condition is achieved due to the 3 h of boiler blow down practice after each start-up.

Total iron analysis

By looking at Figures 10 and 11, due to the thermal stress on the protective oxide layer and the access of

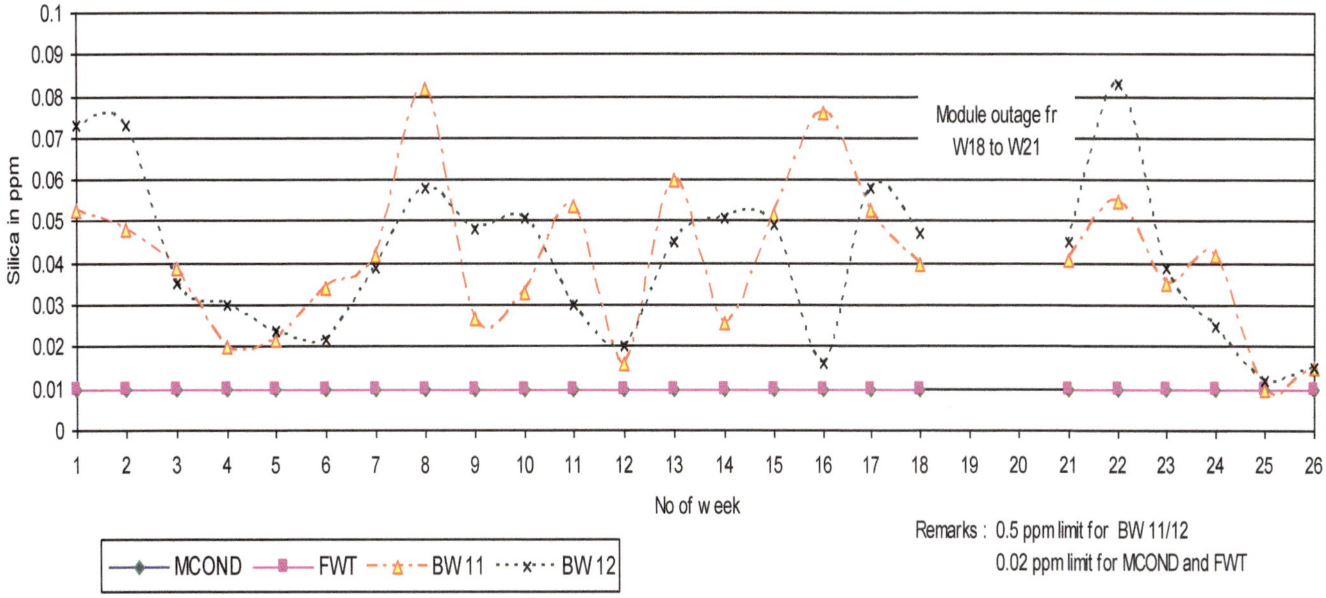

Figure 12. Weekly averages of silica concentrations in water cycle with blowdown.

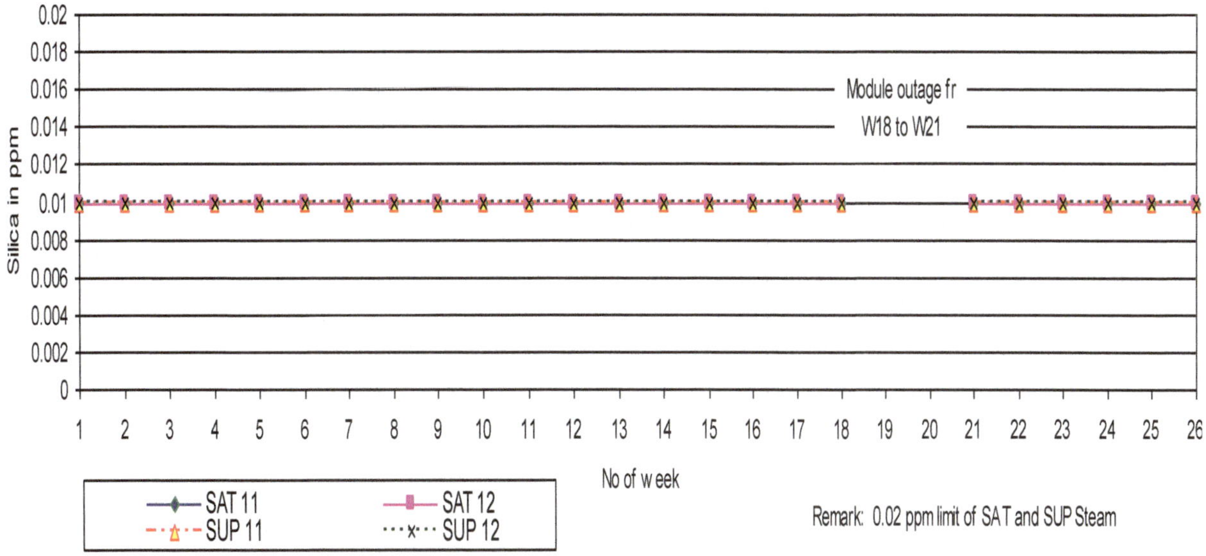

Figure 13. Weekly averages of silica concentrations in steam cycle with blowdown.

free oxygen in the make-up water during boiler start-up, the concentration of total iron in the water and steam cycle was found increasing especially in the boiler drums. The Iron concentration for boiler water HRB11 and HRB12 have increased to 0.06 to 0.045 ppm respectively during initial start-up of the boiler after a scheduled week-end shut down. However after a longer outage, the Iron concentration in the boiler water did not show any sign of increase and remained below 0.01 ppm. This was due to fresh feed water which was used to fill up the boiler prior to start-up and the 3 h of blowdown practice for every start-up of the boiler. The Iron concentration in the steam cycle only show a slight increased during each start-up. In general, the value was less than 0.005 ppm. However, even on a longer term, the analysis on the blowdown practices did not show any sign of deterioration on the water and steam quality.

Silica analysis

Figures 12 and 13 show the trends of silica concentration

in the steam and water cycle respectively. The weekly average of silica concentration in the boiler fluctuated between 0.02 to 0.08 ppm but still far below the limit of 0.5 ppm. The silica concentration can be neglected in the main condensate and feedwater system. This is because the make-up water used has already been treated in the ion-exchanger to become de-mineralized water. At the same time, the Silica (SiO_2) concentration in the steam cycle remained constant at 0.01 ppm throughout the test period.

Conclusions

The study was conducted in order to understand the behavior of the chemical condition for the water-steam cycle by reducing the normal boiler blowdown frequency of once per day. Several conclusions that can be made are:

1. The chemical condition of the water-steam cycle can still be maintained even though the frequency of the blowdown is reduced.
2. Boiler blowdown must be carried out if one of the chemical parameters in the water and steam cycle is trending upwards and reaching the limit value.
3. After a boiler startup, blowdown shall be carried out for approximately 3 h or until the feed water line indicates stable and low concentration of Iron (Fe) and Silica (SO_4), and the conductivity is within the specified range.
4. With the implementation of the new blowdown procedures, saving on the gas, water and chemicals would be achieved for the whole water and steam cycle.
5. The failure of the blowdown valves can be minimized and this again can save a lot of money and down time for the necessary repair work.

REFERENCES

Ahmad Z (2006). Principles of Corrosion Engineering and corrosion control, first ed. Boston, Massachusetts. p. 131

Bahadori A, Vuthaluru HB (2010). A Method for Estimation of Recoverable Heat from Blowdown Systems during Steam Generation. Energy, 35: 3501-3507.

Jiamin Y, Zhansong W (2009). Corrosion Behavior of TP316L of Superheater in Biomass Boiler with Simulated Atmosphere and Deposit. Chinese J. Chem. Eng., 15(5): 849-853.

Khajavi MR, Abdolmaleki AR, Adibi N, Mirfendereski S (2007). Failure Analysis of Bank Front Boiler Tubes, Engine. Failure Anal., 14: 731-738.

Klien HA, Rice JK (2000). A Research Study on Internal Corrosion in High Pressure Boilers, J. Engine. of Power, 3: 5-12.

Kritzer P (2004). Corrosion in High-temperature and Supercritical Water and Aqueous Solutions: A Review, J. Supercritical Fluids, 29: 1-29.

Otsuka N (2002). Effects of Fuel Impurities on the Fireside Corrosion of Boiler Tubes in Advanced Power Generating Systems: A Thermodynamic Calculation of Deposit Chemistry. Corros. Sci., 44: 265-283.

Sandberg J, Chrster K, Rebei BF (2011). A 7 Years Long Measurement Period Investigating the Correlation of Corrosion, Deposit and Fuel in a Biomass Fired Circulated Fluidized Bed Boiler. Appl. Energy, 1: 99-110.

Sidhu BS, Prakash S (2006). Erosion-Corrosion of Plasma as Sprayed and Laser Remelted Stellite-6 Coating in a Coal Fired Boiler. Wear, 260: 1035-1044.

Smiyan OD, Grigorenko GM, Vainman AB (2002). Effect of Hydrogen on Corrosion Damage of Metal of the High-Pressure Energetic Boiler Drum. Inter. J. Hydrogen Energy, 27: 801-812.

Srikanth S, Ravikumar B, Swapan KD, Gopalakrishna K, Nandakumar K, Vijayan P (2003). Analysis of Failures in Boiler Tubes due to Fireside Corrosion in a Waste Heat Recovery Boiler. Engine. Failure Anal., 10: 59-66.

Uusitalo MA, Vuoristo PMJ, Mantyla TA (2004). High Temperature Corrosion of Coatings and Boiler Steels below Chlorine-Containing Salt Deposits. Corros. Sci., 46: 1311-1331.

13

Hydromagnetic forced flow between a rotating disc and a naturally permeable stationary porous disc saturated with fluid

Bhupendra Kumar Sharma[1]*, Tara Chand[2] and R. C. Chaudhary[2]

[1]Department of Mathematics, Birla Institute of Technology and Science- Pilani, -333031, India.
[2]Department of Mathematics, University of Rajasthan, Jaipur-302004, India.

This paper considers the flow of a conducting viscous incompressible fluid between two non-conducting parallel discs, when the magnetic field was applied perpendicularly to the discs. The upper disc is in steady rotation, while the lower one is a stationary porous disc. The whole flow is divided into two regions: the free fluid region (between two parallel discs) and the porous region (this flow is of porous material). The approximate solutions are obtained by solving the Navier-Stokes equations in the free fluid region, and the Darcy's equations in the porous region with suitable boundary conditions at the interface. The effects of rotation, Hartmann number and forced parameter have been considered on the flow characteristics and are illustrated by graphs. The flow is essentially dominated by rotational effect as well as by the forced parameter.

Key words: Magneto-hydrodynamic (MHD), incompressible fluid, forced flow. MSC (2000): 76D10, 76W05, 80A20.

INTRODUCTION

Flow of a viscous fluid through and past a porous medium has been the subject of intensive studies in recent years because of its many engineering and scientific applications. The study of viscous flow near stationary or rotating discs has significant relevance to many applications for industrial devices. Many important applications have motivated studies involving complex geometries, often with through flow and heat transfer, cooling of gas turbines, turbo machinery, boundary layer control, cooling of turbine blades, cooling the skins of high speed aircraft designs, in extraction process of fluid from the porous ground and in lubrication of porous bearings. Probably for the first time, the flow due to an infinite plane disk, rotating with constant angular velocity was discussed by Karman (1921). Cochran (1934) integrated numerically the equations obtained by Karman and compared his results with that of Karman. Batchelor (1951) and Stewartson (1953) applied these equations to the problem of steady flow between two infinite parallel plane discs, rotating at a finite distance apart. The flow due to a rotating disk of infinite radius with uniform suction at the disc has been discussed by Stuart (1954) and he obtained numerical solutions for small values of suction and asymptotic solutions for large values of suction. Rizvi (1962) examined the magneto hydro-dynamic flow over a single disk in the presence of weak magnetic field. The effects of an axial magnetic field on the flow about a rotating disk were studied by Kakutani (1962). Pande (1972) analysed a series solution for the effects of an axial magnetic field and suction (or injection) on the flow about an insulated rotating disk, when there is strong suction and a weak magnetic field. And Nath (1984) developed unsteady rotating flow over an infinite rotating disk with an applied magnetic field. Purohit and Bansal (1995) considered the flow of a viscous income-pressible electrically conducting fluid between a rotating and a stationary naturally permeable disk. Ariel (2002) discussed the numerical behavior of MHD flow near a

*Corresponding author. E-mail: bhupen_1402@yahoo.co.in.

rotating disk. Attia (2003) considered time varying rotating disk flow and heat transfer of a conducting fluid with suction or injection. Darcy (1937) initiated the theory of the flow through a porous medium. Joseph and Tao (1966) has analysed the coupled flow induced by the steady rotation of a naturally permeable disk saturated with fluid. The flow field is divided into two regions, namely (I) free fluid region, and (II) porous region, where the fluid flows through a porous medium. To link flows in the two regions, certain matching conditions are required at the interface of the two regions. This type of couple flows, with different geometries and with several kinds of matching conditions, has been examined by several authors. Khoo et al. (1998) discussed the flow between a rotating and a stationary disc. Steady flow between a rotating and a stationary naturally permeable disc had been studied by Verma and Bhatt (1975). Srivastava and Sharma (1992) studied the MHD flow and heat transfer of a porous medium of finite thickness. Steady viscous flow between two rotating naturally permeable discs had been discussed by Chauhan and Gupta (1999). Srivastava (1999) studied the flow in a porous medium induced by torsional oscillation of a disk near its surface. The flow of viscous incompressible fluid confined between a rotating disk and a porous medium was analyzed by Chaudhary et al. (2004). Sharma et.al (2007) studied forced flow of a conducting viscous fluid through a porous medium induced by a rotating disk with applied magnetic field. Recently, Dufour and Soret effects on unsteady MHD convective heat and mass transfer flow due to a rotating disk, has been investigated by Maleque (2010).

A few investigations have been reported in literature on the MHD flow of a viscous incompressible electrically conducting fluid between a rotating and a stationary naturally permeable disc. Hence, in the present analysis, it is proposed to study the flow of a conducting viscous incompressible fluid between two non-conducting parallel discs, when the magnetic field applied perpendicularly to the discs, is considered. The upper disc is in steady rotation while the lower one is a stationary porous disc.

MATHEMATICAL FORMULATION

We consider the motion of a viscous incompressible electrically conducting fluid confined between two parallel discs of infinite radius. They are placed at a distance h apart. The upper disc is rotating with uniform angular velocity Ω, while the lower disc is stationary and made up of a porous material upto a depth $z = -h$ with an impermeable surface at the bottom. The whole region ($-h \le z \le h$) is divided into two regions namely free fluid region ($0 \le z \le h$) and the region made up of porous material ($-h \le z \le 0$). The cylindrical polar coordinates (r, θ, z) are being used with the origin at the centre of the lower disc and z-axis normal to the disc. A magnetic field of constant intensity $\vec{B}_0$ is applied perpendicular to the discs. The velocity components (u, v, w) in the free fluid region and (U_P, V_P, W_P), in the porous region, are taken to be in the directions of (r, θ, z) respectively. The slip conditions suggested by Beavers and Joseph (1965) have been applied to the radial and transverse velocity components at the interface ($z = 0$).

The governing equations by Navier-Stokes equations for the steady magneto hydrodynamic flow in the free fluid region $0 \le z \le h$ are:

$$u\frac{\partial u}{\partial r}+w\frac{\partial u}{\partial z}-\frac{v^2}{r}=-\frac{1}{\rho}\frac{\partial P_1}{\partial r}+\nu\left[\frac{\partial^2 u}{\partial r^2}+\frac{1}{r}\frac{\partial u}{\partial r}+\frac{\partial^2 u}{\partial z^2}-\frac{u}{r^2}\right]-\frac{\sigma B_0^2 u}{\rho} \quad (1)$$

$$u\frac{\partial v}{\partial r}+w\frac{\partial v}{\partial z}+\frac{uv}{r}=\nu\left[\frac{\partial^2 v}{\partial r^2}+\frac{1}{r}\frac{\partial v}{\partial r}+\frac{\partial^2 v}{\partial z^2}-\frac{v}{r^2}\right]-\frac{\sigma B_0^2 v}{\rho} \quad (2)$$

$$u\frac{\partial w}{\partial r}+w\frac{\partial w}{\partial z}=-\frac{1}{\rho}\frac{\partial P_1}{\partial z}+\nu\left[\frac{\partial^2 w}{\partial r^2}+\frac{1}{r}\frac{\partial w}{\partial r}+\frac{\partial^2 w}{\partial z^2}\right] \quad (3)$$

and the equation of continuity is:

$$\frac{\partial u}{\partial r}+\frac{u}{r}+\frac{\partial w}{\partial z}=0 \quad (4)$$

The flow in porous region $-h \le z \le 0$ is governed by the Darcy's equations which are"

$$U_P=-\frac{K^*}{\mu}\frac{\partial P_2}{\partial r} \quad (5)$$

$$V_P=0 \quad (6)$$

$$W_P=-\frac{K^*}{\mu}\frac{\partial P_2}{\partial z} \quad (7)$$

and the equation of continuity is:

$$\frac{\partial U_P}{\partial r}+\frac{U_P}{r}+\frac{\partial W_P}{\partial z}=0 \quad (8)$$

where P_1, P_2 are the pressures in the free fluid and porous regions, respectively : ρ is the density; μ is the coefficient of viscosity; ν is the kinematic viscosity and K^* is the permeability of the porous medium. The corresponding boundary conditions are:

$$\left.\begin{aligned} &z=h\,;\, u=ar,\, v=r\Omega,\, w=0,\, \frac{\partial P_1}{\partial z}=0\\ &z=0\,;\, P_1=P_2, w=W_P, e_{r\theta}=\gamma V_P, e_{rz}=\gamma(u-U_P)\\ &z=-h\,;\, W_P=0\,,\, \frac{\partial P_2}{\partial z}=0 \end{aligned}\right\} \quad (9)$$

where

$$e_{r\theta} = r\frac{\partial}{\partial r}\left(\frac{v}{r}\right),\ e_{rz} = \frac{\partial w}{\partial r} + \frac{\partial u}{\partial z},\ \gamma = \frac{\alpha}{\sqrt{K^*}},$$

α is a constant depending upon the structure of porous material and a is a forced parameter. Following Batchelor (1951), we seek the solution of the equations (1) to (8) under the boundary conditions (9), in the following form:

$$\left.\begin{aligned} &u = r\Omega F(\zeta),\ v = r\Omega G(\zeta),\ w = (\nu\Omega)^{1/2} H(\zeta) \\ &\frac{P_1}{\rho} = \nu\Omega P_{10}(\zeta) + \frac{1}{2}K\Omega^2 r^2 \end{aligned}\right\} \quad (10)$$

and

$$\left.\begin{aligned} &U_P = r\Omega\bar{F}(\zeta),\ W_P = (\nu\Omega)^{1/2}\bar{H}(\zeta) \\ &\frac{P_2}{\rho} = \nu\Omega P_{20}(\zeta) + \frac{1}{2}\bar{K}\Omega^2 r^2 \end{aligned}\right\} \quad (11)$$

On substituting equations (10) and (11) into the equations (1) to (8) of continuity and motion, we obtain the following set of equations, in non-dimensional form. In the free fluid region:

$$2R^{1/2}F + H' = 0 \quad (12)$$

$$F'' - FM^2 = R^{1/2}HF' + R(F^2 - G^2 + K) \quad (13)$$

$$G'' - GM^2 = R^{1/2}HG' + 2RFG \quad (14)$$

$$H'' = R^{1/2}(HH' + P_{10}') \quad (15)$$

and in the porous region:

$$\bar{F} = -\beta R\bar{K} \quad (16)$$

$$\bar{H} = -\beta R^{1/2} P_{20}' \quad (17)$$

$$2R^{1/2}\bar{F} + \bar{H}' = 0 \quad (18($$

where

$\zeta = \frac{z}{h}$; $R = \frac{\Omega h^2}{\nu}$, the rotational Reynold number; $\beta = \frac{K^*}{h^2}$, dimensionless permeability; $m = \frac{a}{\Omega R}$, dimensionless forced parameter; $M^2 = \frac{\sigma B_0^2 h^2}{\rho}$, the Hartmann number; and primes denotes differentiation with respect to 'ζ'. The corresponding boundary conditions become:

$$\left.\begin{aligned} &\zeta = 1: F = mR,\ G = 1,\ H = 0 \\ &\zeta = 0: H = \bar{H},\ P_{10} = P_{20},\ K = \bar{K},\ G = 0 \\ &\qquad F' = h\gamma(F - \bar{F}) \\ &\zeta = -1: \bar{H} = 0 \end{aligned}\right\} \quad (19)$$

SOLUTION

The Reynolds number R which is defined in terms of the angular velocity of the disc, is assumed to be small. Since the solution for a given ratio of angular velocities of the two discs is not unique for sufficiently high Reynolds number, therefore, the unknown functions can be expanded in ascending powers of R, in the following form:

$$\begin{aligned} &[F(R,\zeta), \bar{F}(R,\zeta)] = R[F_1(\zeta), \bar{F}_1(\zeta)] + R^3[F_3(\zeta), \bar{F}_3(\zeta)] + \ldots \\ &G(R,\zeta) = G_0(\zeta) + R^2 G_2(\zeta) + \ldots \\ &[H(R,\zeta), \bar{H}(R,\zeta)] = R^{1/2}[R\{H_1(\zeta), \bar{H}_1(\zeta)\} + R^3\{H_3(\zeta), \bar{H}_3(\zeta)\} + \ldots \\ &K = K_0 + R^2 K_2 + R^4 K_4 + \ldots \\ &[P_{10}(R,\zeta), P_{20}(R,\zeta)] = R[P_{101}(\zeta), P_{201}(\zeta)] + R^3(P_{103}(\zeta), P_{203}(\zeta)] + \ldots \end{aligned} \quad (20)$$

Substituting equation (20) into equations to (12) to (18) and collecting the coefficients of the like powers of R, we obtain the following set of equations, in the free fluid region : R^0 zeroth order term as:

$$\left.\begin{aligned} &F_1'' - F_1 M^2 = K_0 - G_0^2 \\ &G_0'' - M^2 G_0 = 0 \end{aligned}\right\} \quad (21)$$

R^1 First order term as:

$$\left.\begin{aligned} &G_2'' - M^2 G_2 = 2F_1 G_0 + H_1 G_0' \\ &P_{101}' = H_1'' \\ &H_1' = -2F_1 \end{aligned}\right\} \quad (22)$$

R^2 second order term as:

$$F_3'' - F_3 M^2 = F_1^2 - 2G_0 G_2 + K_2 + H_1 F_1' \quad (23)$$

R^3 Third order term as:

$$\left.\begin{aligned} &H_3'' - H_1 H_1' = P_{103}' \\ &H_3' = -2F_3 \end{aligned}\right\} \quad (24)$$

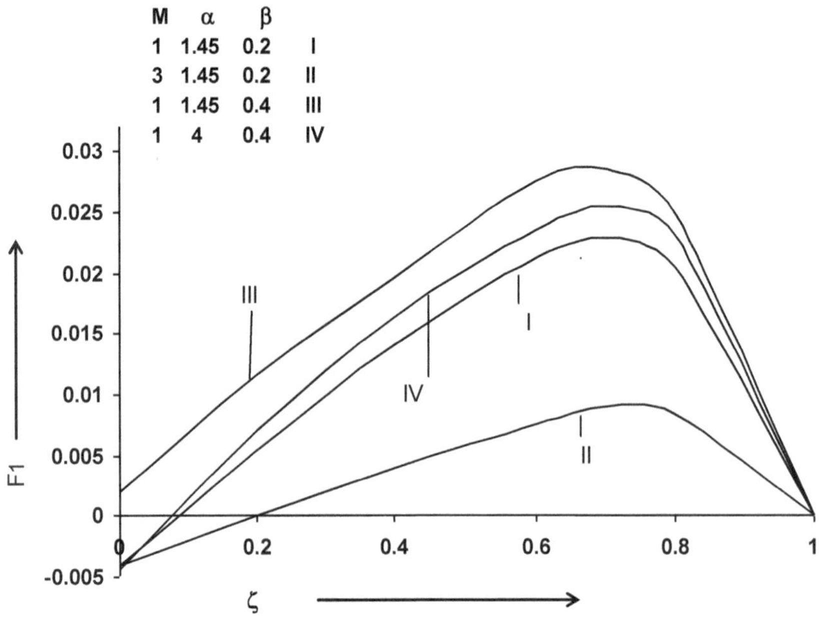

Figure 1. Radial velocity component $F_1(\zeta)$ vs. ζ for m=0.

and the equations in the porous region are

$$\left.\begin{aligned}\overline{F}_1 &= -\beta K_0\\ \overline{H}_1 &= -\beta P'_{201}\\ \overline{H}'_1 &= -2F_1\end{aligned}\right\} \qquad (25)$$

$$\left.\begin{aligned}\overline{F}_3 &= -\beta K_2\\ \overline{H}_3 &= -\beta P'_{203}\\ \overline{H}'_3 &= -2\overline{F}_3\end{aligned}\right\} \qquad (26)$$

The corresponding boundary conditions are reduced to:

$$\left.\begin{aligned}&\zeta=1: F_1 = m, H_1 = 0, F_3 = H_3 = 0, G_0 = 1, G_2 = 0\\ &\zeta=0: G_2 = 0, H_1 = \overline{H}_1. H_3 = \overline{H}_3, P_{101} = P_{201}, P_{103} = P_{203},\\ &\quad F'_1 = h\gamma(F_1 - \overline{F}_1), F'_3 = h\gamma(F_3 - \overline{F}_3)\end{aligned}\right\} \quad 27$$

$$\zeta = -1 : \overline{H}_1 = 0 , \overline{H}_3 = 0$$

The solutions of the ordinary differential equations (21) to (27) are worked out subjected to the boundary conditions (28). We are not including solutions here for the sake of brevity of the paper. The important flow characteristics of the problem are further discussed.

RESULTS AND DISCUSSION

In the present paper, the forced flow of a viscous income-pressible electrically conducting fluid between a rotating and a stationary naturally permeable disc, under the application of a magnetic field acting perpendicular to the discs has been investigated. The whole flow field is divided into two regions; (i) free fluid region, and (ii) porous region. The flow in the free fluid region is governed by Navier-Stokes equations in the presence of magnetic field, while the flow in porous region is governed by Darcy's equations. The Reynolds number defined in terms of the angular velocities of the discs is assumed to be small. The effects of rotation, forced parameter and Hartmann number has been considered on the flow characteristics and illustrated by graphs.

The flow field behavior in the free fluid region and in the porous region under the presence of an applied magnetic field has been considered. The radial velocity component F_1 (zeroth order term) versus distance ζ is shown in Figures 1 and 2 for m = 0, 0.1 and other parameters respectively. An examination of Figure 1 shows that the radial velocity component F_1 decreases in magnitude with the increase in α or Hartmann number M, whereas it increases by increasing β. The magnitude of the radial velocity component F_1 increases with increase in distance from lower disc, until it attains its maximum value, after which it decreases and it becomes zero at the upper disc. The radial velocity component F_1 takes its maximum value near the upper disc. Figure 2 shows that the magnitude of the radial velocity component F_1 decreases with the increasing α or M, where as it increases by increasing β. It increases when we move from lower disc to upper disc and it takes its maximum value at the upper disc. Transverse velocity is shown in Figure 3. The transverse velocity increases with increase in Hartmann number M, where as it decreases if we move towards the

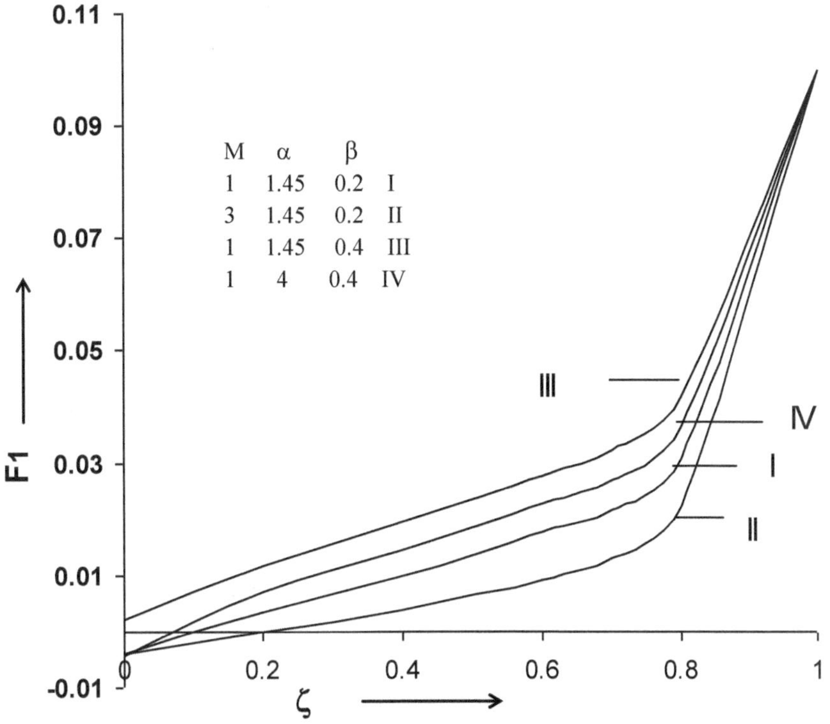

Figure 2. Radial velocity component F1(ζ) vs. ζ for m=0.1.

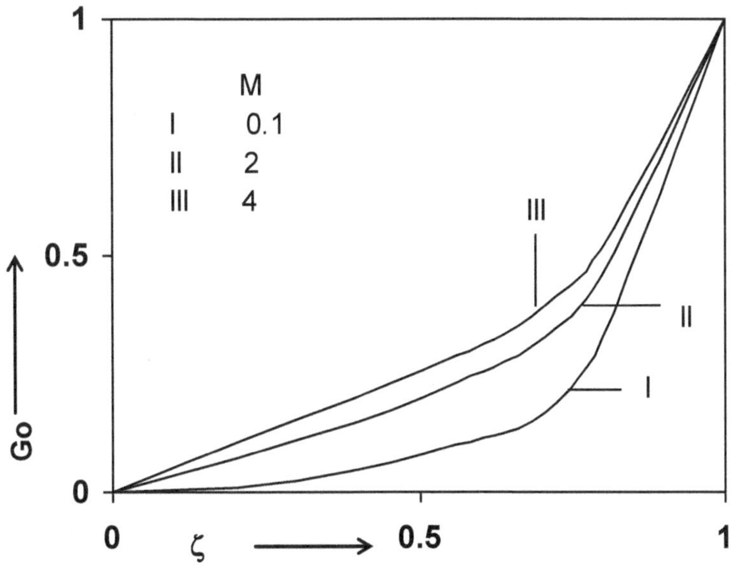

Figure 3. Transverse velocity component Go vs. ζ.

lower disc from upper disc and it becomes zero at the lower porous disc. The axial velocity component H_1 are drawn for different values of the parameters α, β, m and M in Figure 4. It is found that the axial velocity component H_1 decreases in magnitude with the increase in α or M where as it increases by increasing β or m. The magnitude of the velocity component H_1 increases with increase in distance in porous medium, until it attains its maximum value, after which it decreases and it becomes zero at upper disc. It is symmetrical about the axis ζ = 0 (interface). As Reynolds number and forced parameter increase and Hartmann number decreases, more and

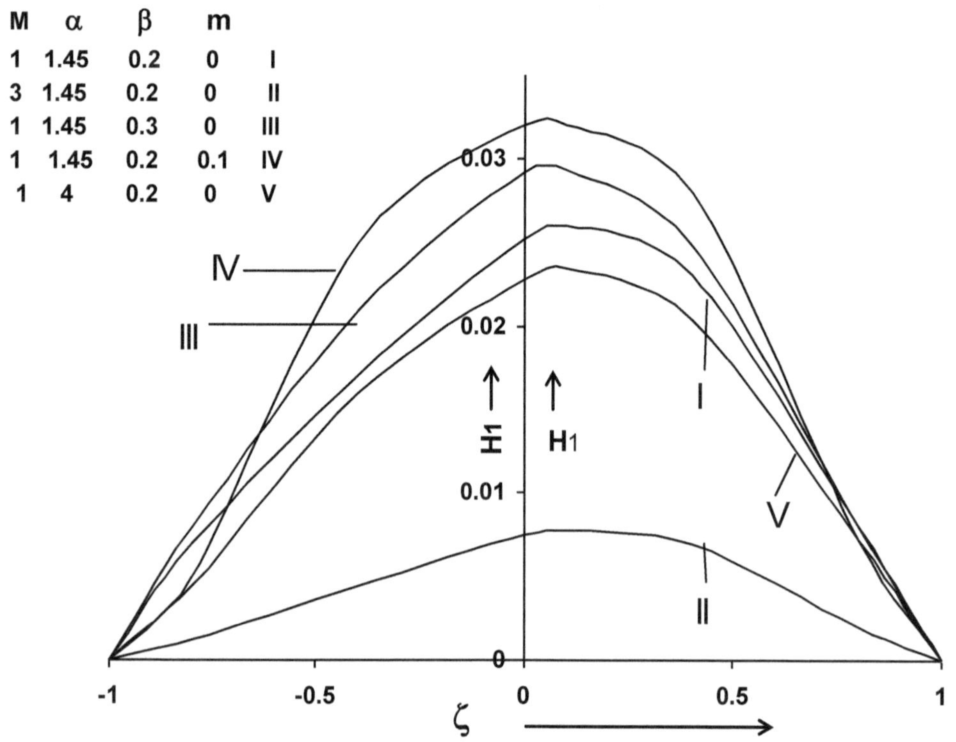

Figure 4. Axial velocity component $H_1(\zeta)$ vs. ζ.

more fluid is thrown out in the neighborhood of the upper disc and radial and axial velocity increases with the increasing of Reynolds number and forced parameter and decreasing of Hartmann number. Thus, magnetic field has a sobering effect on velocity distribution.

Stream functions of the flow

The stream functions ψ_1 and ψ_2 for the free fluid region and porous region respectively are given by:

$$\psi_1^*(\eta,\zeta) = \frac{\psi_1(\eta,\zeta)}{\frac{1}{2}h^2(\nu\Omega)^{1/2}} = \eta^2\, H(\zeta) \qquad (28)$$

$$\psi_2^*(\eta,\zeta) = \frac{\psi_2(\eta,\zeta)}{\frac{1}{2}h^2(\nu\Omega)^{1/2}} = \eta^2\, \overline{H}(\zeta) \qquad (29)$$

The streamlines are drawn in Figure 5, for R = 0.2, β = 0.2 and α = 1.45. We find that the fluid is thrown radially outwards due to the centrifugal forces, hence to fill the gap, the fluid rushes from infinity towards the axis in the stationary lower porous disc and comes out of the porous region to keep the continuity, consequently, as compensation, the fluid is pumped out from the lower disc to maintain the flow. By introducing the forced flow, it is observed that the flux thrown radially outwards is more. As might be expected there is symmetry about the axis ζ = 0.

Skin-friction and torque on both disks

The coefficients of skin-friction are given by:

$$(C_f)_{\zeta=1} = \frac{\tau_1}{(\mu r_0 \Omega/h)} = -\left(\frac{r}{r_0}\right) F'(1) \qquad (30)$$

and

$$(C_f)_{\zeta=0} = \frac{\tau_0}{(\mu r_0 \Omega/h)} = \left(\frac{r}{r_0}\right) F'(0) \qquad (31)$$

Where $\tau_1 = -\left(\frac{\mu r\,\Omega}{h}\right) F'(1)$, is the shearing stress on the upper disc, $\tau_0 = \left(\frac{\mu r\,\Omega}{h}\right) F'(0)$ is the shearing

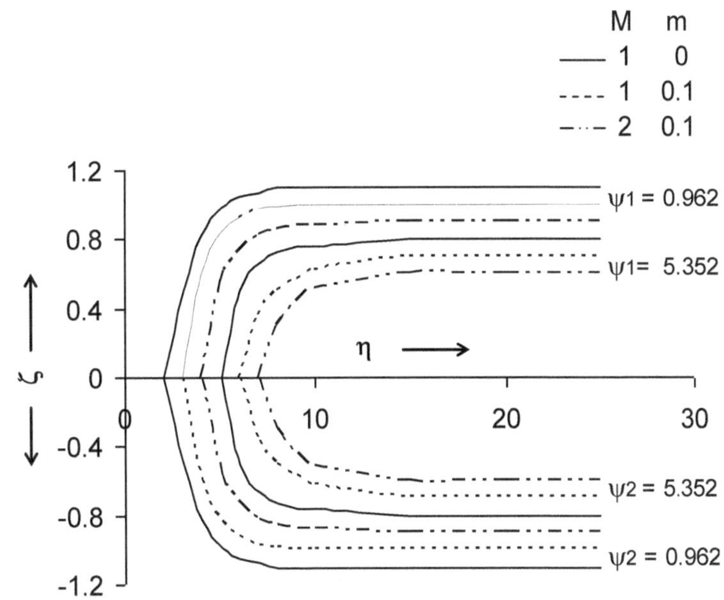

Figure 5. Streamline pattern for R=0.2, α=1.45 and β=0.2.

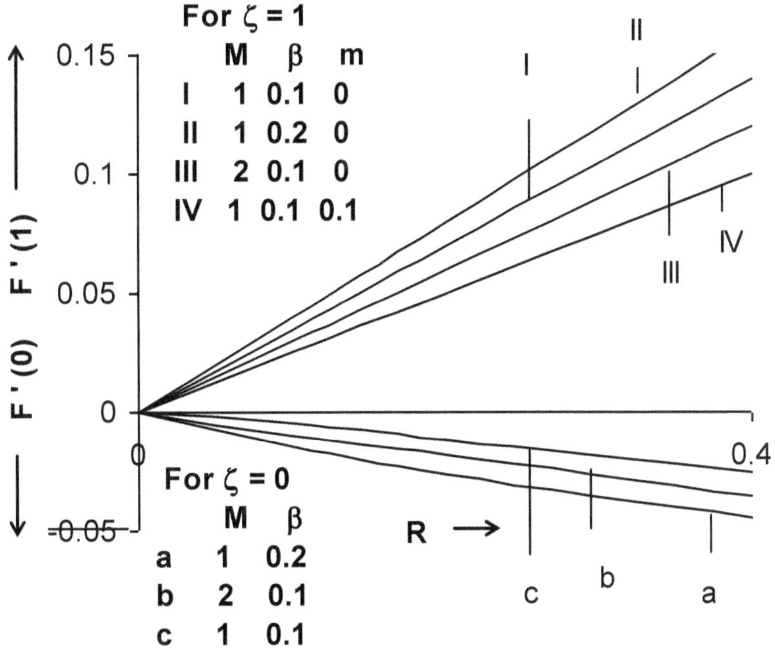

Figure 6. Coefficients of skin friction F' (0) and F'(1) vs. R, for =1.45.

stress on the lower disc and r_0 is a certain distance in the radial direction from the centre of the disc.

Figure 6 shows the variations of coefficients of skin-friction. It is noted that the coefficient of skin-friction at the upper disc increases in magnitude with increase in R and decreases with increasing m or M. It increases with increasing β at the upper disc but it decreases with increasing β or R or M at the lower disc.

Conclusion

In this paper, the forced flow of a viscous incompressible electrically conducting fluid between a rotating and a stationary naturally permeable disc, under the application of a magnetic field acting perpendicular to the discs is studied. The following conclusions can be drawn as a result of the computations:

i. The flow is essentially dominated by rotational effect and as well as by the forced parameter.
ii. The radial velocity component F_1 decreases with the increasing α or M.
iii. The transverse velocity increases with increase in Hartmann number M.
iv. Axial velocity increases with the increasing of Reynolds number and forced parameter.

Nomenclatures: $\mathbf{B_0}$, Uniform magnetic field; **K***, permeability parameter; **M,** magnetic field parameter (Hartmann number); **R,** rotational Reynold number; **m,** dimensionless forced parameter; ν, kinematic viscosity; μ, viscosity; τ, skin-friction (shearing stress); σ, scalar electrical conductivity; ψ_1, ψ_2, stream unctions; $\mathbf{C_f}$, coefficient of skin-friction; ρ, density of the fluid; $\mathbf{P_1}$, $\mathbf{P_2}$, pressures in the free fluid and porous regions respectively; α , a constant of structure porous material; **h,** distance; **u, v, w,** velocity components in the free fluid region in the **r,** θ, **z** directions; $\mathbf{U_P}$, $\mathbf{V_P}$, $\mathbf{W_P}$, velocity components in the porous region in the r, θ, z-directions; **r,** θ, **z,** cylindrical polar coordinates; Ω, uniform angular velocity; β, non-dimensional permeability parameter; ζ, similar distance variable; **a,** forced parameter.

REFERENCES

Ariel PD (2002). On computation of MHD flow near a rotating disk. ZAMM, 82(4): 235-246.
Attia HA (2003). Time varying rotating disk flow and heat transfer of a conducting fluid with suction or injection. Int. Comm. Heat Mass Transfer, 30(7): 1041-1049.
Batchelor GK (1951). Note on a class of solutions of the Navier-Stokes equations representing steady non-rotationally symmetrical flow. Q. J. Mech. Appl. Math., 4: 29-41.
Chaudhary RC, Sharma PK (2004). The flow of viscous incompressible fluid confined between a rotating disk and a porous medium. Inter. J. Appl. Mech. Eng., 9(3): 607-612.
Chauhan DS, Gupta S (1999). Steady viscous flow between two rotating naturally permeable disks. Ganita Sandesh, 13(2), 53-64.
Cochran WG (1934). The flow due to a rotating disk. Proc. Camb. Phil. Soc., 30: 365-375.
Darcy H (1937). The flow of fluids through porous media, McGraw-Hill Book Co., New York.
Joseph DD (1965). The coupled flow induced by the steady rotation of a naturally permeable disk saturated with fluid. Q. J. Mech. Appl. Math., 18: 325-330.
Joseph DD, Tao LN (1966). Lubrication of porous bearing Stokes Solution. J. Appl. Mech. Trans. ASME Series E, 88 (3): 753-760.
Kakutani T (1962). Hydromagnetic flow due to a rotating disk. J. Phys. Soc. Jpn., 17: 1496-1506.
Karman TV (1921). Uber laminare und turbulente reibung. ZAMM, 1: 233-252.
Khoo BC, Chew YT, Lim CP (1998). The flow between a rotating and a stationary disc. Meas. Sci. Technol., 9: 650-658.
Maleque KA (2010). Dufour and Soret effects on unsteady MHD convective heat and mass transfer flow due to a rotating disk. Lat. Am. Appl. Res., 40(2): 105-111.
Mathews JLR, Nath G (1984). Unsteady rotating flow over an infinite rotating disk with an applied magnetic field. Indian J. Pure Appl. Math., 15(3): 279-288.
Pande GS (1972). On the effects of uniform high suction on the steady MHD flow due to a rotating disk. Appl. Sci. Res., 11: 205-212.
Purohit GN, Bansal P (1995). MHD flow between a rotating and a stationary naturally permeable porous discs. Ganita Sandesh, 9(2): 55-64.
Rizvi SAT (1962). Magnetohydrodynamic flow over a single disc. Appl. Sci. Res., 10: 662-669.
Sharma BK, Jha AK, Chaudhary RC (2007). Forced flow of a conducting viscous fluid through a porous medium induced by a rotating disk with applied magnetic field. Ukr. J. Phys., 52(7): 639-645.
Srivastava AC, Sharma BR (1992). The MHD flow and heat transferof a porous medium of finite thickness. J. Math. Phys. Sci., 26: 539-547.
Srivastava AC (1999). Flow in a porous medium induced by torsional oscillation of a disk near its surface. ZAMP, 50: 529-545.
Stewartson K (1953). On the flow between two rotating coaxial disks. Proc. Camp. Phil. Soc., 49: 333-341.
Stuart JT (1954). On the effects of uniform suction on the steady flow due to a rotating disk. Q.J. Mech. Appl. Math., 7: 446-457.
Verma PD, Bhatt BS (1975). Steady flow between a rotating and a stationary naturally permeable discs. Int. J. Eng. Sci., 13: 869-876.

System identification for experimental study for ultra high molecular weight for polyethylene process by using Bi-supported SiO_2 / $MgCl_2$ / $TiCl_4$ catalyst

Ahmmed S Ibrehem[1]* and Hikmat S. Al-Salim[2]

[1]Department of Chemical Engineering, University of Technology Mara (UiTM), 40450, Shah Alam Malaysia.
[2]Chemical and Petroleum Engineering Department, UCSI - University, Cheras, Kuala Lumpur, 56000, Malaysia.

In this work, system identification method is used to capture the reactor characteristics of production rate of polyethylene (PE) based on published experimental data. The identification method is used to measure the percentage effect on the production rate of PE by measuring the effect of input factors of temperature of reaction, hydrogen concentration, and [Al]/[Ti] molar catalyst ratio. Temperature of reaction has big effect which is equal to 52.4% on the output of the system as compared to the other two factors. Also, hydrogen concentration has big effect which is equal to 45.66% on the output of the system. [Al]/[Ti] molar catalyst ratio has big effect which is equal to 1.94%, but less than the reaction temperature and hydrogen concentration. All these results depend on experiment results and these results are very important in industrial plants.

Key words: Catalysts, Ziegler-Natta polymerization, ultracentrifugation, identification.

INTRODUCTION

Fluidized catalytic bed continuous reactors (FCR) entail very complex flow behavior (Ibrehem et al., 2008). Moreover, the presence of four phases in the reaction media complicates the transport phenomena involved which dictates the need for a thorough understanding of parameters effect on production rate and crystallization characteristics.

System identification method (Ibrahem and Hikamt, 2009) should have sufficient accuracy of prediction, wide applicability, the minimum assumption about the most parameters effects like temperature of reaction, hydrogen concentration, and [Al]:[Ti] molar ratio. In the area of production rate for polyethylene system, identification of most input variable effects on the production rate by using experiment work was very little. In 2009, Gholamhossein Zohuri group made a very good experiment work design technique for polymerization process based on experimental data only. We depend on this important work to specify which is the most active input parameters whose effect on the system depends on experimental results and new identification method to specify the percentage effect of production of high molecular weight of polyethylene.

High molecular weight polymers and their morphological development are the most important achievements of heterogeneous catalyst polymerization of the Ziegler-Natta type (Peacock et al., 2000; Zohuri et al., 2008; Dasilva et al., 2002; Shiraishi et al., 1990). $MgCl_2$ and SiO_2 are the most common bi-supports used to heterogenize olefin polymerization catalysts (Pater et al., 2003; Jamjah et al., 2006; Zohuri et al., 2006; Moore, 2005), and SiO_2 is also used to support metallocene and late transition metal catalysts (Galli et al., 2004; Barbe et al., 1986). The morphological development of the catalyst originally obtained from its support could replicate the polymer in controlled conditions. Heterogeneous catalyst particles break up during the early stages of the polymerization, possibly due to the chemical reaction of the catalyst component, mechanical stress and also polymer growth, respectively (Silva et al., 2005; Shaotian et al., 2001). In Ziegler-Natta polymerization, the rate/time profiles are mainly a decay type with very high polymerization activity at the early stage of the reaction. Therefore, the fast growth of the polymer into the pore and channel

*Corresponding author. E-mail: ahmadsaadi1@yahoo.com.

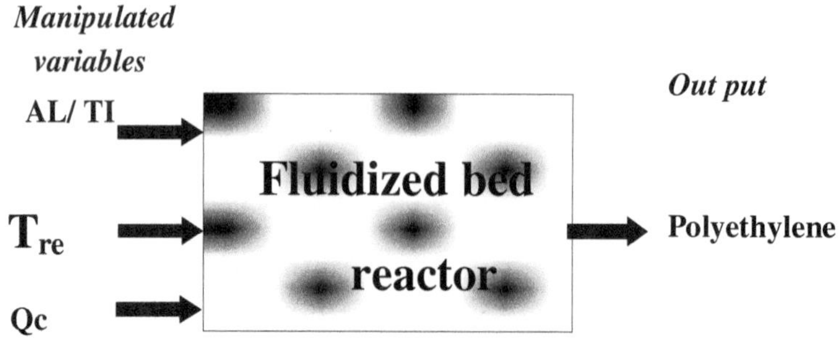

Figure 1. Representation of variables in polymerization system as a MISO system.

of the catalyst could fragmentize the catalyst particles. To achieve a reasonable morphology, the fragmentation must be controlled, since uncontrolled fragmentation may produce fine particles. Using SiO_2 and $MgCl_2$ components of a solid, support not only raised the activity of the catalyst but also improved its morphology. SiO_2 also improved the mechanical and thermal stability of the catalyst (Kono et al., 2001; Yamahiro et al., 1999).

Morphological development of support, catalyst and polymer is the major area of study nowadays. Ultra high molecular weight polyethylene has many desirable physical and mechanical characteristics comparable to some engineering plastics (Fukuda et al., 2003; Zohuri et al., 2003; Zohuri et al., 2001). The present work studies the effect of catalyst spherical Ziegler-Natta flow rate, temperature input and hydrogen concentration on production rate of polyethylene.

DYNAMIC ANALYSIS AND IDENTIFICATION FOR MATHEMATICAL MODEL

System identification involves building a dynamical model from an input/output data and without use of any laws concerning the fundamental nature and properties of the nonlinear system. The batch reactor process involves many variables, which contribute to its operation, and this makes it a single-input single-output (SISO) process. These variables are classified as follows: input or manipulated variables (MVs) are chosen from those variables that have direct effect on the process performance, and practically easy to actuate. In our case, these are: percent ratio of AL/ TI as a catalyst (Qc), temperature input, and concentration of hydrogen. The controlled variables (CVs) are selected from the process outlet streams that are usually of foremost importance such as the production polyethylene (PE). A schematic diagram representing the previously mentioned variables is shown in Figure 1.

So, the identification steps can be defined for PASA as follows (Ibrahem and Hikamt, 2009);

1. To make actual calculations for the system from nominal conditions xn,i to calculate yn,i.
2. To repeat step one with +H% perturbed parameter input vector (H × xn,i) to produce the + H % perturbed output yi matrix.
3. To subtract each element from matrix in first step from the corresponding element in matrix in second step and divide the difference by yn,i. (yi- yn,i) / yn,i.
4. Similarly, calculate the change in parameters (xi- xn,i) / xn,i.
5. Divide each element in third step by fourth step to produce the sensitivity matrix k that depends completely on scale matrix without using any a proper factor.
6. Calculate k matrix for negative direction which mean calculate k+ H % and k- H % so as, we can check the effect of each parameters on output as seen in Table 2;
7. Then use the kaverage slope to find the average angle (θ) that represents the overall effect of each parameter on the measured output (θ1 represents the effects of AL/TI, θ2 represents the effects of reaction temperature and θ3 represents the effects of hydrogen concentration on the system) as shown in Figure 2 . From parameters average slope analysis, we can see a preliminary partitioning of estimating into different groups that depend on slope angle (θ) of kaverage and these groups can be specified as follows:

$(\theta) \geq 20^{\circ}$: Large effects on the system.
$20^{\circ} > (\theta) > 15^{\circ}$: Middle effects on the system.
$15^{\circ} \geq (\theta) \geq 10^{\circ}$: Weak effects on the system.
$10^{\circ} < (\theta)$: Cannot be established.

$$\theta \text{ total} = \theta\text{in the system} + \theta\text{interaction} \quad (1)$$

$$\text{total} = \theta 1 + \theta 2 + \theta 3$$

EXPERIMENTAL STUDY

SiO_2 (PQ3050) was acquired from PQ Corporation (USA). The compound was calcinated at about 400°C for nearly 5 h before use (Zohuri et al., 2001). Spherical adducted $MgCl_2$ $.nC_2H_5OH$ was prepared according to literature (Shiraishi et al., 1990). Slurry polymerization of ethylene was carried out in a 1 L stainless-steel reactor of Buchi (bds300), a semibatch type, equipped with a stirrer speed control, temperature and pressure control. The reactor was purged with nitrogen. A volume of 400 ml heptane was charged into the reactor and degassed at least three times. When the polymerization temperature reached the required temperature, the catalyst

Table 1. Calculation of the percent difference for catalyst ratio against output profile.

Ratio Al/Ti (xi)	(xi - xn,i) / xn,i	PE-mmol(yi- yn,i) / yn,i	Slope (k)
300	0.538	0.117	0.217
350	0.461	0.096	0.208
400	0.384	0.088	0.229
450	0.307	0.074	0.241
500	0.23	0.071	0.308
550	0.153	0.065	0.424
700	0.076	0.053	0.697
750	0.153	0.292	1.908
800	0.23	0.289	1.256
850	0.307	0.205	0.667
900	0.384	0.107	0.278
950	0.461	0.086	0.186
1000	0.538	0.079	0.146
Kaverage			0.52
θ Average			27.49°
%effect (θ average×100)/90°			30.54

Table 2. Calculation the percent difference for hydrogen concentration against output profile.

Hydrogen concentration (ml)	(xi - xn,i) / xn,i	PE-mmol(yi- yn,i) / yn,i	Slope (k)
0	1	1.275	1.275
10	0.8	1.225	1.531
20	0.6	1.125	1.875
30	0.4	0.9	2.25
40	0.2	0.5	2.5
60	0.2	0.3	1.5
70	0.4	0.4	1
80	0.6	0.5	0.833
90	0.8	0.512	0.64
100	1	0.525	0.525
Kaverage			1.3929
θ Average			54.32°
% effect(θ Average×100)/90°			60.36

components were added in the following order; TEA, the solid catalyst, and hydrogen (if any). All the reagents were added as slurry in n-heptane or a gas. The pressure inside the reactor was maintained constant with the monomer gas. Therefore, the required amount of the monomer to feed into the reactor is equal to the consumption of the monomer. At the end of the reaction time, mainly one hour, the monomer feed was stopped. The polymerization waste ruminated by draining the slurry polymer into a small volume of acidified methanol. The polymer obtained was filtered and dried at 70°C overnight (Zohuri et al., 2001).

Catalyst preparation

The SiO_2 /$MgCl_2$/$TiCl_4$ catalyst was prepared using the adducted spherical $MgCl_2$ .nC_2H_5OH. The ethanol was chemically removed during the catalyst preparation procedure. The calcinated SiO_2 (2.5 g) and $MgCl_2$ .nC_2H_5OH (2.5 g) were added to a catalyst preparation reactor containing a sinter glass heated with a jacket. The chemicals were suspended in toluene (100 ml), and then $TiCl_4$ (40 ml) was added drop wise, while the contents of the reactor were stirred at 5°C. The temperature was raised to 115°C in increments of 20°C, while stirring at each step for at least one hour. The product was filtered, and washed with n-heptane (100 ml). Toluene (100 ml) and $TiCl_4$ (40 ml) were added at 40°C. The temperature was raised to 115°C incrementally, as before. The final catalyst was filtered, washed with n-heptane to completely remove unreacted $TiCl_4$, and dried. All the steps were carried out under an atmosphere of dried N_2 (Zohuri GH et al., 2001).

RESULTS AND DISCUSSION

Figure 2 shows the relation between Al/Ti and production

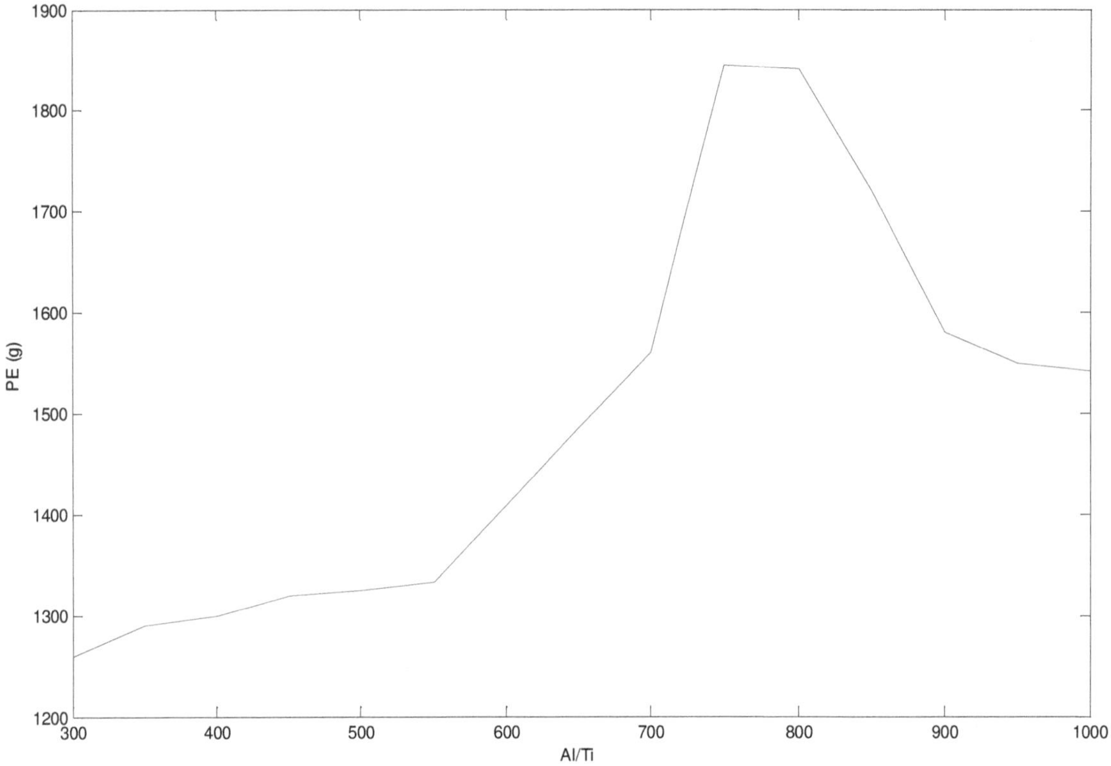

Figure 2. Effect of Al/Ti molar ratio on production rate of polyethylene, temperature 60℃

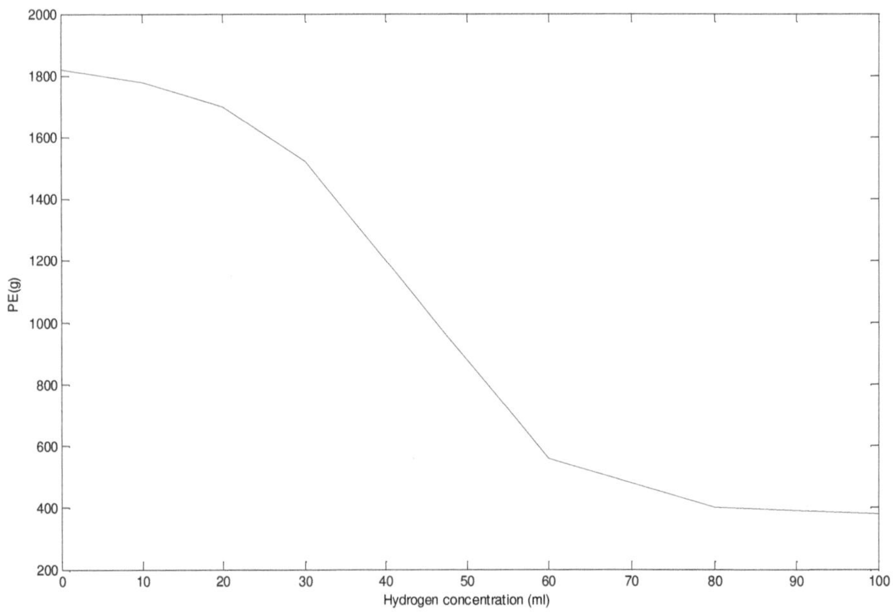

Figure 3. Effect of H_2 concentration on production rate of polyethylene, temperature 60℃, Al/Ti=770:1.

rate of PE. The optimum value for Al/Ti = 770/1 molar ratio gives production rate of PE = 1845 at constant reaction temperature of 60℃. As regard to Figure 2, all calculations are in Table 1. The average slope = 0.52 and θ average = 27.49°, that is, mean Al/Ti ratio has large effects on the system = 30.54%. Figure 3 shows the

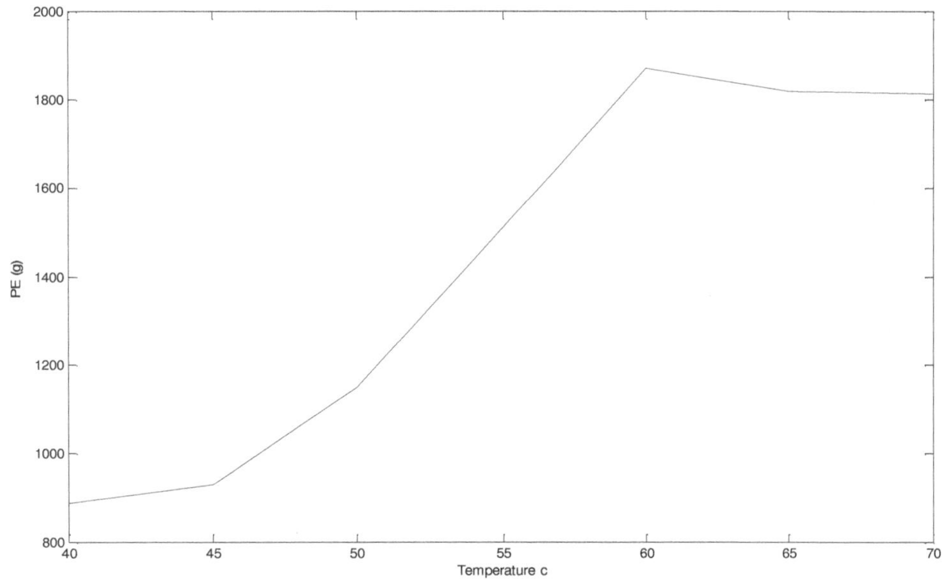

Figure 4. Effect of temperature on production rate of polyethylene, Al/Ti=770:1.

Table 3. Calculation of the percent difference for reaction temperature against output profile.

Temperature (℃)	(xi - xn,i) / xn,i	PE-mmol(yi- yn,i) / yn,i	Slope (k)
40	0.272	0.512	1.882
45	0.181	0.489	2.701
50	0.09	0.368	4.088
60	0.09	0.022	0.244
65	0.181	0.001	0.005
70	0.272	0.003	0.011
Kaverage			1.4885
θ Average			89.99o
% effect(θaverage×100)/90o			99.9%

relation between hydrogen concentration and production rate of PE which is an inverse relation; when the concentration of hydrogen increase, the production rate will be decreased because addition of H_2 closes the rate of live reaction and make dead polymer to stop polymerization by competing with monomer in adsorption on active center that reduce polymerization. Also, in Table 2, average slope = 1.3929 and θ average = 54.32°, that is, mean H_2 concentration has large effects on the system at 60.36%. Figure 4 shows the relation between reaction temperature and production rate of PE. The optimum value for reaction temperature (60℃) gives production rate of PE = 1845 at Al/Ti =770/1 molar ratio. In Table 3, the average slope = 1.4885 and θ average = 89.99°, that is, mean reaction temperature has largest effects on the system at 99.9%. Figure 5 represents all three effects inside the system that are represented by average angle slope. From equation (1), θ total = 89.99° + 54.32° + 27.49° = 171.8°; θ in the system = 90°; θ interaction = θ total – θ in the system; θ interaction = 171.8° – 90° = 81.8°; θ interaction / θ total ≈ θ2 / θ1

Table 4 shows the percentage effects of a temperature, hydrogen concentration and catalyst ratio in the system and interaction between these inputs variables.

Conclusion

Using this active identification method gives us very flexible analysis of polymerization system to reach important results that can be used and which depend on industry. The most active input parameter is reaction temperature, compared to other factors because it is affected by 52.4% in the system for production PE and 47.6% for interaction effects between the two factors.

Also, hydrogen input factor has big effect on the system because it is effected by 45.66% directly in the rate of live and dead polymer for production PE and 1 4.7% for

Table 4. Calculation of the percent effects for in the system and interaction between the inputs variables.

Action	Temperature	Hydrogen concentration	Ratio Al/Ti
Percentage effects inside the system	52.4	45.66	1.94
Percentage interaction effects	47.6	14.7	28.6

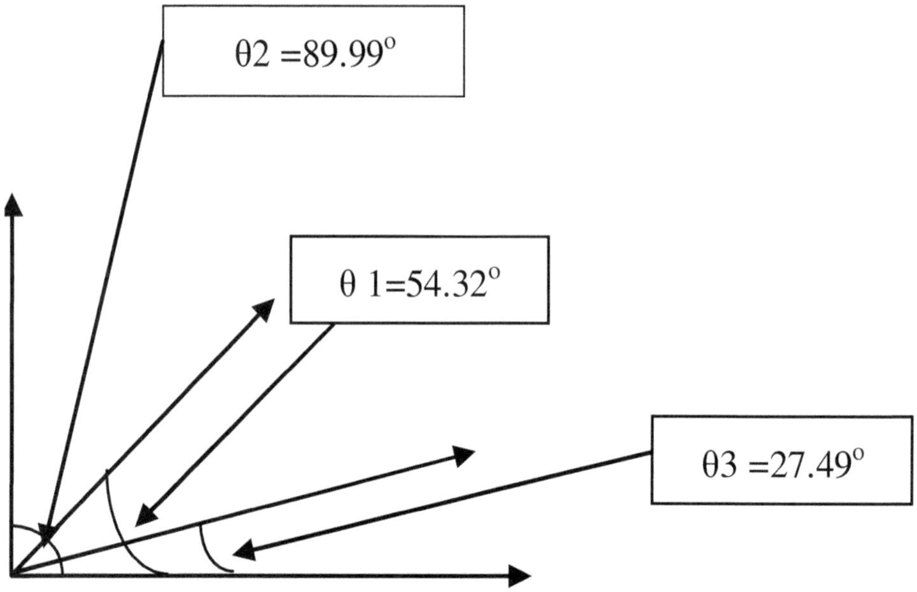

Figure 5. Representation of the average angle slope.

interaction effects between the two factors. Al/Ti ratio has small effect (1.94%) in the system but has big effect (28.6%) in interaction sides between the two factors (reaction temperature and hydrogen concentration). Reaction temperature and hydrogen concentration mainly affect all active sites of catalyst surface reaction. All these results are very important in industrials of polymerization system. The initial stage of the polymerization, including the reaction of the catalyst with the co-catalyst and the high polymerization activity, broke up the catalyst particles into smaller particles of irregular shape and size. The morphology of the polymer particles was almost the same as that of the catalyst, which reacted with the co-catalyst at the early stage of the polymerization. In addition, the morphology of the particles may be affected by polymerization conditions, such as temperature and monomer pressure. However, the parameters of the initial stages may also be critical.

REFERENCES

Ahmmed SI, Hikamt SI (2009). "New Dynamic Analysis and System Identification Of Bio diesel Production Process From Palm Oil. Bull. Chem. React. Eng. Catal., 4(2): 61-68.

Barbe PC, Cecchin G, Noristi L (1986). The catalytic system Ti-complex/$MgCl_2$. Adv. Polym. Sci., 81: 1-81.

Dasilva JC, Defigueiredo GMO (2002). Spherical catalyst, process for preparing a spherical polyethylene of ultra-high molecular weight, US Patent 6,384,163.

Fukuda K, Liu B, Nakatani H, Nishiyama I,Yamahiro M, Terano M (2003). Significant variation of molecular weight distribution (MWD) of polyethylene induced by different alkyl-Al cocatalysts using a novel surface functionalized SiO_2-supported Ziegler-Natta catalyst. Catal Commun., 4: 657-662.

Galli P, Vecellio G (2004). Polyolefins: the most promising large-volume materials for the 21st century. J. Polym. Sci. Part A Polym. Chem., 42: 396-415.

Ibrehem AS, Hussain MA, Ghasem NM (2008). "Mathematical model and advanced control for gas-phase olefin polymerization in fluidized-bed catalytic reactors". Chin. J. Chem. Eng., 16(1): 84-89.

Jamjah R, Zohuri GH, Nekomanesh M, Javaheri M, Ahmadjo S, Farhadi A (2008). Synthesizing UHMWPE by a Ziegler-Natta catalyst based on MgCl2(ethoxide type)/TiCl4/tri-isobutylaluminum. Macromol. Symp., 274: 48-153.

Jamjah R, Zohuri GH, Vaezi J, Ahmadjo S, Nekoomanesh M, Pouryari M (2006). Morphological study of spherical MgCl2.nEtOH supported $TiCl_4$ ene. J. Appl. Polym. Sci., 101: 3829-3834.

Kono H, Mori H, Terano M (2001). Novel olefin block copolymer, polypropene-block-poly(methylene-1,3-cyclopentane-co-propene), synthesized from propene and 1,5-hexadiene by a modified stopped-flow method. Macromol. Chem. Phys., 202: 1319-1326.

Moore EP (2005). Polypropylene Handbook. Pasquini N. (Ed.), 2nd ed., Hanser, Munich, Ch 5.

Pater JTM, Weickert G, Van Swaaij WPM (2003). Polymerization of liquid propylene with a fourthgeneration Ziegler-Natta catalyst: influence of temperature, hydrogen, monomer concentration, and prepolymerization method on powder morphology. J. Appl. Polym. Sci., 87: 1421-1435.

Peacock AJ (2000). Handbook of Polyethylene, Marcel & Dekker, New

York.
Shaotian W, Liu JC, Make MP, Lee CC (2001). Ethylene polymerization process, US Patent 625,541,5
Shiraishi T, Uchida W, Matsuura K (1990), Process for preparing ultra-high molecular weight polyethylene,US Patent, 4: 962-64.
Silva FM, Broyer JP, Novat C, Lima EL, Pints JC, Mckenna TF (2005). Investigation of catalyst fragmentation in gas-phase olefin polymerisation: a novel short stop reactor. Macromol Rapid Commun., 20: 1846-1853.
Yamahiro M, Mori H, Nitta KH (1999). Terano M, Synthesis and basic characteristics of polypropene-block-poly(ethene-co-propene) by modified stopped-flow polymerization with an MgCl2-supported Ziegler catalyst. Macromol. Chem. Phys., 200: 134-141.
Zohuri GH, Ahmadjo S, Jamjah R, Nekoomanesh M (2001). Structural study of mono- and bi-supported Ziegler-Natta catalysts of MgCl2/SiO2/TiCl4/donor systems. Iran Polym J., 10: 149-155.
Zohuri GH, Jamjah R, Ahmadjo S (2006). Comparative study of propylene polymerization using monosupported and bisupported titanium-based Ziegler-Natta catalysts. J. Appl. Polym. Sci., 100: 2220-2226.
Zohuri GH, Azimfar F, Jamjah R, Ahmadjo S (2003). Polymerization of propylene using the high-activity Ziegler-Natta catalyst system SiO2/MgCl2 (ethoxidetype)/TiCl4/Di-n-butylphthalate/ triethylaluminum/dimethoxy methyl cyclohexyl silane. J. Appl. Polym. Sci., 89: 1177-1181.

Process parameters for maximum power of a proton exchange membrane fuel cell

Mojtaba Tafaoli-Masoule*, Mohsen Shakeri and Arian Bahrami

Department of Mechanical Engineering, Fuel Cell Research Technology Group, Babol University of Technology, P. O. Box 484, Babol, Iran.

Pressure and temperature of the cell are the effective parameters in performance of proton exchange membrane (PEM) fuel cell. In this study, a genetic algorithm is applied to determine operating conditions that result in maximum power density of a proton exchange membrane fuel cell (PEMFC). A quasi two dimensional, isothermal model is presented for PEM fuel cell and the polarization curve of the system could be achieved. The maximum power density of the system is used as fitness function of genetic algorithm. An experimental setup is developed and the result of this analytical model is compared to experimental result and has a good agreement.

Key words: Proton exchange membrane (PEM) fuel cell, genetic algorithm, analytical model, optimization.

INTRODUCTION

Polymer electrolyte membrane fuel cell creates high power density at low temperature. It is considered as being one of the most applicable technologies to produce no emission and environmentally friendly energy for different electrical applications (Wahdame et al., 2006). The performance of proton exchange membrane fuel cell (PEMFC) getting more attentions in recent years, is known very well to be influenced by large numbers of parameters such as demanded current, air and hydrogen humidity, fuel cell and humidifiers operating temperatures, anode and cathode pressures, and so forth (Xuan et al., 2010). For improving the performance of PEMFC, it is necessary to know these parameters effect on the PEMFC operations very well (Wang et al., 2003). There are other variables out of the operating parameters such as: different type and thickness of membrane, different catalyst, geometrical parameters of the flow field, different type of the gas diffusion layer and so on. However, these parameters are very important, but are not operating variables and they are not changeable during the fuel cell use. Large numbers of research papers discuss about different parameters affecting the performance of PEMFC. Xuan et al. (2010) developed a dynamic model of fuel cell system that included mass flow model, membrane hydration and electro-chemistry model for optimization of fuel cell. Chen et al. (2011) investigated the influence of gas diffusion layer (GDL) with carbon coated on PEMFC performance. Tamayol et al. (2011) also proposed a new analytical approach for evaluating permeability of GDL and they revealed permeability of GDL as a function of porosity and the fiber diameter. Wang et al. (2003) studied the effects of different operating parameters on the performance of PEMFC using experimental setup and Kumar et al. (2010) focused on numerical simulation of the effects operating conditions and design variables such as stoichiometry of the reactant gases, area of the cell and membrane thickness on cell performance. The main focuses of the past works were in discrete and certain value of parameters because those researches were done by experimental approach but this research work considers these parameters in low discrete step domain by analytic model and genetic algorithm. There are so many available papers for applying genetic algorithm and other optimization algorithms for PEMFC (Sedighizadeh et al., 2011; Li et al., 2011) but certainly, combination of genetic algorithm and valid analytical model as fitness function can be a good tool to the investigation of these

*Corresponding author. E-mail: mojtaba.tafaoli@yahoo.com.

Abbreviations: GA, Genetic algorithm; **MEA,** membrane electrode assembly; **PEMFC,** proton exchange membrane fuel cell.

Table 1. Values of genetic algorithm parameters.

Variable	Value
Number of generations	150
Population size	25
Number of parameters	3
Crossover rate (%)	90
Mutation rate (%)	10

Table 2. Upper and lower values of the parameters to be optimized (search space).

Parameter	Lower limit	Upper limit
Cathode pressure	1 bar	5 bar
Anode pressure	1 bar	5 bar
Cell temperature	50°C	80°C

operating parameters.

Genetic algorithm descriptions

Genetic algorithm (GA) method creates artificial system according to the Darwinian mechanisms of evolution. This evaluation is based on high probability for powerful solutions and low probability for weak cases. The properties of this artificial system are similar to human genetic system and the base of GA is random search. Genetic algorithm is a parallel mathematical algorithm that transforms a set of population namely chains of chromosomes using genetic operations, in to a new population namely next generation based on fitness of each chromosome (Martin-del-Campo et al., 2004).The principle concepts of GA are simple and it can be described as follows:

1. First generation: By concern on the number of parameters that must be optimized, the first generation is created randomly at definite domain.
2. Evaluation of each solution: Each solution or chromosome is evaluated using fitness function.
3. Tournament selection: Each of both cases is selected randomly and according to the fitness function, the better one goes to the mating pool.
4. Crossover: For the purpose of searching for the best solutions at definite domain, each of both cases is combined with certain probability. At the present algorithm, it was used the real digits systems and the crossover are applied to following approach that R is random number between 0 and 1:

$$child_{1,2} = (\frac{parent_1 + parent_2}{2}) \pm | parent_1 + parent_2 | \times R$$

5. Mutation: To achieve the absolute maximum value and escape of the local maximum value and searching in far away locations, mutation is applied with certain probability. Following relation is used for this operator:

$$child = (\frac{parent + \frac{1}{n}\sum_{i=1}^{n} parent_i}{2})$$

A modified GA with elitist concept was applied to keeping the best part in each population and for not missing the best chromosomes. This program was written by manuscript file of MATLAB software that is very simple and user friendly. The parameters value of genetic algorithms used are given in Table 1. Table 2 gives the upper and lower limits of the optimized parameters. The flowchart of the genetic code used to get the optimal parameters is shown in Figure 1.

GA fitness function

The main task in the use of genetic algorithm is fitness function, that is, a gauge of chromosomes. Fitness function in the present model is maximum power density, in other words, the goal is to find the best value of anode and cathode pressure and temperature that have highest output power. For fitness valve of the chromosome, maybe more than a thousand cases must be calculated. Unfortunately, experimental method needs large numbers of experiments; this manner is very time-consuming and costly; to overcome this challenge, we used an analytical model of PEMFC to calculate fitness function of each chromosome.

PEMFC analytical modeling

Two one-dimensional models were used to simulate flow behavior in channels and MEA. These models solve the governing equations in the direction of channels to calculate concentration, velocity and pressure distribution and also the equations in the direction normal to the channel, to find fuel cell polarization curve. Governing equations are given subsequently.

Channel

Variations of velocity and concentration through the channel were assumed one dimensional and so for single phase flow, we have:

$$\frac{d}{dx}[\rho U] = -\frac{M\,N}{H} \qquad (1)$$

where U is average velocity in the channel and N is

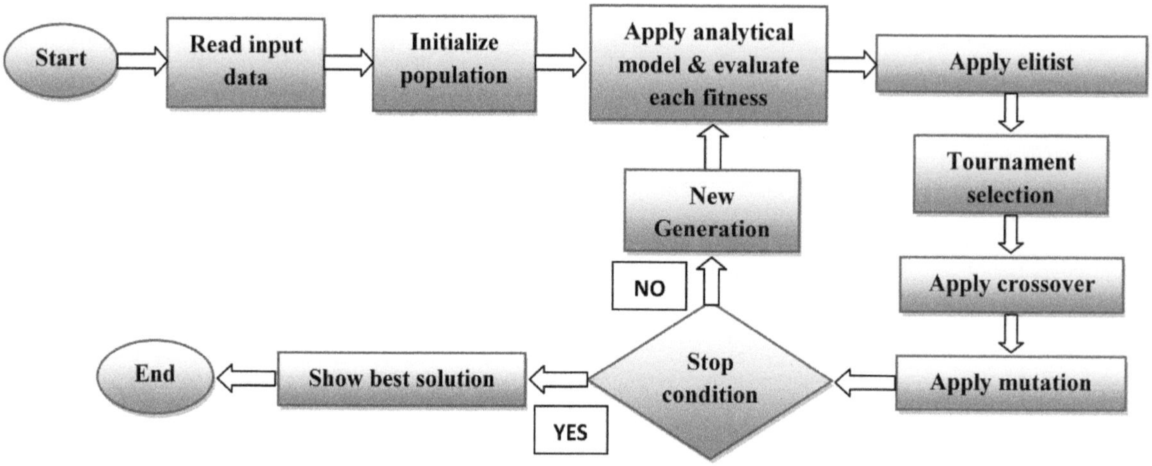

Figure 1. The genetic code to obtain the fuel cell optimal parameters.

molar flux between the channel and MEA. Also, H and ρ are channel's height and gas mixture's density, respectively. By the conservation of species, equation of single phase flow for each component is:

$$\frac{d}{dx}[U_i C_i^k] = -\frac{h_m^k (C_i^k - C_i^k|_{y=H_{ch}})}{H} \quad (2)$$

where C is molar concentration and k is an identifier for the type of species in each side (that is, hydrogen at anode and oxygen and water vapor at cathode side). The right hand side of Equation 2 shows the effect of diffusion at channel/backing layer interface. Coefficient h_m can be estimated (Wang and Wang, 2003) by:

$$h_m^k = Sh\frac{D_{eff}^k}{H} \quad (3)$$

Where H is channel's height and Sh is Sherwood number. Also, D_{eff}^k is effective diffusion coefficient of fluid from channel to backing layer.

PEM modeling

1-D model was used to simulate gases flow and concentration variations through PEM. So, assuming single phase flow, the continuity equation leads to:

$$d(\rho u) = S \quad (4)$$

S is rate of consumption or production in catalyst layers and equal to:

$$S_a = S_{H_2} + S_{H_2O} \text{ at the anode side} \quad (5)$$

$$S_c = S_{O_2} + S_{H_2O} \text{ at the cathode side} \quad (6)$$

Where rate of consumption of each species is:

$$S_{H_2} = -\frac{I}{2F} \quad (7)$$

$$S_{O_2} = -\frac{I}{4F} \quad (8)$$

$$S_{H_2O} = -\frac{\alpha I}{2F} \text{ at the anode side} \quad (9)$$

$$S_{H_2O} = -\frac{(1+\alpha)I}{2F} \text{ at the cathode side} \quad (10)$$

α is net water transport coefficient across the membrane. And so, the mass flow rate per unit area is:

$$\rho u = cte \quad (11)$$

Also, from Darcy's equation we have:

$$\nabla p = -\frac{\mu}{k}u \quad (12)$$

Which k and μ are permeability of cathode backing layer and viscosity of gas, respectively. Distribution of spices at cathode side (oxygen, water vapor and carbon dioxide) can be estimated by Maxwell's equation of distribution as:

$$\nabla x_i = \sum_j \frac{x_i N_j - x_j N_i}{C_T D_{ij}^{eff}} \quad (13)$$

Where x and N are the mole fraction and molar flow rate of each species, respectively, and D_{ij} is the diffusion

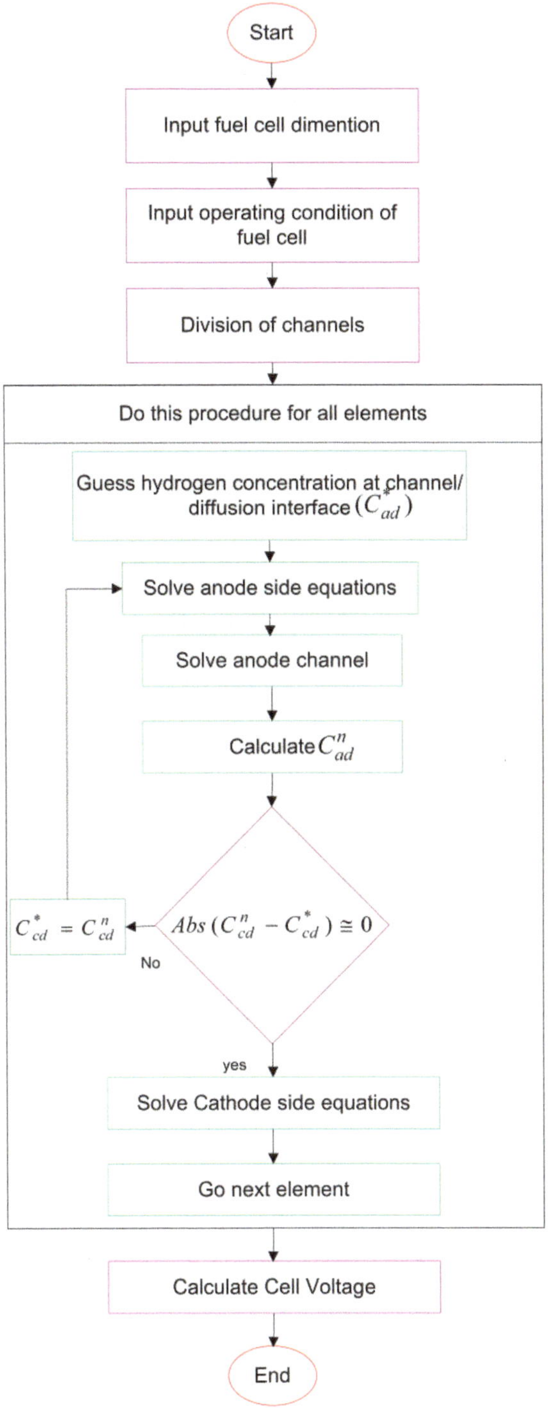

Figure 2. Solution algorithm.

coefficient.

In catalyst layers, transports of species in each side are given by:

$$\nabla(\rho v x_i) = -\nabla N_i + S_i \qquad (14)$$

i is H_2 or H_2O vapor at anode side and O_2 or H_2O vapor at cathode side.

Table 3. MEA properties.

Variable	Value
MEA thickness	2 mills
Electrode active area	5cm*5cm
MEA total area	8cm*8cm
Amount of catalyst in cathode side	0.4 mg Pt/cm^2
Amount of catalyst in anode side	0.4 mg Pt/cm^2
Type of gas diffusion layer	Carbon cloth

Current density, j, is related to H_2 and O_2 concentrations by simplified Butler-Volmer equation:

$$j_a = A_{av}\, j^{ref}_{0,a} \left(\frac{C_{H_2}}{C_{H_2,ref}}\right)^{1/2} \left(\frac{\alpha_a + \alpha_c}{RT} F\eta_a\right) \quad \text{at the anode side} \qquad (15)$$

$$j_c = A_{av}\, j^{ref}_{0,c} \left(\frac{C_{O_2}}{C_{O_2,ref}}\right) \exp\left(-\frac{\alpha_c}{RT} F\eta_c\right) \quad \text{at the cathode side} \qquad (16)$$

Where η and α are over-potential and charge transfer coefficient, respectively. A_{av} and j_o are active area and reference current density at each side. Cell voltage can be obtained as follows:

$$V = E_0 - \eta_a - \eta_c - E_{ohm} \qquad (17)$$

where E_0 is open circuit voltage and E_{ohm} is ohmic voltage losses, η_a and η_c are anode and cathode overpotential, respectively.

METHODOLOGY OF SOLUTION

For solving governing equations, first, channel must be divided into small parts and then by coupling PEM and channels equations for each element, polarization curve of fuel cell can be calculated. By following this process, current density and cell voltage can be estimated. Figure 2, gives an overview of the solution procedure.

Validation of modeling

For this purpose, an experimental set up was developed. Mono cell proton exchange membrane fuel cell with 5 × 5 cm active area was used to experiment validation of present analytic model. A standard MEA consisting of a Nafion N112 membrane in combination with platinum and gas diffusion layers made of carbon cloth was purchased from E-TEK and used. Total specifications of MEA are given in Table 3. Active area was obtained by grooving gas

Figure 3. PEMFC experimental set up

channels having 1 × 1 mm^2 channel's cross section and with 1 mm for channel rib. A schematic of PEMFC experimental set up for the experimental validation is shown in Figure 3. In Figures 4 and 5, the results of analytical model at various input pressure and temperature respectively, was compared by experimental result.

Constant primitive process parameters

The physical characterization and the constant values required for analytical model are extracted from fabrication company documents and are shown in Table 4. After validation of the presented analytical model by experimental work, this analytical model is ready to be used for calculating fitness function of genetic algorithm. Since algorithm. Since the better agreement between analytical model and experimental approach in case of the fuel cell with parallel channels, this type of channel was used for optimization in modeling approach. Input flow rate on both sides and the constant of geometrical parameters are shown in Table 5.

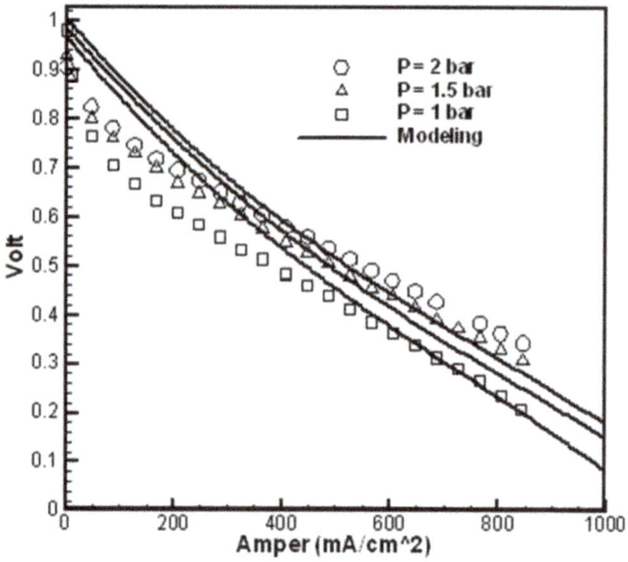

Figure 4. Modelling and experimental polarization curves at different anode and cathode equal pressure. Fuel cell temperature is 60°C.

RESULTS AND DISCUSSION

By executing the MATLAB optimization code, solutions were generated in maximum power densities that are

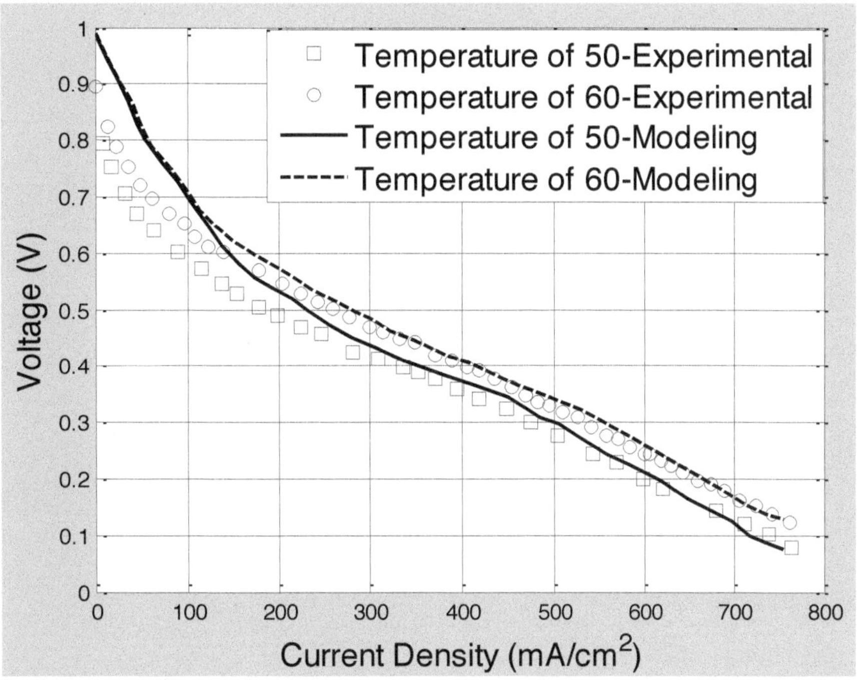

Figure 5. Modelling and experimental polarization curves at different temperature. Anode and cathode equal pressure is 1 bar.

Table 4. Parameters.

Condition and parameters in use	Value	Reference
Anode and Cathode relative humidity (%)	100	Assumed
Anode and cathode diffusion layer thickness(cm)	0.02	Sui et al. (2008)
Anode and cathode catalyst layer thickness(cm)	0.00287	Al-Baghdadi and Al-Janabi (2007)
Anode and cathode diffusion layer porosity	0.4	Al-Baghdadi and Al-Janabi (2007) and Cheddie and Munroe (2007)
Anode and cathode catalyst layer porosity	0.4	Al-Baghdadi and Al-Janabi (2007)
Anodic transfer coefficient (α_a)	0.5	Sui et al. (2008) and Al-Baghdadi and Al-Janabi (2007)
Cathodic transfer coefficient (α_c)	0.5	Sui et al. (2008)
Anode and cathode specific area(1/cm)	6	Assumed
Anode and cathode hydraulic permeability	1	Assumed
MEA thickness(cm)	0.023	Al-Baghdadi and Al-Janabi (2007)
MEA Ionic conductivity (s/cm)	4.25	Cheddie and Munroe (2007)
MEA porosity	0.1	Sui et al., 2008)

Table 5. Geometrical and flow rate parameters.

Parameter	Value
Number of channel	10
Number of cells in stack	1
Anode and cathode width of channel (cm)	0.1
Anode and cathode height of channel (cm)	0.1
Anode and cathode length of channel (cm)	5
Input flow rate in cathode side (kg/s)	1.5e -5
Input flow rate in anode side (kg/s)	1e -4

Table 6. Parameters value that produce maximum power.

Cell temperature (°C)	Anode pressure (bar)	Cathode pressure (bar)
74	5	5

shown in Table 6. Figure 6 illustrates the convergence of the genetic code towards the best solution. Figure 7 shows the maximum power density of analytical model in

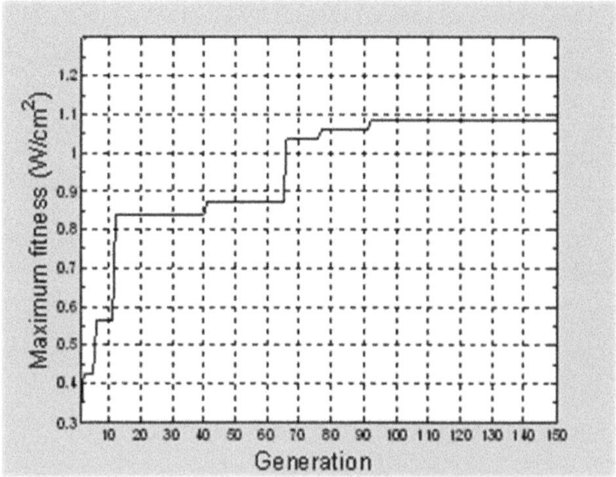

Figure 6. Maximum fitness during the generations.

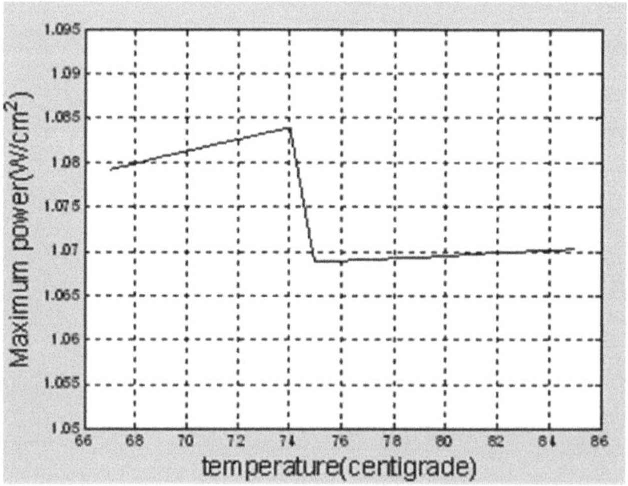

Figure 7. Maximum power in different temperature. The fuel cell pressure on both anode and cathode sides are 5 bar.

different cell temperature and constant cathode and anode side pressure.

When the temperature goes above 50°C, the fuel cell performance gets better, but when it goes above 74°C, it drops due to the reason that at temperatures higher than the humidifier tank, MEA loses its moisture dramatically as a result of cell body warmth, therefore, the proton's conductivity in MEA would be decreased. This results in the enhancement of the cell internal resistance and the drop in cell performance. The reasons behind cell's enhancement efficiency in regard to the increase of temperature from 50°C are as follows:

i. By the temperature increase, the exchange current density will increase; hence, there would be lesser amount of drops concerning cell's activation.

ii. In higher temperatures, the gas capacity for holding water would rise, on the other hand, in lower temperatures, the vapor alteration's possibility in gases in to the water drops exists.

iii. The rise in temperature, providing moisturized-environment in MEA, will result in the decrease of MEA resistance against proton's conductivity through it, consequently, this will lead to the increase of current density and MEA's proton conductivity.

iv. In accordance with the temperature increase, the hydrogen high permeability would be increased.

Maximum power density for different cathode and anode pressure are shown in Figures 8 and 9 respectively. Cell temperature is constant and equal to 74°C; moreover, another pressure side is constant and equal to 5 bar. In these two pictures, by increase of cathode and anode pressure, the cell performance increased monotonously, but by paying attention to these two figures, it is obvious that anode pressure is more

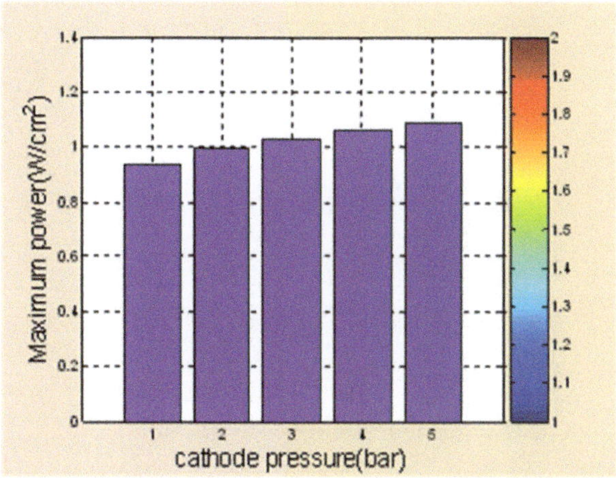

Figure 8. Maximum power at different cathode pressure (The fuel cell temperature is 74°C and anode pressure is 5 bar).

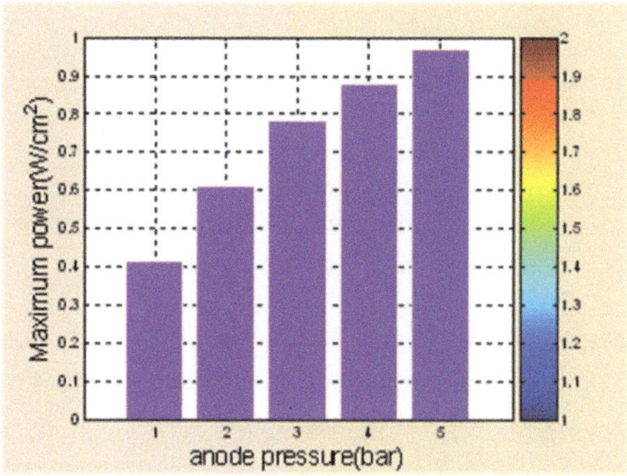

Figure 9. Maximum power at different anode pressure. The fuel cell temperature is 74°C and cathode pressure is 5 bar.

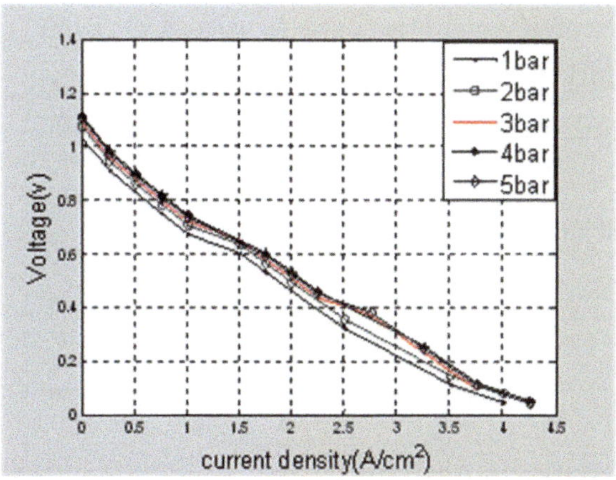

Figure 10. Polarization curves for different cathode pressure (the fuel cell temperature is 74°C and anode pressure is 5 bar).

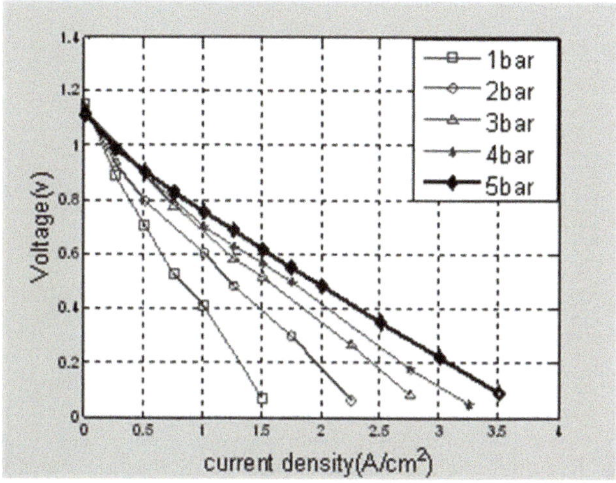

Figure 11. Polarization curves for different anode pressure (the fuel cell temperature is 74°C and cathode pressure is 5 bar).

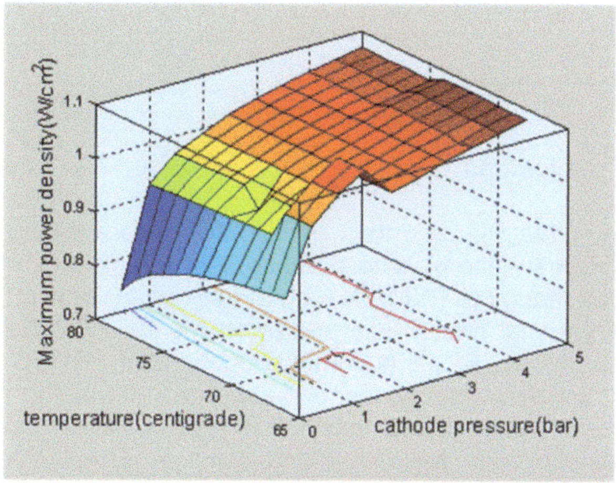

Figure 12. Maximum power value according to temperature and cathode pressure (anode pressure is 5 bar).

sensitive than cathode pressure and it is because of fuel consumption importance and this is the reason why the base reaction takes place at anode side.

Polarization curves for different cathode and anode pressure are shown in Figures 10 and 11 respectively. Cell temperature and another pressure side are constant.

Figures 12 and 13 show maximum powers according to the temperature and one pressure side that another pressure side is 5 bar. Unfortunately, present experimental set up can not suffer this loading condition test because the sealing system has no ability to pass this pressure condition and the heater has no ability to create this loading temperature, but expressed optimal result for maximum power density are in concordance with Kaytakoglu and Akyalcin (2007) who reported 75°C and 5 bar for anode and cathode pressure.

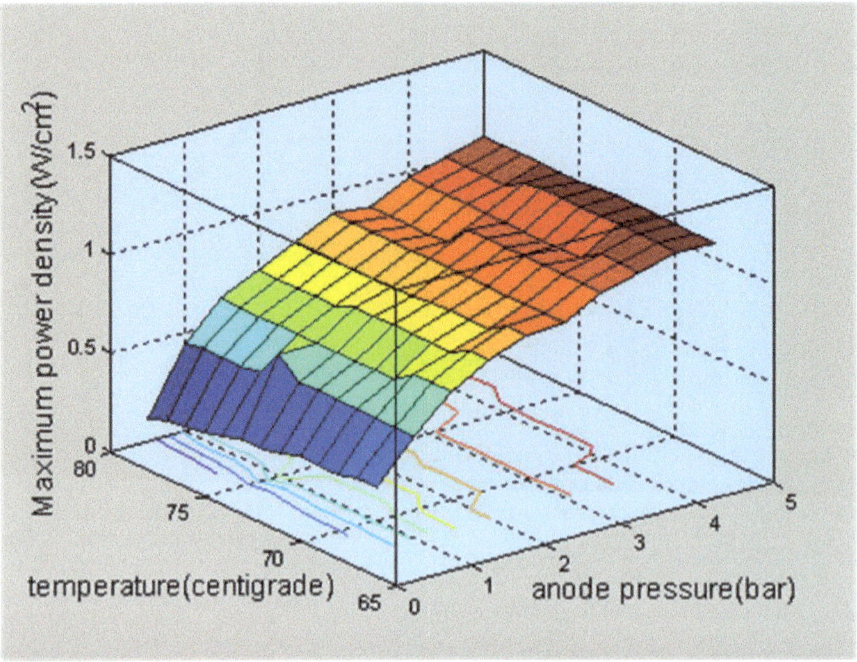

Figure 13. Maximum power value according to temperature and anode pressure (cathode pressure is 5 bar).

Conclusions

In this study, the genetic algorithm was applied to determine the parameters of process for maximum power of a mono cell PEMFC. A quasi two dimensional (1D-1D), isothermal model was presented for the PEM fuel cell. The model solves governing equations of PEMFC in the direction of the channel and a direction normal to it. The present model was used as fitness function for GA. A genetic code was developed using MATLAB. The best working conditions at maximum power density of a PEMFC were attained at 75°C and 5 bar for anode and cathode pressure. The presented values obtained with the presented model are valid and so close in comparison with experimental results reported in literatures.

Nomenclature: A, Area (cm^2); **C,** molar concentration ($mol\ cm^{-3}$); **D,** diffusion coefficient ($cm^2\ s^{-1}$); **E,** electromotive force (V); **F,** Faraday's constant (96,487 C mol^{-1}); **H,** channel's height (cm); **I,** current density ($A\ cm^{-2}$); **J,** current density ($A\ cm^{-2}$); J_o, reference current density ($A\ cm^{-2}$); **K,** permeability; **M,** molecular weight ($kg\ mol^{-1}$); **N,** molar flow rate ($kg\ cm^{-2}\ s^{-1}$); **P,** pressure (Pa); **R,** universal gas constant (8.314 $J\ mol^{-1}\ K^{-1}$); **S,** rate of consumption ($mol\ s^{-1} cm^{-2}$); **T,** temperature (K); **U,** channel's velocity ($cm\ s^{-1}$); **μ,** velocity in *y* direction ($cm\ s^{-1}$); **V,** cell voltage (V); **x,** channel's direction coordinate (cm), mole fraction; **y,** MEA normal direction coordinate (cm).

Greek symbols: **α,** Net water transport coefficient, charge transfer coefficient; **μ,** viscosity of gas ($kg\ m^{-1}s^{-1}$); **η,** overpotential (v); **ρ,** density ($kg\ m^{-3}$); **ν,** kinetic viscosity ($cm^2\ s^{-1}$).

Superscript: **eff,** Effective value; **k,** identifier; **ref,** reference value.

Subscript: a, Anode side; **ad,** anode diffusion; **av,** active; **c,** cathode side; **cd,** cathode diffusion; **ch,** channel; **eff,** effective value; H_2, hydrogen; H_2O, water; O_2, oxygen; **ohm,** Ohmic losses; **ref,** reference state; **T,** total.

REFERENCES

Al-Baghdadi M, Al-Janabi H (2007). Modeling optimizes PEM fuel cell performance using three-dimensional multi-phase computational fluid dynamics model. J. Energy Convers. Manage., 48: 3102-3119.

Cheddie DF, Munroe NDH (2007). A two-phase model of an intermediate temperature PEM fuel cell. J. Hydrogen Energy, 32: 832-841.

Chen HW, Hao KT, Hsiang LJ, Han LC, Wei SC, Hsun WC (2011). Influences of Gas Diffusion Layers with Pitch-based Carbon Coated in Polymer Electrolyte Membrane Fuel Cell. J. Electrochem. Sci., 6: 2192–2200.

Kaytakoglu S, Akyalcin L (2007). Optimization of parametric performance of a PEMFC. J. Hydrogen Energy, 32: 4418-4423.

Kumar N, Khare PS, Swarup A (2010). Optimization of Proton Exchange Membrane Fuel Cell at Different operating and design Variables Using Genetic Algorithm. J. Engine. Sci. Technol., 2: 6720-6730.
Li X, Yan Q, Yu D (2011). PEMFC Model Parameter Optimization Based on a Hybrid PSO Algorithm. J Comput. Inform. Syst., 7: 479-486.
Martin-del-Campo C, Francois JL, Avendano L, Gonzalez M (2004). Development of a BWR loading pattern design system based on modified genetic algorithms and knowledge, J. Annals of Nuclear Energy, 31: 1901-1911.
Sedighizadeh M, Rezazadeh A, Khoddam M, Zarean N (2010). Parameter optimization for a pemfc model with particle swarm optimization. J Eng. Appl. Sci., 3: 102-108.
Sui PC, Kumar S, Djilali N (2008). Advanced computational tools for PEM fuel cell design Part 1. Development and base case simulations. J. Power Sources, 180: 410-422.
Tamayol A, Bahrami M (2011). In-plane gas permeability of proton exchange membrane fuel cell gas diffusion layers. J. Power Sour, 196: 3559–3564.
Wahdame B, Candusso D, Kauffmann JM (2006). Study of gas pressure and flow rate influences on a 500W PEM fuel cell, thanks to the experimental design methodology. J. Power Sources, 156: 92-99.
Wang L, Husar A, Zhou T, Liu H (2003). A parametric study of PEM fuel cell Performances. J. Hydrogen Energy, 28: 1263-1272.
Wang ZH, Wang CY (2003). Mathematical Modeling of Liquid-Feed Direct Methanol Fuel Cells. J. Electrochem. Soc., 150: 508-519.
Xuan D, Li Z, Cheng T, Shen Y (2010). Optimization of PEM fuel cell system using dynamic model, J Appl. Mech. Mater., 26-28:1019-1026.

A prediction to the best artificial lift method selection on the basis of TOPSIS model

[1, 3, 5] Mehrdad Alemi, [3, 5]Hossein Jalalifar, [4]Gholamreza Kamali and [1, 2]Mansour Kalbasi

[1]Department of Petroleum Engineering, Amirkabir University of Technology, Tehran, Iran.
[2]Department of Chemical Engineering, Amirkabir University of Technology, Tehran, Iran.
[3]Department of Petroleum Engineering, Shahid Bahonar University of Kerman, Iran.
[4]Department of Mining Engineering, Shahid Bahonar University of Kerman, Iran.
[5]Environmental and Energy Research Center, Shahid Bahonar University of Kerman, Iran.

Artificial lift as a system adding energy to the fluid column in a wellbore to initiate and enhance production from the well is necessary when reservoir drives do not sustain acceptable rates or cause fluids to flow at all in some cases which use a range of operating principles, including pumping and gas lifting. Technique for order preference by similarity to ideal solution (TOPSIS) model or method is one of the most prevalent multi criteria decision making methods to solve problems involving selection from among a finite number of criteria and specify the attribute information in order to arrive at a choice. In this paper, a novel software method on the basis of technique for order preference by similarity to ideal solution model has been enabled to present the best artificial lift method selection for different circumstances of oil fields.

Key words: Artificial lift, technique for order preference by similarity to ideal solution, multi criteria decision making.

INTRODUCTION

The system that adds energy to the fluid column in a wellbore to initiate or enhance production from the well is called an Artificial Lift. When a reservoir lacks sufficient energy for oil, gas and water to flow from wells at desired rates, supplemental production methods can help. Lift processes transfer energy down hole or decrease fluid density in wellbore to reduce the hydrostatic load on formation. Major types of artificial Lift are Gas Lift (GL) design (Continuous gas lift, intermittent gas lift) and pumping (electrical submersible pump (ESP), progressive cavity pump (PCP), sucker rod pump (SRP), hydraulic jet type pump (HP).

As the well is produced, the potential energy is converted to kinetic energy associated with the fluid movement. This dissipates the potential energy of the reservoir, thereby causing the flow rate to decrease and the flow to eventually cease. It may be economical at any point in the life of a well to maintain or even to increase the production rate by the use of artificial lift to offset the dissipation of reservoir energy.

MCDM refers to making decisions in the presence of multiple, usually conflicting criteria. The problems of MCDM can be broadly classified into two categories: multiple attribute decision making (MADM) and multiple objective decision making (MODM), depending on whether the problem is a selection problem or a design problem. MODM methods have decision variable values that are determined in a continuous or integer domain, with either an infinitive or a large number of choices, the best of which should satisfy the decision maker's constraints and preference priorities. MADM methods, on the other hand, are generally discrete, with a limited number of predetermined alternatives.

By now, the usage of each of the Artificial Lift methods throughout of the world has been in a manner that for GL, ESP, SRP, PCP, HP as different Artificial Lift methods has been equal to 50, 30, 17, > 2 and < 2% respectively.

The most studies in this field have been on the basis of only experiential calculations by now and not based on the scientific MCDM methods, despite its great importance which implies one of the artificial lift selection

*Corresponding author. E-mail: jalalyfar@yahoo.com.

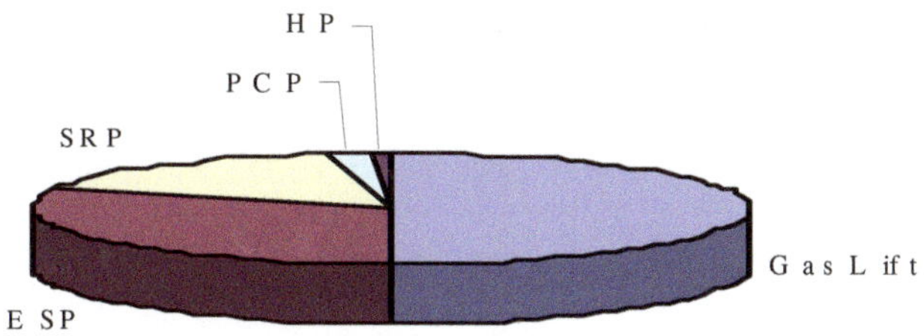

Figure 1. The recent usage of Artificial Lift methods throughout of the world.

previous procedures major imperfections.

However, about the previous Artificial Lift selection procedures, it can be said that some researchers have studied on this matter briefly expressed as the following:

Neely et al. (1981), considered the geographical and environmental circumstances as the dominant factors for artificial lift selection.

Valentine et al. (1988), used optimal pumping unit search (OPUS), a smart integrated system possessing the characteristics of artificial lift methods, for artificial lift selection.

Clegg et al. (1993), studied on some of the operational and designing factors based on artificial lift methods overall capability comparison and design.

Espin et al. (1994), used SEDLA, a computer program possessing the characteristics of artificial lift methods, for artificial lift selection.

Heinze et al.(1995), used "the decision tree" for artificial lift selection, mostly based on a longtime economic analysis.

The paper objective is to specify technique for order preference by similarity to ideal solution (TOPSIS) model as a predicted method for Artificial Lift selection for different circumstances of oil fields.

MATERIALS AND METHODS

The usage of artificial lift methods throughout the world by now has been recently reported (Figure 1), (Weatherford Com.).

It is necessary to mention that sucker rod pump (SRP) is a positive displacement pump that compresses liquid by the motion of a piston. The piston is actuated by a string of sucker rods that extends from the bottomhole pump to the pumping unit at the surface. The rod or structure may limit rate at depth (Gholinegad, 2007).

A progressive cavity pump (PCP) is a kind of pump which transfers fluid by means of a sequence of small, fixed shape, discrete cavities, that move through the pump as its rotor turns.

The cavities taper down toward their ends and overlap with their neighbors, so that, in general, no flow pulsing is caused by the arrival of cavities at the outlet (Gholinegad, 2007).

An electrical submersible pump (ESP) is a dynamic displacement, multistage centrifugal turbine pump coupled by a short shaft to a downhole electrical motor. The motor is supplied with electrical power by a cable extending to the surface. ESP systems have a wide range of applications and offer an efficient and economical lift method. Even if sand production, high gas oil ratio (GOR) and viscosity are concerns, we can find the right ESP for our well and improve production. From onshore high water cut applications to complex offshore, deepwater or subsea applications, we have a system to meet our needs (Gholinegad, 2007).

A hydraulic jet type pump (HP) is an ejector type dynamic displacement pump operated by a stream of high pressure power fluid converging into a jet in the nozzle of the pump.

Downstream from the nozzle, the high velocity, low pressure jet is mixed with the well's fluid. The stream of the mixture is then expanded in a diffuser and as the flow velocity is dropped the pressure is built up. The fluid flow can carry some corrosive additives into wellbore and function as a maintenance material. The constraints to use HP are related to high GOR or contamination in the fluid flow bringing about low efficiency of pump at last (Gholinegad, 2007).

As well, about gas lift (GL), gas is injected into the tubing string to lighten the fluid column and allow the well to flow when it does not flow naturally.

The injected gas is mixed with produced fluid, decreases the flowing gradient in the production string and thus lowers the bottomhole flowing pressure. Basic objective of gas lift design is to equip wells in such a manner as to compress a minimum amount of gas to produce a maximum amount of oil (Gholinegad, 2007).

Previous artificial lift selection procedures

Neely et al. (1981) designated some Artificial Lift methods such as: SRP, ESP, HP, GL and studied about the application circumstances, advantages, disadvantages and constraints of each method. The geographical and environmental circumstances as the dominant factors for Artificial Lift selection and also some other subordinate factors such as: reservoir pressure, productivity index, reservoir fluid properties and inflow performance relationship were considered by him (Neely et al., 1981).

Valentine et al. (1988) used optimal pumping unit search (OPUS) for Artificial Lift selection. Indeed OPUS was a smart integrated system possessing the characteristics of artificial lift methods.

OPUS had the capability to control the technical and financial aspects of Artificial Lift methods. It can be said that the production system was consisted of the downhole pump up to the surface facilities (stock tank). The technical and financial evaluation of this procedure was done by means of some specific computer algorithms. Therefore, knowing the primary required investment value, costs (maintenance, equipment) and technical ability of each Artificial Lift method, Artificial Lift selection was done (Valentine et al., 1988).

Clegg (1988) mentioned some economic factors such as: revenue, operational and investment costs as the basis for Artificial Lift selection. He believed that the selected Artificial Lift method could have the best production rate with the least value of operational costs (Clegg, 1988).

Clegg et al. (1993), studied on some of the operational and designing characteristics of Artificial Lift methods categorized into 3 types based on Artificial Lift methods overall capability comparison and design, some specific operational factors and Artificial Lift methods factors probably causing some specific problems respectively (Clegg et al., 1993).

Espin et al. (1994) used SEDLA for artificial lift selection. Indeed SEDLA was a computer program possessing the characteristics of artificial lift methods. It was composed of 3 modules based on an information bank of human activities, the theoretical knowledge of artificial lift methods and the economic evaluation of artificial lift methods respectively. Therefore, the artificial lift selection was done on the basis of profit value (Espin et al., 1994).

Heinze et al. (1995), used "the decision tree" for artificial lift selection. The most major factor in it was based on a longtime economic analysis. Also, the artificial lift methods evaluation was based on operational costs, primary investment, lifetime cost and energy efficiency. Ultimately, considering these factors besides the decision maker, the Artificial Lift selection was done (Heinze et al., 1995).

Some engineering applications of TOPSIS model used up to now

Application of TOPSIS model as a data classifier, the proposed model could provide additional efficient tool for comparative analysis of data sets. TOPSIS model has been applied in Multiple Criteria Decision Analysis based on Wu's data mining model. It has been applied in supply chain complexity evaluation and simulation has been used to validate the proposed model (Jiang et al., 2007).

Application of TOPSIS model as a new model for mining method selection of mineral deposit based on fuzzy decision making, the fuzzy decision making (FDM) software tool has been employed to develop a fuzzy TOPSIS based model. Application of this model with various values (crisp, linguistic and fuzzy) of the deposit eliminated the existing disadvantages of other methods (Samimi, 2008).

Application of TOPSIS model in initial training aircraft evaluation under a fuzzy environment, the study has applied the fuzzy MCDM method to determine the weights of evaluation criteria and to synthesize the ratings of candidate aircraft. Aggregated the evaluators' attitude toward preference; then TOPSIS has been employed to obtain a crisp overall performance value for each alternative to make a final decision (Wang and Chang, 2007).

Application of TOPSIS model as a Multi Criteria Decision Analysis of alternative fuel buses for public transportation, the result has shown that the hybrid electric bus has been the most suitable substitute bus for Taiwan urban areas in the short and median term. But, if the cruising distance of the electric bus extends to an acceptable range, the pure electric bus could be the best alternative (Tzeng et al., 2005).

Some other certain scientific programs based on MCDM models are listed as below, but because TOPSIS model has been validated with several certain oil fields Artificial Lift selection operations results and a considerable accordance between their final results has been gained, this model has been chosen for Artificial Lift selection. As well, this model gives an appropriate solution that is not only the closest to the best alternative, but also the farthest from the worst alternative:

- TOPSIS (technique for order preference by similarity to ideal solution) model
- SAW (simple additive weighting) model
- ELECTRE (elimination et choice in translating to reality) model
- WPM (weighted product model)
- VIKOR (VIsekriterijumsko KOmpromisno Rangiranje), compromise ranking model.

DISCUSSION

In this paper, a software method (with Visual Basic.net code) based on TOPSIS model has been presented for Artificial Lift selection in oil industry that has been essential to mention to the mathematical and logical strategy of this method.

Technique for order preference by similarity to ideal solution (TOPSIS) model

This model was developed by Hwang and Yoon (1981). This model is based on the concept that the chosen alternative should have the shortest Euclidean distance from the ideal solution and the farthest from the negative ideal solution (Hwang and Yoon 1981; Pimerol et al., 2000).

The ideal solution is a hypothetical solution for which all alternatives relative to criteria attribute values (V_{ij}) correspond to the maximum attribute values in the database comprising the satisfying solutions; the negative ideal solution is the hypothetical solution for which alternatives relative to criteria attribute values (V_{ij}) correspond to the minimum attribute values in the database. TOPSIS thus gives a solution that is not only closest to the hypothetically best, that is also the farthest from the hypothetically worst alternative (Rao, 2007). The main procedure of TOPSIS model for the selection of the best alternative from among those available has been described thus:

At first it was required to allocate suitable quantities scaled from 0 through 10 for the alternative relative to the criteria qualities, (higher each of their qualities, more its value out of 10), (Rao, 2007). The relative scores of different methods relative to Production, Reservoir and Well constraints as well as Produced fluid properties and Surface infrastructure constraints, (Figure 2), have been based on Schlumberger Company certain practical reports. The value of 1 (good to excellent) has been considered as 7 out of 10, the value of 2 (fair to good) has been considered as 5 out of 10 and the value of 3 (not recommended and poor) has been considered as 3 out of 10 in the following (Schlumberger Com.). On the whole, it is believed that the calculations results and the related figures shown as (Figure 3) through (Figure 6) vary in different circumstances of oil fields.

But here, (Figure 3) through (Figure 6) have been related to the condition that ESP application has been the best choice as the shown result in (Figure 3).

In the matrix the number of the alternatives and the number of the criteria have been considered as the number of matrix rows and matrix columns respectively.

Then, the normalizing of the resulted alternatives relative to the criteria quantities matrix had to be done (Rao, 2007).

This resulted normalized matrix had to be weighted by means of a specific weights calculating mathematical method such as Entropy method, (Rao, 2007). Multiplying the normalized matrix by the alternatives relative to the criteria resulted weights diametrical matrix, the normalized weighted matrix has been obtained (Rao, 2007).

Then, the positive ideal (best) and the negative ideal (worst) of each criterion had to be obtained in this step, (Rao, 2007).

$$V^{+} = \{ \overset{\max}{\underset{i}{\Sigma}} V_{ij} \,/\, j \in J), (\overset{\min}{\underset{i}{\Sigma}} V_{ij} \,/\, j \in J') / i = 1,2,...N \},$$
$$= \{V_1^{+}, V_2^{+}, V_3^{+}, ... V_M^{+}\} \quad (1)$$

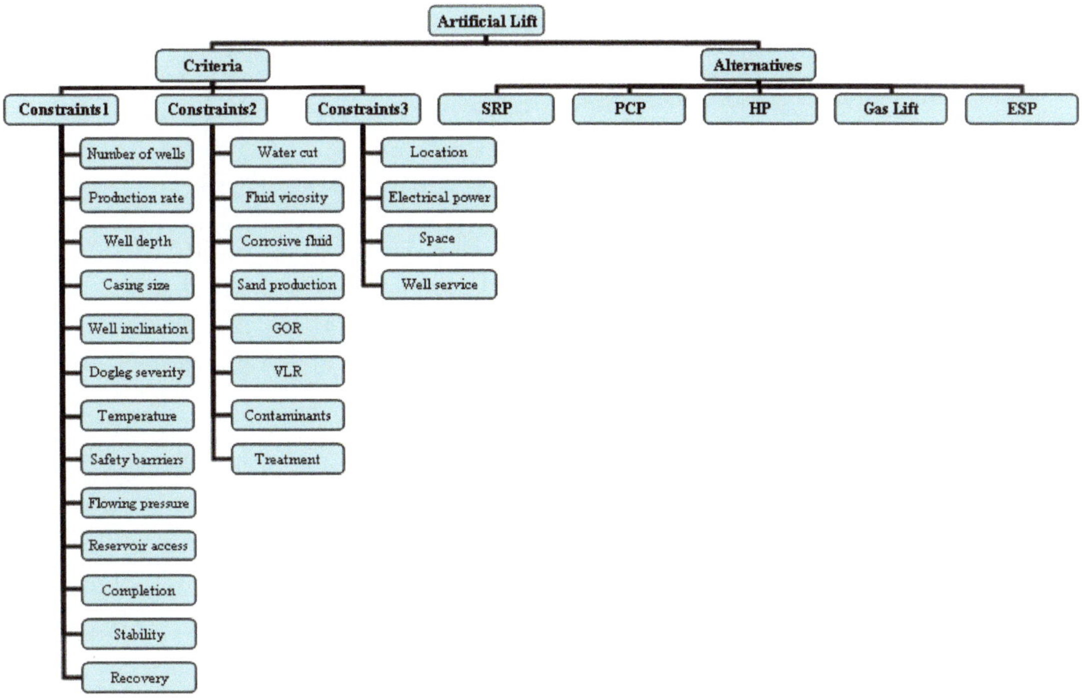

Figure 2. The Alternatives versus the Criteria for Artificial Lift selection

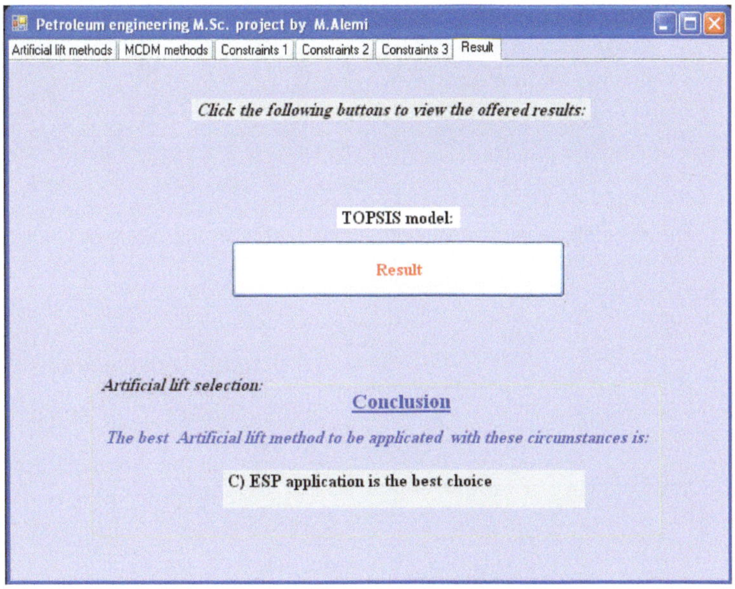

Figure 3. Artificial Lift selection result by the designed computer program (The obtained final result due to the sample input data).

$$V^- = \{ \sum_i^{\min} V_{ij} / j \in J), (\sum_i^{\max} V_{ij} / j \in J') / i = 1,2,...N\},$$
$$= \{V_1^-, V_2^-, V_3^-, ...V_M^-\} \quad (2)$$

Where; J = (j =1, 2, M)/j is associated with beneficial attributes. J'= (j =1, 2, M)/j is associated with non-beneficial attributes.

The separation of each alternative from the ideal one has been given by the Euclidean distance in the following equations (Figure 4 and 5), (Rao, 2007).

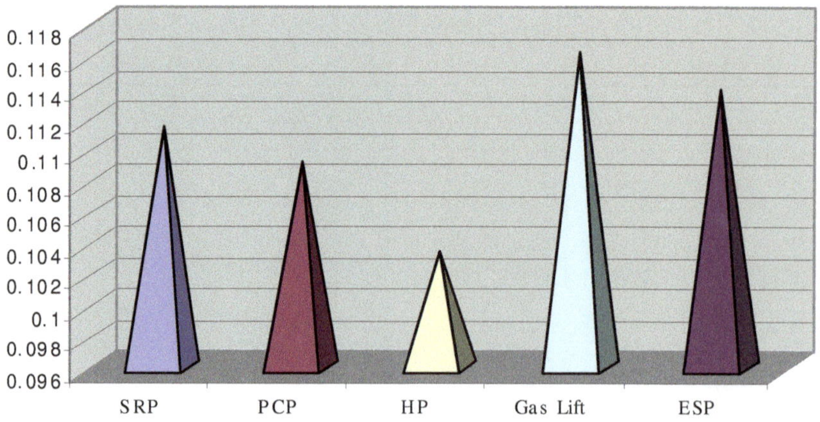

Figure 4. The resulted separation of each alternative from the positive ideal given by the Euclidean distance (obtained from the sample input data).

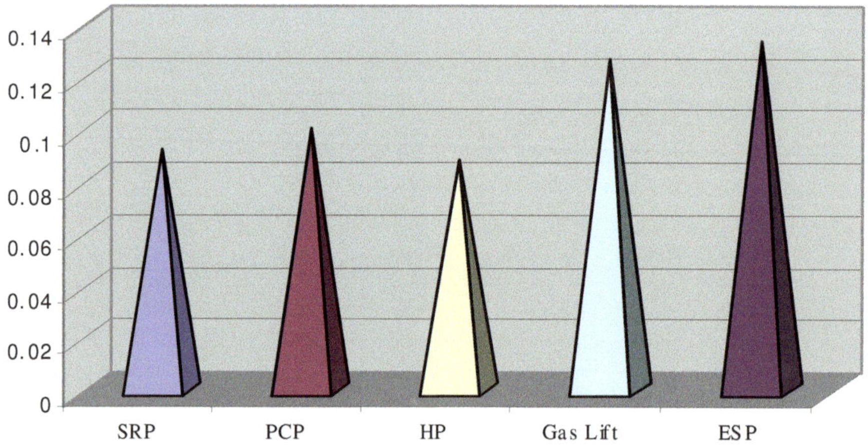

Figure 5. The resulted separation of each alternative from the negative ideal given by the Euclidean distance (obtained from the sample input data).

$$S_i^+ = \{\sum_{J=1}^{M} (V_{ij} - V_j^+)^2\}^{0.5},$$

$$S_i^- = \{\sum_{J=1}^{M} (V_{ij} - V_j^-)^2\}^{0.5}, \qquad (3)$$

$$S_i^+ = \{\sum_{J=1}^{M} (V_{ij} - V_j^+)^2\}^{0.5},$$

$$S_i^- = \{\sum_{J=1}^{M} (V_{ij} - V_j^-)^2\}^{0.5}, \qquad (4)$$

$i = 1, 2, ..., N$

The relative closeness of a particular alternative to the ideal solution, P_i, could be expressed in this step as follows:

$$P_i = S_i^- / (S_i^+ + S_i^-) \qquad (5)$$

The higher value for P_i shows the better choice to be selected (Figure 6), (Rao, 2007). The (Table 1) shows the sample input data too.

Conclusion

In this paper, a novel software method based on TOPSIS model was presented for artificial lift selection in different circumstances of oil fields. The validity of this scientific designed program has been checked and validated with several certain oil fields artificial lift operations results by

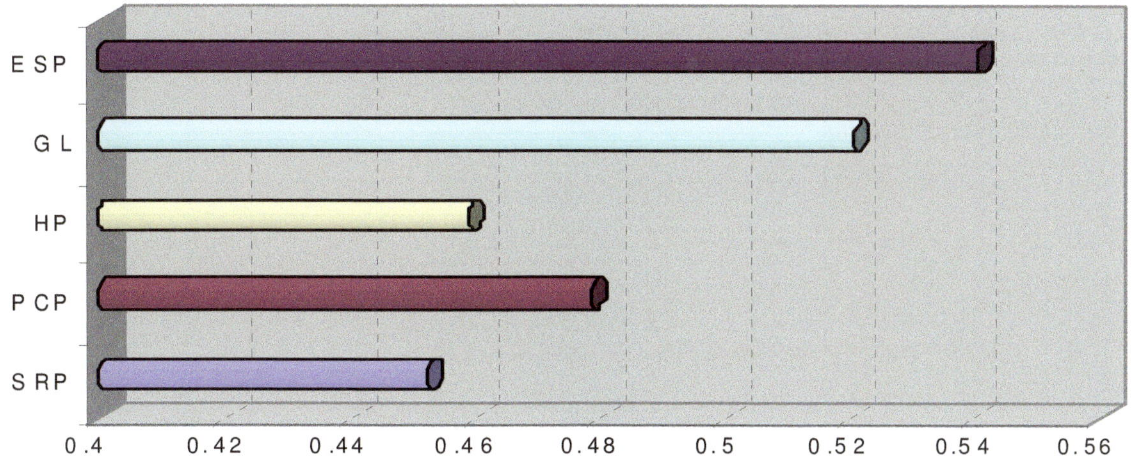

Figure 6. Artificial lift selection result (for the sample input data).

Table 1. The designed software sample input data

Condition	Oil field
Production, reservoir and well constraints	
Number of wells	50
Production rate	56000 STB
Well depth	8000-11000 ft
Casing size	9 5/8"
Well inclination	All of cases
Dogleg severity	0-10 per feet
Temperature	180-210 F
Safety barriers	1
Flowing pressure	Greater than 1000 psi
Reservoir access	Required
Completion	Simple
Stability	variable
Recovery	Secondary waterflood
Produced fluid properties	
Water cut	70%
Fluid viscosity	Less than 100 cp
Corrosive fluid	No
Sand and abrasives	Less than 10 ppm
GOR	650
VLR	Less than 0.1
Contaminants	Scale
Treatment	Acid
Surface infrastructure	
Location	Offshore
Electrical power	Utility
Space restriction	Yes
Well service	Pulling unit

means of the comparison between them. Finally it has resulted that between this designed programs final results and the related oil fields Artificial Lift operations results, a considerable accordance has been available. Indeed, the paper objective has been a prediction study for Artificial Lift selection (with Visual Basic.net code) in oil industry.

ACKNOWLEDGEMENTS

It is considerable to appreciate and thank to the Petroleum Engineering and Development Company (PEDEC Research and Development) and the Iranian Offshore Oil Company (IOOC Research and Development) Tehran, Iran backing up to study on this scientific matter in oil industry.

REFERENCES

Clegg JD (1988). High-rate Artificial Lift. J. Petroleum Technol. SPE#17638.

Clegg JD, Bucaram SM, Heln J (1993). Recommendations and comparisons for selection of Artificial Lift methods. J. Petroleum Technol. 45(12): 1128 -1167.

Espin DA, Gasbarri S, Chacin JE, Intevep SA (1994). Expert system for selection of optimum Artificial Lift method. Argentina, SPE#26967.

Gholinegad J (2007). Artificial Lift. Sharif University of Technology Workshop, Tehran, Iran.

Heinze LR, Herald W, Lea JF (1995). Decision Tree for selection of Artificial Lift method. Oklahoma, SPE#26510.

Hwang CL, Yoon K (1981). Multiple Attribute Decision Making: a state of the art survey. Springer-Verlog.

Jiang W, Zhong X, Chen K, Zhang S (2007). TOPSIS model as a data classifier. Fourth International Conference on Fuzzy Systems and Knowledge Discovery.

Neely B, Gipson F, Capps B, Clegg J, Wilson P (1981). Selection of Artificial Lift method. Dallas, Texas, SPE#10337.

Pimerol JC, Romero SB (2000). Multi Criteria Decision in management: Principles and Practice, Kluwer Academic Publishers.

Rao RV (2007). Decision making in the manufacturing environment: Using graph theory and fuzzy Multiple Attribute Decision Making methods. Springer-Verlog.

Samimi NF, Shahriar K, Ataeepour M, Dehghani H (2008). TOPSIS model as a new model for mining method selection of mineral deposit based on Fuzzy Decision Making. J. S. Afr. Inst. Min . Metall. 108:

381-391.
Tzeng GH, Lin CW, Opricovic S (2005). TOPSIS model as a Multi Criteria Decision Analysis of alternative fuel buses for public transportation. Energy Policy 33: 1373–1383.
Valentine EP, Hoffman FC, Francais du petrole (1988). OPUS: An expert adviser for Artificial Lift, SPE#18184.
Wang TC, Chang TH (2007). TOPSIS model in initial training aircraft evaluation under a fuzzy environment. Int. J. Expert Syst. Appl. 33: 870-880.

Effect of inertia in the study of two-phase oil-gas flow in horizontal pipelines

A. C. Pacheco[1], A. F. Souza[1], D. P. Costa[2], D. A. G. Aranda[1] and L. D. T. Câmara[2]*

[1]Departamento de Engenharia Química, Escola de Química da Universidade Federal do Rio de Janeiro - UFRJ, Rio de Janeiro –RJ, Brazil.

[2]Departamento de Engenharia Mecânica e Energia -DEMEC, Instituto Politécnico da Universidade do Estado do Rio de Janeiro - IPRJ-UERJ, Nova Friburgo-RJ, Brazil.

The flow of oil and gas in the production pipelines is a crucial step for the oil industry, which always seeks to maximize the production of these valuable hydrocarbons. The modeling of the flow patterns allows a better analysis of the behavior of the mixture. In this work, the mechanical energy balance is applied to represent the multiphase oil-gas flow being evaluated; the contribution of the terms of kinetic energy and friction pressure drop along the pipe. The simulation results of pressure drop profiles showed that the kinetic effects due to the inertia of the gas are equivalent to those from compressible flow by gas expansion. The friction can also be significant to the overall result of the pressure drop that will depend on the flow pattern and contact conditions of the gas phase with the pipe wall.

Key words: Oil-gas flow, horizontal flow, kinetic effect, friction effect.

INTRODUCTION

The analysis and study of flow patterns are of paramount importance for the understanding and prediction of the kind of mixture of oil seeping into pipes. There are many scientific studies to better understand the mechanisms that determine the patterns of multiphase flow in oil pipelines (Hapanowicz, 2008; Ruder and Hanratty, 1990; Hewitt et al., 1990; Guang-yao et al., 2007) since they are fundamental in calculating a series of engineering properties related to the oil-gas flow from the production areas to the refining units. Currently, there is interest in the directional/horizontal drilling for achieving higher recovery factors in horizontal reservoirs due to the larger contact area with the production well (Pinto et al., 2003; Mohiuddin et al., 2007). Thus, the study of the behavior of oil and their mixtures through the horizontal multiphase flow in pipelines is necessary in the quest for better routines for calculating the properties concerned.

In multiphase flow of oil, we can find different phase configurations in which the oil-gas is a typical flow pattern in pipelines. In the literature, different configurations of phase were studied each one with its own purpose (Spedding et al., 2005; Guang-Yao, 2007; Jepson and Taylor, 1993). It was observed that the greater the number of phases present, the greater the complexity in the analysis and consequently the modeling. The modeling of multiphase flow in horizontal pipelines has been a constant in the search for models and mechanisms that lead to better understanding of the phenomena related to the dominant flow patterns. In the study of Guang-Yao et al. (2007), both the experimental and modeling study was performed in the horizontal flow of two phase oil-gas. The authors found a critical surface velocity of the liquid phase in the transition from stratified to the slug pattern. Numerical simulations were also performed on the basis of experimental data with the application of the volume of fluid technique (VOF). In this technique, a set of equations is applied to represent the two phases being utilized as an average of properties instead of using separate conservation equations for both phases. Garcia et al. (2007) developed correlations for calculating friction factor in two phase horizontal flow, taking into account the effect of relative velocity of

*Corresponding author. E-mail: dcamara@iprj.uerj.br.

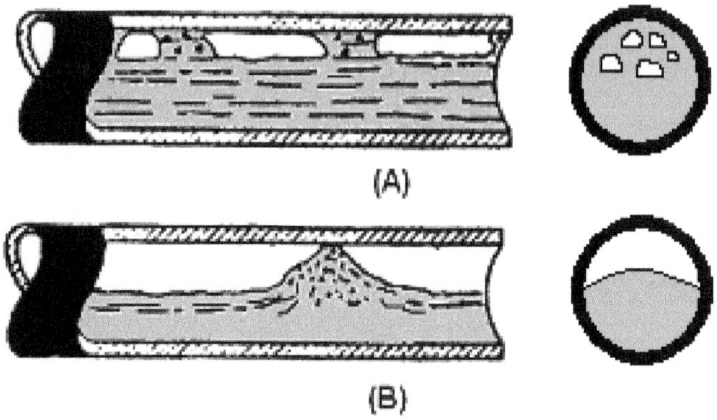

Figure 1. Examples of horizontal flow patterns: (A) bubbles dispersed (B) segregated.

phases. These models were based on data from 2560 experiments of gas-liquid flow in horizontal pipelines.

In this work the horizontal two-phase oil-gas flow is modeled by a mechanical energy balance that assumes independent contributions of the terms of oil and gas phase acceleration in the global balance of energy. The contribution of kinetic and friction terms is analyzed by profiles of pressure drop in wells subject to compressible and uncompressible flows.

MODELING

The modeling of the horizontal flow is assumed as one-dimensional variation of the properties along the length L with the gravitational term neglected. Equation 1 shows the balance of mechanical energy used in which the first term represents the variation of pressure (P) with the length (L) of the pipe. The second and the third terms represent respectively the friction loss, which is related to the friction factor (f), fluid density (ρ), velocity (v) and the pipe diameter (D); and the acceleration, which is related to the fluid velocity and density and the variation of velocity with the pipe length.

$$\frac{dP}{dL} = -f\frac{\rho v^2}{2D} + \rho v\frac{dv}{dL} \tag{1}$$

It is assumed that the gas can increase your speed linearly along the pile length through a type expression.

$$v = v_0 + a.L \tag{2}$$

Where v_o and a correspond to the initial velocity and the angular parameter, respectively. Thus, the increase in speed means that we have an increase in the holdup of the liquid (H_L) due to reduction of velocity of the liquid phase with the consequent increase in sectional area of tube occupied by liquid. In the calculations it is considered a constant volumetric flow of oil and gas along the pipe length.

In calculating the friction factor the model of Drew et al. (1930) is considered.

$$f = 0.0056 + 0.5.\mathrm{Re}^{-0.32} \tag{3}$$

Where the Reynolds number is calculated through the density, velocity and viscosity of the mixture. The latter is calculated using the additive contributions of the liquid and gas holdup. In the acceleration term of the balance of mechanical energy (equation 1) there is the independent contribution of the gas and oil flow.

Figure 1 shows two examples of horizontal flow patterns. According to the type of pattern we can have different distributions of liquid and gas in contact with the pipe wall which leads to different values of energy loss from friction. In the case B the gas promotes a greater contact with the pipe wall than in the case A, since the gas is distributed over the net in the latter. The case A is more related to a pattern of bubbles dispersed in which an increase in the holdup of the liquid does not necessarily lead to an increase in liquid contact with the wall. The case B is be more related to a segregated pattern in which an increase of liquid holdup leads to an increase in liquid contact with the wall.

RESULTS AND DISCUSSION

In the calculations of pressure drop from mechanical energy balance (equation 1), a pipeline discretization of 10 (ten) control volumes along the total length is assumed. In the previous calculations, the speed variations of the gas and liquid phase were not considered. Figure 2 presents the results of pressure drop versus volumetric flow increasing the flow rate of gas-oil mixture. As expected, the increase of the gas-oil mixture flow leads to an increase in pressure drop, as can be seen from this result of available pressure curve. Also, Figure 2 shows that the lowest pressure drop corresponds to the case of larger fraction of the area occupied by gas (Case of lowest gas Holdup which is equal to 0.01). The case of intermediate holdup (H_L = 0.5) led to an intermediate pressure profile and that with higher fraction of liquid (H_L = 0.99) to a pressure drop which is equivalent to those typical experimental field results presented by Oliveira (2002). In the results of Figure 2, the following properties were used:

ρ_{gas} = 2 kg/m^3, ρ_{liq} = 900 kg/m^3, μ_{gas} = 0.02 cP, μ_{Liq} = 7 cP, D = 0.1 m and L = 1000 m

Figure 3 presents the results of the behavior of pressure drop with the pipe length. The simulation results in blue were compared to the experimental data of Oliveira (2002) in red showing that it is not necessary to increase the gas velocity to represent well the experimental points (no slip case). As can be seen, there is a good fit between the experimental data and the simulation showing a linear pressure profile which is equivalent to the data presented by Oliveira (2002). These results are expected as there is no change in the energy term of acceleration due to the constant holdup of liquid which keeps the friction loss of the mechanical energy balance constant. Therefore the pressure drop along the pipe is

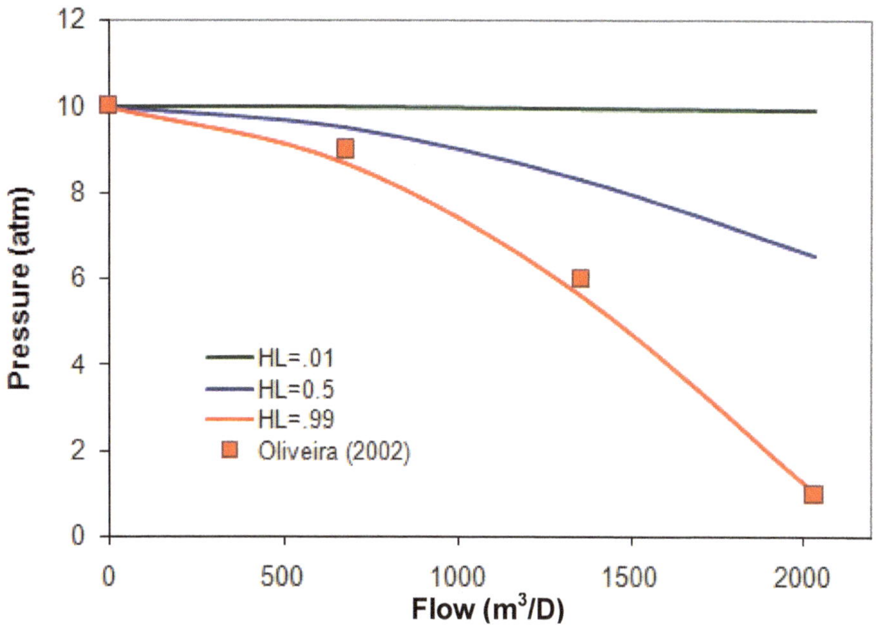

Figure 2. Comparison with field experimental data of available pressure from Oliveira (2002).

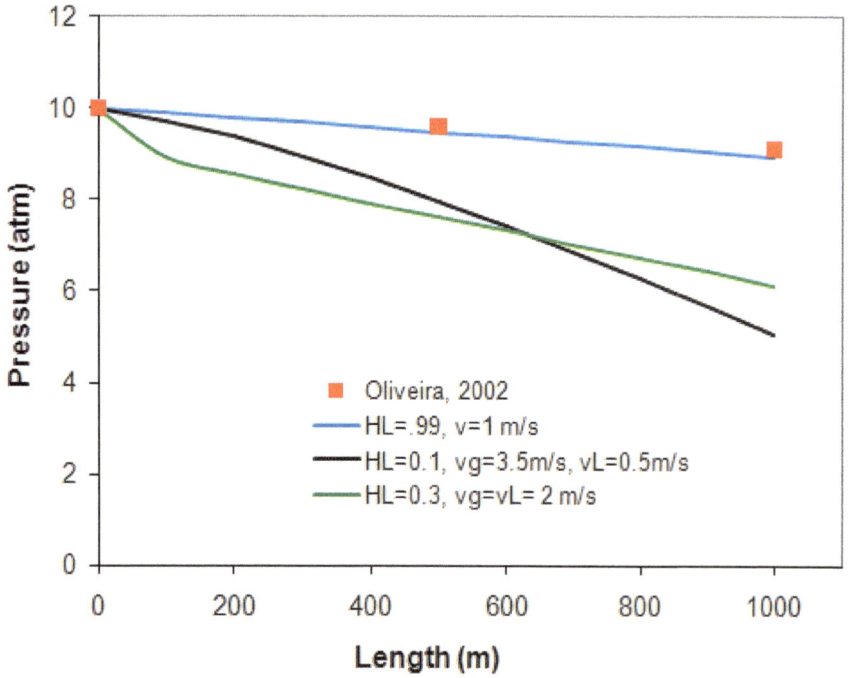

Figure 3. Comparison with experimental data of pressure drop with length from Oliveira (2002).

linear. In such simulation results, the following properties were utilized:

ρ_{gas} = 2 kg/m^3, ρ_{liq} = 900 kg/m^3, μ_{gas} = 0.01 cP, μ_{Liq} = 3 cP, D = 0.1 m and L = 1000 m

According to Oliveira (2002), the black curve corresponds to a typical profile of pressure drop in compressible and

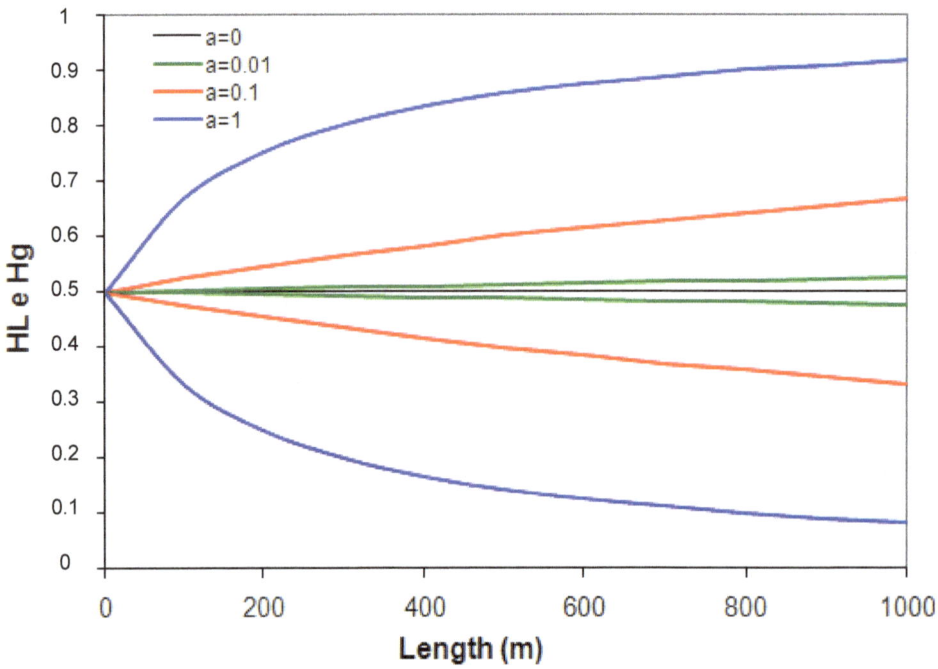

Figure 4. Holdup of liquid (H_L) and gas (H_g) in accordance with increasing gas velocity.

horizontal flow. This profile of pressure has a negative curvature in which the pressure drop increases by increasing the pipeline length. It should be noted that the pressure drop curve in question (black one) was obtained through simulations with lower liquid holdup and higher gas velocities (H_L = 0.1, v_g = 3.5 ms and v_L = 0.5 m/s) with an angular coefficient of gas velocity equal to one (= 1). Thus, such negative pressure drop profiles can be obtained in the conditions of increased acceleration of the gas phase. In the simulations represented here, the increase of the gas velocity is considered because the inertia effect is not responsible for its expansion (the flow of liquid and gas was kept constant with the alteration of the holdup of liquid and gas along the pipe). An exponential pressure profile (green curve) is obtained under no slip conditions (equal velocities) and with an intermediate value of liquid holdup.

In Figures 4, 5, 6 and 7, there is an analysis of sensitivity due to the contribution of the friction factor and the acceleration term in the expression of global balance of mechanical energy.

The conditions utilized are:

$$\rho_{gas} = 2\ \text{kg/m}^3,\ \rho_{liq} = 900\ \text{kg/m}^3,\ \mu_{gas} = 0.01\ \text{cP},$$

$$\mu_{Liq} = 3\ \text{cP},\ v_{og} = v_{oL} = 2\ \text{m/s},\ D = 0.1\ \text{m and } L = 1000\ \text{m}$$

Figure 4 shows the results of the holdup of the liquid (H_L, upper curves) and gas holdup (H_g, lower curves inversely equivalent to H_L) with the pipe length at different values of the angular coefficient of gas velocity (a = 0, 0.01, 0.1 and 1). Note that the holdup of liquid and gas is inversely equivalent. When there is no increase in the angular coefficient of the gas velocity, the gas and oil holdup does not change; it remains constant (black curve). For small increase in the angular coefficient, both holdup presents almost linear profile as in green and red curves. Already, a significant increase in gas velocity (as angular coefficient) leads to both asymptotic holdup profile which can significantly alter the energy loss by friction due to greater contact between the liquid phase with the tube wall.

The friction factors presented in Figure 5 are related to the simulation results from modeling approach implemented. This approach assumes that the increase of gas velocity changes the holdup of the liquid considering the properties of the mixture in the calculation of the friction factor. This is related to a more homogeneous pattern in which the gas bubbles are dispersed in the liquid phase.

It is observed from Figure 5 that, in case of no-slip flow (a = 0), there is no change in the holdup so the friction factor remains constant along the pipe length. For higher speed of the gas velocity, the friction factor tends to an exponential behavior with the length of the pipe. It should be noted that even with the increase of liquid holdup along the pipe, there is a more significant decrease in the friction factor which comes from the reduction in the velocity of the liquid flow.

The friction factors from segregated pattern (case B, Figure 1) were obtained in order to be compared to those from dispersed bubbles (case A, Figure 1). In the latter,

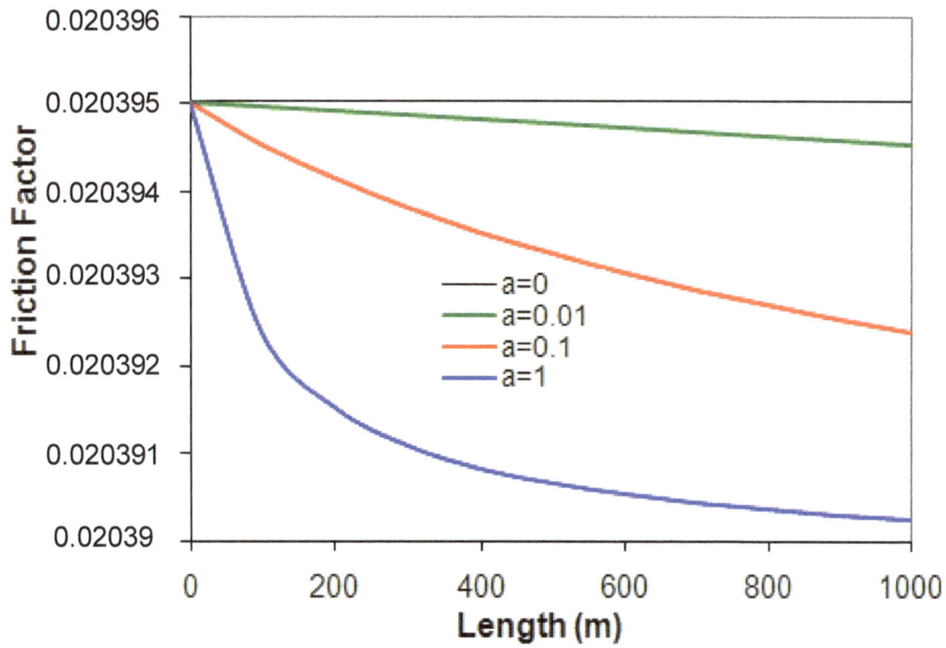

Figure 5. Values of friction factor from dispersed bubbles pattern.

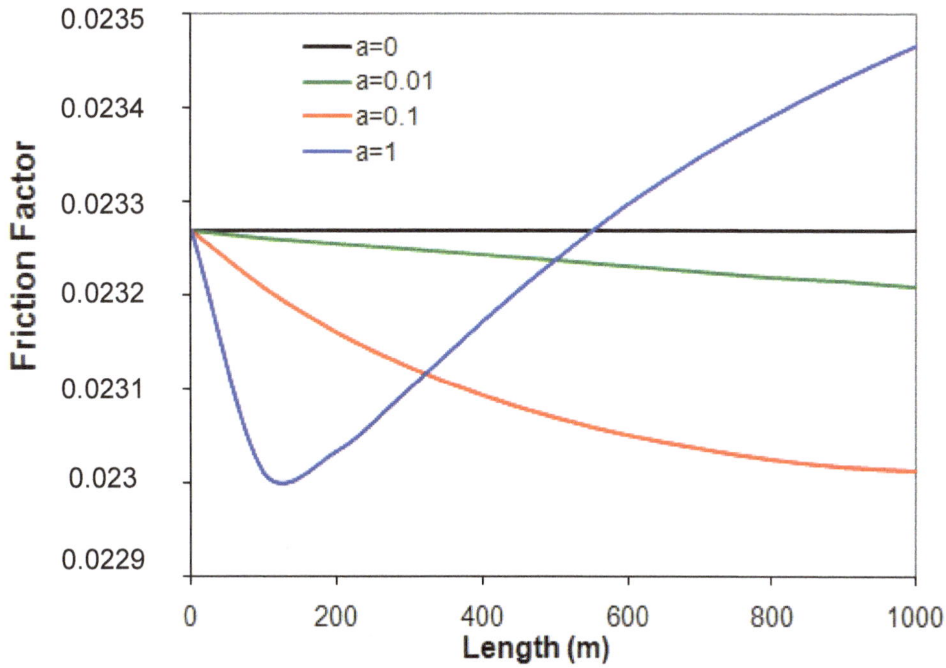

Figure 6. Values of friction factor from segregated pattern.

an increase in the holdup of the liquid leads to a further increase in liquid contact with the pipe wall. This approach assumes that the hydraulic diameter of the liquid and gas phases is proportional to the areas occupied respectively by each one.

$$\frac{\pi D_{HL,Hg}{}^{2}}{4} = A_{L,g} \tag{4}$$

Thus, the Reynolds number is calculated for each phase

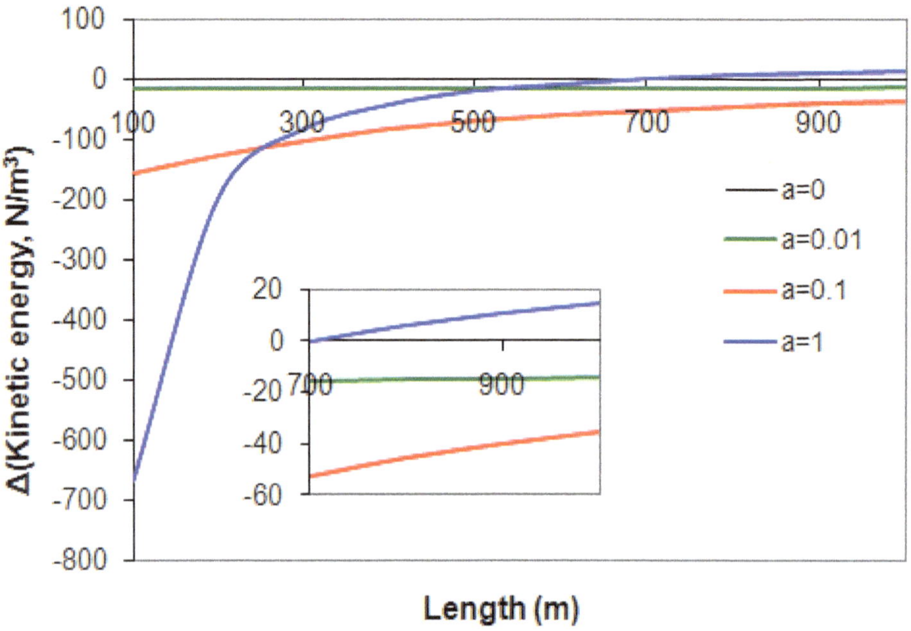

Figure 7. Effect of the kinetic energy term according to the pipe length.

according to the hydraulic diameter of each one. The friction factor was first calculated for each phase according to the respective Reynolds numbers and finally the global friction factor was obtained from the sum of the friction factor of each phase multiplied by each respective holdup. Figure 6 presents the simulation results of friction factor according to the segregated pattern. Note the behavior similar to the approach of bubbles dispersed in relation to small coefficient of gas phase velocity (up to 0.1). For higher values of gas velocity (a = 1) the friction factor shows a drop at the beginning of the pipe presenting a significant increase after this point. This increase is related to the increased contact of the liquid phase with the pipe wall under these conditions of speed.

Figure 7 shows the profiles of change of kinetic energy versus pipe length according to different velocities of gas phase (increasing the angular coefficient). As expected, there is no kinetic energy variation with no-slip between the phases (a = 0). Remember that the delta of kinetic energy is the sum of the portion of the gas and liquid phases. In most of the simulation cases presented in Figure 7, the contribution of kinetic energy term is negative in the balance of mechanical energy showing an increase along the length. In the simulation conditions of Figure 7, the delta of kinetic energy becomes positive after the position length of 700 m which corresponds to the highest value of gas phase velocity (a = 1). The positive variation of the term of kinetic energy is related to the increase of liquid holdup that has a greater share of contribution due to the higher density of the liquid compared to gas phase, despite the slow velocity of liquid phase.

Conclusion

The use of the overall balance of mechanical energy associated with expressions of speed related to the inertia of gas motion in oil-gas two-phase flow can be a good possibility of representation of these systems under such conditions. The modeling routine implemented showed simulation results which are very close to those observed experimentally in the work of Oliveira (2002).

The inertia of gas phase, under the conditions here studied, led to pressure drop along the pipe length which is similar to those observed for compressible flow. In the compressible flow, the pressure drop increases with the length of pipe in horizontal flow due to speed increase of gas associated with its own expansion (Oliveira, 2002). Thus, profiles of such pressure drop (increase with the length) can not exclusively be associated with compressible flow since the speed increase due to the inertia of the gas phase also led to similar results. A previous analysis of the pattern of two-phase flow is of paramount importance as the friction factor can change significantly by variations in the holdup that leads to different contacts of the liquid phase with the tube wall. In those situations of holdup variations due to gas inertia it was observed a positive change in the kinetic energy term that is strongly influenced by the increase in gas velocity.

ACKNOWLEDGEMENTS

The authors acknowledged the financial support from UFRJ, UERJ, CNPq, FAPERJ and CAPES.

Nomenclature

P - Pressure (*atm*)
L - Length (*m*)
f - Friction factor
ρ - Density (*kg/m3*)
v - Velocity (*m/s*)
μ - Dynamical viscosity (*cP*)
v_0 - Initial velocity of gas phase (*m/s*)
a - Angular coefficient of the gas velocity expression
D - Pipeline diameter (*m*)
Re - Reynolds number
H_L - Liquid *Holdup*
D_{HL} - Liquid hydraulic diameter (*m*)
D_{Hg} - Gas hydraulic diameter (*m*).

REFERENCES

Drew TB, Koo EC, McAdams WH (1930). The Friction Factors for Clean Round Pipes. Trans. AIChE pp. 28-56.

García F, García Jm, García R, Joseph DD (2007). Friction factor improved correlations for laminar and turbulent gás-liquid flow in horizontal pipelines. Int. J. Multiphase Flow 33:1320-1336.

Guang-Yao L, Jing W, Zhi-Hai J (2007). Experimental and numerical investigation on horizontal oil-gas flow. J. Hydrodyn. 19(6):683-689.

Hapanowicz J (2008). Slip between the phases in two-phase water-oil flow in a horizontal pipe. Int. J. Multiphase Flow 34:559-566.

Hewitt Gf, Jayanti S, Hope CB (1990). Structure of thin liquid films in gas-liquid horizontal flow. Int. J. Multiphase Flow 16:951-957.

Jepson WP, Taylor RE (1993). Slug flow and its transitions in large-diameter horizontal pipes. Int. J. Multiphase Flow 16:411-420.

Mohiuddin MA, Khan K, Abdulraheem A, Al-Majed A, Awal MR (2007). Analysis of wellbore instability in vertical, directional, and horizontal wells using field data. J. Pet. Sci. Eng. 55:83-92.

Oliveira G (2002). Formation courses in the production operations - Natural and artificial lift, Petrobrás, March.

Pinto CAC, Branco C, Ceciliano MM (2003). Offshore heavy oil in campos basin: the petrobrás experience. Offshore Technology Conference.

Ruder Z, Hanratty TJ (1990). A definition of gas-liquid plug flow in horizontal pipes. Int. J. Multiphase Flow 16:233-242.

Spedding PL, Donnelly GF, Cole JS (2005). Three phase oil-water-gas horizontal co-current flow. Chem. Eng. Res. Des. 83(A4):401-411.

Real-time expert seismology by non-linear oil reserves expert system

E. G. Ladopoulos

Interpaper Research Organization, 8, Dimaki Str., Athens, GR - 106 72, Greece. E-mail: eladopoulos@interpaper.org.

A non-linear 3-D elastic waves real - time expert system is proposed for the exploration of land and marine oil reserves, according to the new theory of "*Real-Time Expert Seismology*". This Generic Technology will work under Real Time Logic for searching the land and marine petroleum developed on the continental crust and on deeper water ranging from 300 to 2000 m, or 3000 m, or even more. Hence, the proposed real - time expert system will be the best device for the exploration of the continental margin areas (shelf, slope and rise) and the very deep waters, too. Furthermore, the above expert system will be suitable for the exploration of land oil reserves, as well. The objectives for exploration of marine and land oil resources are to locate, characterize and evaluate the size of such reserves. Hence, there is a research and development responsibility for the acquisition and analysis of geophysical, geological and reservoir engineering data for the land and the seas all over the world, in order to be explored their petroleum and gas reserves. Thus, through the new technology of *"Real-time Expert Seismology"*, will be effected the exploration of a significant part of marine and land oil reserves very fast and by a low cost.

Key words: Real-time expert seismology, land oil reserves, marine oil reserves, non-linear real - time expert system, real- time logic, generic technology.

INTRODUCTION

As the energy demand for oil and gas is increasing worldwide yearly at a pace of 1.5 to 2.0%, then it will increase up to 2030 by 50 to 60%. The total estimated petroleum all over the world in place stands today approximately at 1.5 trillion barrels, and with current petroleum consumption at 90 million barrels per day, the hydrocarbons in place are predicted to last for the next 40 years. Consequently, there is an absolute need by oil companies all over the world to increase their stock by finding new petroleum and gas reserves. For this reason the big oil companies should be looking into other alternatives, like to drill to ever deeper horizons and in the relatively unexplored ocean depths.

The research and development aspects of marine and land oil reserves can be divided into three main areas:

(a) The acquisition and analysis of geophysical, geological and reservoir engineering data to enable an appreciation to be made of the reserves.
(b) The determination of all-necessary standards and data for the safety to offshore and land operations.
(c) To assist the development of the offshore supplies industry, and to enable it to play a full part in the development of the marine hydrocarbon resources in worldwide markets in the future.

There is a considerable feeling by oil companies and scientific institutes that the unexplored marine and land petroleum and gas resources should be too many, many times more than the terrestrial reservoir of Middle - East. On the other hand, many recent test drillings in the Continental Shelf of Europe and rest world have shown the existence of petroleum resources.

Furthermore, the usual probability considerations indicate that marine oil reserves will be found in areas of thick sedimentary sequences developed on the continental crust. Hence, by such understanding the continental margin areas (shelf, slope and rise) offer good prospects of containing hydrocarbon reservoir. However, up to present the potential of sediments both on and off the shelf remains unquantified. Furthermore, there are deep waters prospects around the seas all over the world, but because of the paucity of the available information it is not possible at present to quantify the amounts that may be recoverable from them.

Since 1920, the basic and prevalent theory on oil reserves exploration has been *"Reflection Seismology"*. The basic idea of this method is to collect reflections of elastic (seismic) waves and through various mathematical operations, by using Snell's law and Zoeppritz equations, convert them to maps of the earth's structure Aki and Richards (1980), Alkhalifah and Tsvankin (1995), Dellinger et al. (1993), Gaiser (1997), Hale (1984), Harrison and Stewart (1993), Thomsen (1988), Thomsen (1999), and Tsvankin and Thomsen (1994) Infer location of oil and gas from structural maps, sometimes of character of reflections, too. Thus, the method of *"Reflection Seismology"* was used almost for a century, with several improvements in the industry of oil reserves exploration.

On the other hand, our proposal on marine and land oil reserves exploration is based on the new theory of *"Real-time Expert Seismology"*, which was very recently proposed by Ladopoulos (2011a, 2011b, 2012). Furthermore, is an extension of the non-linear methods for fluid mechanics. As proposed by Ladopoulos and Zisis (1997, 2000). This is a very "innovative" and "groundbreaking" method on oil reserves exploration. According to the new theory a non-linear 3-D elastic waves real - time expert system is proposed for the exploration of several oil reserves including marine oil reserves, of the seas all over the world. Such Generic Technology will work under Real Time Logic (Emnis et al., 1986; Fritz et al., 1988; Haase, 1990; Jahanian and Mok, 1985, 1986) for searching marine reserves developed on the continental crust and on deeper water ranging from 300 to 2000 m, or even more. The proposed real - time expert system will be therefore the best device for the exploration of the continental margin areas (shelf, slope and rise) and the very deep waters, too. Also, this expert system will be suitable for the exploration of land oil reserves, as well.

The benefits of the new theory of *"Real-time Expert Seismology"* in comparison to the existing theory of *"Reflection Seismology"* are the following:

1. The new theory is based on the special form of the geological anticlines of the bottom of the sea, in order to decide which areas of the bottom have the most possibilities to include petroleum.

On the other hand, the existing theory is only based to the best chance and do not include any theoretical and sophisticated model. So oil companies all over the world by using the existing method of *"Reflection Seismology"* must do many expensive test drillings in big areas of seas, if they want to have a chance to find oil and gas reserves. As every deep drilling cost at least 100 mill.

EURO, then every dry drilling would cost a lot of money to the oil companies.

2. The new theory of elastic (sound) waves is based on the difference of the speed of the sound waves which are traveling through solid, liquid, or gas. In a solid the elastic waves are moving faster than in a liquid and the air, and in a liquid faster than in the air.

Existing theory is based on the application of Snell's law and Zoeppritz equations, which are not giving good results, as these which we are expecting with the new method.

3. The new theory is based on a Real-time Expert System working under Real Time Logic, that gives results in real time, which means every second. Existing theory do not include real time logic.

From the above three points it can be well understood, the evidence of the applicability of the new method of *"Real-time Expert Seismology"*. Also its novelty, as it is based mostly on a theoretical and very sophisticated Real-time Expert model and not to practical tools like the existing method. So, the new method will be the best technology for searching the very deep waters (2000 to 3000 m, or even more) and the very deep depths of the basement rock of earth (20,000 to 30,000 m).

REAL-TIME EXPERT SEISMOLOGY

Generally, marine operations consist of 90% of all data collected worldwide for oil and gas reserves exploration. The depth of the drillings are usually up to 6000 m, but in order to find big petroleum and gas reserves they must be extended to 10,000 m or even from 20,000 to 30,000 m. Beyond the above, by studying geological surveys all over the world indicate that petroleum reserves do not necessarily end at the edge of the continental shelf. So, there is a serious expectation that main resources will be found in areas of thick sedimentary sequences developed on the continental crust. According to the new sophisticated model the oil reserves should occur in areas of geological anticlines. Moreover, there are good possibilities for finding off-shore petroleum and gas reserves in deeper waters, too, ranging up to 2500 to 3000 m, or even more and in very deep drillings ranging from 20,000 to 30,000 m.

The behavior of a reservoir depends not only on the properties of the liquid and gas, but also on a series of factors that may be termed as the "properties of the environment". Amongst these are such items as capillary - pressure effects, the reaction of rock when subjected to high stress, pressure and temperature gradients at the shallower levels in the Earth's crust and influences of the compressibility as pressure are reduced by fluid withdrawals.

There are four conditions that must be satisfied so that a geological formation, or a part thereof, should form a suitable reservoir, for example for the accumulation of oil.

These are porosity, permeability, seal and closure. The first defines the pore space in the rock - space in which the oil may collect. Permeability is the attribute of the rock that permits the passage of fluid through it. Generally, it is a measure of the degree interconnectedness of the pore space, but some reservoir (for example, in the massive limestone deposits, or in igneous intrusions) depends for fluid flow on a network of fractures within the rock.

Furthermore, the seal is the "cap" of the reservoir and prevents the oil from leaking away, while closure is a measure of the vertical extent of the sealed trap or, in the case of resources accumulation bounded below by a moving body of water, of the "height" of the sealed trap where that height is measured along a line perpendicular to the oil - water contact.

Almost all resources occur in sedimentary basins, in porous standstones or limestones and that seal or cap rock is often a clay or shale, or massive unfractured limestone having little or no permeability. On the other hand, three general categories of resources can be mentioned for marine reserves: structural traps, stratigraphic traps and combination traps.

Elastic waves are sound waves, generally three - dimensional and they may be transmitted through matter in any phase - solid, liquid, or gas. Any body vibrating in air gives rise to such waves, as it alternately compresses and rarefies the air adjacent to its surfaces. A body vibrating in a liquid, or in contact with a solid, likewise generates similar longitudinal waves. The frequency of the waves is of course the same as the frequency of the vibrating body that produces them.

The distance between two successive maxima (or between any two successive points in the same phase) is the wavelength of the wave and is denoted by *l*. Since the waveform, travelling with constant velocity *u*, advances a distance of one wavelength in a time interval of one period, it follows that the velocity of sound waves *u* as following:

$$u = l\,v \tag{1}$$

where *v* denotes the frequency.

As it is obvious the velocity *u* differs when the sound waves are travelling through solid, liquid, or gas. In a solid the elastic waves are moving faster than in a liquid and the air, and in a liquid faster than in the air. Therefore, if somebody is searching for example for oil marine resources over the sea, by transmitting sound waves, then there will be a difference in the velocity of the waves in the air, the sea, the solid bottom and in a potential reservoir.

In order to better explain our method, consider the example of Figure 1. In this case consider that in the bottom of the sea there is a potential oil reservoir. Then, the speed of the elastic waves in the air (u_{air}), will be different from the speed in the water (u_{water}), and different from the speed in the solid bottom (u_{solid}) and different from the speed in the potential reservoir (u_{oil}), while the frequency of the elastic waves remaining the same when transmitted through every different matter.

A real-time non-linear 3-D plane-polarized elastic waves expert system is proposed in order to explore the marine oil resources, for the several closed seas all over the world, according to the new theory of *"Real-time Expert Seismology"*, in contrast to the old theory of *"Reflection Seismology"*. Such Generic Sound Waves Technology will work under Real Time Logic for searching marine hydrocarbon reservoir developed on the continental crust and on deeper waters ranging from 300 to 3000 m, or even deeper (Figure 2). There are many deeper water prospects around the seas all over the world, but because of the paucity of the available information it is not possible at present to quantify the amounts that may be recoverable from them. For this reason the proposed real - time elastic waves expert system will be the best device for the exploration of the continental margin areas (shelf, slope and rise) and the very deep waters, too. So, through the new technology of *"Real-time Expert Seismology"*, will be effected the exploration of a significant part of marine and land oil reserves very fast and by a low cost.

According to the new theory of *"Real-time Expert Seismology"* the average velocity of the sound waves is calculated by providing important information about the composition of the materials through of which passed the sound waves. For example the velocity of the sound waves through the air is 331 m/s, through liquid 1500 m/s and through sedimentary rock 2000 to 5000 m/s. Also, the law of Reflection states that the angle of reflection equals the angle of incidence (Figure 3). According to the new technique the arrival times of the seismic waves are analyzed. After the sensor measures the precise arrival time of the wave, then the velocity of the wave can be calculated by using the following method.

The travel time *T* of the seismic waves is given by the formula:

$$T = \frac{2\left(d^2 + x^2/4\right)^{1/2}}{v} \tag{2}$$

where *d* denotes the depth, *x* the distance between source of wave and the geophone or hydrophone detector and *v* is the average speed.

From (2) follows Equation (3):

$$T^2 = \frac{4d^2 + x^2}{v^2} \tag{3}$$

Furthermore, the normal incident time T_o is given as following:

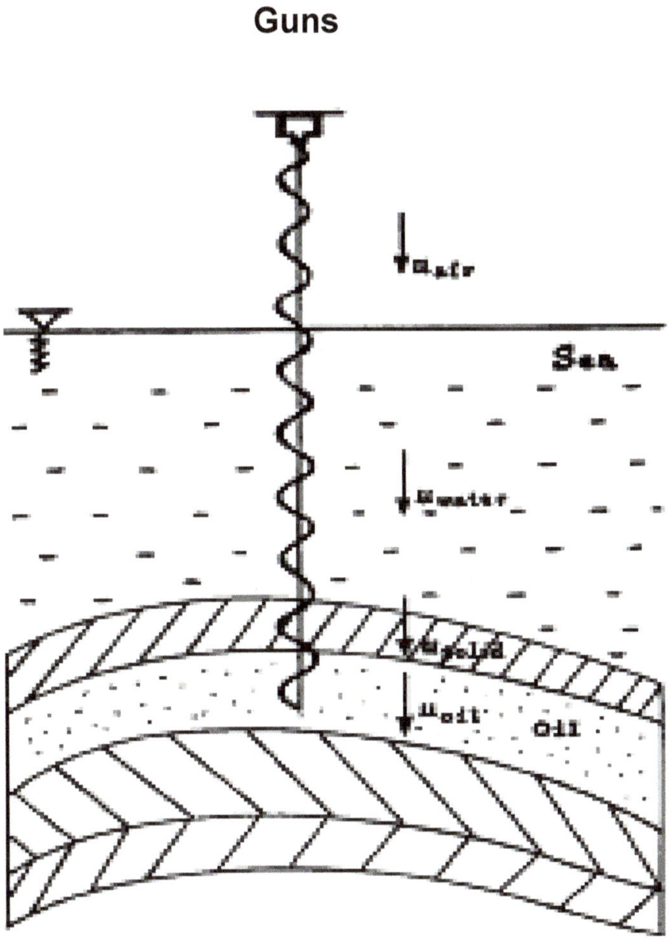

Figure 1. Elastic waves method for the exploration of oil reserves.

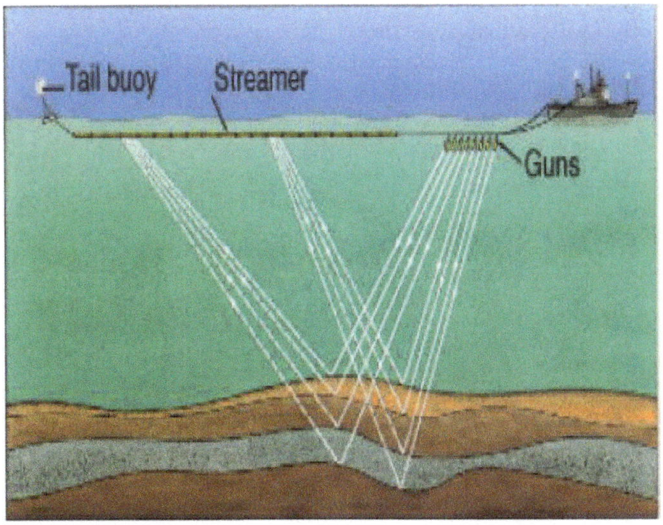

Figure 2. Real-time Expert Seismology.

$$T_o = \frac{2d}{v} \tag{4}$$

From Equations (3) and (4) follows:

$$T^2 - T_o^2 = \frac{x^2}{v^2} \tag{5}$$

and finally from (5) the mean velocity is equal to:

$$v = \left(\frac{x^2}{T^2 - T_o^2} \right)^{1/2} \tag{6}$$

Thus, a real time expert system is used and the apparatus permitted excitation of any combination of elements and reception of any other, visual analysis of the responses, and transfer of the signals to the PC for post processing. The sequencing of transducer excitation, digitiser configuration and subsequent data analysis was performed by a rule based Real-Time Expert System. From the information gathered, the Expert System applies knowledge via a series of software coded rules and provides any one of the following conditions: speed in the air (u_{air}), speed in the water (u_{water}), speed in the solid bottom (u_{solid}) and speed in the potential reservoir (u_{oil}).

REAL-TIME LOGIC FOR OIL RESERVES EXPLORATION

Real-time logic (RTL) is a reasoning system for real-time properties of computer based systems. RTL's computational model consists of events, actions, causality relations, and timing constraint. This model is expressed in a first order logic describing the system properties as well as the systems dependency on external events. The Real-Time Logic system introduces time to the first logic formulas with an event occurrence function, which assign time values to event occurrences. Such real-time logic systems were studied by Jahanian and Mok (1985, 1986).

Beyond the above, real-time computing in common practice is characterized by two major criteria: deterministic and fast response to external stimulation, and both human and sensor and actor based interaction with the external world. Real-time is an external requirement for a peace of software; it is not a programming technology. There are some special software tools for the implementation of real - time systems. Such real-time programming languages were investigated by several scientists like Emnis et al. (1986), Fritz et al. (1988) and Haase (1990).

In general, Real-Time Logic uses three types of constraints:

1. Action constants may be primitive or composite. In a composite constant, precedence is imposed by the event-action model using sequential or parallel relations

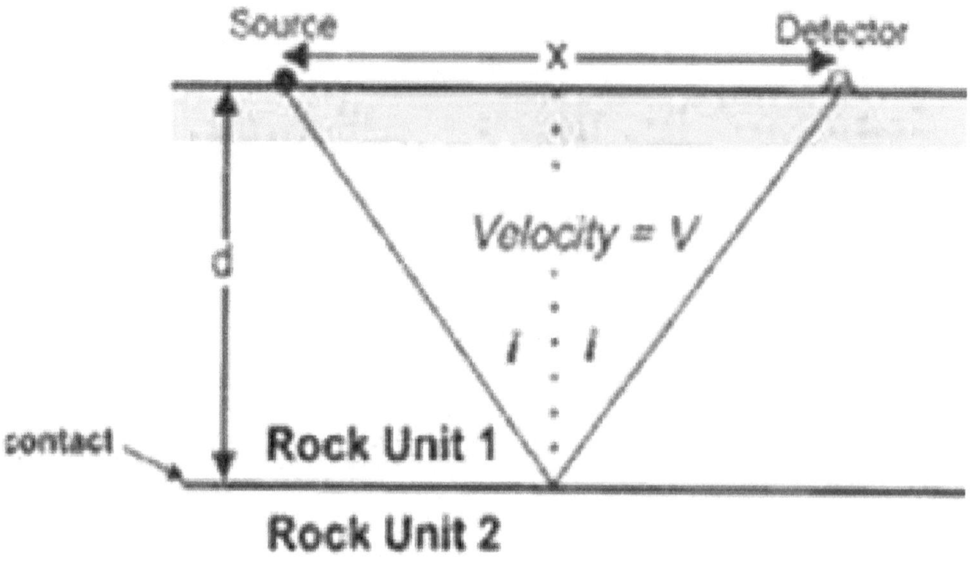

Figure 3. Law of reflection.

between actions.
2. Event constants are divided into three cases. Start/stop events describe the initiation/termination of an action or sub-action. Transition events are those which make a change in state attributes. This means, that a transition event changes an assertion about the state of the real-time system or its environment. The third class, which is the external events, includes those that can be impact system behavior, but cannot be caused by the system.
3. Integers assigned by the accurance function provide time values, and also denote the number of an event occurance in a sequence.
Moreover, the Real-Time Logic System introduces time to the first order logic formulas with an event occurance function denoted by e. The mechanism to achieve a timing property of a system is the deduction resolution.

Let us further consider the following example: Upon pressing button ≠ 30, action TEST is extended within 300 time units. Dusting each execution of this action, the information is sampled and subsequently transmitted to the display panel. Also, the computation time of action TEST is 90 time units.

This example can be further translated into the following two formulas:

∀ x: e (Ω button 30, x) ≤ e (↑ TEST, x) ∧
e (↓ TEST, x) ≤ e (Ω button 30, x) + 300
∀ y: e (↑ TEST, y) + 90 ≤ e (↓ TEST, y)

SEISMIC WAVES

Suppose that a seismic wave of any sort travels from left to right in a medium, then the equation of the travelling wave may be written as following:

$$y(\zeta,t) = A\cos(\omega t - k\zeta) \tag{7}$$

with k the wave number equal to :

$$k = 2\pi / \lambda \tag{8}$$

and λ the wavelength, t the time, A the amplitude of the motion and $\omega = 2\pi v$, v the frequency. A sinusoidal transverse wave travelling toward the right, at intervals of period 1/8 can be seen in Figure 4.

It is further important to distinguish between the motion of the waveform, which moves with constant velocity u along the string, and the motion of a particle of the string, which is simple harmonic and transverse to the string. In order to be understandable the mechanics of seismic waves, consider Figure 5.

Furthermore, by differentiating the wave equation (7) one obtains:

$$\partial y / \partial z = kA\sin(\omega t - kz) \tag{9}$$

$$\partial y / \partial t = -\omega A\sin(\omega t - kz) \tag{10}$$

By combining eqs (11) and (12) and as the velocity is equal to:

$$u = \omega / k \tag{11}$$

Then we obtain:

$$\frac{\partial y}{\partial \zeta} = -\frac{1}{u}\frac{\partial y}{\partial t} \tag{12}$$

which by a second differentiation takes the form:

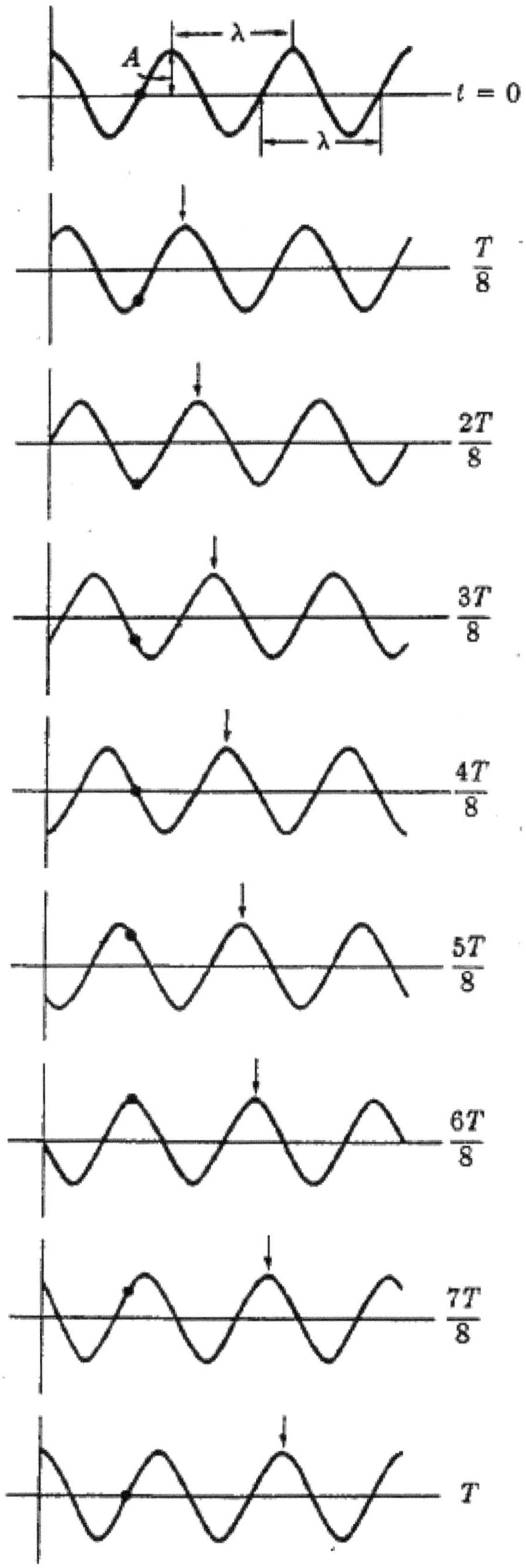

Figure 4. A sinusoidal transverse wave travelling toward the right, at intervals 1/8 of a period.

$$\frac{\partial^2 y}{\partial t^2} = u^2 \frac{\partial^2 y}{\partial \zeta^2} \tag{13}$$

In a transverse wave motion the individual particles vibrate in a direction perpendicular to the direction of propagation of the wave. But there are many such directions - indeed, they are infinitely many. In Figure 6 three transverse waves are illustrated, all travelling in the same direction, but lying in different planes from one another: A in a vertical plane, B in a horizontal plane, C in a plane inclined at an angle of 45° to each of these. In each case the motion of each individual particle is restricted to a single straight line, and the entire wave to a single plane. Each of these waves is said to be plane - polarized. Also, because of their non - linear behavior, are called non - linear plane - polarized. But we could also generate more complex seismic waves by moving one end in any periodic manner, not restricted to a single straight line. In such cases, each particle has two - dimensional motion (it moves in a plane) and the entire wave is three - dimensional. Even then the wave is not necessarily un-polarized, while if the vibrations are ordered in any case, the wave is to some degree polarized.

CONCLUSIONS

In the present investigation the new theory of *"Real-time Expert Seismology"* has been introduced and investigated for the exploration of land and marine oil reserves. As the test marine drillings are very expensive with costs of many millions (or billions) of EURO or USD and many times without any success, which means a big loss of funds, then the highly innovative and groundbreaking technology *"Real-time Expert Seismology"* has been proposed. By this new basic technology for energy applications a strong scientific and technical base will be established for the science and technology all over the world in the emerging areas of oil reserves exploration in the energy field. So, through the new technology of *"Real-time Expert Seismology"* will be effected, the exploration of a significant part of marine and land oil reserves very fast and by a low cost.

The oil market is a market of many billions every year all over the world. So this contribution requires a worldwide approach, rather than a local approach, as it is referred to a market all over the world with value of many billions of EURO or USD. We expect therefore in order for the big oil companies to keep and to improve its leading role in the science and technology in the whole world, to get involved in the new and groundbreaking technology in the area of energy, which we are proposing.

Thus, the potential areas and markets of application of the proposed technology will be the oil market all over the world. The method's results will be applicable to all oil

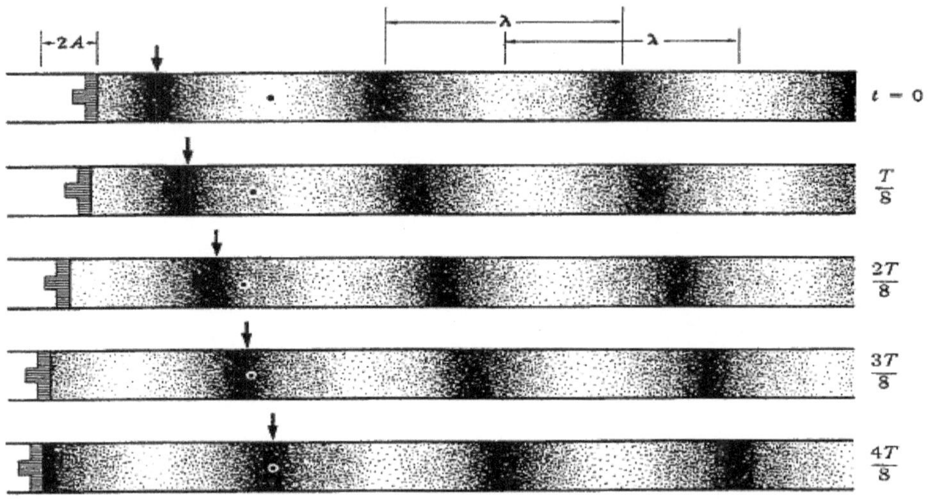

Figure 5. Sinusoidal longitudinal waves, travelling toward the right, at intervals 1/8 of a period.

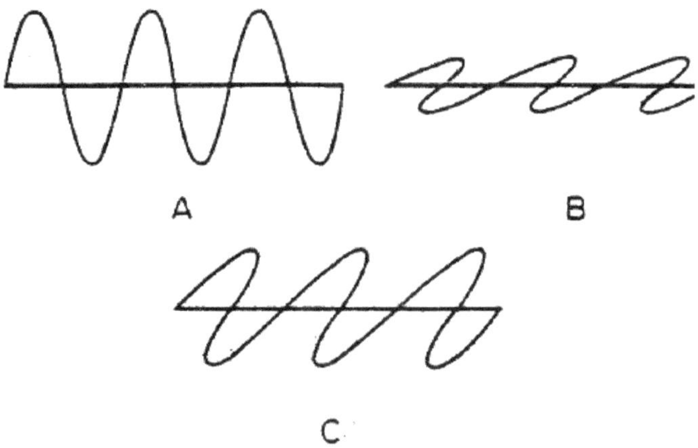

Figure 6. Transverse waves, travelling in the same direction, but lying in different planes from one another.

companies and scientific organizations working on oil exploration in the whole world. We are expecting through our proposed new technology the cost for oil exploration to be reduced too much, as there will be no need for so many test drillings, as today. The test drillings cost many million (or billion) of EURO or USD, and our method will reduce them in the possible minimum.

As our proposed new technology *"Real-time Expert Seismology"*, is based on a non-linear 3-D elastic waves real - time expert system, working under real logic, then is expected to give the best results. Furthermore, our proposed high technology method is based on a very sophisticated model by checking the geological anticlines of the bottom of the sea, in order to decide which areas of the bottom have the most possibilities to include petroleum. So it is not necessary to make test drillings in the half ocean in order to find petroleum, like the existing methods.

REFERENCES

Aki K, Richards P (1980). "Quantitative Seismology, Theory and Methods". Freeman Press, San Francisco.

Alkhalifah T, Tsvankin I (1995). 'Velocity analysis for transversely isotropic media'. Geophysics 60:1550–1566.

Dellinger JA, Muir F, Karrenbach M (1993). 'Anelliptic approximations for TI media'. J. Seis. Explor. 2:23–40.

Emnis R. et al (1986). 'A Continuous Real-Time Expert System for Computer Operations'. IBM J. Res. Dev. 30(1):14-28.

Fritz W, Haase VH, Kalcher R (1988). 'The use of standard software in real time programming - An example demonstrating the integration of ADA, Oracle and GKS', in Puente J. (ed) Proc. 15th IFAC/IFAP Workshop on Real Time Program, Pergamon.

Gaiser JE(1997). '3-D converted shear wave rotation with layer stripping'. U.S. Patent 5, 610, 875.

Haase VH (1990). 'The use of Al-Methods in the implementation of real-time software products'. IFAC 11th Triennial World Con., Tallium, Estonia.

Hale D (1984). 'Dip-move out by Fourier transform'. Geophysics 49:741–757.

Harrison M, Stewart R (1993). 'Poststack migration of P-SV seismic data'. Geophysics 8:1127–1135.

Jahanian F, Mok A (1986). 'A Graph-Theoretic Approach for Timing Analysis in Real-Time Logic'. Proc. Real-Time Systems Symp. (IEEE), pp. 98-108, New Orleans, LA.

Jahanian F, Mok A (1985). 'Safety Analysis of Timing Properties in Real Time Systems', Department of Computer Science, University of Texas, Austin.

Ladopoulos EG (2011a). 'Non-linear singular integral representation for petroleum reservoir engineering'. Acta Mech. 220:247-253.

Ladopoulos EG (2011b). 'Petroleum reservoir engineering by non-linear singular integral equations'. Mech. Eng. Res. 1:1-10.

Ladopoulos EG (2012). 'Oil reserves exploration by non-linear real-time expert telematics system'. Oil Asia J. 32:30-35.

Ladopoulos EG, Zisis VA (1997). 'Non-linear singular integral approximations in Banach spaces'. Nonlin. Anal. Theor. Math. Appl. 26:1293–1299.

Ladopoulos EG, Zisis VA (2000). 'Non-linear finite-part singular integral equations arising in two-dimensional fluid mechanics'. Nonlin. Anal. Theor. Meth. Appl. 42:277-290.

Thomsen L (1988). 'Reflection seismology over azimuthally anisotropic media'. Geophysics 51:304–313.

Thomsen L (1999). 'Converted-wave reflection seismology over inhomogeneous, anisotropic media'. Geophysics 64:678-690.

Tsvankin I, Thomsen L (1994). 'Nonhyperbolic reflection moveout in anisotropic media'. Geophysics 59:1290–130.

Comparison between spreadsheet and specialized programs in calculating the effect of scale deposition on the well flow performance

Khulud M. Rahuma*, F. Sondi, M. Milad and M. AbuGheit

University of Tripoli, Tripoli-Libya, P. O. Box 13126, Libya.

Formation damage is an undesirable operational and economic problem that causes a reduction in the productivity, skin damage, and decrease in well performance. Formation damage occurs during the various phases of oil and gas recovery from subsurface reservoirs including production, drilling, hydraulic fracturing, and work over operations. Formation damage assessment, control, and remediation are among the most important issues to be resolved for efficient exploitation of hydrocarbon reservoirs. Such damage is caused by various adverse processes, including chemical, physical, biological, and thermal interactions of the formation and fluids, deformation of formation under stress and fluid shear. This paper systematically presents an analytical model developed for predicting productivity of reservoir with incidence of scale deposition in the vicinity of horizontal and vertical wellbore. The objective of this study is comparison among the spreadsheet, Kappa Ecrin Saphir program, and Halliburton Wellford Pan System software in calculating the effect of scale deposition on the flow performance for several types of wells. The results of vertical, horizontal wells were close using Kappa Saphir and PanSystem, but far using the Excel program.

Key words: Productivity, scale, skin, well, software.

INTRODUCTION

Scale deposition, unlike other types of deposition, is a complex crystallization process (Urayet, 2000). The formation rate of an initial scale layer and its subsequent rate of growth are determined by the interaction of several rate processes: nucleation, diffusion, chemical reaction, and molecular ordering of the scale crystal lattices, etc. Most, though not all, mineral scale-forming constituents are inversely soluble (their solubility's tend to decrease with increase in temperatures) (Urayet, 2004; Schlumberger, 1998). Therefore, when these super-saturated solutions come into contact with heat transfer surfaces, they precipitate solids due to their lower equilibrium solubility. There are two types of scale deposition organic and inorganic (Ramey, 1992). Once a solution becomes supersaturated and nucleation occurs, the conditions best suited for crystal growth of the particular scale constituent will be the conditions most ideal for scale formation.

Formation damage or skin effect can be found from the analysis or well testing results (pressure build-up "and draw down tests) the damage can also be investigated by analyzing Production logging surveys. Production logging is a useful tool for detecting zones that are not contributing to the total flow systems (Economides and Notle, 1898). Formation damage can be detected by comparing the average permeability for the well, determined by draw down tests (Matthews and Russell, 1987), with the undamaged permeability away from the wellbore determined from pressure build up analysis. Two straight-line portions obtained from a pressure build up test is an indication of wellbore damage surrounding the well (Unneland and Statoil, 1995).

MATERIALS AND METHODS

The oil companies use several methods to calculate the skin factor: 1- spreadsheet is usually used, but it takes a lot of time and accuracy of less 2- programs designed to reduce time and increase accuracy of the results (Ahmed and McKinney, 2005).

Spreadsheet (MS Office Excel)

A modern spreadsheet file consists of multiple worksheets (usually called by the shorter name sheets) that make up one workbook, with each file being one workbook. A cell on one sheet is capable of referencing cells on other different sheets, whether within the same workbook or even, in some cases, in different workbooks. A spreadsheet is essentially just one table, whereas a database is a collection of many tables with machine-readable semantic relationships between them. Spreadsheets are often imported into databases to become tables within them, but it lacks the relational structure of a database.

Kappa Saphir

The Saphir methodology has always been based on the Bourdet derivative as the main diagnostic tool; matching the measured data to the model, taking into account the detailed production history. The ever-increasing processing power of PCs has enabled KAPPA to aggressively expand the technical capability of Saphir. This has resulted in the development of fast numerical models, that has naturally extended to the rigorous solution of nonlinear problems. Saphir is now available only in the non-linear variant known as Saphir NL. When actively maintained under the current pricelist, Saphir NL also includes the well performance analysis module Amethyste as a free add-on. In a minifrac analysis component was developed and the slug-test option was reintegrated. Substantial improvements in the analytical and numerical models were added to better address unconventional formations. An improved Perrine method addresses multiphase flow and material balance correction has been added to the de-convolution. The Rubis sector now allows the Saphir NL numerical model to start from a dynamic state of a Rubis simulation as next: data loading and editing, Data QA/QC and datum correction, Test design, Extracting ΔP and de-convolution, Specialized plots, Analytical models, Numerical models, Use of Rubis sectors, Multilayer analysis, Optimization and sensitivity analysis, AOF / IPR, Minifrac analysis, Slug and Pulse, Formation tests, Reporting and exporting.

EPS PanSystem

PanSystem is a powerful tool for: Preparing and editing Well Test Data from conventional gauges and wire line formation testers. Analyzing and history matching Transient Well Test Data using Analytical and Numerical methods. Analyzing and history matching long-term Flowing Pressure and Production Data using Pressure Decline Analysis (PDA) methods. Analyzing and history matching wireline formation tester probe or packer data simultaneously with observation probe data Computing Well Deliverability/ Injectivity and Forecasting long-term production.

The Well and Reservoir is initialized via the Well and Reservoir Description (Numerical) sub-menu option from the PanSystem Dataprep Menu. The actual Numerical Simulation is performed within PanMesh; this is accessed via the Numerical sub-menu option from the PanSystem Simulate Menu. Deliverability in PanSystem provides the following two methods for calculating the Deliverability/Injectivity of a Reservoir: Using the fitting to measured Test Point Data method for Oil and Water Fluid Types involves the program fitting the Vögel IPR relation to up to three measured flowing/injection Pressures and Rates. For Gas and Condensate Fluid types, the Deliverability curve is computed either using the Darcy (B) and Non-Darcy (F) Flow Coefficients estimated by an LIT Analysis of measured flowing/injection Pressures and Rates, or using the C-coefficient and n-exponent obtained from a simplified C&N Analysis of measured data.

For semi-theoretical derivation using results from Transient Well Test Analysis (k, S) and Extended Drawdown Analysis (A, CA), for Oil and Water Fluid Types, the Productivity Index (J) is calculated from the semi-steady-state inflow equation, and the Vogel IPR relation applied. For Gas and Condensate Fluid Types using the LIT Method, the Darcy Flow Coefficient (B), is calculated from the semi-steady-state inflow equation and the Non-Darcy Flow Coefficient (F), is derived from a Rate versus Skin analysis of Transient Well Test Data to allow the Deliverability curve to be calculated. For the C and n Method the "stabilized deliverability" is estimated from a theoretical stabilized Flow Rate, derived from the C-coefficient and n-exponent . Production Forecasting for Oil, a semi-theoretical approach is used. For Gas, a theoretical approach is used (for example, LIT and C&n). This predicts Flow Rate versus Time for the current Reservoir Model (that is, Infinite, Semi-Infinite or Closed System), assuming constant Bottom-Hole Flowing Pressure (BHFP), or constant Wellhead Pressure (in which case, an Eclipse-compatible VFP File must be imported to model the Tubing Performance. File is the data used by the program is stored in a number of files. Every file must have a name which will allow the operating system to refer to it. There are different types of file, each storing different types of information. From these two methods Flow Regimes can be identified and the various Well and Reservoir parameters derived. During Analysis, results from any particular plot can, if desired, be transferred to the Reservoir Description (that is, Model Parameters) set using the Confirm toolbar option. In this way, the Reservoir Description can be built up and continually updated as the Analysis proceeds. Up to six different Interpretation Models can be stored in a single (*.PAN) file. The latest status of every plot invoked (that is, lines fitted, results, Flow Regimes, etc.) is written to the (*.PAN) file by the File Save or File Save As options. Subsequent recall of the file will reinstate all plots so that the Analysis can be resumed from where it was left off. If users have more than one Pressure and/or Flow Rate channel to choose from (for example, two Pressure Gauges were run), they can change their selection at any time by returning to Dataprep Gauge Data (TPR) Preparation and selecting a different Column Name in the Master File/Columns section.

RESULTS AND DISCUSSION

The aim of this paper is the comparison of spreadsheet and two other specialized programs Kappa Ecrin Saphir and Halliburton Wellflo PanSystem in the calculation of the Effects of scale deposition on well flow performance for several types of wells and compares these results to the results of the company. Ten wells were selected of the total wells in the field where it was a choice of five

Table 1. All results of well N1 and company result.

Software	Skin (value)
Kappa Saphir	54.213
EPS PanSystem	56.8612
MS Office Excel	77.43
Company	**Skin (value)**
Mellitah OIL and GAS B.V.	57.33

Table 2. All results of wells and company result.

Well	Type	Skin			
		Kappa Saphir	EPS Pansystem	MS Office Excel	Mellitah B. V.
N6	HORI. OIL	39.15	37.77	44.48	16.80
N7H	HORI. OIL	12.92	13.21	18.34	12.50
N8H	HORI. OIL	6.84	7.02	19.21	7.00
N9H	HORI. OIL	4.13	4.77	11.70	4.13
N10H	HORI. OIL	2.59	1.96	9.25	2.40
N14	VERT. OIL	281.43	281.74	317.40	287.60
N18	VERT. OIL	192.78	192.16	181.44	193.50
N15	VERT. OIL	138.23	140.84	154.65	140.00
N1	VERT. OIL	54.21	56.86	77.43	57.30
N16	VERT. OIL	45.65	45.01	49.37	42.00

vertical and five horizontal wells.

The following workflow has been followed to perform the interpretation:

(i) Data preparation for each well.
(ii) Import gauge data from the excel file (pressure).
(iii) Remove noisy data.
(iv) Define flow periods (for softwares only).
(v) Define well, layer and fluids parameters (for softwares only).
(vi) Define reservoir thickness (for softwares only).
(vii) Define PVT to find PVT fluid parameters (for softwares only).
(viii) Plot flow periods and assign flow rate for each period.
(ix) Analyze the buildup using the pressure derivative on the log-log plot.
(x) Analyze the buildup using the semi-log pressure plot (Horner plot) and define or confirm the following parameters:

(a) Extrapolated pressure P*.
(b) Kh permeability thickness product.
(c) Radius investigation.
(d) Skin pressure drop.
(e) Skin factor.

(xi) Verify the analysis made on build-up data on the draw down periods.

The results obtained by software's calculation are shown in Table 1 and Figure 9.

We calculate the skin of the several wells using the three programs and compared the results given by the company, which was calculated using manual methods (Excel); the results are in Table 2 for all wells.

Conclusion

Calculated results of the skin using the programs and manual methods (Excel) for ten wells compared with company results. The skin obtained from vertical wells show that the results are similar between the company's results and that obtained from Kappa Saphir and PanSystem, while the horizontal wells results were close in four wells using Kappa Saphir and PanSystem to company results, and far using the Excel program. Generally high values result obtained from the Excel program in most wells Figures 1 to 8.

ACKNOWLEDGEMENT

The authors would like to express their deep appreciation to Libyan Oil Company for their help and provision of the necessary data.

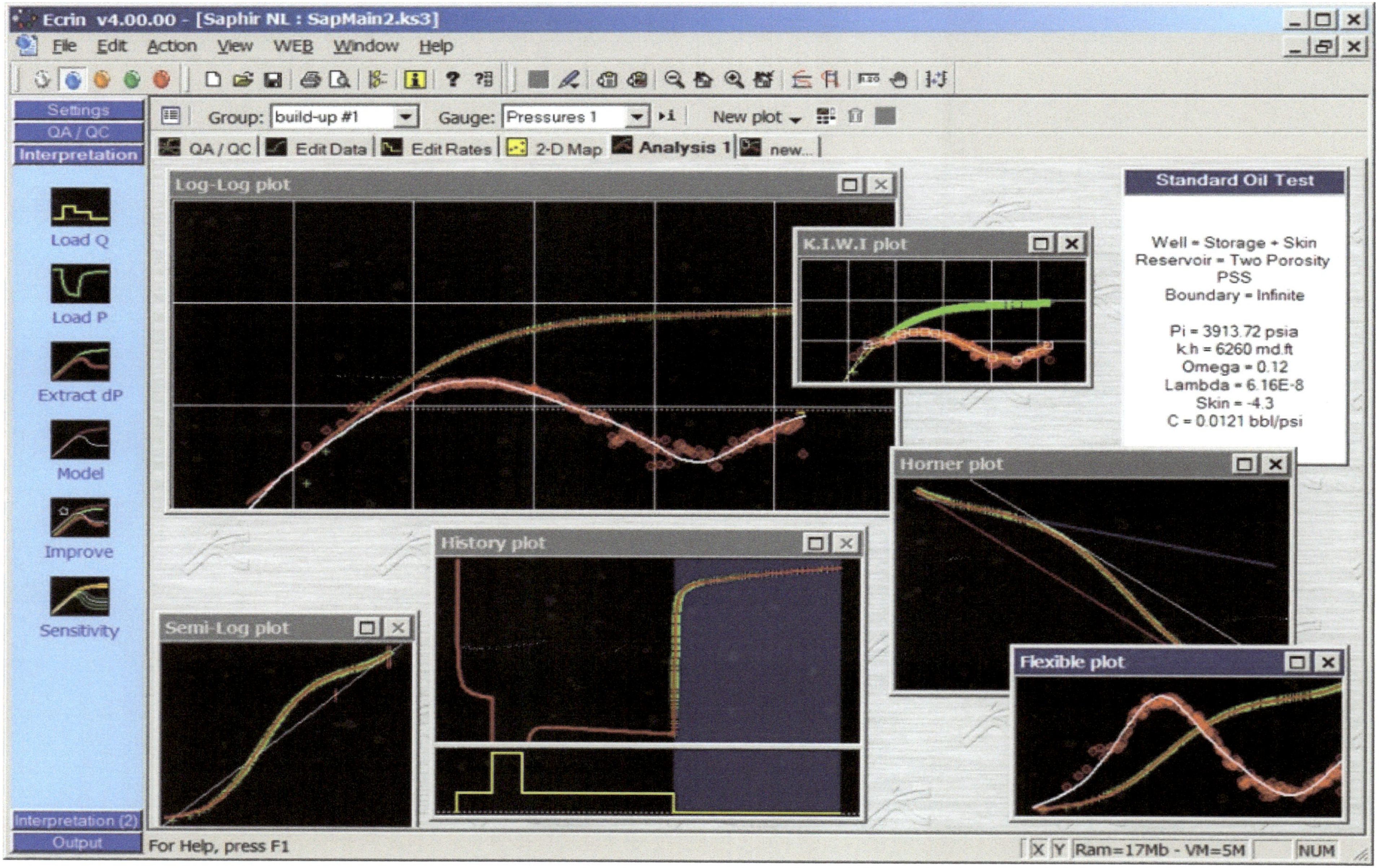

Figure 1. Saphir main window.

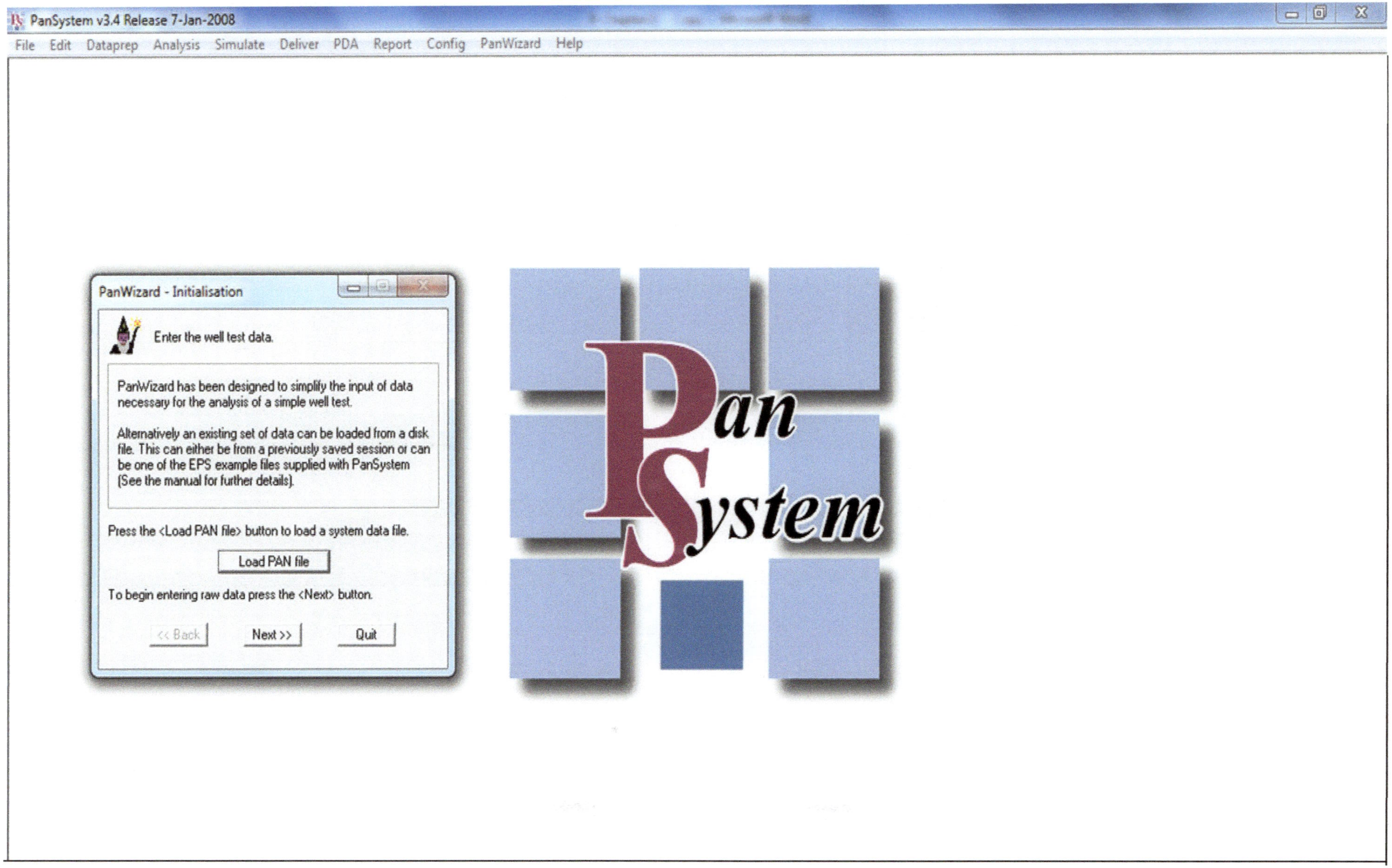

Figure 2. EPS PanSystem main windows.

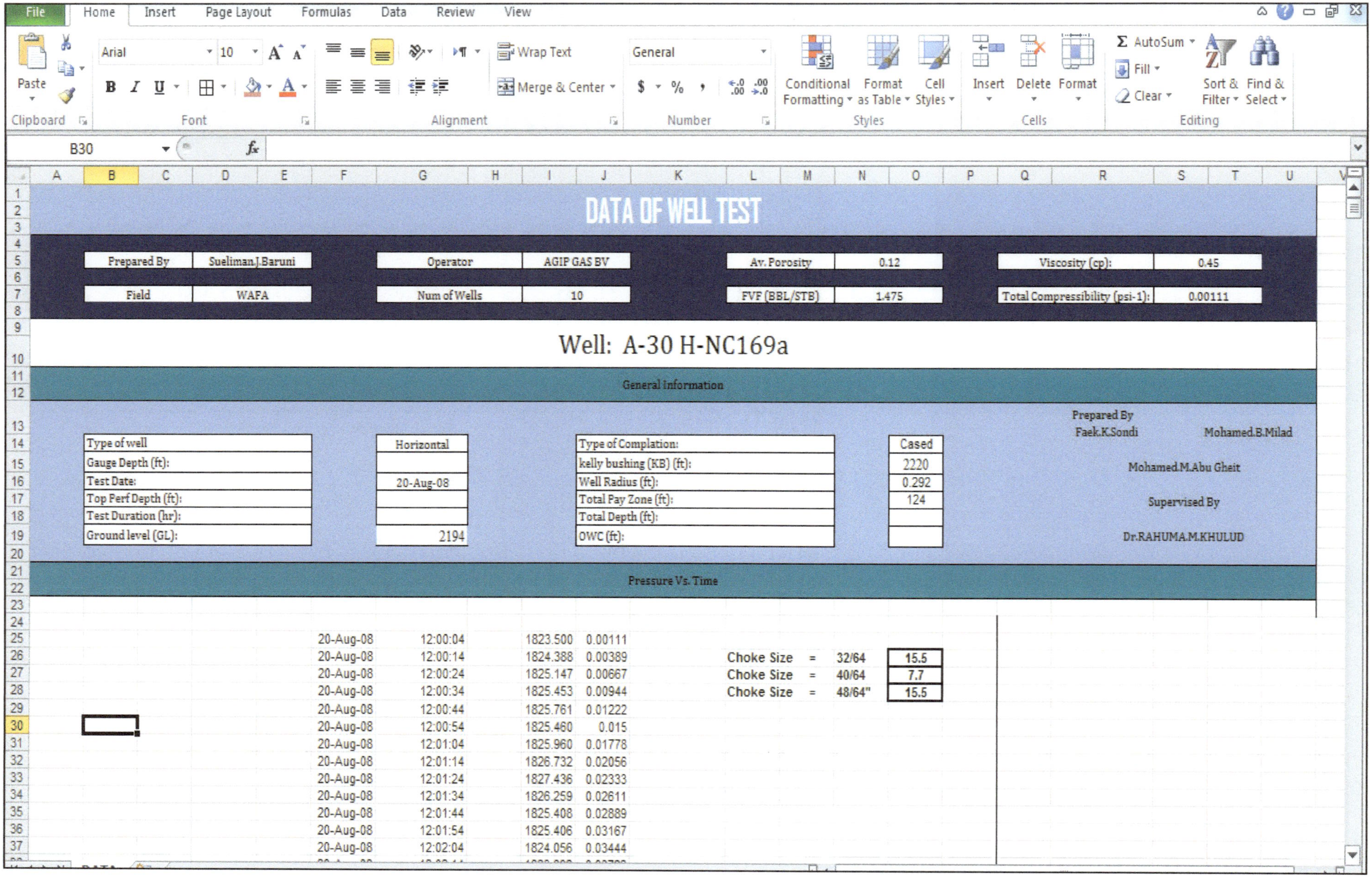

Figure 3. The data preparation.

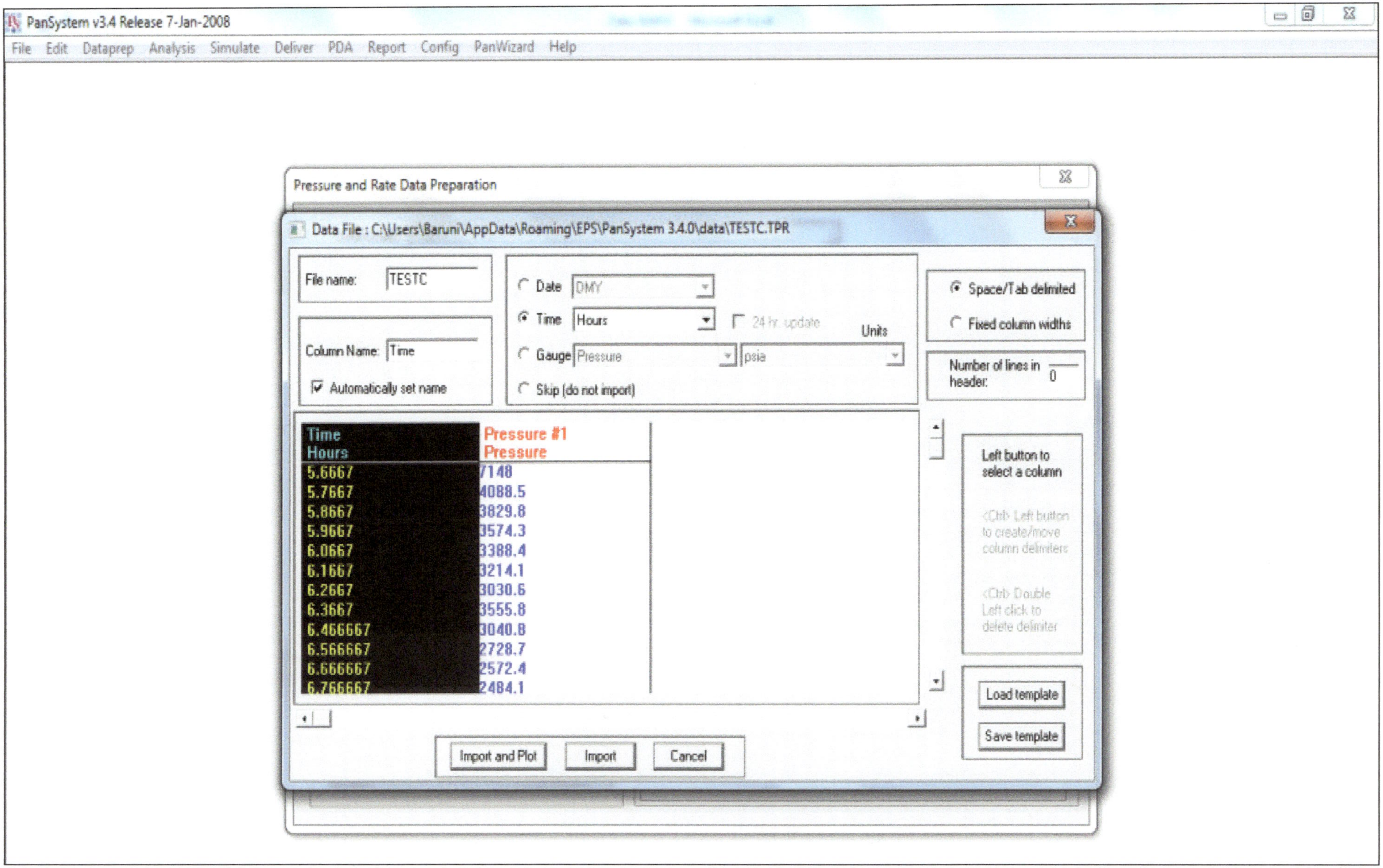

Figure 4. Define data (Pansystem).

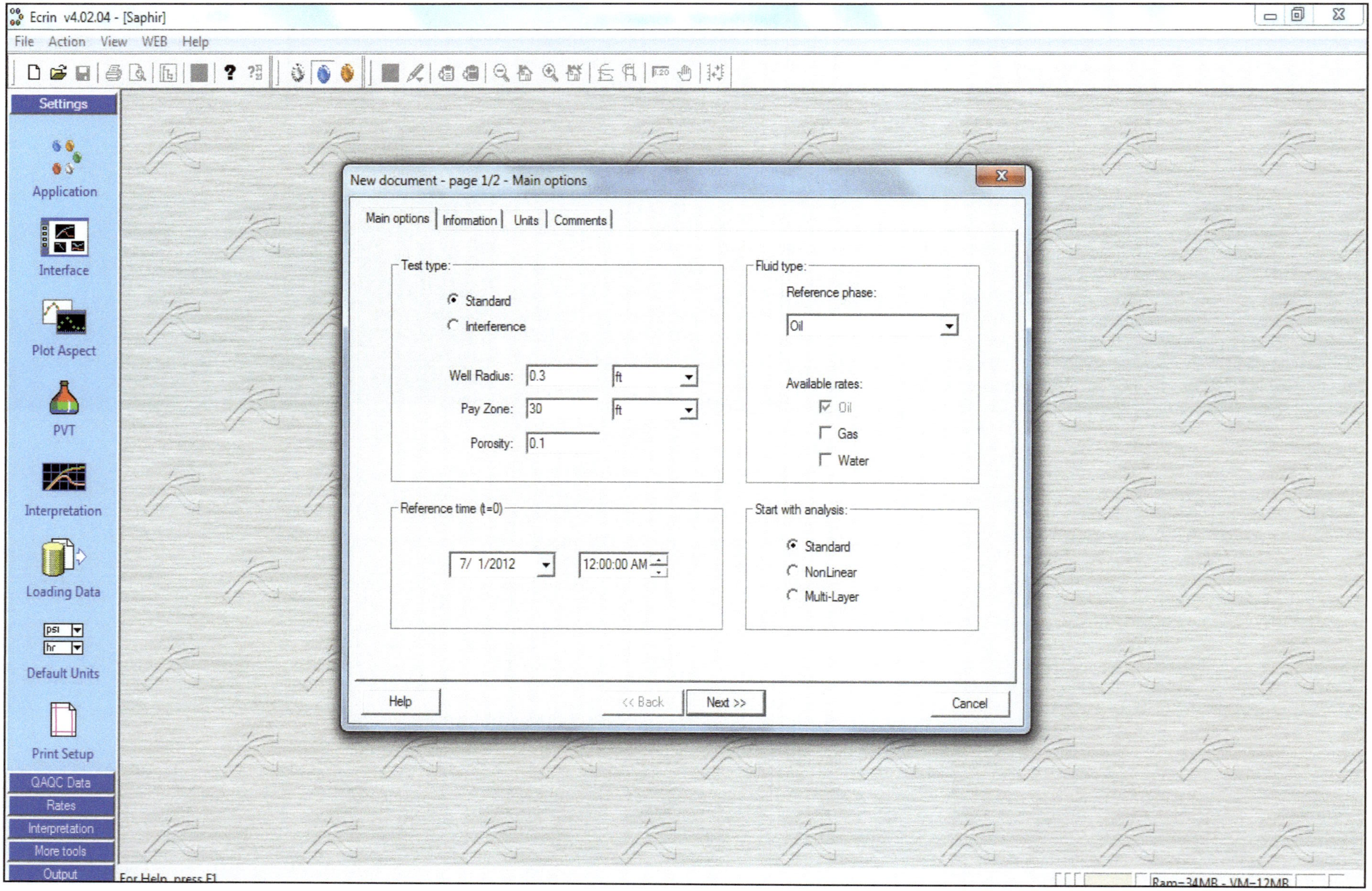

Figure 5. Define data (Ecrin Saphir).

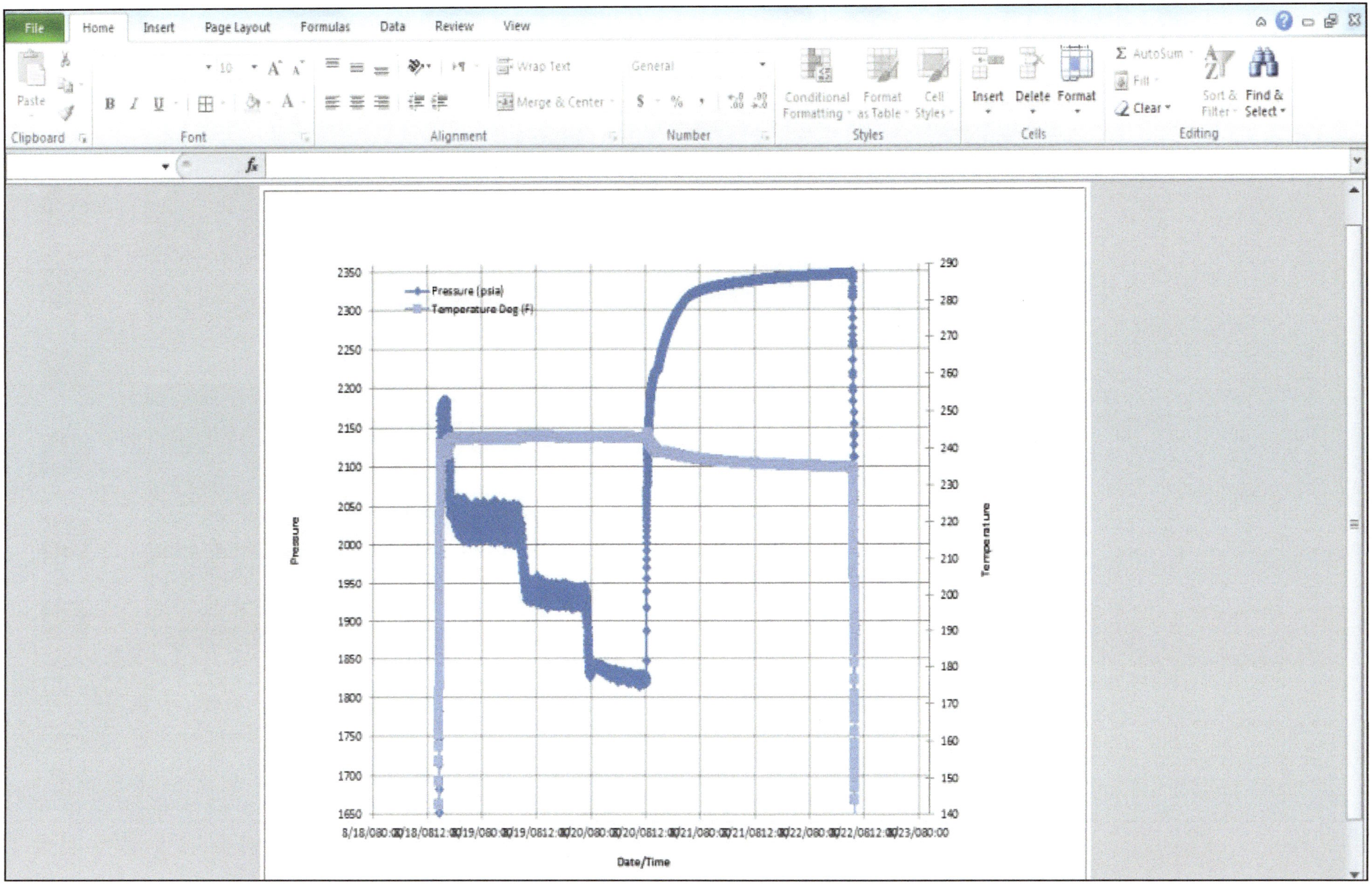

Figure 6. The buildup using the semi-log pressure plot (MS. Excel).

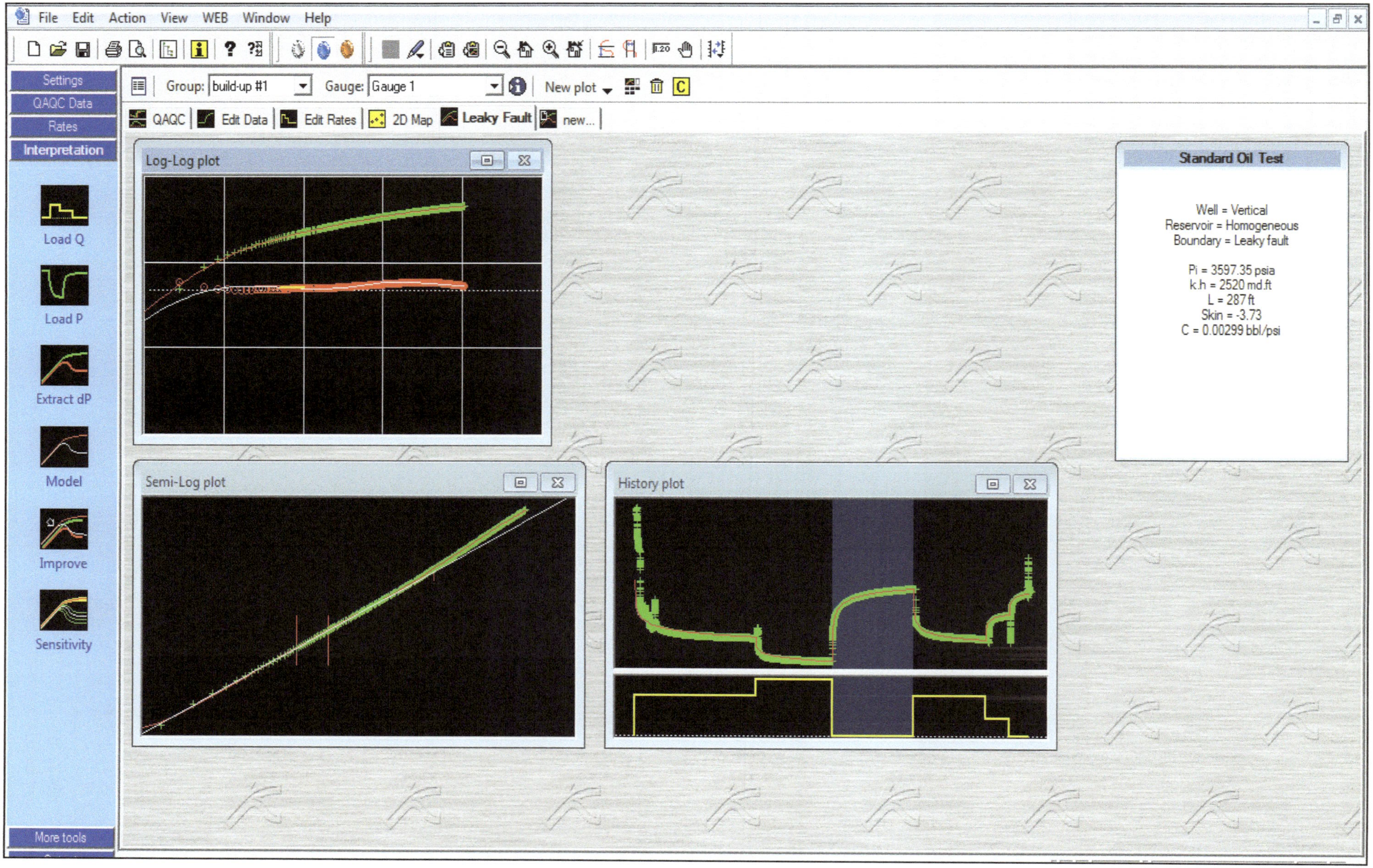

Figure 7. The buildup using the semi-log pressure plot (Ecrin Saphir).

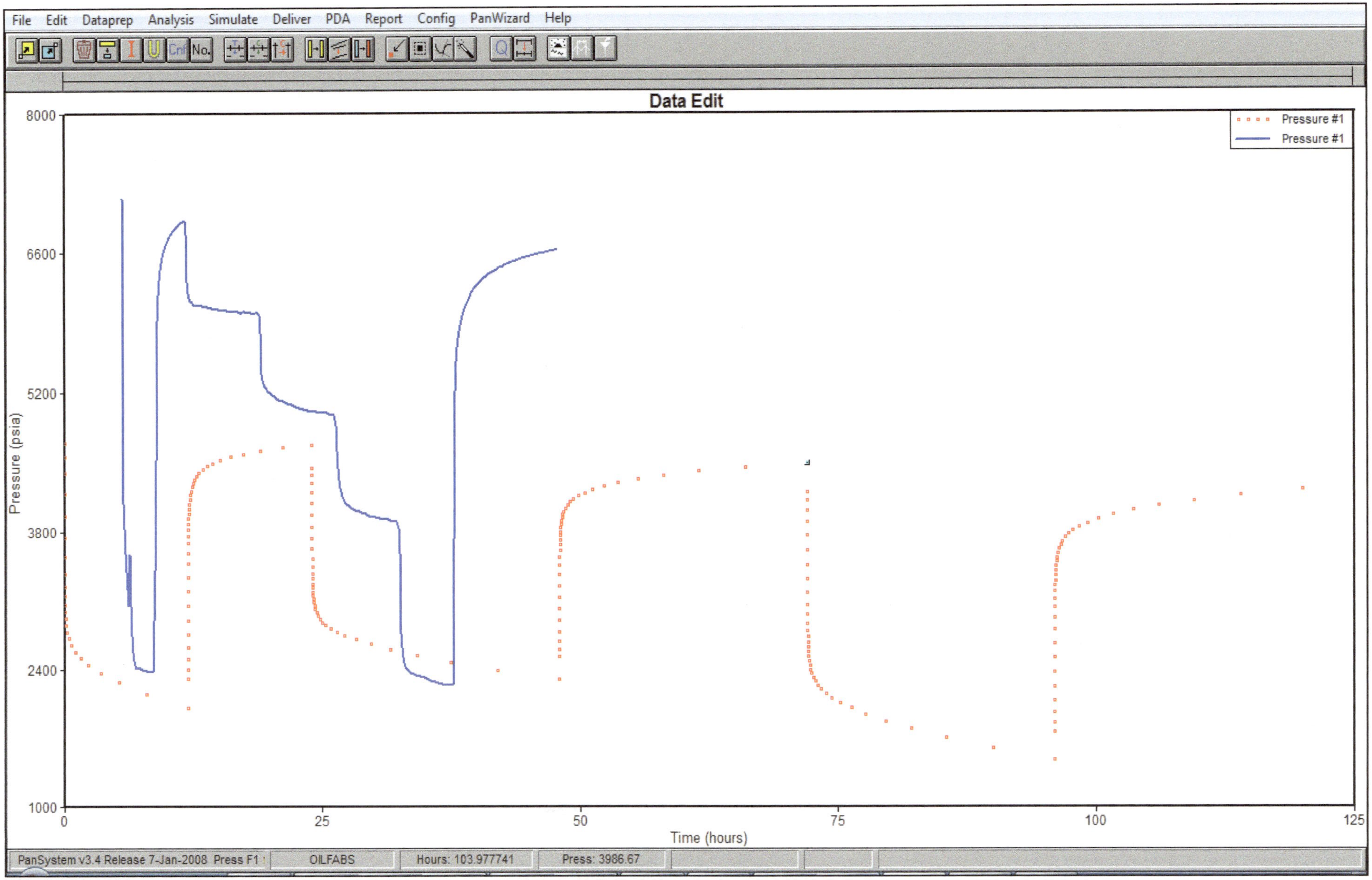

Figure 8. The buildup using the semi-log pressure plot (Pansystem).

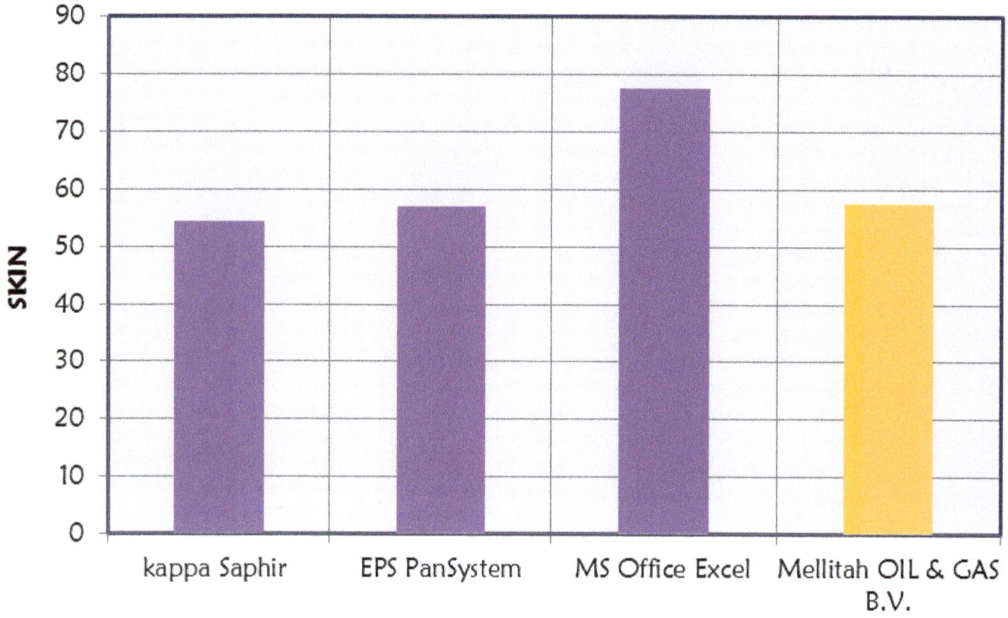

Figure 9. All results of well N1 and company result.

REFERENCES

Urayet AA (2000). "Transient Pressure Analysis". Part of the Technical Program Organized for the Petroleum Engineering Department, University of Tripoli, Tripoli.

Urayet AA (2004). "Advanced Topics in Transient Pressure Analysis". Part of the Technical Program Organized for the Petroleum Research Center, Tripoli.

Schlumberger W (1998). "introduction to well testing". England.

Ramey HG Jr (1992)."Advances in practical well test analysis". SPE 20592.

Economides MJ, Notle KG (1898). Reservoir simulation. Second edition, Texas.

Matthews CS, Russell DG (1987). Pressure Buildup and Flow Tests in Wells". SPE, Dallas.

Unneland T, Statoil LL (1995). "Limitations of the Skin Concept and Its Impact on Success Criteria Used in Sand Control". SPE.

Ahmed T, McKinney PD (2005). "Advanced Reservoir Engineering", pp. 5-14.

20

Identifying the flow dimension in fractured rock using an interference test

H. S. Liang[1]*, S. L. Lin[1], T. K. Huang[1], D. L. Chen[1] and M. C. Tom Kuo[2]

[1]Exploration and Development Research Institute, Chinese Petroleum Corporation (CPC), Miaoli, Taiwan.
[2]Department of Mineral and Petroleum Engineering, National Cheng Kung University, Tainan, Taiwan.

The actual flow dimension in fractured rock may be between 2-D cylindrical flow and 1-D linear flow. This paper demonstrated the importance of the method used to identify the flow dimension in fractured rock correctly. A case study of a long-duration interference test was presented to illustrate the application of Barker's type curves and to identify a flow dimension in fractured rock. It was difficult to judge the flow dimension based solely on the type-curve matching. Both geological and production data could also be taken into account to help select the appropriate flow dimension. To investigate the sensitivities to pumping time for each observation well, the σ function was defined for an observation well equals the sum of the square relative-errors between the observed and predicted drawdowns. For a long-duration interference test with sufficient pumping time, the σ function of an observation well is sensitive to the flow dimension. An empirical rule, θ (dimensionless time) > 100, was developed for estimating the minimum pumping time required in an interference test to identify the flow dimension between the production well and an observation well in fractured rock.

Key words: Interference test, fractured rock, flow dimension.

INTRODUCTION

Analytical solutions and type curves for 2-D cylindrical flow and 1-D linear flow dimensions developed by Theis (1935) and Jenkins and Prentice (1982), respectively, are widely used for analyzing interference test data for estimating the aquifer parameters (Fan et al., 2005). To determine the aquifer parameters, it is usually assumed that the flow dimensions are predefined along with assumptions of homogeneity and isotropy before analyzing drawdown data. However, it is usually the circumstance that no presumption about the dimension of the flow system can be made (Chakrabarty, 1994). In addition, the calculated aquifer parameters would be varied with flow dimensions. Thus, hydrologically based approaches for determining the flow dimensions and aquifer parameters are greatly needed.

Walker and Roberts (2003) indicated that the flow dimension is not necessarily a simple function of radial distance; the flow dimension and heterogeneity are inter-changeable when interpreting the flow dimension based on the assumption that hydro-geologic properties are function of radial distance. Chen and Liu (2007) pointed out the determination of apparent flow dimensions should consider all other knowledge of the system to construct a meaningful conceptual model of the system when commenting on the article by Walker and Roberts (2003).

While 2-D cylindrical flow dimension is most common, linear flow has been recognized in some fractured aquifers (Muskat, 1937; Jenkins and Prentice, 1982). Many studies recognized linear flow in the vicinity of a production well with buildup or drawdown data collected at the production well during short-duration tests (Gringarten et al., 1975). However, very little information is currently available for interference tests with sufficient pumping

*Corresponding author. E-mail: 155446@cpc.com.tw.

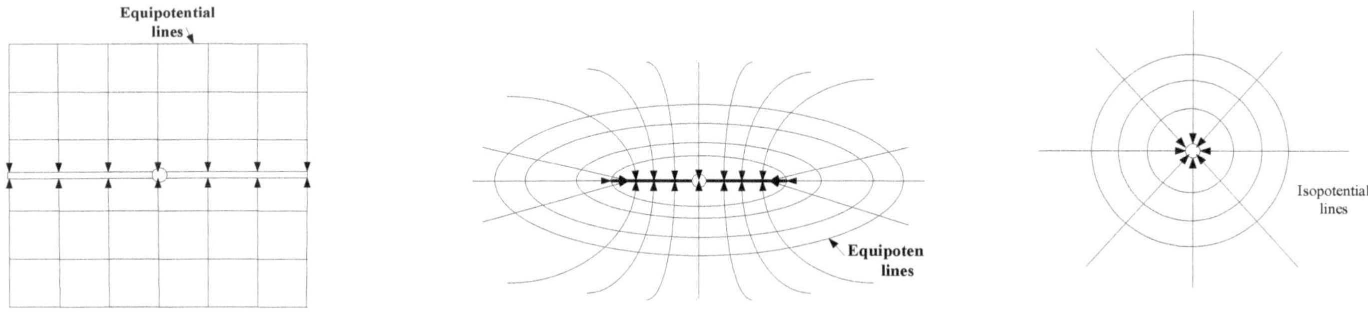

Source Boundary Conditions

$$\lim_{r\to 0}(Kb^2 \frac{\partial s}{\partial r}) = -\frac{\mathbf{Q}}{\alpha_n} = -\frac{Q}{2}$$

$$\lim_{r\to 0}(Kb^{3-n} r^{n-1})\frac{\partial s}{\partial r} = -\frac{\mathbf{Q}}{\alpha_n}$$

$$\lim_{r\to 0}(Kbr \frac{\partial s}{\partial r}) = -\frac{\mathbf{Q}}{\alpha_n} = -\frac{Q}{2\pi}$$

Analytical Solutions ($r_w \to 0$)

Planar source solution (Jenkins and Prentice, 1982)

Generalized Theis solution (Barker, 1988)

Line source solution (Theis, 1935)

(a) Flow dimension (n) = 1 **(b) 1 < Flow dimension (n) < 2** **(c) Flow dimension (n) = 2**

Figure 1. Flow dimensions, source boundary conditions and analytical solutions for fractured rock.

time to identify the flow dimension in fractured rock at an observation well. Jenkins and Prentice (1982) reported an interference test to identify the linear flow dimension in fractured limestone. However, Sen (1986) noted that the pumping time of Jenkins and Prentice's test was too short to identify the linear flow dimension at observation wells.

Barker (1988) developed a generalized radial flow model for hydraulic tests in fractured aquifers by regarding the dimension of the flow as a parameter. The objective of this paper was to demonstrate the application of Barker's generalized solution to identify the flow dimension correctly in fractured rock using a long-duration interference test in Chingshui geothermal field.

For exploring the flow dimension, it is important to determine the hydrogeological parameters with the flow dimension simultaneously. When analyzing drawdown data from the interference test, it is difficult to choose an appropriate flow dimension in a fractured formation system. The flow geometry may be considered as a 3-D spherical flow, 2-D radial flow, or 1-D linear flow according to the fracture density and isotropic/ anisotropic distribution. And the fractional flow dimension would exist with the variations of connectivity of the fracture system, spatial and temporal changes of flow dimension (Leveinen et al., 1998; Leveinen, 2000). To identify the flow dimension between the production well and an observation well, the σ function was utilized to further examine the curve-fitting through the sum of the square relative-errors between the observed and predicted drawdowns. This study focuses on 2-D areal flow; both geological and production data could also be taken into account to help select the appropriate flow dimension. The case study was further investigated to develop an empirical rule for determining the minimum pumping time required for an interference test using observation wells to identify the flow dimension in fractured rock at an observation well.

MATERIALS AND METHODS

The two flow dimensions of practical interest for fractured aquifers are radial flow and linear flow. In radial flow dimension, the flow lines are straight and converge in two dimensions (n = 2) toward a common center, for example, a well (Figure 1c). In linear flow dimension, the flow lines are parallel in one dimension (n = 1) and the cross section exposed to flow is constant (Figure 1a). Barker (1988) recognized that the actual flow dimension may be between radial and linear in fractured rock (1< n < 2; Figure 1b) and developed generalized Theis solution.

Theis solution – line source solution

Theis (1935) investigated unsteady radial flow problems in a confined aquifer with the following equation in terms of drawdown, s:

$$\frac{\partial^2 s}{\partial r^2} + \frac{1}{r}\frac{\partial s}{\partial r} = \frac{S}{T}\frac{\partial s}{\partial t} \qquad (1)$$

subject to the following boundary and initial conditions:

$$s(r,t) = 0 \qquad for \quad r \to \infty$$

$$\lim_{r\to 0}(r\frac{\partial s}{\partial r}) = -\frac{Q}{2\pi T} \qquad for \quad r \to 0$$

and

$$s(r,o) = 0 \qquad for \quad t = 0$$

The well-known Theis solution written in terms of the aquifer drawdown, s, is

$$s = \frac{Q}{4\pi T}[-Ei(-u)] = \frac{Q}{4\pi T}W(u) \tag{2}$$

where Q is the constant well production rate; T is the aquifer transmissivity; $-Ei(-u)$ is the exponential integral; and $W(u)$ is the well function; and $u = (r^2 S)/4Tt$; where r is the distance between the pumping well and observation well, S is the storage coefficient, and t is the pumping time. Both u and $W(u)$ are dimensionless. Well function, $W(u)$, represents an exponential integral, $-Ei(-u)$, that is,

$$W(u) = -Ei(-u) = \int_u^\infty \frac{e^{-\theta}}{\theta}d\theta \tag{3}$$

Generalized Theis solution

In order to extend Theis solution to a general flow dimension (1≤n≤2), Barker (1988) defined an area factor of equipotential surfaces, α_n, for any flow dimension, n. Using Theis assumptions, Barker (1988) derived a generalized flow equation in term of drawdown, s, expressed as follows:

$$S_s\frac{\partial s}{\partial t} = \frac{K}{r^{n-1}}\frac{\partial}{\partial r}\left(r^{n-1}\frac{\partial s}{\partial r}\right) \tag{4}$$

where S_s is the specific storage of the fracture system; K is the hydraulic conductivity; n is the dimension of the fracture flow system; r is the radial distance from the centre of the source; t is the well production time. For the constant-rate condition, equation (4) could be subjected to the following boundary and initial conditions:

$$s(r,t) = 0 \qquad for \quad r \to \infty$$

$$\lim_{r\to 0}(Kb^{3-n}r^{n-1}\frac{\partial s}{\partial r}) = -\frac{Q}{\alpha_n} \qquad for \quad r \to 0$$

and

$$s(r,o) = 0 \qquad for \quad t = 0$$

$$s(r,o) = 0 \qquad for \quad t = 0$$

where Q is the constant well production rate; $\alpha_n = 2\pi^{n/2}/\Gamma(n/2)$; b is the extent of the flow region; $\Gamma(x)$ is the gamma function.

The generalized Theis solution of drawdown, s, at any radial distance, r, from the center of the source and at any time, t, was solved using Laplace transform by Barker (1988) as follows:

$$s(r,t) = \frac{Qr^{2v}}{4\pi^{1-v}Kb^{3-n}}\Gamma(-v,u) \tag{5}$$

or

$$s(r,t) = \frac{Qr^{2v}}{4\pi^{1-v}Tb^{2-n}}\Gamma(-v,u) \tag{5a}$$

$$v = 1 - \frac{n}{2} \tag{6}$$

$$u = \frac{S_s r^2}{4Kt} \tag{7}$$

or

$$u = \frac{S\ r^2}{4\ T\ t} \tag{7a}$$

where S is storage coefficient; T is aquifer transimissivity; $\Gamma(-v,u)$ is the incomplete gamma function and can be calculated as follows:

$$\Gamma(a,u) = \Gamma(a) - \sum_{m=0}^{\infty}\frac{(-1)^m u^{a+m}}{m!(a+m)} \qquad for \quad a \neq 0,-1,-2... \tag{8}$$

In metric units, equations (5a) and (7a) can be written as follows:

$$\Gamma(-v,u) = W(\frac{n}{2}-1,u) = \frac{4\pi^{1-\upsilon}Tb^{2-n}}{Qr^{2\upsilon}}s \tag{9}$$

$$u = \frac{Sr^2}{4(60)Tt} \tag{10}$$

where Q is in m³/min; T is in m²/min; s is in meter; r is in meter; b is in meter; S is dimensionless; and t is in hour. $\Gamma(-v,u)$ was regarded as generalized well function,

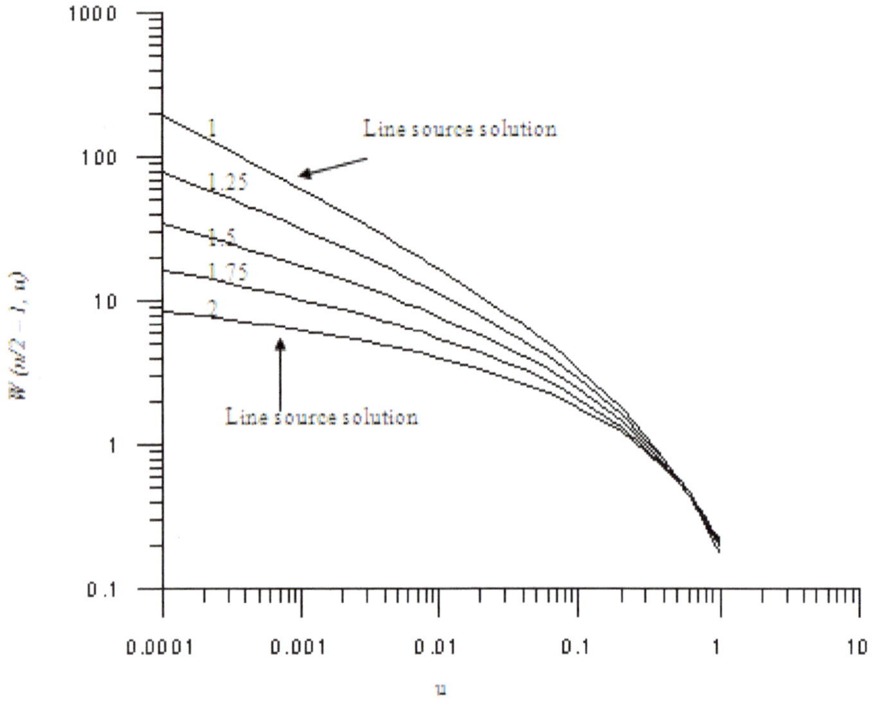

Figure 2. Barker's type curves for various flow dimensions (n = 1, 1.25, 1.5, 1.75, and 2).

$W(\frac{n}{2}-1,u)$ (Barker, 1988).

Using the generalized Theis solution, equations (8), (9), and (10), Barker (1988) also constructed a family of type curves on a log-log plot for various flow dimensions using *n* as a parameter. The flow dimension, *n*, is not necessarily an integer and must be determined empirically (Barker, 1988). Figure 2 shows five type curves with various n values (that is, n = 1, 1.25, 1.5, 1.75, 2). The type curves of n = 2 and n = 1 are line source solution for radial flow dimension (Theis, 1935) and planar source solution for linear flow dimension (Jenkins and Prentice, 1982), respectively.

RESULTS

Case study

Taiwan is located at the western rim of the Circum-Pacific margin, a convergent and compression boundary between the Philippine Sea and Eurasian Plates. The Chingshui geothermal field is located in the Northeast portion of Taiwan. It can be imaged that Chingshui geothermal area is a fracture-dominated aquifer. Production in the liquid-dominated Chingshui geothermal field is largely from a fractured formation. An interference test was conducted for the initial assessment of Chingshui geothermal reservoir to determine the transmissivity and storage coefficient for estimating deliverability and reserves (Chang and Ramey, 1979). Figure 3 is a scale map showing both surface and bottom-hole locations of these wells. Also, it is evident that the subsurface fracture dominated the movement (N-S direction) of borehole while drilling. Furthermore, The Chingshui geothermal area is situated on a monocline structure, which is cut internally by numerous thrust faults that essentially trend parallel to the bedding (NE-SW) and are lightly curved; the most important ones are the Tashi, Hsiaonanao and Hanhsi faults. To identify the flow dimension, the geological data shows a possible linear flow dimension that the fractured formation dominated the flow system in subsurface.

During the aquifer test, the well 16T was produced, and pressure responses were observed in wells 4T, 5T, 9T, 12 T, 13T, and 14T. A complete set of aquifer data is presented in Table 1. Table 2 summarizes the well capacity and completion for wells 4 T, 9 T, 12 T, 14 T, and 16 T. The hot-water production rate of the well 16T measured by weir ranged from 80 to 84 tons/hour during the eleven-day aquifer test. Equivalent to 80 tons/hour of hot-water, the total production rate of well stream was 1.89 m^3/min. The test was conducted by observing wellhead pressures at the observation wells. The aquifer data for wells 5 T and 13 T did not appear to be reliable because of some malfunction of equipment.

The test data can be analyzed by means of curve fitting using a family of Barker's type curves for various flow dimensions (n = 1, 1.25, 1.5, 1.75, and 2). Table 3 shows

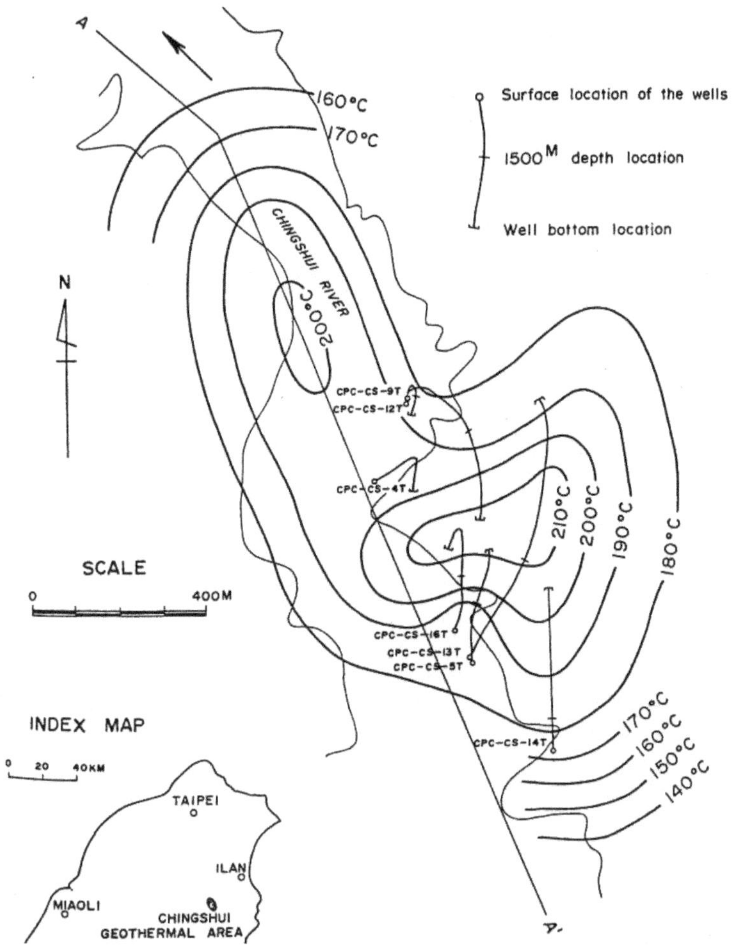

Figure 3. Location of the wells and the isotherms of presumed formation temperature for 1500 m depth in the Chingshui geothermal area (Chang and Ramey, 1979).

that the calculated values of aquifer transmissivity and storage coefficient (T and S) obtained for each observation well depended on the flow dimension selected for type-curve matching. However, it is difficult to judge the flow dimension based solely on the type-curve matching. Both geological and production data must also be considered to help select the flow dimension.

The geological data also strongly indicate a possible linear flow dimension on a field-wide scale. Since the porosity and permeability of slates (aquifer rock) are low, faults, joints, and other extensive fractures provide the conduits for the fluid flow. Figure 4 shows the rose diagram for 67 joints measured at an outcrop of the aquifer rock located near the Chingshui geothermal field (Tseng, 1978). The most predominant set of joints strikes a linear direction of Northwest.

In addition, the capacity of a well depends on a number of factors such as transmissivity, well skin, and well completion. According to Darcy's law, under similar conditions of well completion and skin, well capacity is proportional to the transmissivity. The transmissivity estimated for various flow dimensions was compared and correlated to well capacities to select the appropriate flow dimension. An idea of linear correlation between well capacity and transmissivity was presented in this work to demonstrate the positive relationship. A production test was strongly suggested for each well to identify the well capacity. The well capacities could be further utilized to verify the flow dimension.

Figure 5 illustrates regression lines of well capacity versus transmissivity using data from wells 4T, 9T, 12T, and 14T for five various flow dimensions (n = 1, 1.25, 1.5, 1.75, and 2). The smallest value of the sample correlation squared regression coefficient (that is, R^2 = -0.4319) corresponding to the radial dimension (n = 2) indicates that the two parameters are not well correlated. On the other hand, coefficient is the highest (i.e., R^2 = 0.9013) for the linear flow dimension (n = 1). According to the results

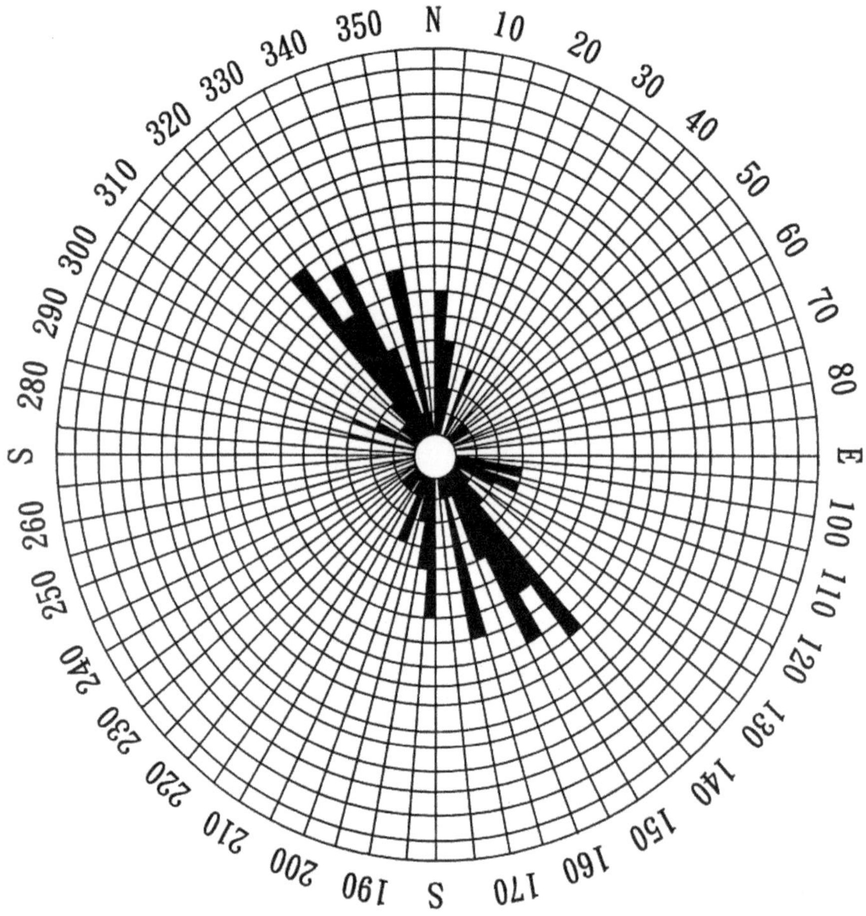

Figure 4. Rose diagram of 67 joints in the Chingshui geothermal area (Tseng1978).

shown in Figure 4, the transmissivity estimated using the linear flow dimension (n = 1) appeared to best correlate with well capacity, whereas the correlation was the worst for the radial flow dimension (n = 2).

DISCUSSION

Method to identify the flow dimension in fractured rock

To identify the flow dimension and investigate the sensitivities to pumping time for each observation well, the σ function for an observation well was proposed as follows:

$$\sigma = \sum_{i=1}^{M}\left[\left(s_i - \frac{Qr^{2v}W(\frac{n}{2}-1,u_i)}{4\pi^{1-v}Tb^{2-n}}\right)\Bigg/ s_i\right]^2 \quad (11)$$

where s_i is the observed drawdown; $\frac{Qr^{2v}W(\frac{n}{2}-1,u_i)}{4\pi^{1-v}Tb^{2-n}}$ is the predicted drawdown; $u_i = Sr^2/4Tt_i$ for the i th data point; and M is the total number of data points at an observation well.

Figure 2 shows that the type curves of various flow dimensions are not discernible for a large value of u. Therefore, the σ function is not sensitive to the flow dimension for an observation well located too far away from the production well, or, for a short-duration interference test with insufficient pumping time. However, for a small value of u, the type curves of various flow dimensions separate from each other. For a long-duration interference test with sufficient pumping time, the σ function of an observation well is sensitive to the flow dimension. The correct flow dimension can be selected at the minimum value of the σ function.

Figure 6 shows the variations of the σ function with

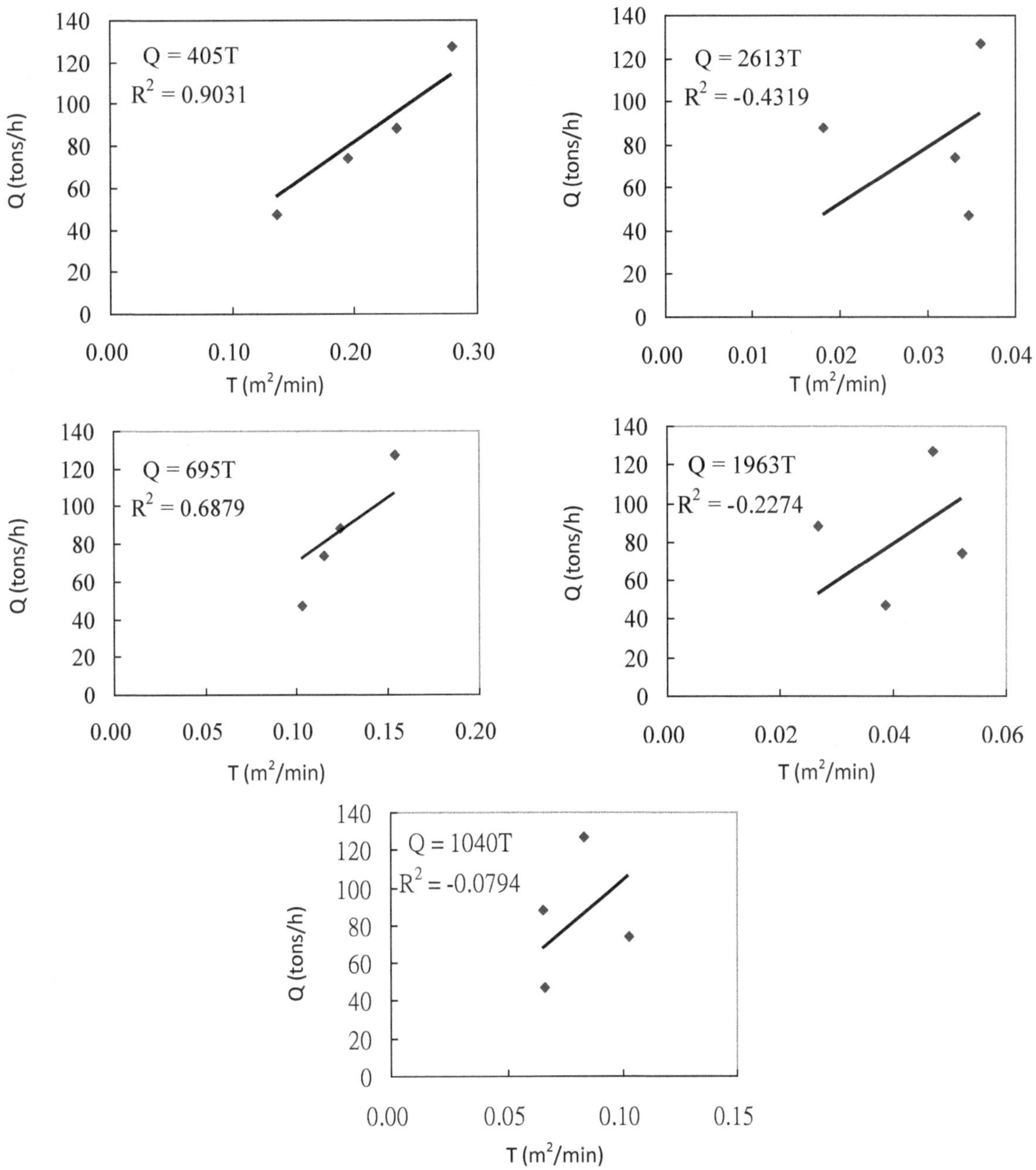

Figure 5. Regression lines of well capacity versus aquifer transmissivity for various flow dimensions.

the flow dimension, n, for four observation wells (4T, 9T, 12T, and 14T) in Chingshui field. For the observation wells 9T and 14T which are 300 m and 330 m from the pumping well 16T, the σ function is not sensitive to the flow dimension. The total pumping time of the well 16T in Chingshui interference test was 258.5 h and was not long enough to determine the flow dimension in fractured rock at observation wells 9T and 14T because the distance from both observation wells is too far away from the pumping well 16T.

For the observation well 4T which is 175 m from the pumping well 16T, the σ function shows a little sensitivity to the flow dimension. For the observation well 4T, the pumping time of 258.5 h starts to reveal the flow dimension.

For the observation well 12T, which is the closest observation well and only 90 m from the pumping well 16T, the σ function is very sensitive to the flow dimension. The pumping time of 258.5 h is sufficient to identify the flow dimension between the observation well 12T and the

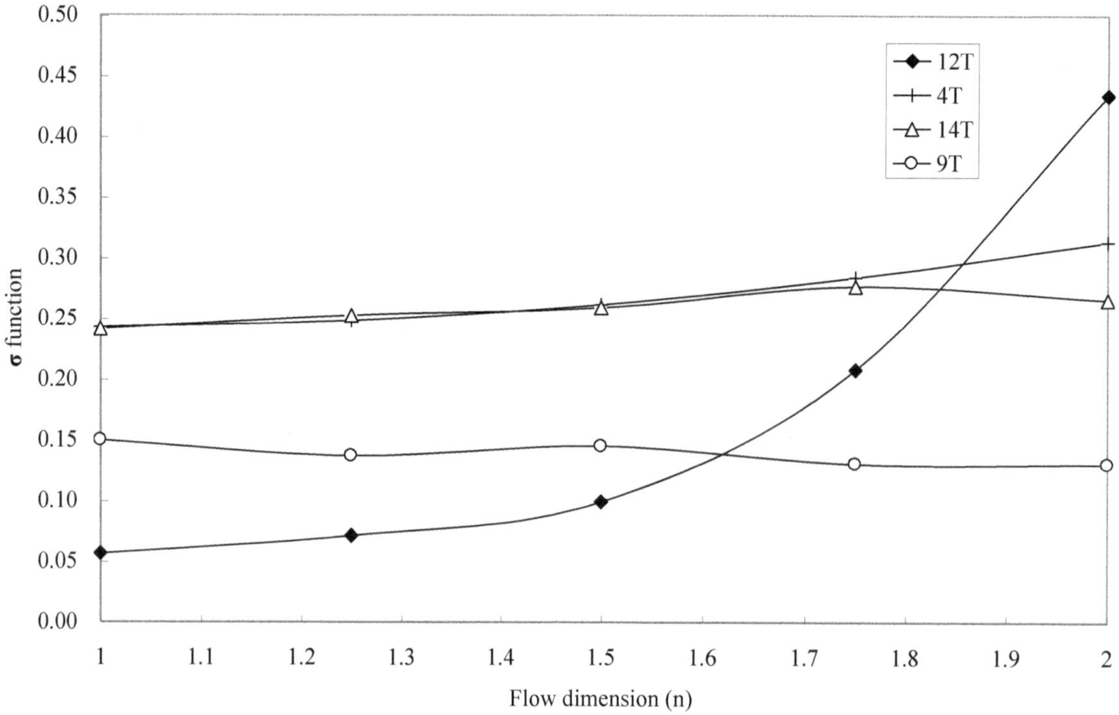

Figure 6. Variations of σ function with flow dimension for four observation wells (12T, 4T, 14T, and 9T).

Table 1. Aquifer test in Chingshui geothermal field (Chang and Ramey, 1979).

Time (h)	Observation wells												Flowing well		
	4T			9T			12T			14T			16T		
	WHP*		s **	WHP		s	WHP		s	WHP		s	WHP		Weir water rate
	kg/cm²	psi	m-H₂O	kg/cm²	psi	m-H₂O	kg/cm²	psi	m-H₂O	kg/cm²	psi	m-H₂O	kg/cm²	psi	(tons/h)
0	12.09	172	0.00	9.70	138	0.00	13.15	187	0.00	9.35	133	0.00	18.14	258	0
18.5	12.02	171	0.73	9.63	137	0.70	13.01	185	1.41	9.35	133	0.00	4.85	69	24
42.5	11.81	168	2.93	9.49	135	2.11	11.39	162	3.52	9.14	130	2.11	4.08	58	83.5
66.5	11.67	166	4.41	9.35	133	3.52	12.80	182	3.52	8.79	125	5.63	3.94	56	83.1
90.5	11.67	166	4.41	9.14	130	5.63	12.66	180	4.92	8.79	125	5.63	3.94	56	83.1
114.5	11.60	165	5.14	9.14	130	5.63	12.59	179	5.63	8.65	123	7.03	3.94	56	82
138.5	11.53	164	5.87	9.14	130	5.63	12.52	178	6.33	8.51	121	8.44	3.94	56	82.4
162.5	11.53	164	5.87	9.07	129	6.33	12.44	177	7.03	8.44	120	9.14	3.80	54	82.4
186.5	11.46	163	6.61	9.00	128	7.03	12.37	176	7.74	8.37	119	9.85	3.80	54	81
210.5	11.39	162	7.35	8.93	127	7.74	12.30	175	8.44	8.37	119	10.55	3.73	53	80
234.5	11.39	162	7.35	8.93	127	7.74	12.30	175	8.44	8.23	117	11.25	3.66	52	80
258.5	11.32	161	8.08	8.86	126	8.44	12.30	175	8.44	8.09	115	12.66	3.66	52	80***

* WHP: Wellhead pressure. ** Δp : Pressure interference. *** Equivalent of well stream total production rate of 105 tons/h, or, 1.89 m³/min.

pumping well 16T in fractured rock.

The variation of the σ function at the observation well 12T as shown in Figure 6 can be used to judge the flow dimension. The values of the σ function at the observation well 12T are 0.434, 0.208, 0.099, 0.071, and0.056 for the flow dimension, n = 2, 1.75, 1.5, 1.25,

Table 2. Well capacity and completion data for wells in Chingshui geothermal field.

Well	Well capacity (tons/h)	Total depth TD (m)	Completion interval (m)	Temperature at TD (°C)	Distance from well 16T (m)
4T	126.7	1505	498 - 1503	201	175
9T	74.0	2079	490 - 2074	205	300
12T	46.9	2003	1048 - 1998	223	90
14T	88.0	2003	947 - 1995	215	330
16T	116.2	3000	830 - 2990	225	-

Table 3. Aquifer transmissivity (T) and storage coefficient (S) estimated for various flow dimensions using type-curve matching.

Flow dimension	Observation wells			
	4T	9T	12T	14T
Transmissivity (m^2/min)				
1	280×10^{-3}	195×10^{-3}	136×10^{-3}	235×10^{-3}
1.25	154×10^{-3}	116×10^{-3}	103×10^{-3}	124×10^{-3}
1.5	826×10^{-3}	102×10^{-3}	65.8×10^{-3}	65.1×10^{-3}
1.75	47.0×10^{-3}	52.1×10^{-3}	38.5×10^{-3}	26.7×10^{-3}
2	36.1×10^{-3}	33.1×10^{-3}	34.6×10^{-3}	18.0×10^{-3}
Storage coefficient				
1	6.72×10^{-3}	5.14×10^{-3}	9.79×10^{-3}	3.21×10^{-3}
1.25	7.40×10^{-3}	3.88×10^{-3}	7.17×10^{-3}	2.98×10^{-3}
1.5	5.95×10^{-3}	2.45×10^{-3}	13.0×10^{-3}	1.88×10^{-3}
1.75	5.87×10^{-3}	2.50×10^{-3}	18.5×10^{-3}	1.47×10^{-3}
2	5.02×10^{-3}	1.75×10^{-3}	1.99×10^{-3}	1.65×10^{-3}

and 1, respectively. The value of the σ function attains a minimum with a linear flow dimension (n = 1). Therefore, the flow dimension between the production well 16T and the observation well 12T can be identified as linear (n = 1). The validity of identifying the linear flow dimension between wells 16T and 12T is strongly supported by the geological data.

An empirical rule for estimating the minimum pumping time

It is also of practical interest to be able to estimate how long the test duration is required to determine the flow dimension between the production well and an observation well in fractured rock. The field data of Chingshui aquifer test could provide opportunities to identify the flow dimension in fractured rock, and be investigated further to develop an empirical rule for determining the minimum pumping time required to identify the flow dimension in fractured rock.

For a long-duration interference test with sufficient pumping time, the type curves of the generalized well function become discernible among various flow dimensions when the value of dimensionless time ($\theta = 4Tt/Sr^2$) gets large enough, or, $u = Sr^2/4Tt$ gets small enough (Figure 2). Table 4 lists the calculated values of dimensionless time (θ) for observation wells 12T, 4T, 9T, and 14T using a pumping time of 258.5 h and a linear flow dimension. Based on Table 4, we tried to develop the following empirical rule for estimating the minimum pumping time required in aquifer tests to identify the flow dimension.

$$\theta_{\min} = \frac{4Tt_{\min}}{Sr^2} > 100 \tag{12}$$

where $\theta_{\min}$ is the minimum dimensionless pumping time; $t_{\min}$ is the minimum pumping time required; r is the distance between the observation well and pumping well; T is the aquifer transmissivity; and S is the storage

Table 4. Calculated values of u for various observation wells in Chingshui aquifer test.

Observation wells	Transmissivity, T*, (m^2/min)	Storage coefficient, S*	Distance from pumping well 16 T r, (m)	Dimensionless time $\theta = {4Tt}/{Sr^2}$
12T	0.136	0.00979	90	106
4T	0.280	0.00672	175	84
9T	0.195	0.00514	300	26
14T	0.235	0.00321	330	41

*Estimated aquifer parameters for flow dimension, n = 1. **production time of well 16T, t = 258.5 h.

coefficient. The proposed empirical rule would be useful for planning an interference test. Very little information is currently available for interference tests with sufficient pumping time to identify the flow dimension in fractured rock at an observation well. It is recommended to improve the concept of equation (12) when additional data sets become available.

Conclusions

1. The actual flow dimension in fractured rock may be between radial flow and linear flow. If the flow dimension in fractured rock is not identified correctly, there would be significant error in aquifer parameters estimated from the analyses of aquifer tests.
2. Barker's type-curves of generalized Theis solution can be employed to analyze drawdown data from an interference test in fractured rock. However, it is difficult to judge the flow dimension based solely on the type-curve matching. Both geological and production data must also be taken into account for selecting an appropriate flow dimension.
3. Very little information is currently available for interference tests with sufficient pumping time to identify the flow dimension in fractured rock at an observation well. The case study of Chingshui demonstrates that a long-duration interference test with sufficient pumping time can provide important information concerning the flow dimension between the production well and an observation well in fractured rock.
4. To identify the flow dimension in fractured rock, the σ function for an observation well to investigate the sensitivities to pumping time for each observation well was proposed. For a long-duration interference test with sufficient pumping time, the σ function of an observation well is sensitive to the flow dimension. The σ function of an observation well is not sensitive to the flow dimension for a short-duration interference test with insufficient pumping time.
5. An empirical rule was developed for estimating the minimum pumping time required for an interference test to identify the flow dimension in fractured rock at an observation well. The proposed empirical rule would be useful for planning an interference test. It is recommended to improve the concept of equation (12) when additional data sets become available.

REFERENCES

Barker JA (1988). A generalized radial flow model for hydraulic test in fractured rock. Water Resour. Res. 9(10):1796-1804.

Chakrabarty C (1994). A note on fractional dimension analysis of constant rate interference tests. Water Resour. Res. (30):2339-2341.

Chang CRY, Ramey HJ (1979). Well interference test in the Chingshui geothermal field. In 5th Geothermal Reservoir Engineering Workshop. Stanford University, Stanford, Calif.

Chen CS, Liu IY (2007). Comment on "Flow dimensions corresponding to hydrogeologic conditions" by Douglas DW and Randall MR. Water Resour. Res. (43):W02601.

Fan KC, Kuo MCT, Liang KF, Lee CS, Chiang SC (2005). Interpretation of a well interference test at Chingshui geothermal field, Taiwan. Geothermics 34:99-118.

Gringarten AC, Ramey HJ, Raghavan R (1975). Applied pressure analysis for fractured wells. J. Pet. Technol. pp. 887-892.

Jenkins DN, Prentice JK (1982). Theory for aquifer test analysis in fractured rocks under linear (nonradial) flow conditions. Ground Water 20(1):12-21.

Leveinen J (2000). Composite model with fractional flow dimensions for well test analysis in fractured rocks. J. Hydrol. (234):116-141.

Leveinen J, Ronka E, Tikkanen J, Karro E (1998). Fractional flow dimensions and hydraulic properties of a fracture-zone aquifer, Leppavirta, Finland. Hydrogeol. J. (6):327-340.

Muskat M (1937). The flow of homogeneous fluids through porous media. McGraw-Hill Book Co. pp. 125-128.

Sen Z (1986). Aquifer test analysis in fractured rocks with linear flow pattern. Ground Water 24(1):72-78.

Theis CV (1935). The relation between the lowering of the piezometric surface and the rate and duration of discharge of a well using ground-water storage. Trans. Am. Geophys. Union 16:519-524.

Tseng CS (1978). Geology and geothermal occurrence of the Chingshui and Tuchang districts. Ilan. Petrol. Geol. Taiwan 15:11-23.

Walker DD, Roberts RM (2003). Flow dimensions corresponding to hydrogeologic conditions. Water Resour. Res. 39(12):1349. doi: 10.1029/2002WR001511.

Prediction of drilling pipe sticking by active learning method (ALM)

Morteza Elahi Naraghi[1]*, Peyman Ezzatyar[2], Saeid Jamshidi[3]

[1]Petroleum and Geosystems Engineering Department, University of Texas at Austin, Austin, TX.
[2]Science and Research Branch, Islamic Azad University, Tehran, Iran.
[3]Chemical and Petroleum Engineering Department, Sharif University of Technology, Tehran, Iran.

Stuck piping is a common problem with tremendous impact on drilling efficiency and costs in oil industry. Generally, the stuck pipe troubles are solved after their occurrences by using some standard techniques; here we attempt to predict the causes of occurrence of such problems to eschew risks and excessive drilling costs. If these risks are identified in advance, better solutions can be provided to reduce the associated consequences. Based on the literature, this problem is caused by numerous parameters, such as drilling fluid properties and the characteristics of the mud cake that is formed while drilling. In this study, an attempt is made to develop a model for stuck pipe prediction. To consider all aspects of pipe sticking and behavior of the involved variables, the fuzzy logic and active learning method (ALM) can be used as a primary predictive tool. Active Learning Method is a robust recursive fuzzy modeling without computational complexity. These methods are broadly used in many industries; including oil and gas. This paper proposes a systematic approach for pipe stuck prediction based on ALM. The results of this method are more accurate than other methods and prediction accuracy is close to perfect either in stuck or non-stuck cases. This study presents a case study in which the ALM is used successfully to estimate pipe sticking. Thus, the proposed method possesses reliable results for prediction of pipe stuck, and can be used in order to minimize the risk of pipe sticking.

Key words: Pipe stuck prediction, active learning method (ALM), artificial intelligence, drilling engineering.

INTRODUCTION

Over several years oil industry is facing troubles associated with the stuck pipes. Differential pipe sticking is one of the stuck pipe mechanisms with a major impact on drilling efficiency and well costs (Adams, 1977a; Weakley, 1990; Wisnie and Zheiwei, 1994). These occurrences are common everywhere in the world and are estimated to cost the industry hundreds of millions of dollars annually. In some areas, events related to differentially stuck pipe can be responsible for as much as 40% of the total well cost. Differential pipe sticking problems generally result in the significant amount of downtime and remedial costs and well cost and time overruns as a non-productive time in terms of loss of rig days either due to stopping of drilling operations or an attempt to free the stuck pipe. This huge loss is always accounted for in the well budget cost as a contingency factor for the risks associated with the stuck pipe problems in the well planning and drilling performance approach (Adams, 1977b; Beigler and Kuhn, 1994; Wisnie and Zheiwei, 1994; Sharif, 1997; Aadnoy et al., 1999). The recent increase in drilling activity, shortage of experienced personnel and equipment, and drilling in higher-risks areas have increased the risk of stuck pipe events in all drilling operations (Yarim et al., 2007).

The concept of differential pressure sticking of drill pipe was first reported by Helmick and Longley (1957) according to laboratory tests. They stated that pipe sticking results when the drill pipe becomes motionless against a permeable bed and a portion of the area of the pipe is isolated by filter cake. Hempkins et al. (1987) analyzed pipe sticking statistically based on drilling parameters. This was done by comparing the properties of non-pipe stuck wells with the ones that had stuck piping. Then drilling operations were planned according to non-pipe stuck wells' characteristics. In that study, the parameters of 221 wells were investigated in 131 stuck pipes' cases in Mexico's wells and the risk of stuck pipe occurrence in others wells were estimated. Biegler and Kuhn (1994) generated a data base including 22 drilling parameters in 73 non-pipe stuck wells and 54 pipe sticking wells in Mexico's gulf. Recently, some research is being conducted in order to determine the characteristics of stuck pipe such as the depth of pipe sticking. Torne et al. (2011) tried to determine the depth of pipe sticking by means of continuous Free-Pipe logs. These studies were the base of primary comparative analysis that could identify the pipe sticking mechanisms in addition to its probability prediction. Howard and Glover (1994) improved the prediction stuck pipes' models by applying statistical techniques in 100 wells of Mexico's gulf. These models were used for prevention of pipe sticking and operation saving. Siruvuri et al. (2006) recently presented an application of Artificial Neural Network (ANN) methods for understanding the causes of differential stuck pipe. Miri et al. (2007) implemented ANN to predict the pipe sticking in Iranian offshore oil fields. Murillo et al. (2009) did a study to predict and avoid pipe sticking based on adaptive fuzzy logic. Al-Baiyat and Heinze (2012) investigated application of ANN and Support Vector Machines (SVM) in stuck pipe prediction. Jahanbakhshi and Keshavarzi (2012) used SVM with Gaussian kernel function to predict differential pipe sticking. Elahi Naraghi et al (2013) did a comprehensive study to compare the performance of different Neural Networks and Neuro Fuzzy Systems in prediction of pipe stuck.

Differential pipe sticking occurs when a part of the drill string, casing, or logging tool becomes embedded in a mud solids filter cake and is held there by a significant amount of differential pressure. This differential pressure is the pressure difference between the hydrostatic pressure of mud and the formation pore pressure. Usually, because of the excessive differential pressure, the sticking takes place across porous and permeable formations such as sandstone or limestone, where a mud filter cake builds up during drilling. It does not occur in very low permeability formations such as shale, where mud filter cakes normally do not form. Stuck pipe is identified as an impedance of drilling mud flow in the annular space and the difficulty of the pipe movement either in the upward or downward direction. In a complete stuck pipe situation, neither circulation nor pipe movement are possible.

Although these symptoms are similar to Key Seat sticking, they usually occur under different drilling conditions. Significant mud overbalance, as well as an exposed permeable section, must also exist for differential sticking to occur. Clearly, as many reservoirs become depleted, a significant number of wells will be drilled with high overbalance pressures, thereby maintaining the industry's concerns over differential sticking.

The likelihood of differential sticking increases further with the length of the permeable section that is open to the drilling fluid. The continued trend towards extended reach and horizontal drilling means that increasing lengths of permeable formations are exposed. Clearly, the nature of the rock formations encountered certainly cannot be altered. Therefore, if those formations carry a high risk of differential sticking, this has to be accepted. Furthermore, high overbalance pressures may be unavoidable if they are needed to maintain well control or wellbore stability in other parts of the open-hole section. However, mud composition and properties can be modified, within limits, in the prevention of differential sticking.

In the past multivariate statistical analysis techniques and simulated sticking testes using different drilling fluids have been performed to identify and modify parameters that lead to differential pipe sticking in order to prevent or minimize sticking. A review of published literature and laboratory data establishes the importance of mud filter cake properties (thickness, shear strength, and lubricity) on the differential sticking tendencies of mud.

Artificial Intelligence methods, such as Neural Networks and Fuzzy logic, have the ability to represent complex stuck pipe situations, which involve several variables. The methodology enables drilling industry personnel to estimate the risk of occurrences of stuck pipe not only during well planning but also during drilling. A proper prediction of the risk of differential pipe sticking will identify the main causes of the problem and consequently, the best techniques to prevent stuck pipe can be done.

This paper is organized as follows. First, an overview of active learning method is presented. The subsequent methodology section discusses input and output parameters, preprocessing step, and evaluation method are explained. Finally, the implementation results and accuracy of prediction are presented and analyzed in the numerical results part of this work.

MATERIALS AND METHODS

As discussed above, in this paper a novel approach for prediction of pipe sticking which is based on active learning method (ALM) is presented. In this method, our system possesses multiple inputs and one output, which is the probability of pipe sticking. Our model is trained by Active Learning Method. Next an overview of Active Learning Method is presented.

Active learning method

Active Learning Method (Bagheri Shouraki and Honda, 1997) is a robust recursive fuzzy modeling without computational complexity. The main idea behind ALM is to break M. I. S. O. system into some S. I. S. O. subsystems and aggregate the behavior of subsystems to obtain the final output (Mita, 2000; Nishino et al., 1999; Sakurai et al., 2003). This idea is the same as the brain activity which stores the behavior of data instead of the exact values of them. Each S. I. S. O. subsystem is expressed as a data plane (called IDS plane) resulted from the projection of the gathered data on each input-output plane (Bagheri Shouraki and Honda, 1997; Bagheri Shouraki et al., 1999; Yuasa et al., 1992). Two types of information can be extracted from an IDS plane. First, the behavior of output respect to each input variable that is illustrated by a curve called narrow path. Second, the level of confidence for each input variable which is proportional to the reciprocal of variance of data around the narrow path. Narrow paths are estimated by applying Ink Drop Spread (IDS) on data points and Center of Gravity (COG) on data planes (Bagheri Shouraki, 2000; Bagheri Shouraki and Honda, 1999). IDS and COG are two main operators of ALM (Bagheri Shouraki and Honda, 1997; Sagha et al., 2008).

It is very difficult for human to memorize the numerical data points but tries to memorize the general behavior function of data points. In addition, for modeling, the human converts a MIMO (Multi Inputs - Multi Outputs) system to some SISO (Single Input – Single Output) systems and then he tries to find the general behavior function in each SISO system and the effects of other inputs are considered as the deviation of data points around of the general behavior function. In addition, human can save the data points on a continuous path which means the general behavior function, but usually fails to save the randomly distributed data points in the space of variable. ALM algorithm uses all of these mentioned constructions of human modeling method. Taheri Shahraiyni (2007) developed new heuristic search, fuzzification and defuzzification methods for ALM algorithm.

In this paper, a fuzzy-based modeling approach is applied. Active Learning Method (ALM) is one of the fuzzy modeling methods which usesa basic level of mathematics. ALM was invented by Bagheri Shouraki and Honda (Bagheri Shouraki and Honda, 1997).ALM has a very simple algorithm that avoids mathematical complexity and its accuracy increases unlimitedly by increasing the number of iterations inthe algorithm.

IDS and COG Method

IDS method breaks down a complicated system into simpler sections the same as the way in which humans act encountering sophisticated subjects. For multi-input single-output (MISO) systems, this is done by dividing the MISO system, $y = f(x_1, x_2, \ldots, x_N)$, into multiple single-input single-output (SISO) systems. From available input-output training data, each SISO builds a pattern which will be utilized in modeling procedure in IDS method. For each MISO system, IDS method creates N 2-dimensional discrete planes which N is the number of inputs in this MISO system. The horizontal and vertical axes of ith plane are respectively xi and y. Then, all the training data is scattered on all of these planes. This step is called "data spread" or "distilling ink drop". As individual data spreads overlap each other, the overlapping regions become exceedingly darker, and eventually result in a pattern on all the planes. By implementing IDS method to constructed pattern image of each plane, two different types of information are elicited. The first one is the narrow path and the other is the deviation of the spread data points around each narrow path. Figure 1 depicts a typical result of implementing IDS method to data by radius equal to 1. This figure is a result of implementing ALM steps on data. This figure is a function between output and measured depth in the last step of dividing. The details of the steps are explained in next sections. Each Narrow path illustrates the relationship between its horizontal and vertical axes. For instance, the narrow path of 4th plane represents the relation existing between the output and the 4th input (x_4). The deviation of spread points from the narrow path reflects the importance degree of the horizontal axes in system behavior. The less deviation of spread data from the narrow path, the higher importance degree the parameter possesses in system behavior. In other words, in this case MISO system can be approximately simplified to the SISO system described by the narrow path in that plane.

The next step is to select a representative point in each column of each plane based on COG method. This point is obtained by calculating the weighted mean of all diffused data points in each column. Figure 2 depicts the extracted path of Figure 1 using COG. In order to determine the corresponding output of any new input vector $x_t = [x_{t1}, x_{t2}, \ldots, x_{tN}]$ by inference, values of narrow paths and spreads are calculated at this point from the pattern images of the planes. These values are then transferred from IDS units to the upper layer for being used for the inferential process in the ALM. ALM utilizes this information and approximates output value.

Heuristic search method

Partitioning of multi-dimensional space is a combinatorial problem. There is no theoretical approach for it; hence, heuristic search methods are used (Taheri Shahraiyni, 2007). The heuristic search is a guided search method.

Consider k inputs ($x_1, x_2, \ldots, x_k$), and a single output (y)system which is the same as our case with seveninputs and one output as mentioned earlier. The algorithm of the new heuristic search method for this system is depicted in Figure 3.

Step 1. The domain of x_1 is divided into two parts (small and big). Using the ALM algorithm, the best continuous path is determined for each part of the x_1 domain. Assume these paths are $f_{11}(x_j)$ and $f_{12}(x_m)$, which are the best paths for the first dividing step and for the small and big parts of the divided variable that are the functions of the jth and m^{th} variables, respectively. Here, the rules for modeling are:

If (x_1 is small) then $y=f_{11}(x_j)$

If (x_1 is big) then $y=f_{12}(x_m)$

Then, the modeling error (e_{11}) is calculated for the above rules. Similarly, the domain of other variables is divided and their modeling errors are calculated and a set of k errors ($e_{11}, e_{12}, \ldots, e_{1k}$) are generated. The variable corresponding to the minimum error is the best one for dividing of space. Suppose $e_{1s}$is the minimum error and it is correspond to x_s, then, the x_s domain is divided into small and big values. If e_{1s} is more than the threshold error, the dividing algorithm should continue.

Step 2. Consider all possible combinations of x_s–x_j ($j=1, 2, \ldots, k$) for each part of x_s and then divide the domain of x_j again into two parts. Thus, 2^kcombinations are generated (k combinations of $x_{s(small)}$–x_j

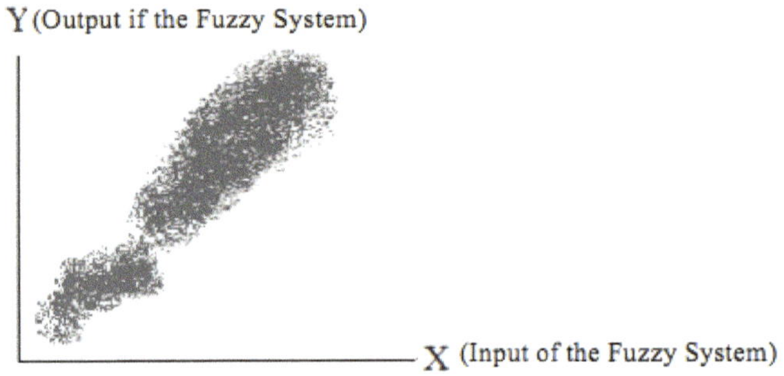

Figure 1. Ink drop spread results for spread radius 1.

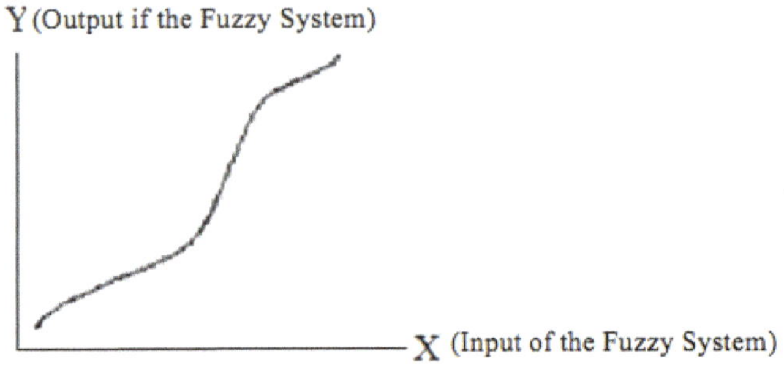

Figure 2. Extracted narrow path by Center of Gravity.

and k combinations of $x_{s(big)}$–x_j). Similarly, the ALM algorithm is applied to each part and the minimum modeling error is calculated for each k–combinations. Suppose these are $e_{2m}$and $e_{'2n}$. They imply that the minimum modeling errors in the second step of dividing the space of variables is related to dividing of mth and n^{th}variables for the small and big parts of x_s, respectively. Based on minimum errors, x_m and x_n are divided and the rules for modeling after dividing are:

If (x_s is small & x_m is small) then ...
If (x_s is small & x_m is big) then ...
If (x_s is big & x_n is small) then ...
If ($x_s$is big & x_n is big) then ...

e_{2m} and $e_{'2n}$ are the local minimum errors. The suitable global error (e_2) can be calculated using minimum local errors (e_{2m} *and* $e_{'2n}$). Dividing continues until the global error is less than the threshold error. In this heuristic search method, the global error decreases simultaneously by decreasing the local errors.

Figure 3 depicts the next step of dividing algorithm which is step 3. This heuristic search method uses an appropriate criterion to select a variable for dividing and the median of data is used as the boundary for crisp dividing. Hence, the number of data points in the subspaces is equal.

Fuzzy modeling in ALM

Although, ALM implements either crisp or fuzzy dividing methods, but fuzzy dividing and modeling methods can enhance the ALM performance by:

1 Satisfaction of the continuity condition,
2 Better knowledge extraction of multi-variable non–linear systems,
3 Decrease of ALM sensitivity to noise.

Fuzzy dividing is similar to crisp dividing. In crisp dividing, the dividing point of a variable is the median. However, in fuzzy dividing, the boundary of small values of a variable is bigger than the median and vice versa. Hence, the regions of small and big values of a variable can overlap.

The fuzzy systems are not too sensitive to the dividing points. Therefore, the appropriate points for fuzzy dividing can be calculated by investigating various alternatives to select the most appropriate one.

Since the presented new heuristic method utilizes a complex dividing method, the typical fuzzification methods are not compatible with it. In this paper, a simple fuzzy modeling method is explained and used which is attuned to the heuristic search method. This fuzzy modeling method has been developed by Taheri Shahraiyni (2007). The details of how fuzzy rules are extracted are as follows.

We denote the membership function of a fuzzy set as $A_{ij}^{ks}(X_k^m)$ in which *i* is the dividing step, *j* is the number of dividing in each *i* which has a value between 1 and 2^{i-1},*s* is the membership function that is related to small (*s*=1) and big parts (*s*=2) of a variable domain.

K denotes the divided variable number and X_k^m is the mth member of the kth variable (X_k) $(x_k^m \in X_k)$ $and\ X_k \in X\,.X = \{X_1, \dots X_n\}$ is a set of n variables. ALM can be implemented by fuzzy modeling with miscellaneous shapes of membership functions and the performance of ALM as a fuzzy modeling method is not sensitive to the shape of membership function. Trapezoidal membership functions are one of the most-used membership functions. Besides, implementation of a fuzzy modeling method using trapezoidal membership functions is very straight forward. As a result, trapezoidal membership functions are applied here.

The truth value of a proposition is calculated by a combination of membership degrees. For example, the truth value of ' x_1^1 is A_{11}^{11} and x_2^1 is A_{21}^{22}' is expressed as:

$$\left(x_1^1\ is\ A_{11}^{11}\ and\ x_2^1 is A_{21}^{22}\right) = \left(A_{11}^{11}\left(x_1^1\right) \cap A_{21}^{22}\left(x_2^1\right)\right) = \left(A_{11}^{11}\left(x_1^1\right) \times A_{21}^{22}\left(x_2^1\right)\right)$$

In this fuzzy method, the general fuzzy rules are defined as below:

$$R_p : if\left(x_{k1}^m\ is\ A_{1j_1}^{k_1S_1}\ \&\, x_{k2}^m\ is\ A_{2j_2}^{k_2S_2}\ \&\ldots\right) then\ y_p^m = f_p\left(x_{k_3}^m\right)$$

Where p is the rule number and has a value between 1 and h (h is total number of fuzzy rules), *Rp* is the pth rule and *fp* is the *p*th one–variable non-linear function for the pth subspace (p^{th} rule). *1/P(fp)* is considered as the weight of the pth rule *(Wrp)* where *P(fp)* is PAE of *fp* (continuous path in the pth rule). Fire strength or membership degree of the pth rule, W_{fp}^m is equal to the truth value of the proposition which is:

$$W_{fp}^m = A_{1j_1}^{k_1S_1}\left(x_{k1}^m\right) \times A_{2j_2}^{k_2S_2}\left(x_{k2}^m\right) \times \ldots$$

Obviously, the summation of truth values of all of the propositions should be equal to 1

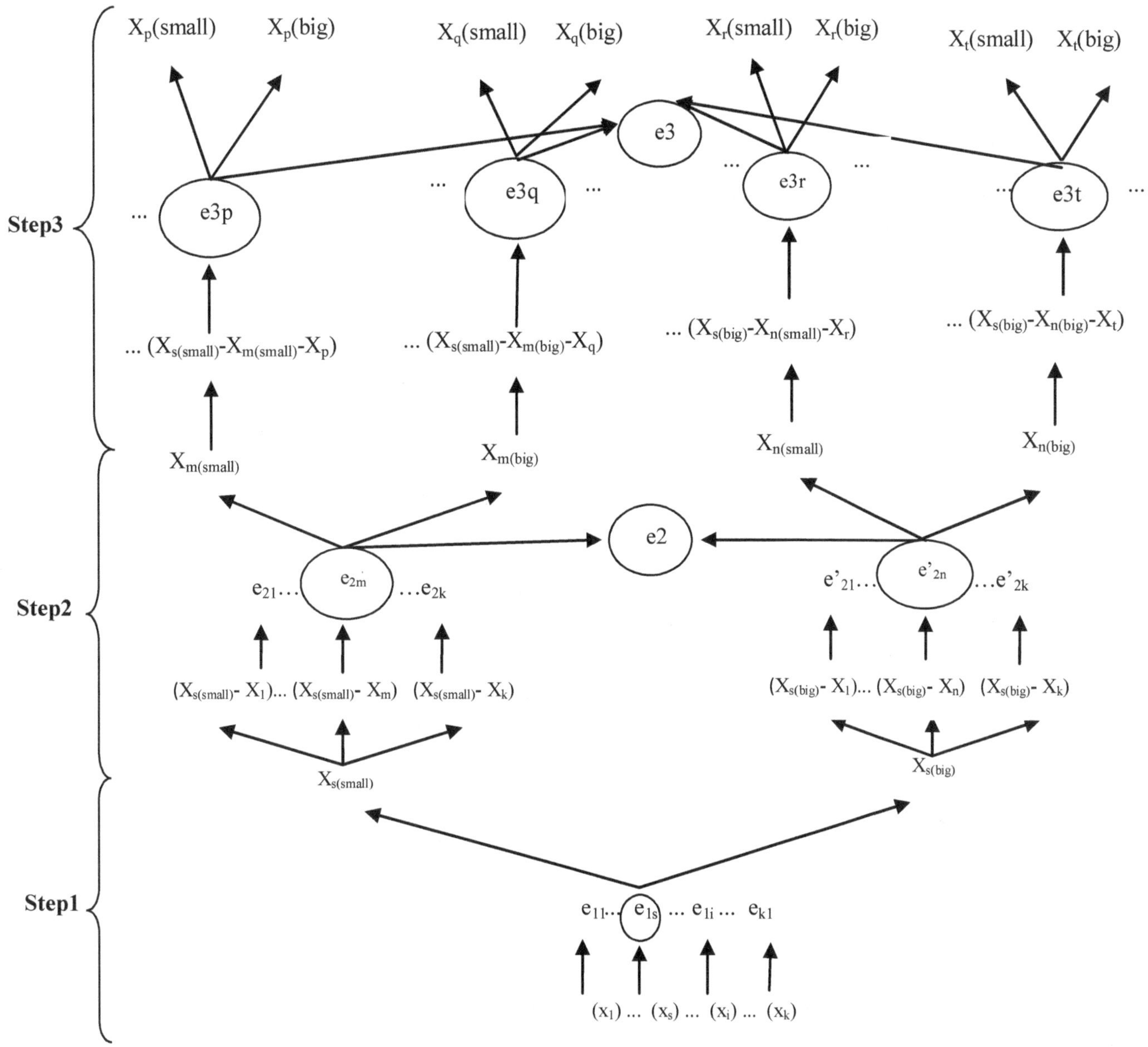

Figure 3. Algorithm of the new heuristic search method for dividing the space.

$$\sum_{p=1}^{h} W_{fp}^{m} = 1$$

Finally, the corresponding output (y^m) to m^{th} set of input dataset $(X_1^m, \ldots X_k^m, \ldots X_n^m)$ is calculated as:

$$y^m = \frac{\sum_{p=1}^{h}\left(y_p^m \times W_{fp}^m \times W_{rp}\right)}{\sum_{p=1}^{h}\left(W_{fp}^m \times W_{rp}\right)}$$

The most prominent target followed in this paper is to develop a modeling system by which the probability of pipe stuck can be estimated. As a result, this interference system possesses multiple inputs which are assumed to have a considerable impact on pipe sticking. The output of this system is the probability of pipe sticking; hence, our system has one output, and is considered as a MISO (Multiple Inputs Single Output) system. As a modeling problem, the first step is gathering data and dividing it in two parts which are train data and test data. The train data is utilized to construct the model, and the train data which is independent of train data in used to evaluate the method. Afterwards, it should be tried to find a numerical relationship between the inputs of train data and the outputs. Then, the numerical relationship ought to be explained mathematically. In the next step, the mathematical method is used in order to calculate the output by conducting inputs of test data to the expression. Then, the calculated and actual outputs are compared to calculate the error. Afterwards, the mathematical expression is modified if required.

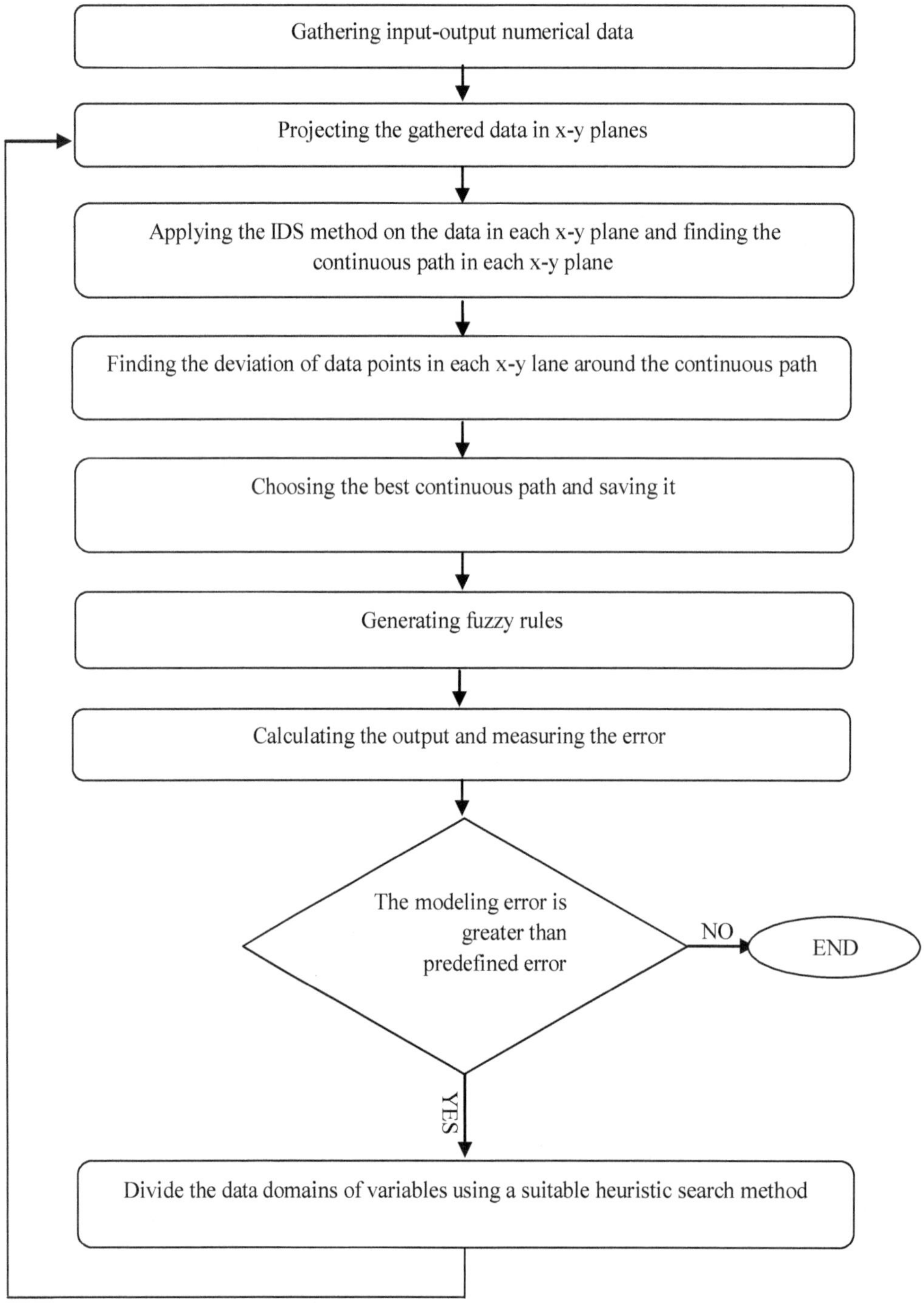

Figure 4. The flowchart of the proposed methodology.

Implemented algorithm for ALM method is applied as follows: The first step in ALM modeling is to gather the input-output numerical data. The inputs are called 'x' and the outputs are called ' y'. Afterwards, the gathered data is projected in x-y planes. As mentioned earlier, the next step in each ALM modeling process is to apply the IDS method on the data in each x-y plane and to find the continuous path which is the general behavior or implicit nonlinear function in each x-y plane. The next step is to extract the fuzzy rules. Afterwards, the test data is conducted to fuzzy model, and the outputs are calculated. Then, the error based on a comparison between actual outputs and calculated ones is calculated. If the error is higher than required, the data domains of variables using a suitable heuristic search method is divided and the process is done again. The flowchart of the methodology is illustrated in Figure 4.

Data gathering

As discussed earlier, the main target of this paper is to develop a system by which the probability of pipe sticking is determined. This system determines the stuck pipe probability based on some properties of drilling mud and drilling operation characteristics. Following the section, an overview of different parameters affecting pipe sticking is presented.

Mud type

A comparison of generic mud types has shown oil-based muds to have the lowest stickance values and gel-water based mud has the highest. Polymer-water-based muds fall between these two extremes. It was found that the sticking potential also varies greatly within a mud type, depending on the precise formulation tested (Reid et al., 2000).

Lubricant

The addition of certain lubricants for water- and oil-based muds will reduce the effect of differential sticking. If sticking still occurs, then reduce the force needed to free the stuck pipe or tool.

Solids level

Type and amount of solids play a role in cake characteristics and affect the degree of pipe sticking and pull out force to get it free (Isambourg et al., 1999). Increasing the solids level in the mud (both weighting agent and drilled solids) has been found to increase the force needed to free the pipe. This effect depends on the type of mud used. For example, salt muds have the lowest sticking tendency until reactive drill solids are added, which resulted in one of the highest measured forces (Bushnell-Watson and Panesar, 1991).

Fluid loss

Improving fluid loss can reduce the stickance tendencies of a mud. Oil-based muds usually have low fluid loss values. However, reducing the fluid loss does not have the same effect on stickance in all mud systems (Isambourg et al., 1999). It is currently not possible to determine accurately the sticking potential of the mud from a single mud property, such as density, fluid loss, solids content, or lubricity. However, laboratory work has shown that several mud treatment options, including adding a lubricant, can reduce the sticking tendencies of a mud.

Mud cake properties

A mud cake is formed on permeable formations if the formation pressure is significantly lower than the hydrostatic pressure of the drilling fluid. As a result, there is an invasion of the liquid phase into the permeable zone and deposition and/or penetration of the corresponding solids inside and against the formation (Courteille and Zurdo, 1985). After a period of time, equilibrium is reached and deposition is balanced by erosion, resulting in a constant cake thickness. However, when the mud is static, erosion then occurs and the cake thickness increases with time. Increasing the time that the mud is not circulating will increase mud cake thickness and the likelihood of differential sticking.

Cake thickness cannot be used alone to predict the sticking tendency of different types of mud. However, for one particular mud formulation, an increase in cake thickness will increase the force required to free the pipe. Darcy's Law predicts that the cake thickness will increase with the square root of time and the increase in the force to free the pipe also follows the same relationship. This suggests that the change in the sticking tendency is a result of increasing contact area. In general, to reduce the chance of differential sticking, the time that the mud is left static in the hole should be minimized (Bushnell-Watson and Panesar, 1991)

Another parameter that influences pipe sticking is the friction factor. The friction between steel and mud cake varies with changes in mud composition. Previous studies have shown that the friction factor increased with increased barite content of the mud. Carboxymethylcellulose had no effect on the friction factor. Emulsification of oil in the mud had the effect of reducing the friction factor. In summary, the mud composition may be altered to reduce the friction between the pipe and mud cake (Annis and Monaghan, 1962).

Several characteristics of the mud cake have an effect on pipe sticking and on the necessary force to pull out the pipe. The sticking tendency of a mud cake depends on more than one parameter. It will vary as a result of cake thickness (contact area) and mud cake properties (friction/adhesion and surface roughness). The combination of these factors means that predicting the sticking tendency of any mud is not simple.

In this research, after analyzing the general properties of well and drilling fluid, seven most effective ones were selected to be used as input variables in the neural network model; these parameters are defined as follow:

Measured depth

The length of the wellbore is as if determined by a measuring stick. At greater depth, more stresses will be imposed on formation and it could be a major stuck pipe variable.

Yield point

YP is the yield stress extrapolated to a shear rate of zero. YP is used to evaluate the ability of a mud to lift cuttings out of the annulus. A high YP implies a non-Newtonian fluid, one that carries cuttings better than a fluid of similar density but lower YP. YP is lowered by adding deflocculant to a clay-based mud and increased by adding freshly dispersed clay or a flocculant, such as lime.

Plastic viscosity (PV)

PV is the slope of the shear stress/shear rate line above the yield point. A low PV indicates that the mud is capable of drilling rapidly because of the low viscosity of mud exiting at the bit. High PV is caused by a viscous base fluid and by excess colloidal solids. To lower PV, a reduction in solids content can be achieved by dilution of the mud.

Gel strength (Initial and 10 min)

The shear stress measured at low shear rate after a mud has set quiescently for a period of time (10 s and 10 min in the standard API procedure). Some drilling fluids are thixotropic, forming gelled structures when stagnant and liquefying when sheared. The specific gel strength of a drilling fluid is described as low-flat (most desirable), progressive or high-flat (both undesirable) according to its measured gel strength versus time.

Weight on bit (WOB)

Weight on bit is an essential factor in the drilling process, which can affect the rate of penetration as well as natural frequencies of the drill string in the bending mode of vibration. The WOB can also be related to the load carrying capacity of the drill string (buckling load). Increasing the weight on bit will bend the drill collars behind the near-bit stabilizer more, so the rate of build will increase.

Revolutions per minute (RPM)

A higher rotary speed will tend to `straighten' the drill collars and hence reduce the rate of build. Increasing the RPM of the bit provides more opportunities to cut the formation in a given amount of time.

In the presented approach, the pipe stuck is assumed dependent to the mentioned parameters, that is, measured depth, Yield Point (YP), Plastic Viscosity (PV), Initial Gel Strength, 10 minutes Gel Strength, WOB and RPM. The output is pipe stuck prediction. The block diagram of the system is depicted in Figure 5.

Data processing

The next step in gathering data is preprocessing data. In our study, 245 daily drilling reports of the field are utilized to train and test the method. 194 of them are used to construct and train the model, and 51 of them are utilized in order to assess the method. At first, the used data is normalized based on following equation in order that all variables are in the interval [0 1].

$$Xn = \frac{X - X_{\min}}{X_{\max} - X_{\min}}$$

Where Xn, X, $X_{\min}$, and $X_{\max}$ are normalized parameter, original parameter, minimum used parameter, and maximum used parameter. After the data is processed, the data is projected on x-y planes, and ALM steps are applied on them, and the inference system is constructed. Then the method ought to be assessed. The next part explains the evaluating process and its criteria.

Evaluating method and error calculation

The next step, after constructing a fuzzy model, is to evaluate the model, and calculate the error. As mentioned earlier, as a modeling problem, the data is divided in two parts. The first part which is the training data has been used to construct the fuzzy rules, and train the ALM model. The test data which is independent of training data is utilized in order to assess the method. The details are as follows. The test data is conducted to the model, and the outputs are calculated based on the constructed model. The next step is to calculate the error.

As discussed earlier, the main target followed by this study is to predict the pipe sticking while drilling. The data can be divided into two parts. Stuck data which is the data that pipe sticking has occurred while drilling, and none stuck data which is the data that no pipe sticking has occurred while drilling operation. In this study, the output of stuck data is assumed 1, and the output of none stuck data is assumed 0. The most important target of this study is to construct a model to determine whether pipe sticking will occur while drilling operation or not. Thus, our study is to classify the data into two parts which are stuck data and none stuck data.

In order to calculate the error, the test data is conducted to trained model, and the output is calculated. The output is a real number in [0 1] interval. If the output is greater than 0.5, the data is considered as stuck data. Otherwise, the data is considered as none stuck data. Then, the error based on a comparison of actual output and determined one is calculated. The error is defined as follows:

$$\text{Error} = \frac{\text{The number of stuck data considered as none stuck} + \text{The number of none stuck data considered as stuck}}{\text{total number of test data}}$$

After the error is calculated, if the error is higher than required the procedures based on Figure 1 are reiterated.

RESULTS

Afterwards, when the model is constructed, it requires being assessed. As discussed in previous section, the test data is conducted to the constructed model in order to assess the proposed method. As explained earlier, in this study, the same as all classification and modeling studies, the data is divided in train and test data. In this study, 150 data of daily reports of The field have been considered as train data, and the drilling reports of well number 129 of The field is used as test data. After the test data is conducted to the model, the output is calculated based on the fuzzy rules. The details of constructing the model and extracting the rules are explained earlier. Then, the error has been calculated according to previous section, and the confusion matrix is calculated. Each array of the confusion matrix is calculated as follows:

$$c_{ij} = \text{number of } i^{th} \text{case that has been identified as } j^{th} \text{ case}$$

where the c_{ij} is the element of row *i*, and column *j*. For example, the array of first row and second column of the Table 1 is the number of none stuck data that has been considered as stuck one. As it can be seen in Table 1, there exist no error in data, and all the test data have been predicted precisely.

As mentioned earlier, the inputs of the method are measured depth, plastic viscosity, yield point, initial gel strength, 10 min gel strength, Weight on Bit (WOB), and Revolution per Minute (RPM), and the output is the probability of pipe sticking. In order to evaluate the

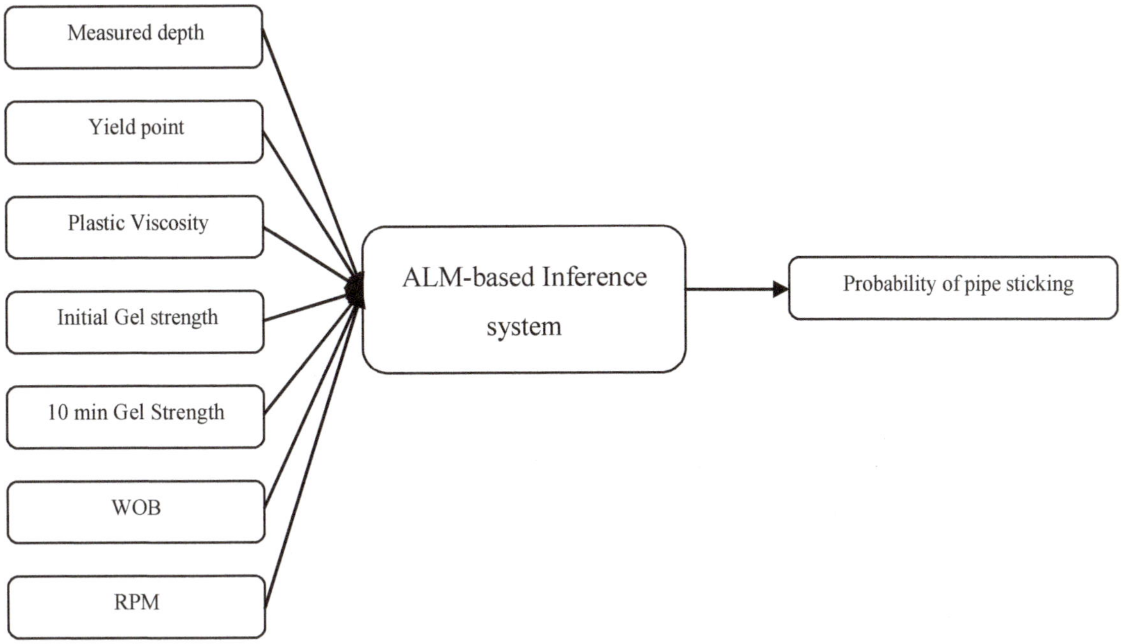

Figure 5. The block diagram of the proposed method.

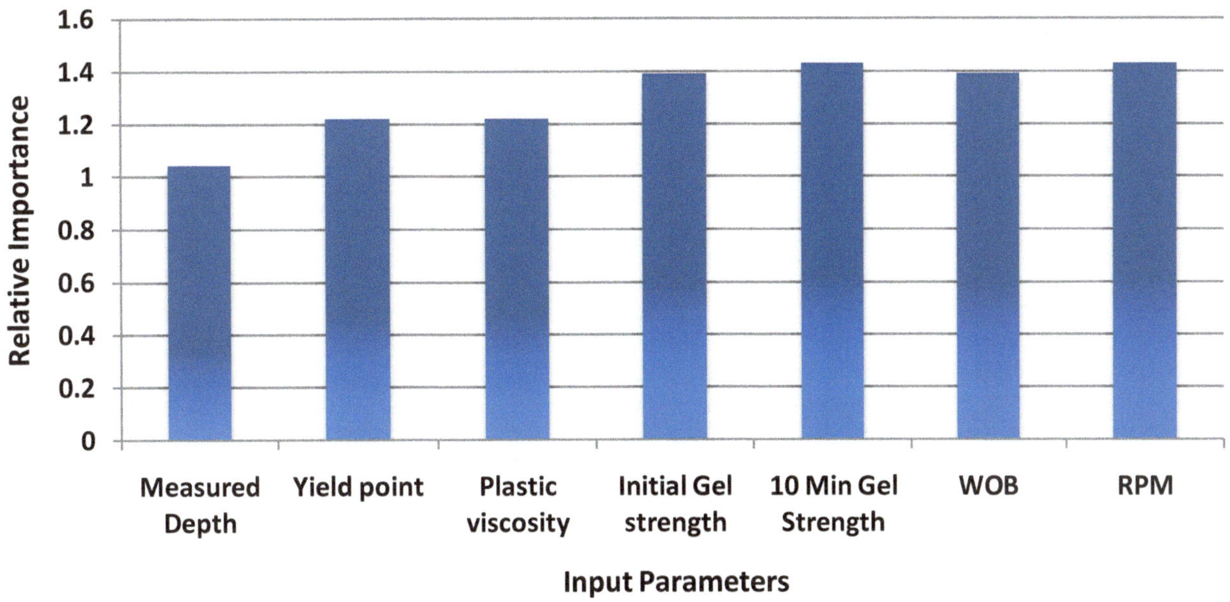

Figure 6. The relative importance of input parameters.

importance of each input parameter in pipe sticking, following steps has been implemented. For each input parameter, a fuzzy ALM model has been constructed based on the new data. The new data has six inputs which are all inputs except the input which its importance is under investigation. All previous steps for constructing, training and testing the model have been conducted on the model, and the classification accuracy of the model is calculated as follows:

$$\text{Accuracy} = \frac{\text{the number of data predicted correctly}}{\text{total number of data}}$$

As mentioned before, artificial intelligence has been used for pipe stuck prediction (Al-Baiyat and Heinze, 2012; Jahanbakhshi and Keshavarzi, 2012; Murillo et al., 2009; Miri et al., 2007). Table 2 shows a comparison between the results of the existing methods and the proposed method.

Table 1. Confusion matrix by ALM method.

Output/desired	
Stuck	**None stuck**
0	44
7	0

Table 2. Comparison between proposed methods and existing ones.

Method	Accuracy (%)
Jahanbakhshi and Keshavarzi	92.19
Al-Baiyat and Heinze	85
Murillo et al.	98.4
Miri et al.	92.86
ALM	100

Afterwards, the relative importance of each parameter in pipe sticking is calculated as follows:

$$\text{Relative importance} = \frac{\text{accuracy of the method}}{\text{accuracy of the method without selected parameter}}$$

The numerator of the previous equation is the accuracy of the method by means of all inputs which is 100% based on Table 1. Figure 3 depicts the relative importance of all input parameters.

Conclusion

In the ALM method, the constructed modeling system accepts seven inputs and one output which is the probability of pipe sticking. The 150 data of drilling daily reports of the field have been considered as train data. The interference system and corresponding rules have been constructed. Afterwards, in order to evaluate the proposed method, the method is implemented on the data of well number 129 of the field. Each case, which has the output less than 0.5, is considered as none stuck, and each case that has the output value greater than 0.5 is considered as stuck. The confusion matrix has been calculated which shows the method is accurate to predict the pipe stuck. In order to find the relative importance of each input parameters on pipe sticking, sensitivity analysis has been done. The result of sensitivity analysis shows that RPM and gel strength are the most important parameters in pipe sticking occurrence.

REFERENCES

Aadnoy BS, Larsen K, Berg PC (1999). "Analysis of Stuck Pipe in Deviated Boreholes", paper SPE 56628 presented at the 1999 SPE Annual Technical Conference and Exhibition, Texas, Oct. 3-6.

Adams N (1977a). Member SPE-AIME, Prentice and Records Enterprises, Inc."A Field Case Study of Differential-Pressure Pipe Sticking." SPE 6716.

Adams N (1977b). "A Field Case Study of Differential Pressure Sticking." Paper SPE 6716 presented at the SPE Annual Technical Conference and Exhibition, Denver, Oct. 12.

Al-Baiyat A, Heinze L (2012). "Implementing Artificial Neural Networks and Support Vector Machines in Stuck Pipe Prediction." SPE Kuwait International Petroleum Conference and Exhibition, Kuwait City, Kuwait.

Annis MR, Monaghan PH (1962). "Differential Pressure Sticking – Laboratory Studies of Friction between Steel and Mud Filter Cake." SPE 151. JPT 15(5):537-543.

Bagheri SS, Honda N (1997). "A New Method for Establishment and Saving Fuzzy Membership Functions." 13th Fuzzy Symposium, Toyama, Japan pp. 91-94.

Bagheri SS, Honda N (1999). "Recursive Fuzzy Modeling Based on Fuzzy Interpolation,"J. Adv. Comput. Intell. 3(2):114–125.

Bagheri SS, Honda S, Yuasa G (1999). "Fuzzy interpretation of human intelligence." Int. J. Uncertainty Fuzziness Knowl.-Based Syst. 7(4):407-414.

Bagheri SS (2000). "A Novel Fuzzy Approach to Modeling and Control and its Hardware Implementation Based on Brain Functionality and Specifications." The University of Electro Communications.

Biegler MW, Kuhn GR (1994). "Advances in Prediction of Stuck Pipe Using Multivariate Statistical Analysis." Presented at the SPE/IADC Conference in Dallas, Texas.

Bushnell-Watson YM, Panesar SS (1991). "Differential Sticking Laboratory Tests Can Improve Mud Design." Paper SPE 22549-MS presented at the SPE Annual Technical Conference and Exhibition, 6-9 October, Dallas, Texas.

Courteille JM Zurdo C (1985). "A New Approach to Differential Sticking." Paper SPE 14244 presented at the SPE Annual Technical Conference and Exhibition, 22-26 September, Las Vegas, Nevada.

Elahi Naraghi M, Ezzatyar P, Jamshidi S (2013). "Adaptive Neuro Fuzzy Inference System and Artificial Neural Networks: Reliable approaches for pipe stuck prediction." Aust. J. Basic Appl. Sci. 7(8):604-618, 2013, ISSN 1991-8178.

Helmick WR, Longley AJ (1957). "Pressure Differential Sticking of Drill Pipe and How It Can Be Avoided or Relieved.', Oil Gas J. 55:132.

Hempkins WB, Kingsborough RH, Lohec WE, Nini CJ (1987). "Multivariate Statistical Analysis of Stuck drill Pipe Situations." SPE DRILLING ENGINEERING.

Howard JA (1994). Enertech Engineering & Research Co., and Glover.S.S., Enertech Europe "Tracking Stuck Pipe Probability While Drilling." Paper SPE 27528 presented for Presentation at the 1994 IADC/SPE Drilling conference held in Dallas, Texas.

Isambourg P, Ottesen S, Benaissa S, Marti J (1999). "Down-Hole Simulation Cell for Measurement of Lubricity and Differential Pressure." Paper SPE 52816-MS presented at the SPE/IADC Drilling Conference, 9-11 March, Amsterdam, Netherlands.

Jahanbakhshi R, Keshavarzi R (2012). Intelligent Prediction of Differential Pipe Sticking by Support Vector Machine Compared with Conventional Artificial Neural Networks: An Example of Iranian Offshore Oil Fields. SPE Drill. Completion J. 27(4):586-595.

Miri R, Sampaio J, Afshar M, Lourenco A (2007). 'Development Of Artificial Neural Networks To Predict Differential Pipe Sticking in Iranian Offshore Oil Fields. Paper SPE 108500 presented at the 2007 international Oil conference and Exhibition in Mexico held in Veracruz, Mexico.

Mita T (2000). "Introduction to Nonlinear Control – Skill Control of Underactuated Robots." (Japanese), Syokodo Pub. Co.

Murillo A, Neuman J, Samuel R (2009). "Pipe Sticking Prediction and Avoidance Using Adaptive Fuzzy Logic and Neural Network Modeling." Paper SPE 120128 presented at Production and Operations Symposium held in Oklahoma City, Oklahoma, USA.

Nishino L, Tagawa A, Shmi H, Odaka T, Ogura H (1999). "Hierarchical Fuzzy Intelligent Controller for Gymnastic Bar Action." J. Adv. Comput. Intell. 3(2):106-113.

Reid P, Meeten G, Way P, Clark P, Chambers B, Gilmour A, Sanders M

(2000). Differential-sticking mechanisms and a simple wellsite test for monitoring and optimizing drilling mud properties. SPE Drill. Completion 15(2):97-104.

Sagha H, Bagheri SS, Khasteh H, Kiaei AA (2008). "Reinforcement Learning Based on Active Learning Method." IEEE Second Int. Symp. Intell. Inf. Technol. Appl. 5:598–602.

Sakurai Y, Honda N, Nisbino J (2003). "Acquisition of Knowledge for Gymnastic Bar Action by Active Learning Method." J. Adv. Comput. Intell. Intell. Inform. (JACIII) 7:1.

Sharif QJ (1997). "A Case Study of Stuck Drill pipe Problems and Development of Statistical Models to Predict the Probability of Getting Stuck and If Stuck, the Probability of Getting Free." University of Engineering & Tech, Lahore, Pakistan.

Siruvuri C, Halliburton Digital and Consulting Solutions, Nagarakanti S, Samuel R (2006). "Stuck Pipe Prediction and Avoidance: A Convolutional Neural Network Approach." Paper SPE 98378 presented at the 2006 IADC/SPE Drilling Conference, Miami, Florida.

Taheri SH (2007). Extraction of water quality parameters using the satellite data and inverse modeling of radiative transfer equation by active learning method, PhD Thesis, Civil Engineering Department, Sharif University of Technology, Tehran, Iran. p. 167 (In Persian).

Torne J, Rourke M, Derouen B, Kessler C (2011). Middle East Case-Study Review of a New Free-Pipe Log for Stuck-Pipe Determination and Pipe-Recovery Techniques, SPE Asia Pacific Oil and Gas Conference and Exhibition, Jakarta, Indonesia.

Weakley RR (1990). Chevron Services Inc. "Use of Stuck Pipe Statistics to Reduce the Occurrence of Stuck Pipe." SPE 20410.

Wisnie AP, Zheiwei Z (1994)."Quantifying Stuck Pipe Risk in Gulf of Mexico Oil and Gas Drilling." SPE 28298.

Yarim G, Uchytil R, May R, Trejo A, Church P (2007). "Stuck Pipe Prevention – A Proactive Solution to an Old Problem." Paper SPE 109914 presented at the SPE Annual Technical Conference and Exhibition, 11-14 November, Anaheim, California.

Yuasa G, Bagheri SS, Honda N, Sakurai Y (1992). "Applying an active learning method to control problems." Asian Fuzzy System Symposium 2000 (AFSS2000), & HA Japan, pp. 572-577.

Improving well productivity in an Egyptian oil field using radial drilling technique

Adel M. Salem Ragab

Department of Petroleum Engineering, School of Sciences and Engineering (SSE), American University in Cairo (AUC), Egypt.
Department of Petroleum Engineering, Faculty of Petroleum and Mining Engineering, Suez University, Egypt.

Radial drilling (RD) technique utilizes hydraulic energy to create several lateral holes in different directions and levels with several lengths. These lateral holes are made by milling the casing with small bit then extending these holes laterally using high pressure hydraulic jetting. This work presents full descriptions and analysis for RD applications in one of the Egyptian oil field. Moreover, it attempts to analyze and then optimize the different means for performing this technique. Therefore, several tests was performed and analyzed. The total depths of these wells vary from 8856 to 8987 ft. The first well was laterally drilled by about 164 ft long by seven laterals and the angle between each two is 90°. The second well was drilled by 6 laterals in two different levels. Five of them extended to 164 ft, and one of them with 295 ft long. In the third well, 4 lateral holes were radially drilled with 165 ft long. After evaluation, the first well gross rate increased by 37.4%, and the net oil rate was improved by more than 31.4%; the second well shows an improvement by about 73.34% increase in gross rate, and 47.3%, an increase in the net oil rate; and the third well shows an improvement by about 14.3% gross rate, and 14.7%, an increase in the net oil rate for very short period. Several experiences have been gained from using this technology which extends the productive life of wells and accordingly of the whole field with reasonable cost.

Key words: Radial drilling, stimulation, lateral holes, hydraulic jetting, URRS.

INTRODUCTION

The well is drilled and then completed to move the oil and gas from its original location in the reservoir to the surface, and due to damage during drilling and completion operation, not all of these fluids can move to the area around the wellbore, therefore, it is always important to think about any unconventional means to reach or communicate these areas with the wellbore. One of the proposed techniques is to bypass the damaged zone which is called radial drilling (RD).

The objective of the radial drilling is to provide an extended wellbore radius by means of multiple radials from a vertical wellbore. It can be applied both in new and old wells. It is mainly used as high pressure jet flow energy to penetrate and elongate a number of lateral holes radiated from the main wellbore in the same layer or different layer. The choice of radial length, number of radials, and radial array is a function of the reservoir properties.

Radial drilling

The radial drilling (RD) is a technique that can create several small diameter drains from the well in a relatively

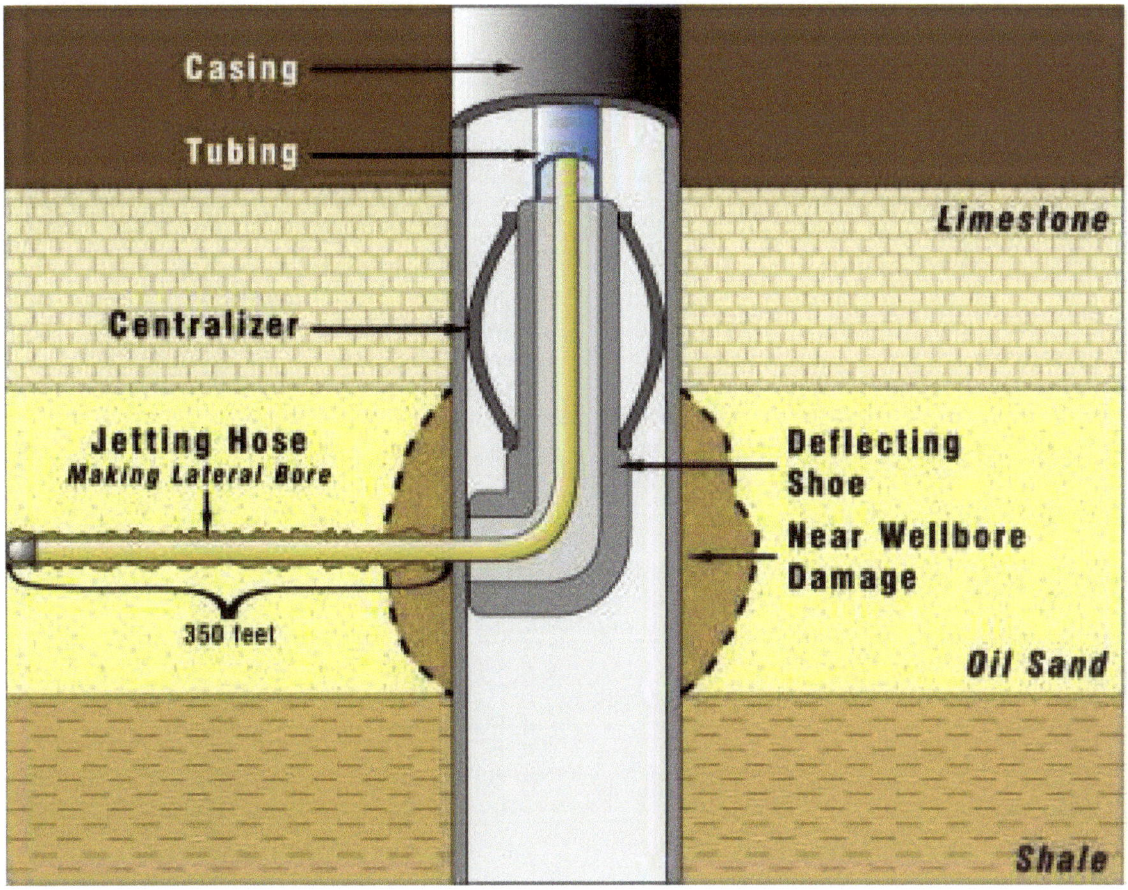

Figure 1. Radial drilling technique.

short time, normally 2 to 3 days per level. The diameter of these holes varies from few millimeters to several tenths of millimeters in the casing using high pressure fluid at selected depths and azimuth, and installed at single or multiple levels. These perforations can be extended radially up to 100 m perpendicular from the main wellbore, as shown in Figure 1. The first application of this technique had been performed in former Soviet Union (Elliott, 2011).

Working principle of radial drilling

The features of the tool are the capability to create and open a hole in the casing and subsequently jetting laterally into the reservoir formation creating a lengthy hole to bypass the damaged zone around the well which is called altered or skin area. To perform this, a special bottom hole assembly (BHA) which consists of a drilling machine; high pressure hose, and jet nozzle have been assembled. The drilling machine involves a drill bit driven by a combination of an electrical motor and a hydraulic piston, making a hole in the casing, its size depends on the bit size used. Then jetting through this hole may extend this hole to 100 m (≈ 330-ft), based on the hose length.

The components of the tool used in radial drilling mainly consist of the drilling machine and the jet drum. In addition, as depicted in Figure 1, it involves the following components:

(i) Tubing end connector
(ii) Controller unit/power pack
(iii) Anchor
(iv) Orienter/Indexer
(v) Steering-tool
(vi) Stroke cylinder.

Applications

The main application is to provide a fast and economical method to recover the remaining hydrocarbons form marginal or mature oil and gas fields.

This technique can be applied in different disciplines in oil well industry such as:

1. Well completions

2. Well stimulations
3. Directed reservoir treatment
4. Improve water injection
5. Improve vertical cleaning
6. Reduce water coning
7. New wells instead of standard completion methods
8. Water disposal and re-injection
9. Steam applications in heavy oil and tar sands
10. Mining applications (leaching).

Radial drilling benefits

Based on what have been performed on some worldwide fields, the radial drilling methods have some economical and technical benefits such as (Bruni et al., 2007; Dickinson-Dickinson, 1985):

1. Enhance production rate and recoverable reserves from marginal wells.
2. Improves injection rates in water disposal/injection wells.
3. Allows directional treatment of wells for example, acid, steam, CO_2, etc.
4. Outperforms conventional stimulation methods at a lower cost, in reduced application time and with higher potential production results.
5. Improved and extended drainage area in productive formations.
6. Radial drilling penetration greatly exceeds conventional (perforation) penetration and can reach substantially beyond the damaged area of the well-bore.
7. Reach beyond the damaged area of the well-bore.
8. Allows multi-layer application in thicker reservoir zones.
9. Most effective on old, low productivity wells.
10. No need for large, expensive rotary rigs.
11. Does not require mud pits that can damage the environment.
12. No casing milling requirement, therefore no need to circulate mud back to the surface.
13. No additional stimulation required.
14. The process is fast, average operation duration is two days per well, so no big loss in production.
15. No logging expense required.
16. No need to change well-bore configuration.

Limitations of radial drilling

Based on the previous operations using radial drilling all over the world, there are some limitations and challenges in applying such technique in oil and gas wells (Abdel-Ghany et al., 2011). From these:

1. Difficulties of penetration under porosity of 3 to 4%.
2. Maximum working depth about 10000 ft.
3. Bottom hole temperature not to exceed 120°C (248°F)
4. Maximum wellbore inclination 30° and no more than 15° at the zone target depth/zone of interest.
5. Maximum tensile strength 100,000 psi.

Radial drilling field operations

Bruni et al. (2007) show a brief description of the operation and bottom hole assembly of radial drilling technology. Radial drilling technique is described as a new coiled tubing convoyed drilling technique, were several new wells or lateral holes are jet drilled perpendicular from the mother well and into the reservoir formation, this technology is targeted for increasing the well productivity of the new and existing wells.

The following is a brief procedure for performing this job in a specific well (Figure 2):

1. Run in hole with deflector sub on drill pipe string correlating its depth with well logging.
2. Orient string with gyro.
3. Run in hole with milling tool (milling bit).
4. Pull out of hole with milling tool.
5. Run in hole with jetting tool.
6. Pull out of hole with hose, nozzle jetting tool.
7. Rotate deflector shoe and repeat operation at each lateral hole for any horizontal layer.

In another way, the radial drilling technology can be performed in three main steps, first step is milling the casing , second step is jetting the formation with high pressure nozzle and the last one is washing out the formation while pull out of hole. Figure 2 illustrates these steps.

The bottom hole assembly for milling the casing are; 1¾" bit connected with flexible shaft, both are rotated by conventional mud motor connected to coiled tubing up to surface, connected to coiled tubing unit with its monitoring system. Figures 3 to 5 show the bottom hole assembly for milling the casing.

The created hole size depends on formation strength, confining strengths and compressive loads from overburden and matrix stress, as well as on the speed of penetration of the jet. From surface tests an average hole size of 4 to 5 cm in diameter was obtained.

MECHANISM OF PENETRATION

It is reported in the literatures (Bruni et al., 2007; Buset et al., 2001) that there are four main penetration mechanisms. These mechanisms are: 1) surface erosion; 2) hydraulic fracturing; 3) poroelastic tensile failure; and 4) cavitation.

The net forces that affect to drive jetting nozzle forward can be derived from three main mechanisms; under

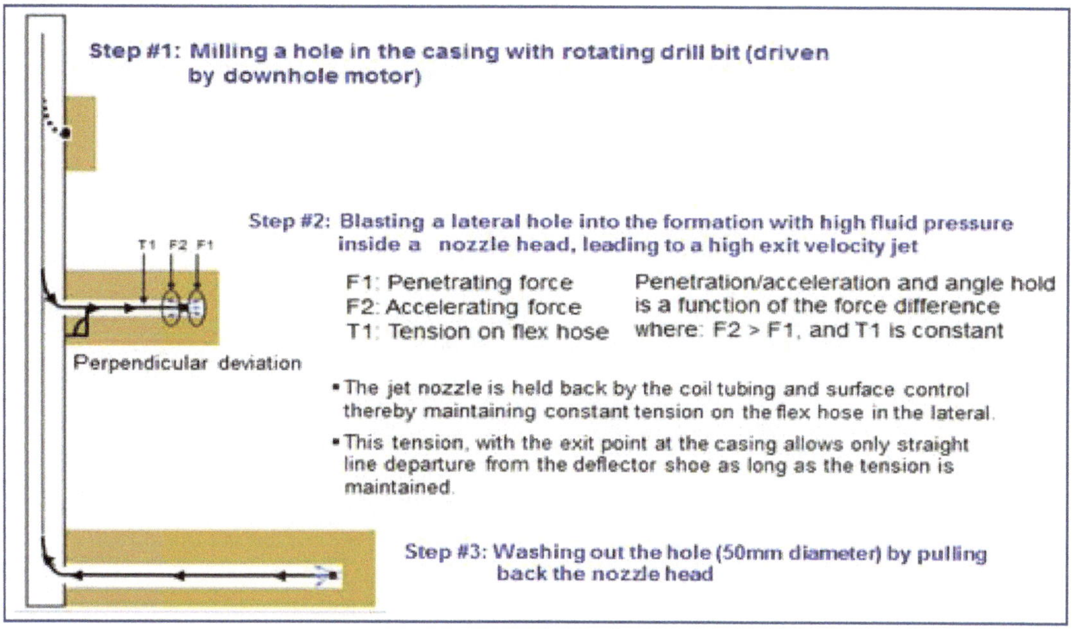

Figure 2. The three steps for radial drilling process.

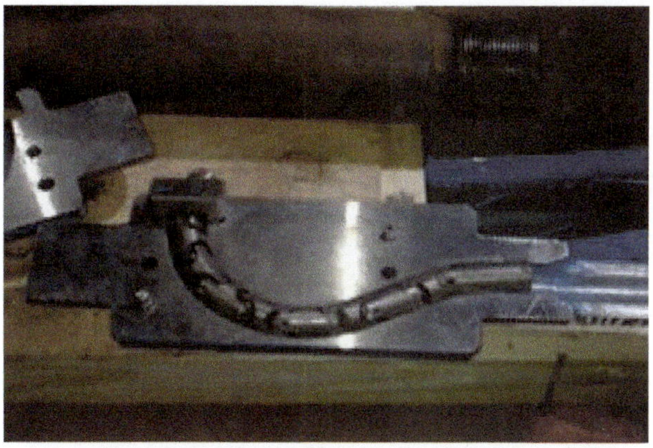

Figure 3. Deflector shoe with flexible shaft inside.

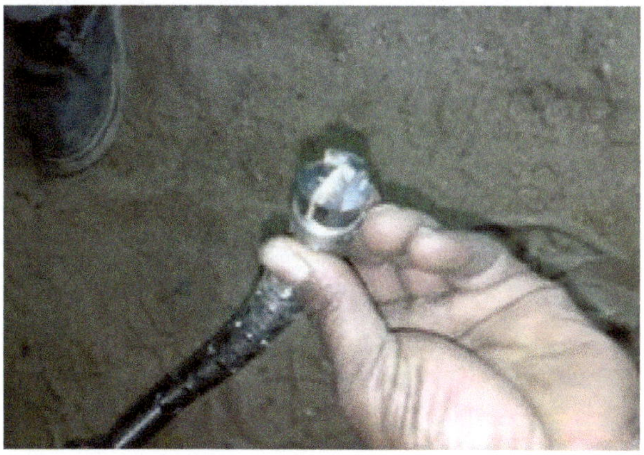

Figure 4. The 1¾" bit connected to flexible shaft

Figure 5. The 1 ¾" bit and the flexible shaft connected to motor.

pressure force, jetting force and ejector force. The main mechanism is jetting force mechanism; Figures 6 and 7 show the driving mechanisms of jetting nozzle and it effect of core sample.

The BHA is gathered and then circulating with an intermediate flow rate; once the BHA is close to the baffle shoe, the flow rate is increased and the tool is slipped allowing for the introduction into the anchor. Once the hose enters the formation, it will move horizontally in the formation due to the force generated by the distribution of

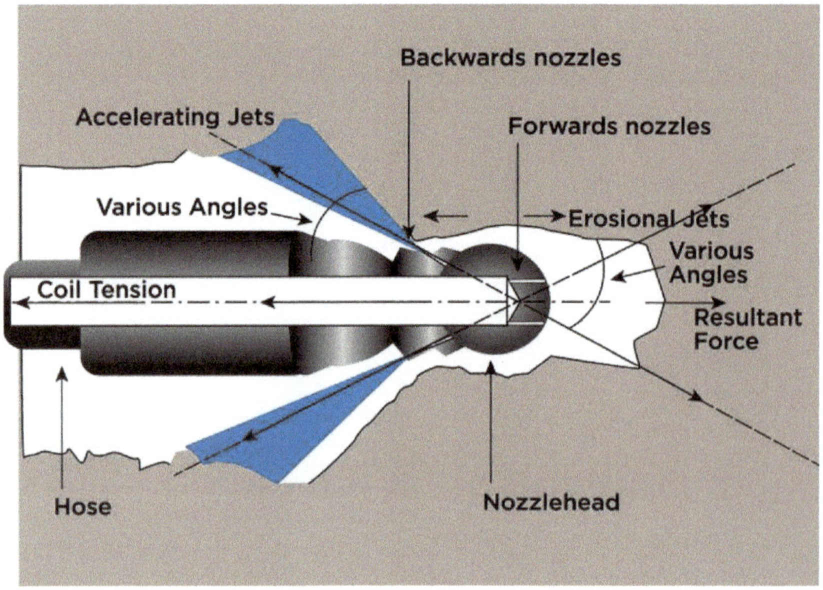

Figure 6. Sketch of the nozzle.

Figure 7. Jet nozzle of 8000 psi and its effect in core sample.

jet nozzles. The force S in the driving direction can be calculated from the following equation:

$$S_{jetting} = \rho u_o^2 A_o - \sum_{i=1}^{6} \rho u_i^2 \cos\varphi_i \, A_i$$

Where:
A_o = Inside hose area (m^2)
A_i = Nozzle area (m^2)
u_o = inside hose velocity (m/s)
u_i = Nozzle velocity (m/s)
ρ = Density of water (kg/m^3)
φ = Angle of the nozzle.

WORLDWIDE FIELD APPLICATIONS

In 1985, Dickenson-Dickenson described a new system to drill horizontal holes of 100 to 200 ft in length, and about 4 inches in diameter, through an unconsolidated formation in Californian oil field. Then placing a 1¼ inch OD production carbon steel tube within the 4 inch diameter bore hole in the formation while drilling is in progress. The drill system uses 8000 to 10000 psi water jet drilling causes a velocity ranges from 6 to 120 ft/min. They mentioned that this drilling system is not limited to horizontal shallow, 100 to 200 foot radials, it is neither limited to a 1¼ inch production tube nor a 4 inch radial bore. The system can work vertically and can be place in very long and large tubes, moreover, it can be applied in offshore reservoirs, consolidated formations, waste disposal and mineral recovery by solution mining.

In 1989, Dickenson et al. extended their previous work not by elongating the depth of penetration but by increasing the number of radials, they named it ultra-short radius radial system (URRS). Multiple radials can be placed at the same level and on multiple levels and horizontal completions can be provided, including 100% gravel packing, *in-situ* electrolytic perforation and cutting, and flexible sand barriers (FSB's). Initial field applications were in unconsolidated formations. The basic URRS uses an erectable whip stock lowered down hole by a 4½ inch work string into an under reamed cavity or hydraulically slotted opening of 22 inch diameter. The surface water drilling fluid pressure ranges from 8000 to 10000 psi which is pumped into the long vertical work string with a conventional fracture pump (Dickinson et al., 1989).

Dickinson et al. (1992) presented for the first time the use of combination of water jet drilling and coiled tubing, URRS and quick radial system (QRS) provide multiple horizontal radials on a single horizon near wellbore. These have been done by milling of casing and under reaming of a cavity. The hardware of the URRS includes whipstock, drilling string, jet drilling, control while drilling, positional survey, cutoff and perforation, gravel packing, slotted liner, and gravity drainage.

Yonghe et al. (2000) described the application and development of using URRS by high pressure jet flow techniques, and applied this technique for two wells. They described the details of the operation in which they performed radial holes with lengths ranged from 4 to 10 m. They concluded first, that this technique can greatly increase oil recovery and oil production rate and second it needs some improvement to solve some problems before the large scale uses.

Marbun et al. (2011) reported herein discusses the application of the URRS. The specific variables like reservoir thickness, vertical and horizontal permeabilities, oil properties, well spacing, outer-boundary reservoir pressure, gravity drainage, thermal and non-thermal processes and impermeable partings within the reservoir has been studied together with drilling operation parameter like casing design, drilling fluid design, radial drilling equipment, Bottom Hole Assembly to optimize the drilling process and completion of the URRS at field.

In the North Urtabulak Oil Field in southern Uzberkistan (Bruni et al., 2007), a trail program was conducted in five existing wells using radial drilling, a coiled tubing workover technique in which lateral holes of 2-in. diameter are drilled up to 330 ft from the original well bore by high pressure water jetting. All but one of the five trials wells displayed significant post workover uplift in production, leading to overall incremental production of more than 17000 bbl in 2011.

Buset et al. (2001) in his paper titled "Jet Drilling Tool: Cost-Effective Lateral Drilling Technology for Enhanced Oil Recovery" addressed a new coiled tubing (CT) conveyed drilling technique, were several new wellbores are jet-drilled perpendicular from the mother well and into the reservoir formation. Their objective is to improve the production profile around the mother well, by penetrating the damaged skin zone, and connecting to possible hydrocarbon pockets left behind in the reservoir. They described mathematically the new CT tool in terms of the Penetration Effect, and pull effect. The first effect which is penetration mechanisms were Identified by Surface Erosion, Hydraulic Fracturing, Poroelastic tensile failure and Cavitation. The second effect, the net pull effect that works the nozzle forward can be derived from three main mechanisms: the under pressure force, the jetting force and the ejector force. Finally, they concluded that further research and testing of the jet nozzle penetration mechanism are required in order to identify the optimal nozzle configurations.

Elshahawi et al. (2001) presented several case studies performed in Belayim Fields of Sinai, Petrobel-Egypt to enhance their production. The main reasons for trying this new technique were the reduction in the wells productivity, and the rapid decline in total field production. The major wells in this field were under artificial lifting and this involved the integration of data from various sources. They showed in their study that the majority of wells in the field were suffering from severe damage. Using nodal analysis and skin modeling, the wells were then categorized based on the value of their completion factors (the ratio of actual to theoretical productivity indices), and the potential sources of damage were identified for each category.

The method used to mitigate this shortage in the wells productivity is to deep penetrating perforating charges, this had been performed after investigating the origin and the development of formation damage in Belayim Fields and how this damage was attributed to different damage mechanisms using a novel combination of nodal and damage analysis techniques.

Elshahawi et al. (2011) addressed several methods for damage removal such as vacuum strings (for perforation cleaning), surging using an atmospheric chamber (failed), acidizing (failed), demulsifiers and other chemicals to break emulsions (failed), and finally Deep-penetrating perforating. The later technique in Petrobel was the most successful productivity enhancement over the last few years. The main reasons for this success are to increase the surface area available for flow, decrease pressure drop and thus reduce the flow velocity across the perforated interval. And present three cases illustrating how deep penetrating perforating has been successfully used to remove well damage and increase well productivity, namely Well BM-30 m, Wells Sidri-3 and Sidri-4, and Well Sinai-02.

FIRST CASE OF RADIAL DRILLING APPLICATION IN EGYPT

This technique was applied for the first time in Egypt in Belayim Land Oil Field in Petrobel Company (Dickinson et al., 1989). The formation of the Belayim Oil Field is located in the central part of the Gulf of Suez along Sinai Peninsula. Belayim oil fields are characterized by multiple layered reservoirs generally formed from sand with inter-bedded shale and anhydrite from different ages. Belayim oil field main production now depend on artificial lift; secondary recovery are used (water injection).

The data of the first job are collected and analyzed in order to decide extending this technique to more other wells in the company. Many considerations had been taken into account for well selection. From these consideration, open hole logs for defining the pay zones, lithology, static bottom hole reservoir pressure, average porosity (about 20%) and permeability, cased hole logs and casing types and grades. Based on all of the mentioned parameters three wells were selected to evaluate radial drilling technique from layered reservoir zones II-A and IV. Currently, Zone IV contains about 23% of Belayim OOIP and contributes about more than 27% of production.

This technique is applied in three wells, in the first one, which was producing a commingle from two zones (Belayim and Sidri), six lateral drains at two levels was performed, five of which are 50 m long and one penetrate about 92 m long, all oriented by gyro tool. In the second well, seven laterals have been performed in Sidri formation, all penetrate 50 m long each at one different depth and oriented like a spiral by rotating the BHA at the surface one and half turn. The last well, four lateral are performed at two different depths, all are 50 m long (Abdel-Ghany et al., 2011).

In 2011, Abdel-Ghany et al. presented some experience gained from using this technique and concluded that; for one well, an increase in the rate of 75 to 130 m^3/day, for the second one, the production rate is increased from 41 to 45 m^3/day, and for the last well, the results showed no change in flow rate.

RESULTS

Case 1: Well # 1

Well #1 is produced from two different zones (zone II-A and zone IV) with an average daily rate of 251 bpd, the current static reservoir pressure is about 900 psi and productivity index is 2 barrel per day per psi, the formation porosity 20% in average, the rock permeability is varied largely, the total depth of this well is about 8856 ft and net pay thickness is 82 ft. Figure 8 illustrates the casing and perforation interval of that well. This well radially drilled on 2010 to evaluate and optimize this technique in the field.

Radial drilling job were performed on this well by milling and jetting seven lateral holes with 164 foot lateral length at different depths from zone IV as shown in Figure 8. The hydraulic jet pressures for all of them were 7000 psi. Table 1 shows the accurate depths and elongations for each lateral.

While performing RD in this well, a vacuum test was performed before and after radial drilling job to evaluate this technique. Production rate shows an increased after stimulating the well with this technique. The gross production rate increased from 252 to 346 bpd, where the net oil rate increased from 220 to 289 bpd, which means 37.5% increase for the gross rate and 31.4% for the net oil rate. Figures 9 to 11 show the results of vacuum tests before and after jobs. Table 2 shows comparison between rate before and after radial drilling job, and Figure 12 displays gross production performance before and after applied radial drilling technique. As shown in that figure, the rate maintained awhile after performing the test and then declined again but still better than before conducting the technique. This is attributed to 'do not fill the hole with any material' and that can keep the channel open in order to guarantee constant production rate.

Case II: Well #2

This well is produced from three zones (zone II, II-A and zone IV) with average daily production rate 471 bpd, its current static reservoir pressure is about 1990 psi at top of perforation 7126 feet sub sea level and productivity index is 1 barrel per day per psi from last vacuum test before applied radial drilling, its rock porosity is 20%, heterogeneous permeability, the total depth of this well is 8134 ft and net pay thickness is 133 ft. Figure 13 illustrates casing and perforation interval for well #2.

Radial drilling job was performed on this well by milling and jetting six lateral holes at two depths with 164 ft lateral length and one of them was 295 ft lateral length from zone II-A. Table 3 shows the full details of the six laterals performed in well #2. In the previous well, the lateral holes were drilled randomly but here in this well the lateral hole selection was based on geological maps and faults orientation so holes were made at two depths and oriented with consideration to north direction.

A vacuum test was performed on this well before and after radial drilling job to evaluate this technique. Production rate obviously show increase after stimulating

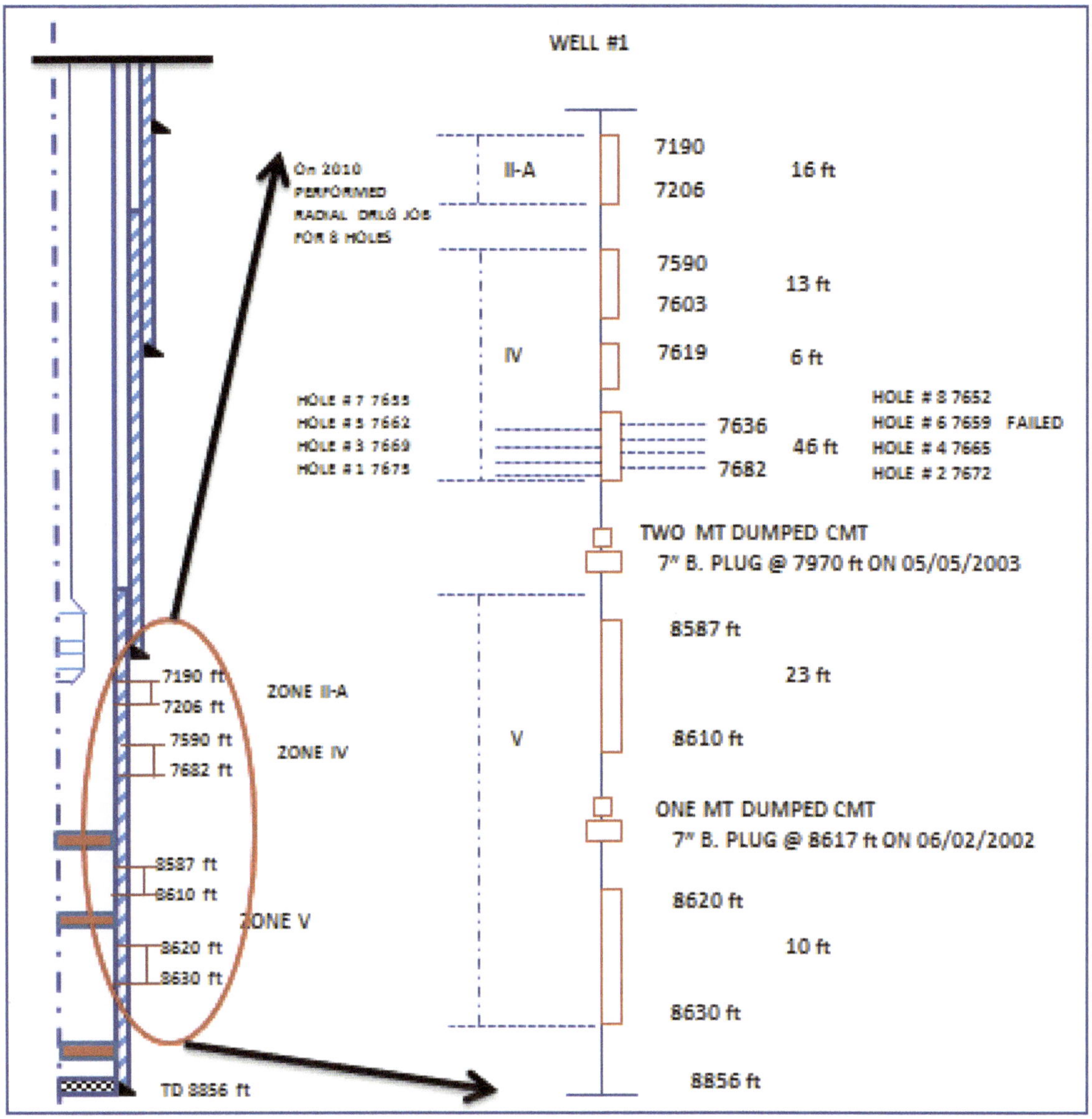

Figure 8. Casing and perforation sketch of well #1.

Table 1. Radial drilling operation performed on well #1.

No. of hole	Hole depth (ft)	Penetration length (ft)	Jetting pressure (Psi)	Remarks
1	7675.20	164	7000	
2	7671.92	164	7000	
3	7668.64	164 (L)	7000	
4	7665.36	164 (L)	7000	
5	7662.08	164 (L)	7000	
6	7658.80	***	7000	Tried two times to drill, no success
7	7655.52	164	7000	During POOH found the bit and flex shaft in side deflector shoe
8	7652.24	164	7000	

(L): Lateral.

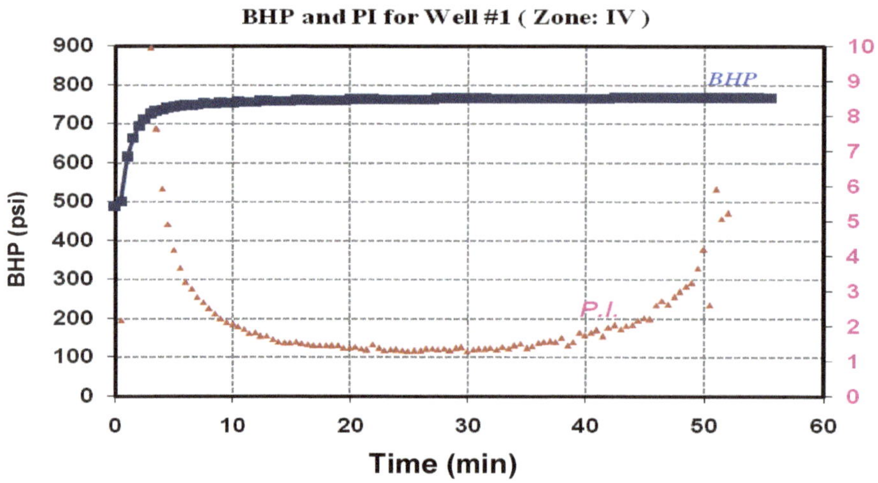

Figure 9. Fill-up test for BHP and P.I test before RD of well #1.

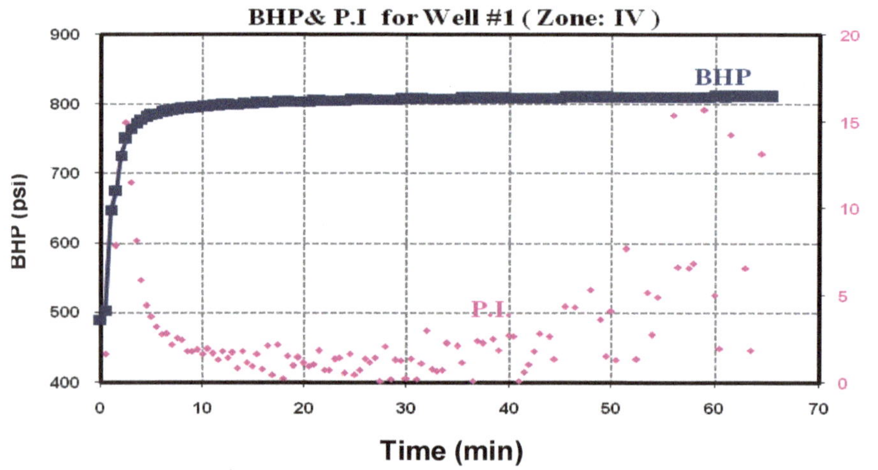

Figure 10. Fill-up test for BHP and P.I test after radial drilling.

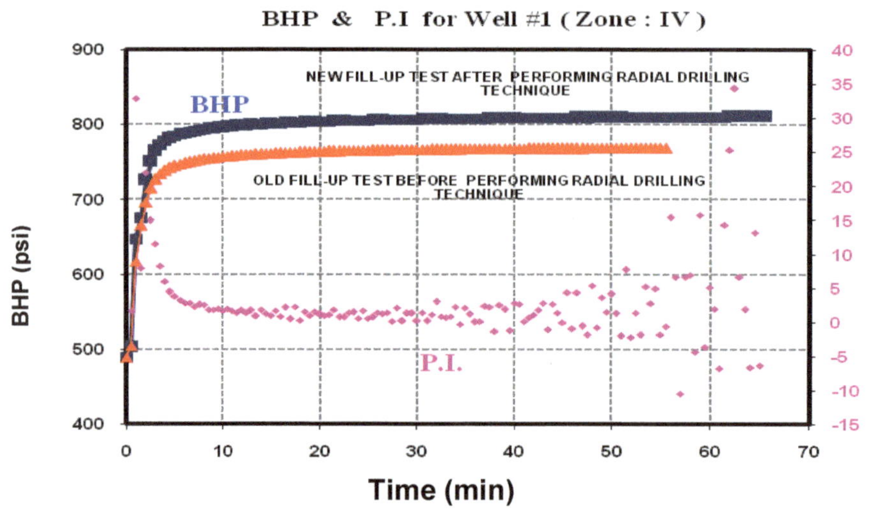

Figure 11. BHP comparison between the two tests before and after RD.

Table 2. Comparison between the rate before and after RD.

Before radial drilling			After radial drilling		
Rate (bpd)	WC (%)	Net oil	Rate (bpd)	WC (%)	Net oil
252	12	220	346	16	289

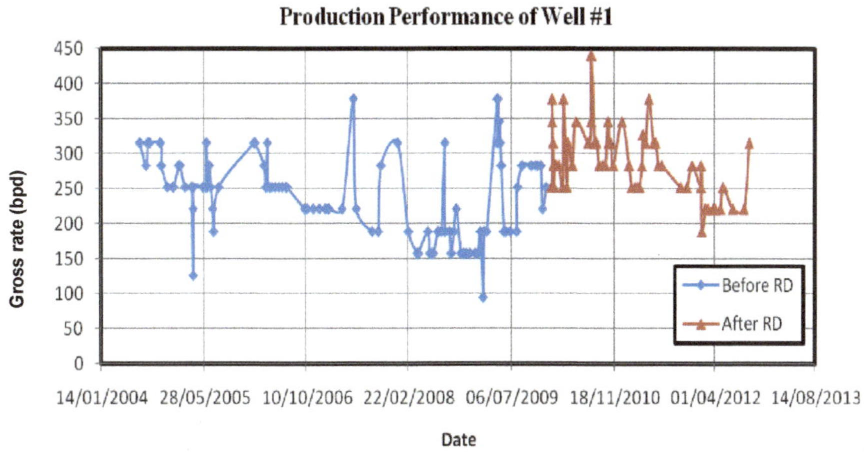

Figure 12. Production performance for well #1.

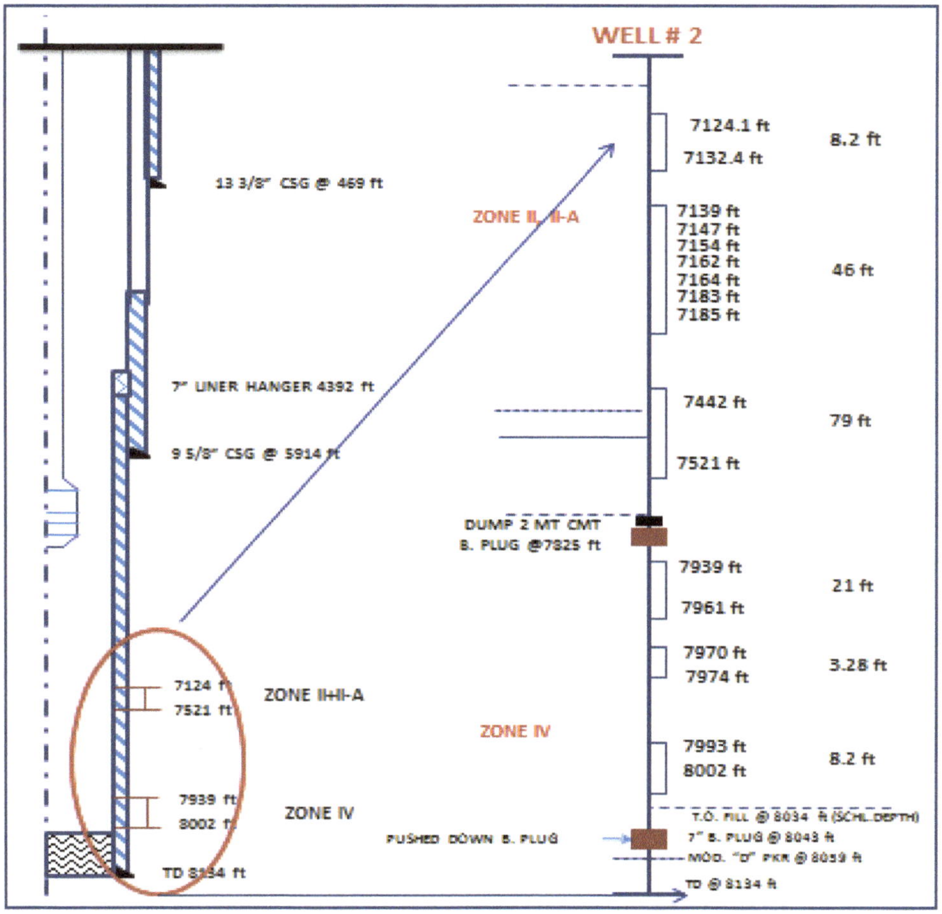

Figure 13. Casing and perforation sketch for well #2.

Table 3. Radial drilling operation performed on well 2#.

Depth (ft)		7462 ft			7449 ft	
No. of holes	1	2	3	4	5	6
Angle from north	20	150	225	20	150	240
Depth of lateral (ft)	164	164	295	164	164	164

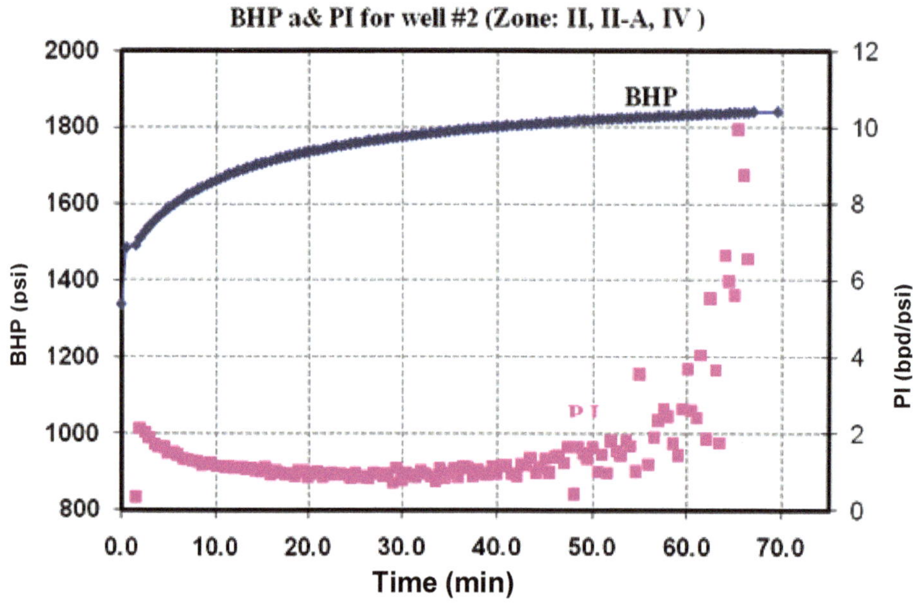

Figure 14. B.H.P and P.I vs. time, vacuum test results before radial drilling.

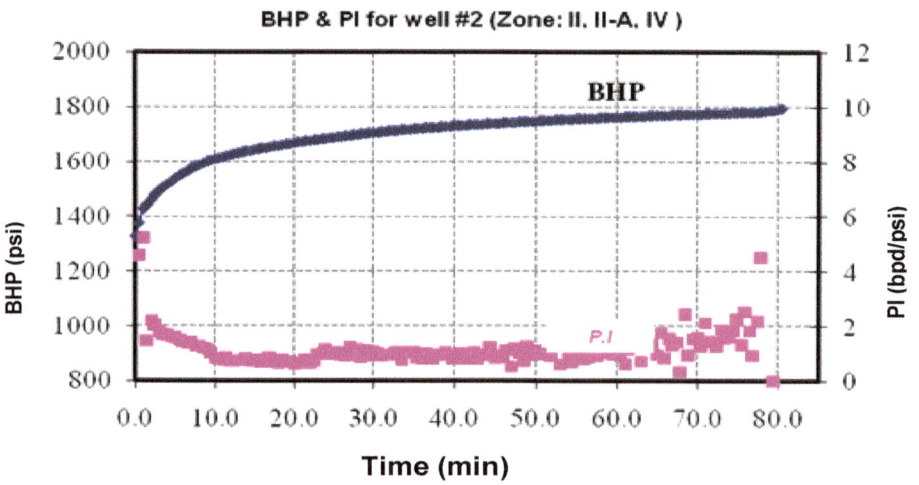

Figure 15. B.H.P and P.I vs. time, vacuum test results after radial drilling.

the well with this technique. The gross rate increased from 472 to 818 bpd immediately after the RD performed, and the net rate increased from 465 to 686 bpd. However the static well pressure decreased which mean a decrease in fluid level, and productivity index still the same. Figures 14 and 15 show the bottom hole pressure before and after RD performed respectively. Figure 16 illustrates overlapped results of well #2 before and after

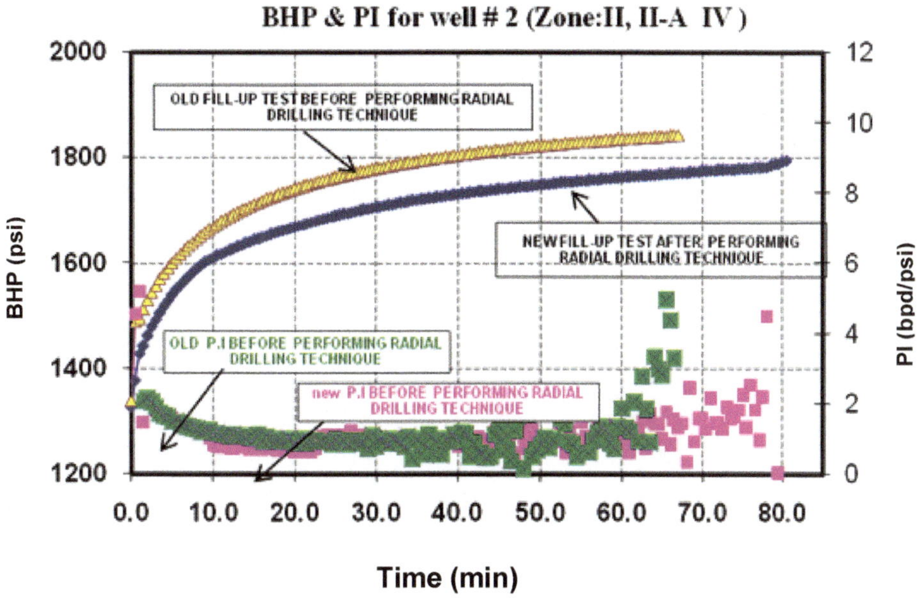

Figure 16. Comparison between the two vacuum tests.

Table 4. Comparison between rate before and after radial drilling for well #2.

Before RD			After RD		
Rate (bpd)	WC (%)	Net oil (bpd)	Rate (bpd)	WC (%)	Net oil (bpd)
472	1.6	465	818	16	686

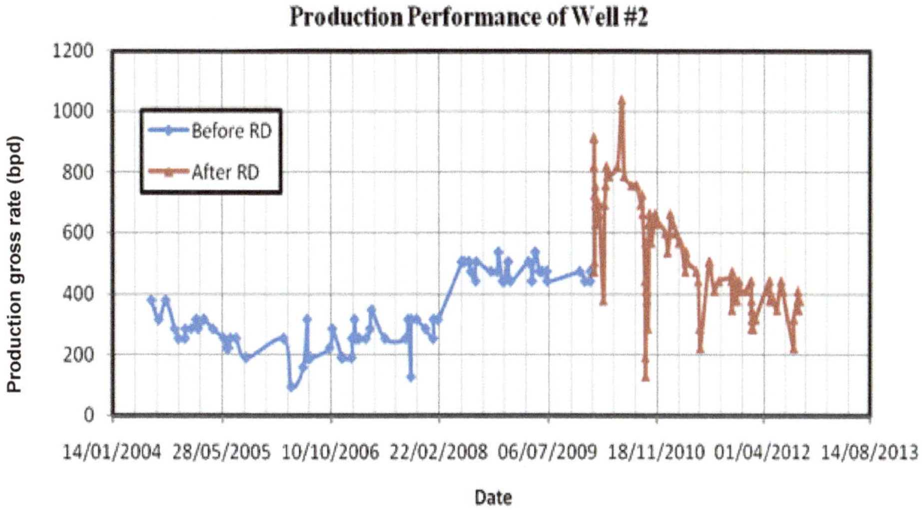

Figure 17. Production performance of well #2 before and after radial drilling.

the vacuum test. Table 4 shows comparison between rate before and after radial drilling job which mean gain in net oil production about 221 bbl/day. Figure 17 shows the production performance of well #2 before and after the radial drilling operation. It shows how RD enhances the productivity from this well and how it declines after a

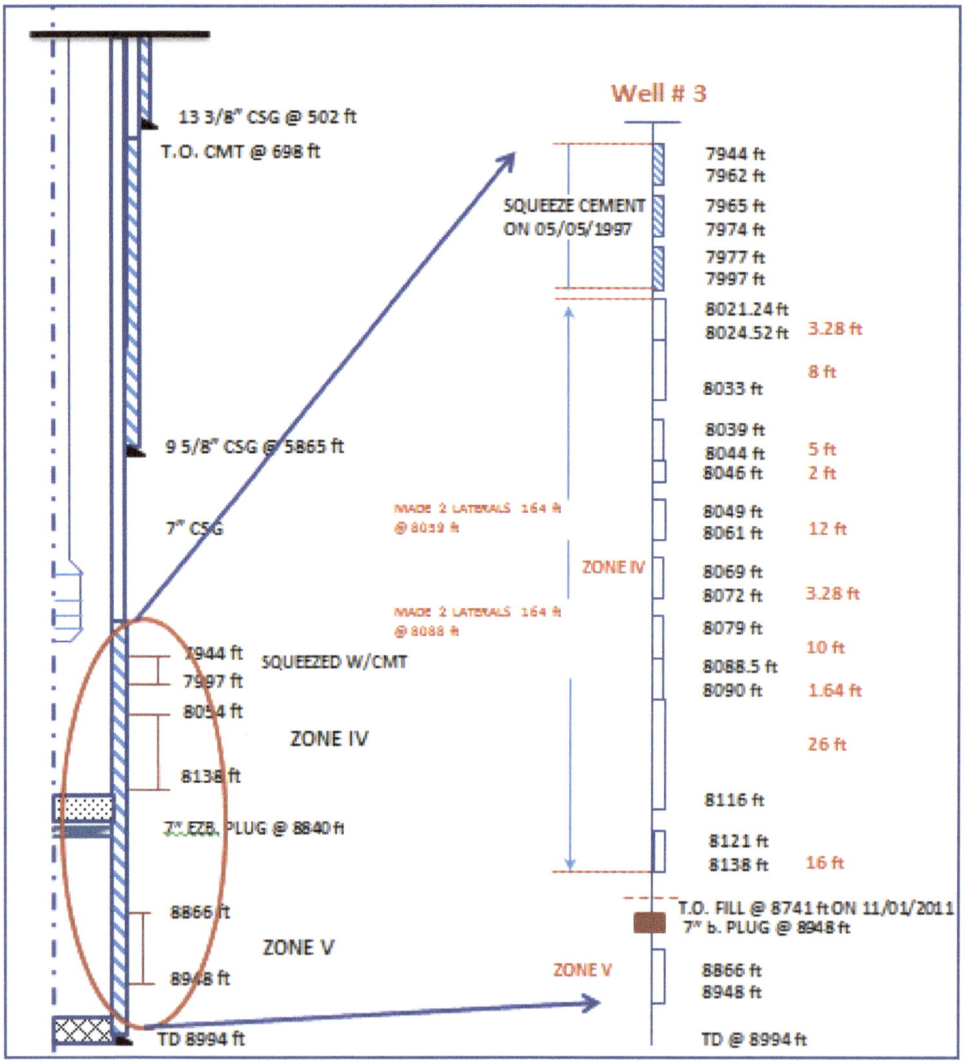

Figure 18. Casing and perforation sketch for well #3.

Table 5. Radial drilling operation performed on well #3.

Depth (ft)	8088 ft		8059 ft	
No. of holes	1	2	3	4
Angle from the north	115	295	115	295
Depth of lateral (ft)	164	164	164	164

short period of time.

Case 3: Well #3

This well is produced from only one zone (zone IV) with average daily rate 189 bpd, the static reservoir pressure is about 970 psi at datum 7900 feet sub sea level and productivity index is 2 barrel per day per psi from last vacuum test before applied radial drilling, the average porosity 20%, heterogeneous permeability, the total depth of this well is about 8994 ft, and net pay thickness is 87 ft. Figure 18 illustrates casing specification and perforation interval of well #3.

Radial drilling job was performed on this well by milling and jetting four lateral holes at two depths with 164 ft lateral length from zone IV. There were a lot of trials to make another two holes at another depth but the holes were canceled due to 1¾" bit and flexible shaft lost in hole several time, and that cause more lost time. Table 5 shows the depths and the length of penetration of each drain holes. In the first well, well #1 the lateral holes

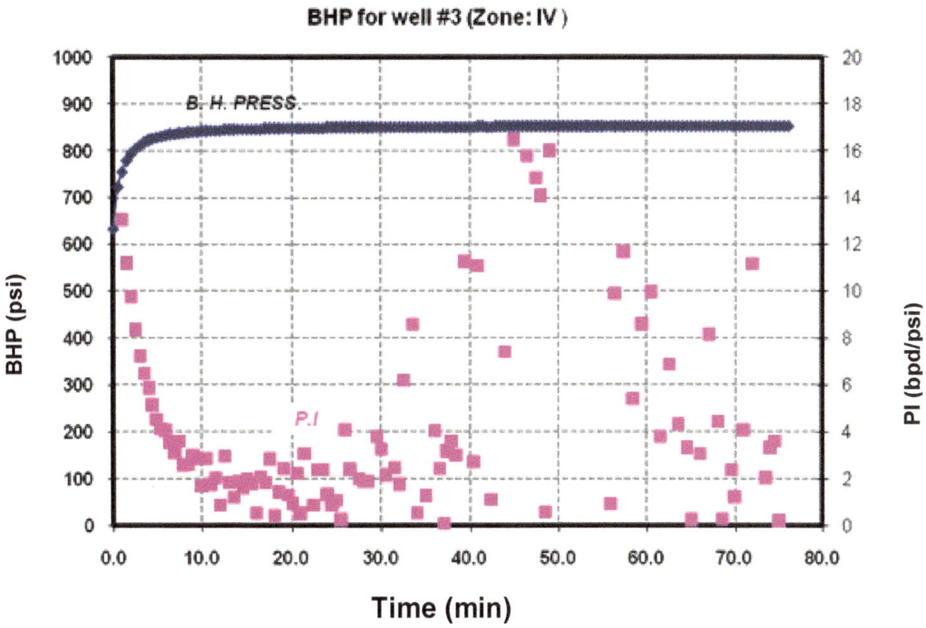

Figure 19. S.B.H.P and P.I vs. time, vacuum test results for well #3 before radial drilling.

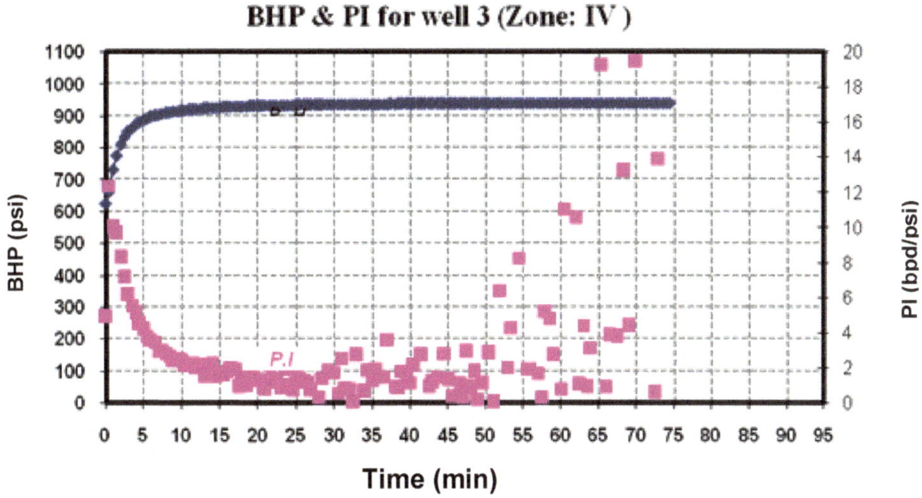

Figure 20. S.B.H.P and P.I vs. time, vacuum test results for well #3 before radial drilling.

were drilled randomly but here in this well the lateral hole selection was based on geological maps and faults orientation so holes were made at two depths and oriented with consideration to north direction.

A vacuum test was performed on this well before and after radial drilling job to evaluate this technique. Production rate show a little increase and then decreased after stimulating the well with this technique; the productivity index still the same. Figures 19 and 20 depict the bottom hole pressure and productivity index for this well before and after RD respectively. Table 6 shows comparison between rate before and after radial drilling job, Figure 21 show gross production performance before and after applied radial drilling technique.

CONCLUSIONS

The evaluation thus reveals that:

1. This technology is succeeded from the mechanical point of view as the holes were drilled successfully.
2. The well which has the deepest lateral hole length show a significant increase in production rate so as the length of lateral hole increase and the results will be

Table 6. Comparison between rate before and after radial drilling for well #3.

Before radial drilling			After radial drilling		
Rate (bpd)	WC (%)	Net oil (bpd)	Rate (bpd)	WC (%)	Net oil (bpd)
220	3.2	214	252	3.2	245

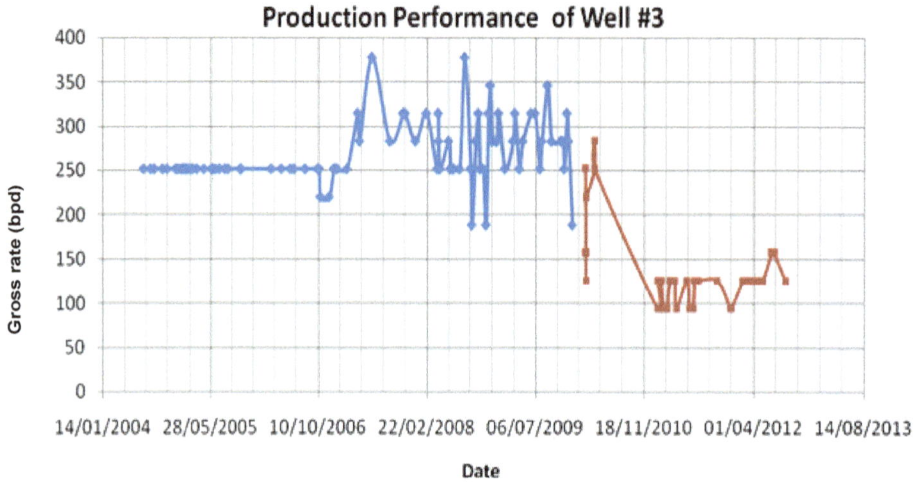

Figure 21. Production performance of well 3 # before and after radial drilling.

more good.

3. Using this technique in consolidated rock better than un-consolidated ones in order to maintain the hole open.

4. The technology has been successfully applied to provide multiple horizontal radials at the same level in a single well.

5. Radial drilling by high-pressure jet flow techniques can greatly increase oil recovery and oil production rate.

6. Before the large scale usage of this technique, rock mechanics must be considered prior to design.

7. Increase in the reservoir contact length (improve drain efficiency), therefore increase the production, control the direction of perforations, and can be low cost when utilized in existing wells.

8. Further research and testing of the jet nozzle penetration mechanism are required in order to identify the optimal nozzle configurations.

9. Easy to be applied and less expensive than other conventional techniques.

10. The experience shows that it is necessary to have petrophysical studies, rock mechanics and pressure tests, prior to intervention, to know the reservoir conditions.

11. An improvement is needed in order to maintain the rate increase in production rate, such as gravel packing or using slim tubes specially in unconsolidated formation.

REFERENCES

Abdel-Ghany MA, Siso S, Hassan AM, Pierpaolo P, Roberto C (2011). New Technology Application, Radial Drilling Petrobel, First Well in Egypt, presented at the 10th Offshore Mediterranean Conference and Exhibition (OMC) in Ravenna, Italy.

Bruni M, Biassotti H, Salomone G (2007). Radial Drilling in Argentina, SPE Paper 68504, Latin American and Caribbean Petroleum Engineering Conference, Buenos Aires, Argentina.

Buset P, Riiber M, Arne E (2001). Jet Drilling Tool: Cost-Effective Lateral Drilling Technology for Enhanced Oil Recovery. Presented at the SPE/ICoTA Coiled Tubing Roundtable held in Houston, Texas.

Dickinson W, Dickinson R, Herrera A, Dykstra H, Nees J (1992). Slim Hole Multiple Radials Drilled with Coiled Tubing. SPE 23639, presented at the Second Latin American Petroleum Engineering Conference held in Caracas, Venezuela.

Dickinson W, Anderson RR, Dickinson RW (1989). The Ultrashort-Radius Radial System. SPE Paper 14804, SPE Drilling Engineering, pp. 247-254.

Dickinson W, Dickenson RW (1985). Horizontal Radial Drilling System. SPE 13949, presented at the SPE 1985 California Regional Meeting, held in California.

Elliott S (2011). Coiled-tubing Method Drills radial Laterals to Improve Oil Production from a Depleted Reservoir. World Oil 232:10.

Elshahawi H, Siso S, Samir M, Safwat M (2001). Production Enhancement in the Belayim Fields: Case Histories. SPE 68688, presented at the SPE Asia Pacific Oil and Gas Conference and Exhibition held in Jakarta, Indonesia.

Marbun BTH, Sinaga SZ, Arliyando L, Putra SK (2012). Review of Ultrashort-Radius Radial System (URRS). IPTC 14823, presented at the International Petroleum Technology Conference held in Bangkok, Thailand.

Yonghe L, Chunjie W, Lianhai S, Weiyi G (2000). Application and Development of Drilling and Completion of the Ultrashort-radius Radial Well by High Pressure Jet Flow Techniques. SPE 64756, presented at the SPE International Oil and Gas Conference and Exhibition in China held in Beijing, China.

Modeling of thermal cracking process in a crude oil refinery

Sepehr Sadighi and S. Reza. Seif Mohaddecy*

Project Manager, Catalysis and Nanotechnology Division, Catalytic Reaction Engineering Department, Research Institute of Petroleum Industry (RIPI) Iran

In a thermal cracking process, the molecular bonds of the liquid are broken to the lighter ones. Recently due to the availability of more heavy oils, the process interest was to yield light and middle distillate products. In this research, thermal cracking of vacuum residue in a commercial soaker-visbreaking plant is studied. The product of the process is characterized to the light gas (C_1, C_2), liquefied petroleum gases (C_3, C_4), gasoline (IBP-180°C), gas oil (180 to 320°C) and fuel (320^+°C). Then to model the visbreaking process, a six-lump kinetic network with fifteen reactions and thirty kinetic parameters is developed. In this model, visbreaking process is modeled as an equal distributed heater, and the soaker is modeled as a complete stirred tank reactor. After evaluating the rate of reactions by estimated kinetic parameters, it is confirmed that a reduced reaction network with seven reaction paths and fourteen kinetic parameters is reliable enough to simulate the performance of the reactor with the absolute average deviation ($AAD\%$) of 4.75%.

Key words: Kinetic model, visbreaking, lumped model.

INTRODUCTION

In the present day scenario the availability of more heavy crude oils has resulted in the increased production of atmospheric and vacuum residues and simultaneous decrease in light and middle distillate fractions (Singh et al., 2012). A visbreaker is a processing unit in oil refinery whose purpose is to reduce the quantity of residual oil produced in the distillation of crude oil and to increase the yield of more valuable middle distillates (heating oil and diesel) by the refinery. A visbreaker thermally cracks large hydrocarbon molecules in the oil by heating in a furnace to reduce its viscosity and to produce small quantities of light hydrocarbons (LPG and gasoline). The process name of "visbreaker" refers to the fact that the process reduces (that is, breaks) the viscosity of the residual oil. The process is non-catalytic (Gary et al., 1984; Speight et al., 2006).

There are two types of commercial visbreaking units: the coil or furnace type, and the soaker process. The coil-visbreaker is operated at high temperature and low residence time whilst in a soaker one by adding an adiabatic drum after the coil furnace, the product is held for a longer time so that the coil is kept at relatively lower temperature. Therefore, the heater duty and, in turn, the fuel consumption is only 70% of that for the coil-visbreaking process (Joshi et al., 2008). Worldwide, about 200 visbreaking units are under operation, and Europe alone accounts for about 55% of the total visbreaking capacity (Joshi et al., 2008).

To effective design and perfect control of any process, a model is needed to predict product yields and qualities versus variables such as space velocity and temperature. However, the complexity of visbreaking feed and product makes it extremely difficult to characterize and describe its kinetics at a molecular level. Modern day rigorous simulators such as Aspen plus or Hysys from Aspen Technology do not have such restrictive limits on the total number of components, and it is possible to use a unique set of pseudo components for every petroleum assay stream (Sadighi et al., 2012). This approach increases the calculation time, however, and characterization of the streams and subsequent reports become unnecessarily complicated (Sadighi et al., 2012).

One approach to simplify the problem is to consider the partition of the species into a few equivalent classes, the so-called lumps or lumping technique, and then assume each class is an independent entity. Developing simple kinetic models (for example, power-law model) for complex catalytic reactions is a common approach as it can give basic information for reactor design and optimization. In this field, many investigations were reported in which visbreaking process was modeled with discrete two-lump (Al-Soufi et al., 1988; Krishna et al., 1988; Di Carlo and Jenis, 1992), three-lump (Benito et al., 1995), 4-lump (Del Bianco et al., 1993; Trauth et al., 1992; Singh et al., 2012), five-lump (Kataria et al., 2004; AlHumaidan et al., 2013) and 7-lump (Xiao et al., 2002) approaches. In all these investigation, the experiments were carried out in a micro or pilot scale reactor. But, the common advantage of all of these presented models is their simplicity, less required computational time and the least needed structural information. Furthermore, the more sophisticated lumping models are continuous one that one of them was recently presented by Shadbahra et al. (2011).

The aim of this research is developing a simple yield predictor model, according to a six-lump reaction approach, to predict the most added value products consists of gas, LPG, gasoline, diesel and fuel oil in a commercial soaker unit. The main advantage of this work is presenting a simple approach for the commercial visbreaking process in which the temperature profile of the furnace is also included in the model.

DATA GATHERING

Feed characterization

An Industrial soker-visbreaker unit was chosen as a case study. This unit was designed to visbreak 20,000 barrel per day of a mixture of vacuum residuum and slop vacuum gas oil which are both taken from the vacuum tower; the composition of the fresh feed can vary slightly with time from start of run (SOR) to end of run (EOR). The specification of the combined feed, which was analyzed during this research, is shown in Table 1.

Process description

The visbreaking feed is charged to the coil furnace at the temperature about 340°C. The visbreaking furnace is included of two sections fired independently. After the coil furnace, the two hot streams are drained into a transfer line; then the mixed product is entered into the soaker drum. The specifications of cells and the soaker drum are presented in Table 2. The output product from the

Table 1. Feed characterization.

Property	Unit	Value
Sp.Gr.	-	1.006
Sulfur Content	wt %	3.19
Va + Ni Content	wt ppm	188

Distillation (ASTM D1160)	
Vol %	**Temperature (°C)**
IBP[1]	303
5	409
10	457
20	503
30	543
50	585

Table 2. Specifications of the cell and soaker of the visbreaking unit.

Coil specification	Unit	Value
Number of tubes	-	128
Number of convection tubes	-	76
Number of radiation tubes	-	52
Tube length	m	18.745
Outside diameter	m	0.114
Soaker specification		
Outside diameter	m	2.405
Length	m	16.5

soaker drum is quenched by the cold recycle stream to stop the cracking reactions, and so to inhibit the coke formation. Finally, the combined stream is transferred to the fractionation tower and side strippers to separate the visbreaking products. The simplified process flow diagram of the described unit is depicted in Figure 1.

During one year of data gathering, nine sets of data including product flow rates, feed inlet temperature and soaker outlet temperature were gathered from the target visbreaking process (Table 3). As it is illustrated in Figure 2, light gases including C_1, C_2 and LPG, gasoline and tar are the output streams from the visbreaking plant. Performing mass balance around the unit proved that the error for all experiments was less than 2%, mainly related to the gross error for the measuring of the gaseous products and maybe related to the coke formation. Moreover, the boiling range of VGO feed, fuel and gas oil samples are analyzed according to the ASTM D1160 standard procedure (ASTM Standard D1160-06, 2009) whilst the one for the gasoline sample is analyzed according to the ASTM D86 method (ASTM Standard D86-08a, 2009).

KINETIC MODEL

This work considers six lumps, that is, vacuum residue (V), fuel (F), gas oil (D), gasoline (N), LPG and gas (G) to match all main

[1] Initial Boiling Point

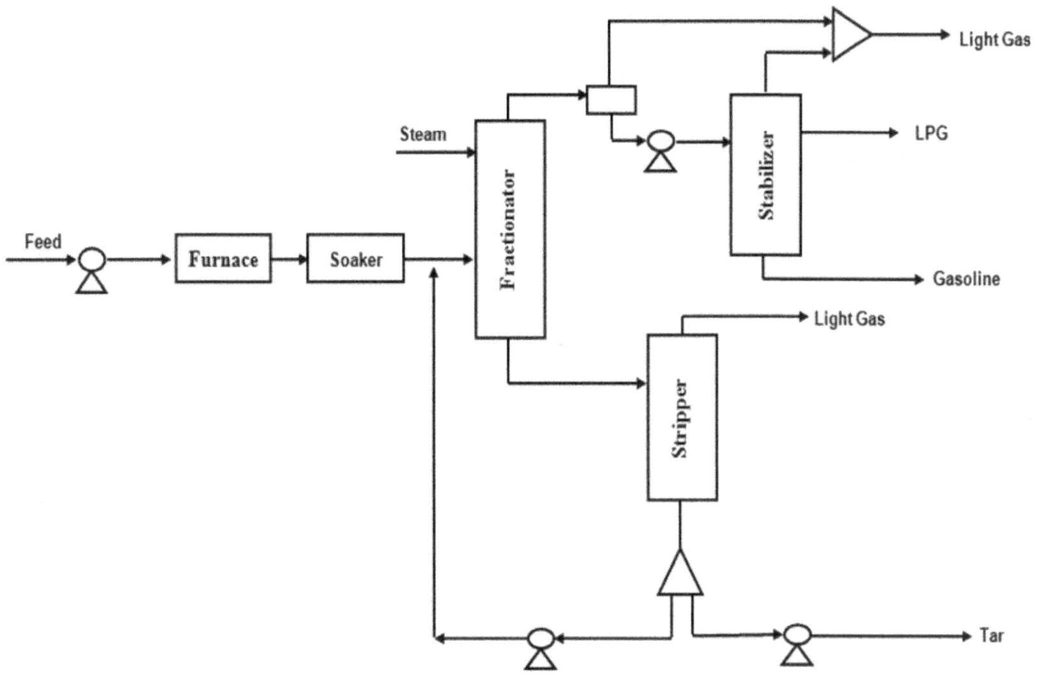

Figure 1. Block flow diagram of visbreaking process.

Table 3. Feed flow rate and reactor operating condition.

Case	Vacuum residue (kg/h)	Inlet temperature (°C)	Outlet temperature (°C)
1	1.243E+05	326.5	439
2	1.286E+05	326	438.5
3	1.346E+05	324.4	440.7
4	1.193E+05	327.4	438.5
5	1.433E+05	324.8	441.3
6	1.313E+05	324.9	440.5
7	1.393E+05	324.8	439.3
8	1.156E+05	328.5	437.5
9	1.325E+05	324.8	440.5

products in the commercial process. The rate of coke formation with time can be considered low for a soaking visbreaker (Dente et al., 1997), and so the coke as a main product can be neglected. Figure 2 shows the fifteen reaction pathways associated with this strategy, illustrating the complexity of the network if all possible pathways are considered. So, the model consists of thirty kinetic parameters which should be estimated using experimental data. However, some considerations can be normally utilized to reduce the model complexity without sacrificing the accuracy (Sadighi et al., 2010b, c).

For each reaction, a kinetic expression (r) is formulated as the function of the mass concentration of the reactants (C), furnace temperature (T) and kinetic parameters (k_0 and E). The reaction of VGO hydrocracking to yield products is assumed first order (Benito et al., 1995).

According to the above assumptions, the kinetic constants of the model are expressed as:

Vacuum residue (V): $$k_{Vj} = k_{0Vj}\exp(\frac{-E_{Vj}}{RT}) \quad (1)$$

where j in Equation (1) represents all products lighter than the Vacuum residue lump;

Fuel (F): $$k_{Fj'} = k_{0Fj'}\exp(\frac{-E_{Fj'}}{RT}) \quad (2)$$

where j' in Equation (2) represents all products lighter than the Fuel lump;

Gas oil (D): $$k_{Dj''} = k_{0Dj''}\exp(\frac{-E_{Dj''}}{RT}) \quad (3)$$

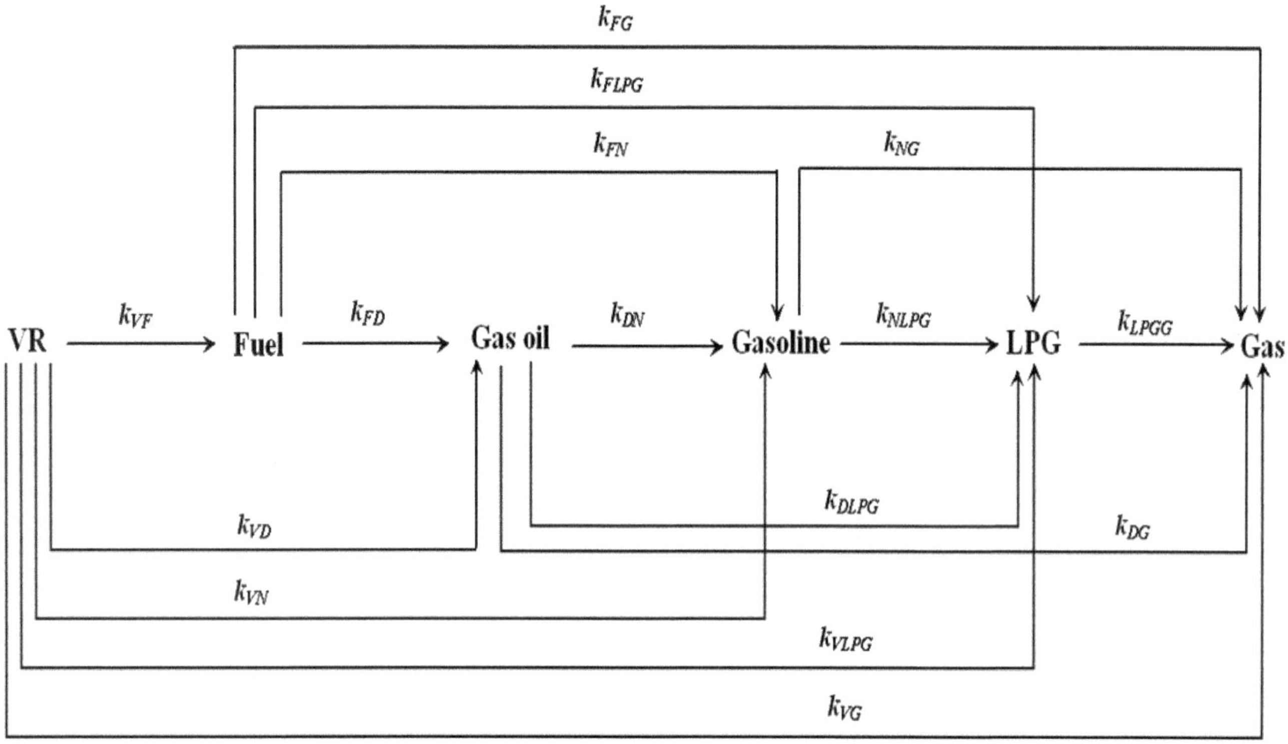

Figure 2. The complete six-lump kinetic model.

where j'' in Equation (3) represents all products lighter than the Gas oil lump;

Gasoline (N): $k_{Nj'''} = k_{0Nj'''} \exp(\frac{-E_{Nj'''}}{RT})$ (4)

where j''' in Equation (4) represents all products lighter than Gasoline; and

LPG (LPG): $k_{LPGG} = k_{0LPGG} \exp(\frac{-E_{LPGG}}{RT})$ (5)

In Equations (1) to (5), T and R are the absolute value of the coil temperature of the visbreaking furnace and the ideal gas constant, respectively. Thus, the reaction rates (r) can be formulated as the following:

Vacuum residue (r_V): $r_V = \sum_{j=V}^{G} k_{Vj} C_V$ (6)

Fuel (r_F): $r_F = k_{VF} C_V - \sum_{j=F}^{G} k_{Fj} C_F$ (7)

Gas oil (r_D): $r_D = k_{VD} C_V + k_{FD} C_F - \sum_{j=D}^{G} k_{Dj} C_D$ (8)

Gasoline (r_N): $r_N = \sum_{j=V}^{D} k_{jN} C_j - \sum_{j=N}^{G} k_{Nj} C_N$ (9)

LPG (r_{LPG}): $r_{LPG} = \sum_{j=V}^{N} k_{jLPG} C_j - k_{LPGG} C_{LPG}$ (10)

Gas (r_G): $r_G = \sum_{j=V}^{G} k_{jG} C_j$ (11)

MASS BALANCE EQUATIONS

A soaker-visbreaking unit is divided to two reactive parts in series. The first one is the coils of the furnace assumed as an ideal plug-flow reactor (Joshi et al., 2008), and the second one is the soaker drum which is considered as complete mixed reactor. So, the mass balance equations for the coil and soaker drum can be given as follows.

For the coil: $\frac{\partial(\upsilon C_j)}{\partial V_C} \pm r_j = 0$ (12)

For the soaker drum: $\frac{\Delta(\upsilon C_j)}{V_D} \pm r_j' = 0$ (13)

In Equations (12) and (13), j ranges from the vacuum residue lump (V) to the gas (G), C is the mass concentration of the lump, V_C is the volume of coil, V_D is the volume of drum; r_j and r'_j are reaction rates; negative sign "-" is used for the feed (or VGO), and a positive sign "+" is used for products.

For the coil: $$\frac{\partial(\rho v)}{\partial V_c} = 0 \quad (14)$$

For the soaker drum: $$\frac{\Delta(\rho v)}{V_D} = 0 \quad (15)$$

$$X_j = \frac{C_j . v}{F_m} \quad (16)$$

$$\frac{1}{\rho} = \sum_{j=V}^{G} \frac{X_j}{\rho_j} \quad (17)$$

In Equations (14) to (17), ρ and v are the stream density and volumetric flow rate through the reactor, respectively; F_m is the mass flow rate of the stream passing through the coil; X_j, and ρ_j are the mass fraction and density of lump *j*, respectively.

After calculating the mass concentration and volumetric flow rate of each lump in the effluent stream, the product yields can be found as the following:

$$Y_j = \frac{C_{j\,out} . v_{out} (1 - F_R)}{F_m} \quad (18)$$

In Equation (18), F_R is the fraction of recycle of the lumps, which is mixed with the fresh feed.

COIL TEMPERATURE MODEL

In this work, it is supposed that there is an equal heat flux throughout the furnace to close the overall heat balance. Therefore, the following expression can be written for the temperature profile through the furnace tubes:

$$\frac{\partial T}{\partial z} = \frac{F_m \left[(\sum_{j=G}^{V} X_j Cp_j T)_{out} - (\sum_{j=G}^{V} X_j Cp_j T)_{in} \right]}{L_t} \quad (19)$$

where T is the fluid temperature flowing the coil (reaction temperature); L_t is the total length of the tubes and Cp_j is the heat capacity of lump *j*; T_{co} and T_0 are coil inlet and outlet temperatures, respectively.

Because the difference between the inlet and outlet temperatures of the soaker drum is negligible, it can be modeled similar to an isothermal reactor.

PARAMETER ESTIMATION

To estimate the kinetic parameters, the sum of squared errors, SQE, as given below, is minimized:

$$SQE = \sum_{n=1}^{N_t} \sum_{j=V}^{G} (Y_{jn}^{meas} - Y_{jn}^{pred})^2 \quad (20)$$

In Equation (20), N_t, Y_{jn}^{meas} and Y_{jn}^{pred} are the number of test runs, the measured product yield and the yield predicted by the model, respectively.

The visbreaking model according to Equations (1) to (19) is coded and solved simultaneously using the Aspen Custom Modeler (ACM) programming environment (AspenTech, 2006) to calculate the product yields (Y_{jn}). ACM is an easy-to-use tool for creating, editing and re-using models of process units. ACM uses an object-oriented modeling language to build the simulation applications; then, the entire process can be simulated by combining these applications on a graphical flow sheet.

To estimate kinetic parameters, Equation (20) is minimized by sequential application of the NL2Sol and Nelder-Mead algorithms which are both found in the Aspen Custom Modeler software.

Additionally, to compare the simulated and measured product values, absolute average deviations (*AAD*) are calculated by the following equation (Sadighi et al., 2010a):

$$AAD\% = 100 \frac{\sum_{n=1}^{N_t} \sum_{j=F}^{G} \sqrt{\frac{(Y_{jn}^{meas} - Y_{jn}^{pred})^2}{Y_{jn}^{meas\,2}}}}{N_t} \% \quad (21)$$

RESULTS AND DISCUSSION

During the field study, nine sets of data consisting of flow rate of products, composition of gaseous products, distillation curve of cuts and soaker temperature were gathered from the target soaker-visbreaking plant. The Petro-sim process simulator was employed to lump the feed and products into components with the specific boiling-point ranges and properties, presented in Table 4, including gas (C_1&C_2), LPG (C_3&C_4), Gasoline (IBP-180°C), Gas oil (180 to 320°C), Fuel(320$^+$°C) and Vacuum residue. Hence, the process flow diagram of the visbreaking simulator can be shown as Figure 3.

The thirty kinetic parameters for the assumed model (Figure 1) were estimated, using measured industrial data, reported in Table 5. In this table, the ratio of all rate constants to the highest one (k_{VF} or vacuum residue to fuel) were calculated. After parameter estimation and simulation, the $AAD\%$ was 4.75% in comparison to the measured data.

Table 4. Average properties of the visbreaking lumps.

Parameter	IBP-FBP (°C)	Sp. g	Heat capacity (kJ/kg.°C)
Gas	C_1&C_2	0.364	1.86
LPG	C_3&C_4	0.55	1.97
Gasoline	IBP-180	0.739	2.4
Gas oil	180-320	0.806	2.6
Fuel	320^+	0.999	2.95

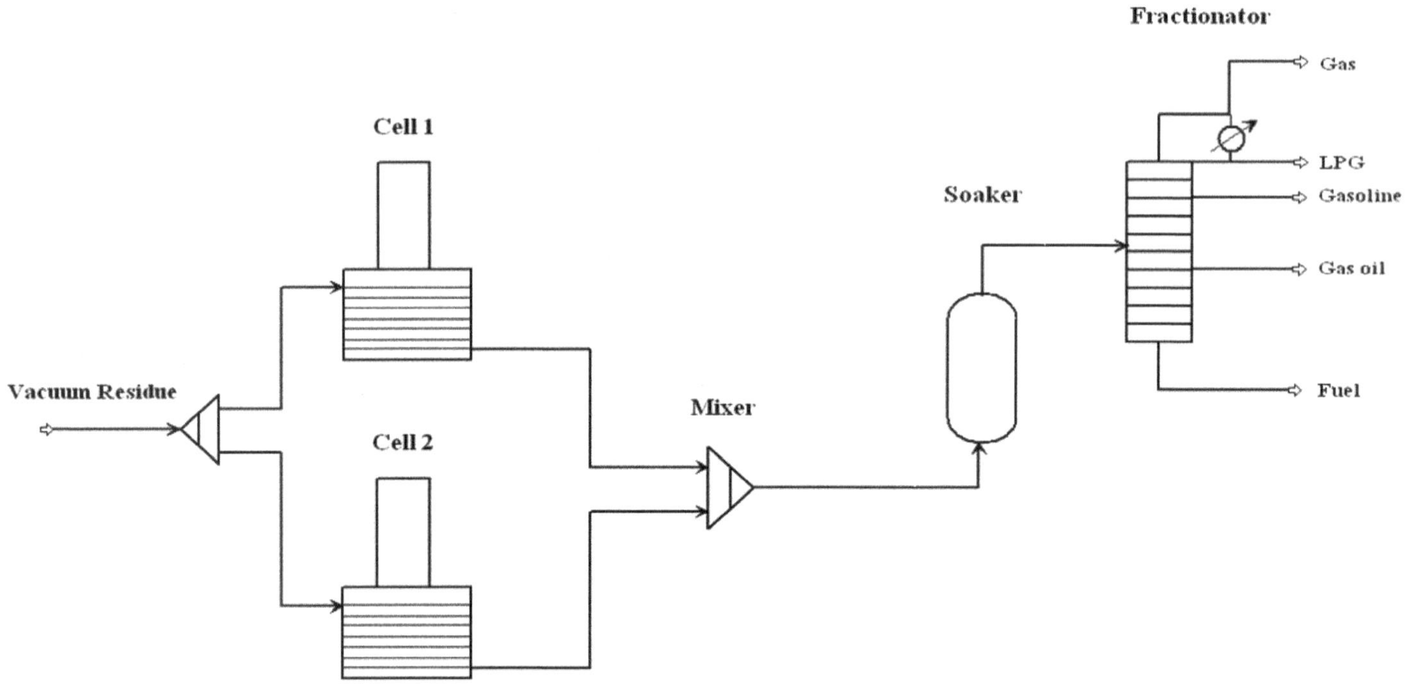

Figure 3. The scheme of the process flow diagram of visbreaking simulator.

Table 5. Kinetic parameters for the reaction network.

Frequency factor k_0 ($m^3.hr^{-1}.m^3\ cat^{-1}$)		Activation energy E (kcal/mol)		Rate $k_o \exp(-E/RT_{mean})$	Order (to k_{VF})
k_{OVF}	243082	E_{VF}	8.70	520.98	1
k_{OVD}	6785.12	E_{VD}	11.10	2.66	5.11E-03
k_{OVN}	0	E_{VN}	31.11	0	0
k_{OVLPG}	0	E_{VLPG}	30.91	0	0
k_{OFG}	3034.89	E_{FG}	31.29	7.63E-07	1.46E-09
k_{OFD}	0	E_{FD}	29.08	0	0
k_{OFN}	91224.183	E_{FN}	19.53	0.093	1.78E-04
k_{OFLPG}	2184.96	E_{FLPG}	31.01	6.70E-07	1.29E-09
k_{OFG}	15776.3	E_{FG}	19.30	0.019	3.63E-05
k_{ODN}	0	E_{DN}	29.32	0	0
k_{ODLPG}	0	E_{DLPG}	29.26	0	0
k_{ODG}	1766.11	E_{DG}	30.66	6.91E-07	1.33E-09
k_{ONLPG}	1344.11	E_{NLPG}	12.12	0.256	4.92E-04
k_{ONG}	1.03799	E_{NG}	16.97	6.45E-06	1.24E-08
k_{OLPGG}	1344.11	E_{LPGG}	31.15	0	0

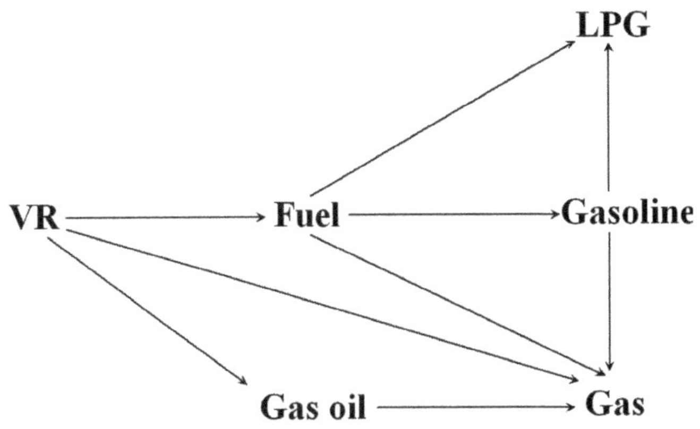

Figure 4. The complete six-lump kinetic model.

From Table 5 it can be concluded that I) the selectivity of the process to convert vacuum residue to fuel is the highest. Moreover, the fuel product is fairly stable (k_{FD} ~0 and k_{FN} is low); therefore these phenomena can justify the highest yield of fuel in the visbreaking process, II) Gas oil is fairly stable in the visbreaking process (k_{DN}, k_{DLPG} ~0), III) most of the gas and LPG of the visbreaking unit are produced from the thermal cracking of gasoline which can be the reason for low yield of gasoline in the visbreaking process, and IV) LPG cannot be converted to gas in the visbreaking process which is rational due to the stability of C_3 and C_4 chains.

After eliminating the low rate reaction paths (~0) and re-predicting the yields, the *AAD%* of reduced model were found to be still 4.75%. This deviation is acceptable thus justifying the removal of the less important reactions.

The simplified reaction-path network for the six-lump visbreaking model is shown in the Figure 4 named as the reduced model.

Figures 5, 6 and 7 show the comparison between the measured and predicted product yields. As it can be observed, acceptable mappings are realized.

The *AAD%* of all lumps is presented in Table 6. As it can be observed, the simulated yields for the nine commercial data, for the vacuum residue, fuel, gas oil and gasoline are in good agreement with the actual data. It was supposed that the observed deviation was caused by assuming the average density for the VGO feed and visbreaking products (Equation 17) reported in Table 4, and also some fluctuations in the property of the feed (such as sulfur, nitrogen and aromatic contents) and also reactor pressure which were not included in the model. Furthermore, we cannot neglect the possibility of error measurements in gathering data obtained with some faults such as signal transmission, calibration and power fluctuation.

It was thought that the high *AAD%* for the LPG and gas lumps were for the reason of the difficulty of their measurement in the commercial unit, creating large gross error. In addition, there are existed several vents in the gas system for which flow rates were not reported in the test runs. Because, the yield of these lumps, especially LPG and gas, were low, a little deviation could make a flagrant *AAD%*.

Conclusions

In this research, a six-lump kinetic model for a commercial vacuum residue visbreaker was proposed. The model consists of vacuum residue, fuel oil, gas oil, gasoline, LPG and light gas as the discrete lumps. Nine sets of industrial data gathered from a soaker-visbreaking unit were used to estimate the apparent activation energies and frequency factors. For the modeling of the visbreaking furnace, it was supposed that there was an equal heat flux throughout the furnace to satisfy the overall heat balance. Moreover, the furnace and soaker drum were simulated as a plug ideal flow and a completely mixed reactors, respectively.

Product yields predicted by this model showed a good agreement with commercial test runs with an absolute average deviation of about 4.75%. Results confirmed that the prediction was more accurate for heavy products than the light ones (gas and LPG). It was thought that the reason for higher deviation of gas and LPG was probably difficulties in measuring these gaseous flows.

Nomenclature

V, Vacuum residue lump.
F, Fuel lump.
D, Gas oil lump.
N, Gasoline lump.
LPG, LPG lump.
G, Light gas lump.
k, Rate constant ($m^3\ h^{-1}\ m^{-3}_{cat.}$).
k_0, Frequency factor ($m^3\ h^{-1}\ m^{-3}_{cat.}$).
E, Activation energy (kcal/mol).
T, Absolute temperature (K).
R, Ideal gas constant (kcal $kmol^{-1}K^{-1}$).
j, All products lighter than the vacuum residue lump.
j', All products lighter than the fuel lump.
j'' All products lighter than the gas oil lump.
j''', All products lighter than gasoline lump.
r, Reaction rate (kg $h^{-1}m^{-3}$).
C, Lump mass concentration (kgm^{-3}).
V_C, Coil volume (m^3).
V_D, Drum volume (m^3).
ρ, Stream density (kg m^{-3}).

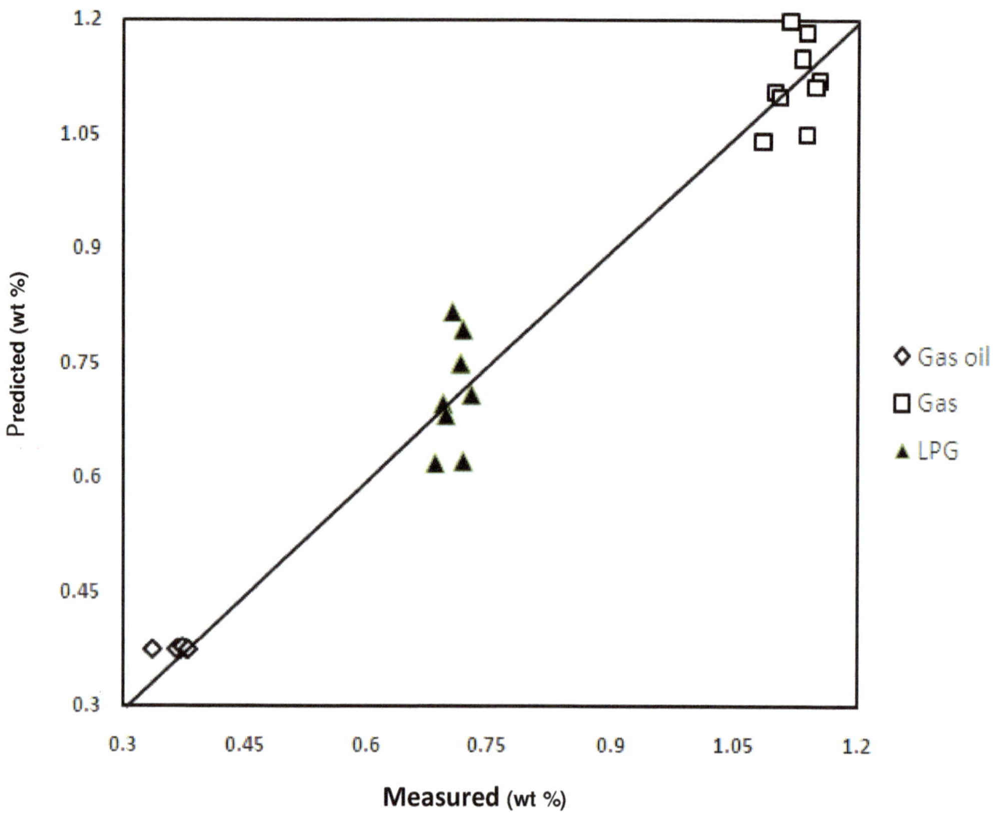

Figure 5. Comparison between the measured yields and the predicted yields of Gas, LPG and Gas oil.

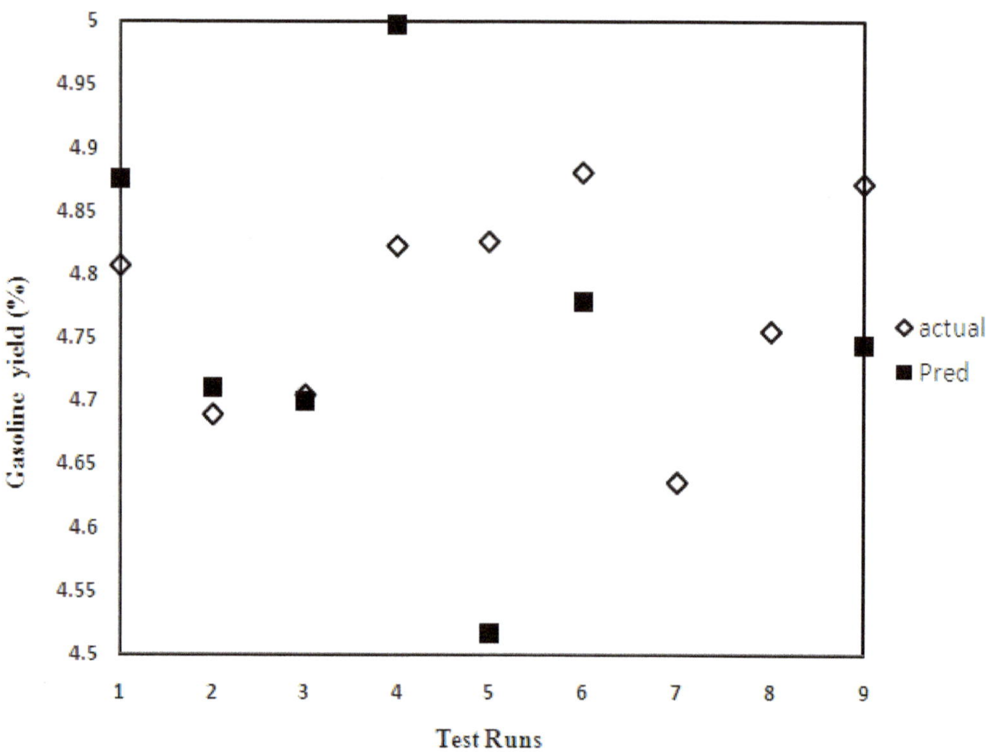

Figure 6. Comparison between the measured yields and the predicted yield of Gasoline.

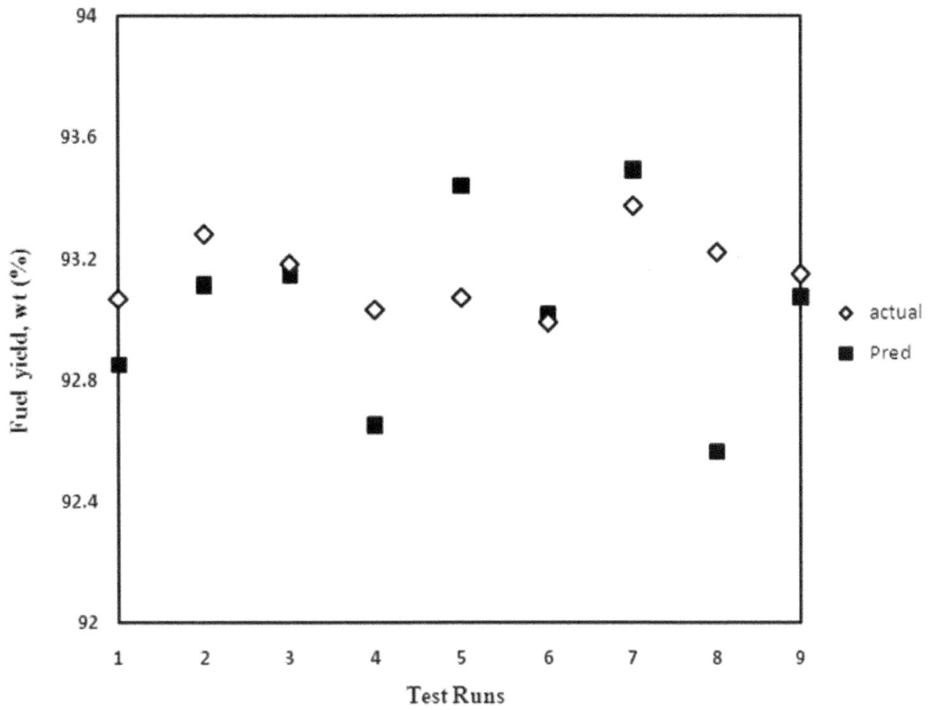

Figure 7. Comparison between the measured yields and the predicted yield of Fuel.

Table 6. AAD% of model prediction in comparison to measured data.

Lump	*AAD%*
Fuel	0.24
Gas oil	2.49
Gasoline	6.24
LPG	10.52
Gas	7.57
Ave%	4.75

V, Volumetric flow rate (m^3 h^{-1}).

F_m, Mass flow rate of the stream passing through the coil (kg h^{-1}).

X_j, Mass fraction of lump.

ρ_j, Density of lump (kgm^{-3}).

F_R, Fraction of recycle of the lumps.

L_t, Total length of the tubes (m).

Cp_j, Heat capacity of lump (kJ/kg.°C).

z, Furnace tubes Length (m).

SQE, Summation of squared errors.

N_t, Number of test runs.

Y_{jn}^{meas}, Measured product yield (wt %).

Y_{jn}^{pred}, Yield predicted by the model (wt %).

$AAD\%$, Absolute average deviations.

REFERENCES

Al Humaidan F, Lababidi MSH, Al-Rabiah H (2013). Thermal cracking kinetics of Kuwaiti vacuum residues in Eureka process. Fuel 103:923-931.

Al Soufi HH, Savaya ZF, Moahmmed HK, Al-Azami IA (1988). Thermal conversion (visbreaking) of heavy Iraqi residue. Fuel 67:1714-1715.

ASTM Standard D1160-06 (2009). Standard Test Method for Distillation of Petroleum Products at Reduced Pressure. ASTM International. West Conshohocken, PA.

ASTM Standard D86-08a (2009). Standard Test Method for Distillation of Petroleum Products at Atmospheric Pressure. ASTM International, West Conshohocken, PA.

Clarke GM, Kempson RE (1997). Introduction to the design and Analysis of Experiments. John Wiley & Sons.

Del Bianco A, Panartili N, Anelli M, Beltrame PL, Carniti P (1993). Thermal cracking of petroleum residues 1. Kinetic analysis of the reaction. Fuel 72:75-80.

Dente M, Bozzano G, Bussani G (1997). A comprehensive program for visbreaking simulation: Product amounts and their properties prediction. Comput. Chem. Eng. 21(10):1125–1134.

Di Carlo S, Janis B (1992). Composition and visbreakability of petroleum residues. Chem. Eng. Sci. 47:2695-2700.

Gary JH, Handwerk GE (1984). Petroleum Refining Technology and Economics. 2nd Ed., Marcel Dekker, Inc.

Krishna R, Kuchhal YK, Sarna GS, Singh ID (1988). Visbreaking studies on Aghajari long residue. Fuel 67:379-383.
Montgomery DC (2001). Design and Analysis of Experiments. New York: John Wiley & Sons.
Sadighi S, Ahmad A (2012). An Optimisation Approach for Increasing the Profit of a Commercial VGO Hydrocracking process. Can. J. Chem. Eng. DOI 10.1002/cjce.21720.
Sadighi S, Ahmad A, Mohaddecy SR (2010a). 6-Lump Kinetic Model for a Commercial Vacuum Gas Oil Hydrocracke. Int. J. Chem Reactor Eng. 8(1):1542-6580.
Sadighi S, Arshad A, Irandoukht A (2010b). Modeling a Pilot Fixed-bed Hydrocracking Reactor via a Kinetic Base and Neuro-Fuzzy Method. J. Chem. Eng. Jap. 43(2):174-185.
Sadighi, S, Arshad A, Irandoukht A (2010c). Kinetic Study on a Commercial Amorphous Hydrocracking Catalyst by Weighted Lumping Strategy. Int. J. Chem. Reactor Eng. 8:A60.
Sawarkar AN, Joshi JB, Pandit AB, Kataria KL, Kulkarni RP, Tandon D, Ram Y, Kumar MM (2007). Petroleum Residue Upgrading via Visbreaking: A Review. Can J. Chem. Eng. 85:1-24.
Shadbahra J, Khosravanib L, Khorasheh F (2011). Development of a continuous kinetic model for visbreaking reactions. Sci. Iranica. 18(3):465-469.
Singh J, Kumar S, Garg MO (2012). Kinetic modeling of thermal cracking of petroleum residues: A critique. Fuel Process. Technol. 94(1):131-144.
Speight JG (2006). The Chemistry and Technology of Petroleum. 4th ed., CRC Press.
Trauth DM, Yasar M, Neurock M, Nasigam A, Klein M, Kukes SG (1992). Asphaltene and Resid Pyrolysis: Effect of Reaction Environment. Fuel Sci. Technol. Int. 10(7):1161-1179.
Xiao J, Wang L, Chen Q, Wang D (2002). Modeling for Product Distribution in Thermal conversion of Heavy Oil. Pet. Sci. Technol. 20(5):605-612.

Variation of petrophysical properties due to carbon dioxide (CO_2) storage in carbonate reservoirs

Adel M. Salem[1]* and Shedid A. Shedid[2]

[1]American University in Cairo (AUC), and Suez University, Egypt.
[2]British University in Egypt (BUE), Cairo, Egypt.

Depleted hydrocarbon reservoirs, deep saline reservoirs, and un-mineable coal seams are considered the best geological sequestration candidates for carbon dioxide (CO_2) geologic storage formations. CO_2 sequestration in carbonate reservoirs provides a good way to reduce CO_2 release to the atmosphere. This work investigates the effect of the temperature, pressure, and brine salinity on the petrophysical changes in the carbonate cores due to CO_2 storing. Two groups of experiments were undertaken; (1) investigating the CO_2 solubility under different salinities, pressures and temperatures, and (2) studying the effect of CO_2 storage duration on porosity and permeability of carbonate rocks. Actual cores saturated with 25 000 ppm NaCl brine were used. The potential of the CO_2 storage capacity and variations in porosity and permeability are evaluated and quantified. The results showed that solubility of CO_2 decreases with increase in brine salinity and/or temperature. The increase of pressure causes an increase in CO_2 solubility. The results also indicated that storing CO_2 more than 150 days increases the porosity and permeability of carbonate rocks. The application of the achieved results is expected to have good impact on design storage process of CO_2 in deep saline water incarbonate reservoirs, and on validation of developed mathematical models.

Key words: Carbon dioxide (CO_2) storage, sequestration, carbonate formation, petrophysical properties, deep saline reservoirs.

INTRODUCTION

Carbon dioxide is used in many industries such as metal industry to enhance their hardness; manufacturing and construction such as in metal inert gas / metal active gas (MIG/MAG) welding for protection and increase of the welding rate; rubber and plastic industry to remove flash with crushed dry ice in a rotating drum; Food and beverages as quick freezing, surface freezing, chilling and refrigeration in the transport of foods; Health as an additive to oxygen for medical uses as a respiration stimulant; Environment as a propellant in aerosol cans; Chemicals, Pharmaceuticals and Petroleum Industry such as CO_2 injection as tertiary recovery to increase the recovery from the reservoirs.

Geologic sequestration of carbon dioxide in aquifers or in hydrocarbon reservoirs offer a promising alternative to reduce the amount of CO_2 released to the atmosphere (McCabe et al., 2007). Among the emerging technologies for large-scale reduction of CO_2 emissions, one of the most promising is the carbon dioxide capture and storage (CCS) in deep geological formations. Geological storage of CO_2 can be undertaken in a variety of forms, including its injection in depleted oil and gas fields and its use to enhance recovery from producing wells (D'Alesio et al., 2011).

Nguyen (2003) summarized four options for the CO_2 geological sequestration as follows; (1) the CO_2 can be

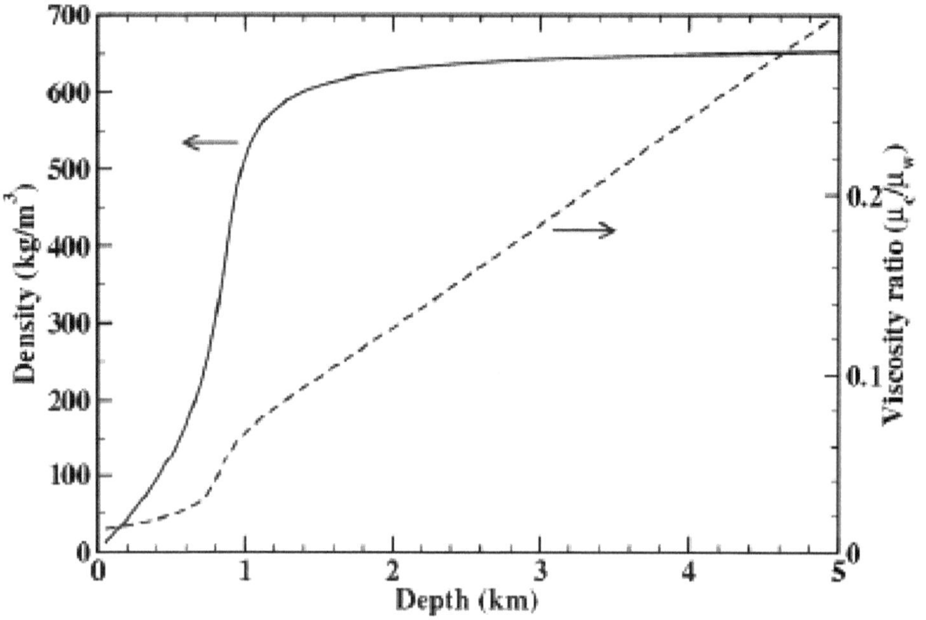

Figure 1. CO_2 density and viscosity at subsurface conditions, surface temperature 15°C/km and 10 MPa/km (Ennis-King and Paterson, 2002).

used in enhanced oil recovery (EOR) processes, (2) the CO_2 can be used to enhance coal bed methane recovery, (3) the CO_2 can be stored in depleted oil and gas reservoirs, and (4) the CO_2 can be stored in deep saline aquifers. Aquifers are considered to be the most widely available.

Physical properties of carbon dioxide (CO_2)

Based on pressure and temperature, CO_2 may exist in three phases under its critical point: solid, liquid and/or gas with a triple point at -56.6°C and 75.1 psi. CO_2 is in supercritical phase at temperature values greater than 31.04°C and pressure values above 1071 psi. Supercritical CO_2 performs like a condensed gas having low viscosity and high density. In order to avoid separation of CO_2 into gas and liquid phases during injection process, disposal of CO_2 is made at supercritical conditions. A minimum depth of 2625 ft (800 m) is required to reach these conditions (Dirik et al., 2004).

Using Peng-Robinson equation of state, the compressibility factors of CO_2 have been calculated by a computer program for different temperature and pressure values. Also, another equation of state for CO_2 has been proposed as a recent research by Span and Wagner (1996) which can be used as an alternative for the calculations of compressibility factor of CO_2.

Ennis-King and Paterson (2002) presented the density and the viscosity ratio of pure CO_2 to typical formation brine (32000 ppm) as a function of depth in Figure 1. This ratio increases with depth, but in the depth range of 1000 to 3000 m, it varies from 0.05 to 0.20. This figure is developed based on a geothermal gradient of 30°C/km, and at mean surface temperature of 15°C and a hydrostatic pressure gradient of 10 MPa/km. Therefore the CO_2 is more mobile than in the formation than the brine. The important conclusion of this figure is that the density of the gas increases rapidly down with the depth of about 1000 m, then levels out around 600 to 700 kg/m^3. This provition of 30 to 40% less dense than typical formation brine and the injected CO_2 will rise and accumulates below the top seal of the formation. In homogeneous formation, this may lead to viscous fingering in the displacement front. The effect of reservoir heterogeneity on the flow of CO_2 was investigated by Chang et al. (1994). The study used a compositional simulator to study the CO_2 flooding under unstable first-contact miscible flow with stochastic permeability fields. The study indicated that the permeability heterogeneity and gravity override - dominate the displacement pattern of the CO_2 flow in underground formation.

Carbon dioxide (CO_2) geological sequestration

The major criteria for the CO_2 storage in the depleted oil and gas reservoirs are the capacity, injectivity, lithology, and cap rock integrity (Taber et al., 1997; Shaw and Bachu, 2002; Cinar et al., 2007). The main problems of the CO_2 sequestration are the overriding of the gas to the top of the formation, the reduction of the CO_2 injectivity over time, and the early breakthrough of the CO_2 during injection due to the heterogeneity of the formation.

Le Gallo et al. (2002) summarized the advantages of the CO_2 storage as (1) available large pressure range for injection, (2) allowing the storage of significant gas volume for a low compression power, and (3) available data of dynamic and geologic characterization of depleted reservoirs. The appropriate selection of a suitable reservoir is crucial to the success of the CO_2 sequestration. Le Gallo et al. (2002) and Izgec et al. (2005) identified three major steps (mechanisms) to achieve the CO_2 storage as follows; (1) Hydrodynamic Trapping; in which the CO_2 is injected in the host formation and flows according to the pressure gradient, (2) Solution Trapping: It is the process in which the CO_2 reaches the thermodynamic equilibrium in all available phases; and (3) Mineral Trapping: it is the kinetic (or geochemical) reaction of the CO_2 with the rock and fluids of the aquifer. These mechanisms lead to storage of the CO_2 as a free-phase gas in pore space, dissolved phase CO_2 in formation water, and CO_2-converted to rock matrix. Kaldi and Bachu (2009) summarized the basic requirements for any carbon dioxide sequestration site to be as follows; (A) sufficient economic capacity to store targeted CO_2 quantity, (B) suitable injectivity to accept CO_2 at the rate at which it is being generated by the target source, and (C) containment assurance for no leakage from the site in long-term.

Phase properties of CO_2-brine systems

There are two major steps for the CO_2 underground sequestration including (1) the dissolution and diffusion of the CO_2 in oil and brine and following the pressure gradient (hydrodynamic trapping), and (2) the injected CO_2 reacts with the formation and induces dissolution/precipitation reactions (mineral trapping), that may impair the well injectivity and/or rock properties.

Natural gas storage in saline formation has been practiced in North America, Europe, and Australia (Coats and Richardson, 1967; Hower at al., 1993; Chang et al., 1994, 1998; Ennis-King and Paterson, 2002; Nghiem et al., 2009). However, a little attention has been paid to the mineral trapping of the CO_2 storage in carbonate formation saturated with brine. Storage of the CO_2 into carbonate formation was investigated by Izgec et al. (2005). The study evaluated the potential of a carbonate aquifer in the south-east of Turkey. The results indicated that permeability was initially increased and decreased later under slower injection rate because of precipitation of calcium carbonate ($CaCO_3$). Porosity of the aquifer was also decreased. The study also showed that orientation of core plugs was influential in rock-fluid-CO_2 interactions. Gupta (2010) presented the results of experimental and simulation work to understand opportunities and challenges in sequestering CO_2 in carbonate rocks. The results indicated that petrophysical properties of core samples had dissolution in some pores but precipitation in other pores of carbonate matrix.

Ülker et al. (2007) proved that the solubility of CO_2 in brine is much higher than that of hydrocarbon components. The CO_2 solubility depends essentially on pressure, temperature, total salinity and brine composition. In general, CO_2 solubility increases with increasing pressure and decreases with increasing temperature. An increase in the salinity of the formation brine decreases the CO_2 solubility significantly.

Reid et al. (2003) investigated the viability of, and optimal locations for sequestering CO_2 in the subsurface require detailed knowledge of the complex interactions among CO_2, rock matrix, and pore fluids under appropriate *in-situ* pressure and temperature conditions.

Many physical and chemical processes are known to occur both during and after geologic CO_2 injection, including diagenetic chemical reactions and associated permeability changes. Although it is commonly assumed that CO_2 sequestered in this way will ultimately become mineralized, the rates of these changes, including CO_2 hydration in brines, are known to be relatively slow. Together with hydrated CO_2, cations from brines may form solid-state carbonate minerals, ostensibly providing permanent sequestration (Reid et al., 2003).

Geochemical reaction of CO_2 and carbonate rocks

In generally, the predominant minerals of carbonate rocks are calcite ($CaCO_3$) and dolomite [$CaMg(CO_3)_2$], which are highly susceptible to dissolution, especially by carbonic acid (H_2CO_3). Dissolution processes are important to understand, as they may significantly alter the porosity and permeability of the rock. The following are the possible chemical reaction that might happen if CO_2 contact the carbonate rocks:

First the CO_2 dissolution into the formation water produces a slightly acidic solution as follow:

$CO_2 + H_2O \leftrightarrow H^+ + HCO_3^- \leftrightarrow H_2CO_3$, and/or
$H_2O \leftrightarrow OH^- + H^+$

As a result of the decreased pH of the reservoir brine, dissolution of calcite and dolomite along with increased alkalinity (HCO^{3-}) occur as represented by the following equations:

Reaction with calcite formation:

$CaCO_3 + CO_2 + H_2O \leftrightarrow Ca^{2+} + 2HCO_3^-$, and/ or
$H^+ + CaCO_3 \leftrightarrow Ca^{2+} + HCO_3^-$ **OR**

$CaCO_3 + H_2CO_3 \leftrightarrow Ca^{2+} + 2HCO_3$

Reaction with limestone formation:

$2H^{+} + CaMg(CO_3)_2 \leftrightarrow Mg^{2+} + Ca^{2+} + 2HCO_3^{-}$ or,
$CaMg(CO_3)_2 + 2H_2CO_3 \leftrightarrow Ca^{2+} + Mg^{2+} + 4HCO_3$

The reaction rate constant of calcite (1×10^{-5} mol./cm^2/s) is larger than the reaction rate constant of dolomite (1×10^{-7} mol/cm^2/s) (Wellman et al., 2003). Therefore, when a formation has a combination of calcite and dolomite, carbonic acid will tend to react with calcite (Mohamed et al., 2011).

Although dissolution can enhance effective porosity and permeability, precipitation of carbonate minerals can significantly reduce those characteristics. Carbon dioxide precipitation in solid form is also important because it represents a relatively secure sequestration mechanism. As such, understanding the conditions in which dissolution and precipitation will occur is important for predicting the evolution of porosity and permeability (and resultant increases or decreases in injectivity), and long-term storage mechanisms.

Moh et al. (2010) mentioned that CO_2 dissolves in the formation brine, generating carbonic acid, which dissolves carbonate rock. Dissolution impacts brine composition, which affects solubility. Calcium carbonate may tend to precipitate with changing concentration of bicarbonates. Precipitation may occur in either EOR operations or during primary CO_2 sequestration. Injectivity changes are a concern during EOR operations, while storage capacity and seal integrity are primary concerns during CO_2 sequestration. From their study, they concluded first that the changes in porosity and permeability were independent of each other, secondly, pressure lower than critical CO_2 pressure resulted in an increase in the core porosity and permeability, thirdly, permeability increased as CO_2: water volumetric ratio was increased.

Vanorio et al. (2010) found that, the injection of CO_2-rich brine into carbonates induces dissolution of the microcrystalline matrix (that is, micrite) leading to porosity enhancement.

Alam et al. (2011) studied the effect of CO_2 sequestration in Ekofisk Formation and Tor Formation chalk (*a form of limestone composed of the mineral calcite*) of the South Arne field formation, North Sea. They observed that, an increase in porosity and decrease in specific surface is due to injection of supercritical CO_2. This indicates that a reaction between CO_2 enriched water and particles takes place which smoothens the particle surface. Accordingly, partial increase in permeability was also noticed. A porosity increase of 1 to 2% was observed in Ekofisk Formation samples. In Tor Formation samples, the increase was typically 2 to 3%. Specific surface measured by nitrogen adsorption technique shows a decrease after CO_2 injection in Tor Formation samples. In Ekofisk Formation any change in specific surface is not evident. A decrease in specific surface of pores can be due to an increase in porosity. The trend of change in specific surface is in accordance with the change in porosity. It indicates that supercritical CO_2 injection smoothens the grain surface, which leads to increase in porosity and permeability.

Taberner et al. (2009) presented a simulated study of supercritical CO_2 injection into a deep saline aquifer from a carbonate formation (calcite and dolomite, with minor anhydrite). They found that the near well-bore porosity reduces by 5 to 17% due to halite precipitation (dryout zone); calcite and dolomite dissolve as the CO_2 plume advances during injection (Zhang et al., 2009).

Ricardo (2012) investigated the effect of CO_2 injection and storage in carbonate rocks, and they investigated experimentally the interactions between CO_2, water and carbonate rock. Their experiments involved saturating dolomite and limestone samples by carbonated water under pressures up to 10 000 psi in a high-pressure vessel for 24 and 48 h. Then they measured the porosity, permeability and mass of the rock. They concluded that the porosity and permeability changes produced improvements or damages on the flow properties, which is related to the fine grains produced during the dissolution process leading to the partial blockage of porous media. Mass reduction was up to 0.95 and 0.35%, respectively for limestone and dolomite. Permeability varied +/-60% (dolomite) and up to 86% (limestone). Porosity varied +/-3% for dolomite and +/-20% for limestone (Ricardo, 2012).

Gupta (2010) studied the capacity and constraints for CO_2 sequestration in aquifers and depleted carbonate reservoirs and found that after aging two of the samples with CO_2 for one and two weeks, the final measured porosity-permeability clearly indicate reduction in both porosity and permeability.

Izgec et al. (2006) performed an experimental and numerical modeling of direct injection of CO_2 Into carbonate formations and they showed using CT monitored core-flood experiments and core scale and field scale simulations that 1) chemical reactions occurred preferentially at the center of the core (at the inlet) where CO_2 injection is performed, 2) The common feature in all experiments was that dissolution occurred at the inlet face. 3) the duration of CO_2 – rock contact and the amount of area contacted by CO_2 seems to have a more pronounced effect compared to rate effect, 4) for the temperature range studied (18 to 50°C) permeability and porosity alteration trends were similar, 5) either a permeability improvement or a permeability reduction can be obtained through the injection of CO_2 into carbonate aquifers. The trend of change in rock properties is very case dependent because it is related to distribution of pores, brine composition and as well the thermodynamic conditions, 6) Precipitation process can impact the permeability drastically while small change in porosity is observed.

It is clear that there is still a debate in the literature about variation in petrophysical properties. Therefore, this study is designed to experimentally investigate (1) CO_2

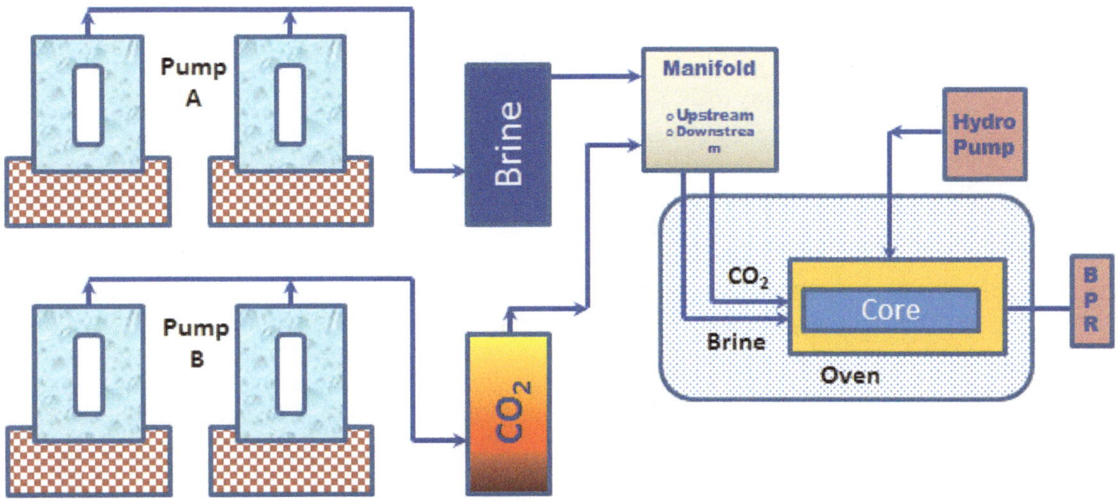

Figure 2. Schematic diagram of the experimental apparatus.

solubility under different brine salinity and also under different conditions of pressure and temperature, and (2) variation in porosity and permeability due to CO_2 storage in carbonate rocks.

EXPERIMENTAL SET UP AND PROCEDURES

Two groups of experiments were carried out. These two groups included; (A) solubility tests for carbon dioxide in different salinities and under different pressure and temperature conditions, and (B) core storage experiments to study variation in porosity and permeability under different time intervals from 7 to 170 days. The procedure for each experimental group is described below;

Solubility tests

The experimental procedure for investigating the solubility in distilled water and in different salinity brines can be summarized as follows;

1. The laboratory equipment, as shown in Figure 2, was calibrated to secure accurate measurements.
2. A brine of a specific salinity concentration or distilled water is pumped into the system according to the required volume.
3. A specific volume of CO_2 was then injected into the distilled water and also into brine of different salinities.
4. For each experiment, the volume of CO_2 was released from solution for each pressure depletion step, removed and measured at atmospheric conditions using a gas flow meter.
5. The procedure was repeated for different conditions of brine salinity, pressure, and temperature.

Variation in rock properties experiments

The experimental apparatus consists of core holder, injection system, and data recording system, as shown in Figure 2. The injection system includes constant rate displacement pumps (A and B), brine tank, CO_2 cylinder, gas flow meter, back pressure regulator (BPR), and pressure transducers.

The experimental procedure for studying the variation in porosity and permeability in carbonate reservoir rocks can be described as follows;

1. Core samples are evacuated for 24 h, and then saturated with brine of 25000 ppm NaCl solution. From weight difference, initial porosity is calculated. The permeability is measured under steady-state Darcy's flow condition.
2. The CO_2 is injected at constant pressure of 1000 psi continuously until irreducible brine saturated is well established.
3. The outlet is closed and the CO_2 is injected at 1000 psi and the pressure along the core sample is monitored till pressure equilibrium is achieved along the core sample.
4. The change in porosity and permeability is calculated by measuring these properties after each experiment.
5. The above procedure is typically repeated for different storage time intervals ranged from 7 to 170 days under 2 400 psi pressure and 200°F temperature.

RESULTS AND DISCUSSION

Tables 1, 2, and 3 presented the experimental results attained for carbon dioxide solubility for different brine salinity and under different pressure and temperature conditions. The graphical presentations of this data are shown in Figures 3, 4, and 5. Comparison of the results listed in Tables 1, 2, and 3, and Figures 3, 4, and 5, indicated that the increase of reservoir pressure decreases the carbon dioxide solubility for different brine salinity and under different temperature conditions.

Effect of pressure on carbon dioxide (CO_2) solubility

Figure 3 for distilled water, indicates that for the same brine salinity and the same reservoir temperature, the increase of pressure increases the carbon dioxide solubility. The same trend is also shown in Figures 4 and

Table 1. Carbon dioxide solubility in distilled water at different temperatures.

Pressure (psia)	Carbon dioxide solubility (SCF/STB) at different temperature		
P / T	70°F	120°F	170°F
0	0	0	0
300	75.8	48.6	29.5
600	126.4	83.4	61.8
900	158.6	123.8	92.6
1200	168.5	137.8	110.3
1500	179.8	147.9	123.7
2000	195.2	163.2	144.2
4000	210.3	186.1	179.3
6000	227.4	196.4	193.1

Table 2. Carbon dioxide solubility in 15 000 ppm NaCl water at different temperatures.

Pressure (psia)	Carbon dioxide solubility (SCF/STB) at different temperature		
P / T	70°F	120°F	170°F
0	0	0	0
300	48.6	34.7	23.5
600	91.4	64.7	47.8
900	128.6	73.8	74.6
1200	132.5	101.8	77.3
1500	138.5	113.6	102.3
2000	153.6	122.5	106.4
4000	162.9	138.4	137.7
6000	177.6	141.6	140.8

Table 3. Carbon dioxide solubility in 25 000 ppm NaCl water at different temperatures.

Pressure (psia)	Carbon dioxide solubility (SCF/STB) at different temperature		
P / T	70°F	120°F	170°F
0	0	0	0
300	25.8	21.8	20.1
600	86.4	56.2	53.3
900	102.6	75.6	70.3
1200	121.5	86.5	73.4
1500	136.2	108.9	84.6
2000	141.6	120.2	100.2
4000	156.3	128.1	125.8
6000	160.2	139.4	138.8

5 for different salinities of 15 000 ppm and 25 000 ppm NaCl brines, respectively. Therefore, a conclusion can be drawn that the increase of pressure increases the CO_2 solubility for different brine salinities and different reservoir temperatures. It is also clear that for temperature equal to or more than 120°F, the temperature effect diminishes and starts to have no effect, especially under pressures equal to or greater than 4000 psia. The reported data has real importance on validation simulation models describing the process of CO_2 storage in fresh and brine aquifers. It is also concluded that the increase of temperature and/or salinity increases the CO_2 solubility.

The results of four storage experiments undertaken studying the variation of porosity and permeability of carbonate reservoir are listed in Table 4 for different storage time intervals. Table 4 shows the petrophysical properties for the core samples used and the CO_2 storage time in carbonate reservoir rock under 2400 psia pressure and 200°F reservoir temperature.

Effect of temperature and salinity on carbon dioxide (CO_2) solubility

Based on the experimental results obtained, it is found that, as the solubility decreases the temperature increases for all prepared solutions at a given pressure, that is, the same results of distilled water, and for any brine salinities. On other hand, at a constant temperature, the CO_2 solubility increases with increase of the applied pressure. These results are depicted in Figures 6, 7, and 8 for distilled water, 15000 ppm, and 25000 ppm salinity respectively. These results are valid for any pressure applied on the system, since the pressure range studied in this work are varied from 300 to 6000 psi.

By comparing these Figures (6, 7, and 8), it is very clear that as salinity increases as the solubility decreases. This can be proved, if the CO_2 solubility at a given constant temperature and pressure have been picked and compared for different salinities. For example, the CO_2 solubility for distilled water at pressure 2000 psi and 120°F is 163.2 SCF/STB, for 15 000 ppm brine-salinity is 122.5 SCF/STB, and for 25 000 ppm brine-salinity is 120.2 SCF/STB.

Variation of petrophysical properties

Figure 9 presents the variation in porosity for the four core samples used under different storage time intervals from 7 to 170 days. Figure 9 shows that the porosity is decreased by almost 20% when the carbon dioxide is stored for 7 days. On the other side, the porosity was increased when the storage time was increased to 150 days and more. This analysis can be proven if the initial and final porosity plotted versus the time duration as shown in Figure 10. Form the previous figure; one can conclude that, the porosity of carbonate rocks is decreased with the aging time from zero day to about 85 days, then the relationship reversed, which means the final rock porosity is increased if the time is greater than

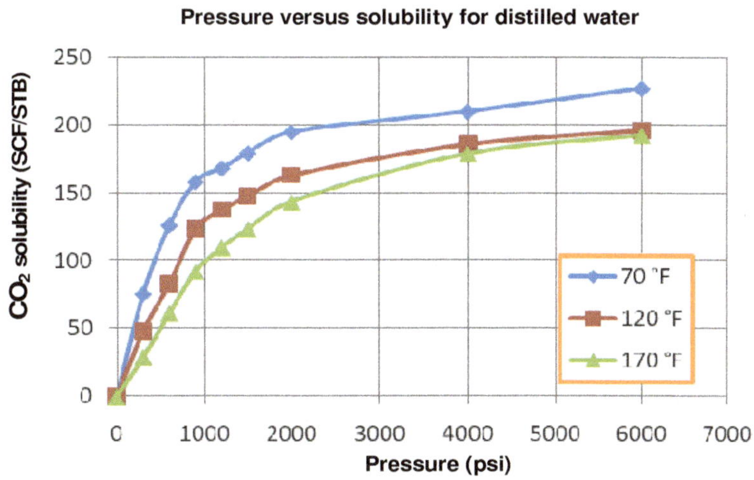

Figure 3. Carbon dioxide solubility in distilled water.

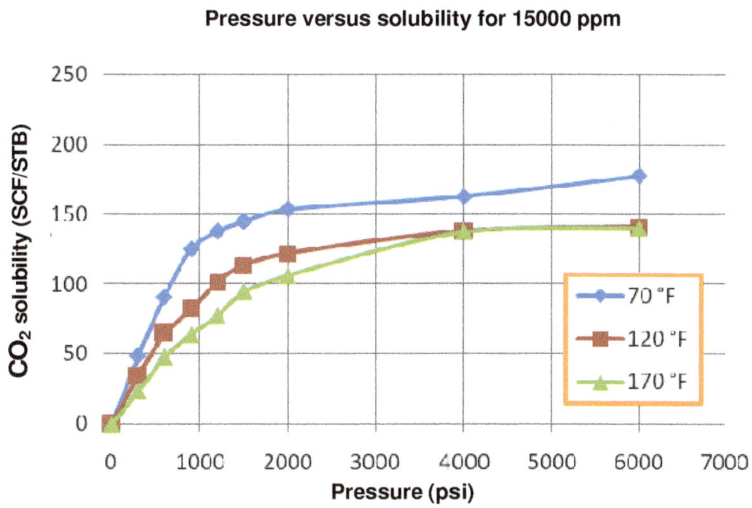

Figure 4. Carbon dioxide solubility in 15000 ppm NaCl brine.

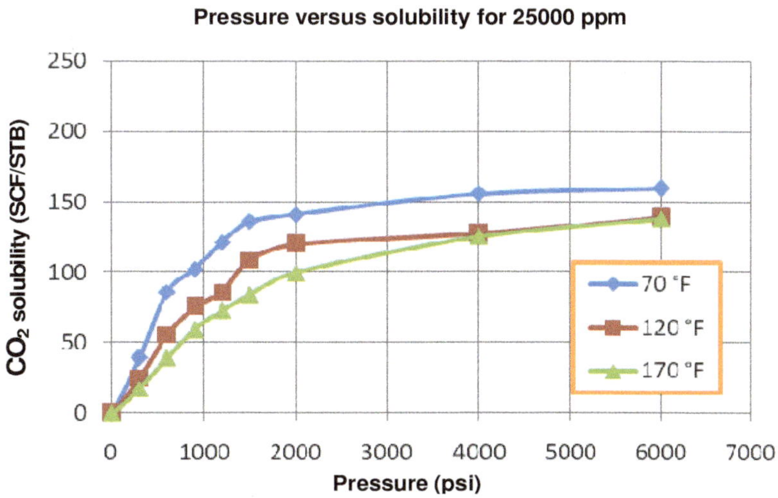

Figure 5. Carbon dioxide solubility in 25 000 ppm NaCl brine.

Table 4. Petrophysical properties of used core samples.

Sample code	PV (cc)	Initial porosity (%)	Final porosity (%)	Initial K (md)	Final K (md)	CO_2 storage time (days)
SF-1	7.35	18.57	14.72	10.8	7.71	7
AC-4	8.21	20.12	23.35	69.12	88.05	150
BK-6	8.86	21.23	25.42	57.31	65.87	160
SD-2	9.34	26.45	29.83	75.42	90.23	170

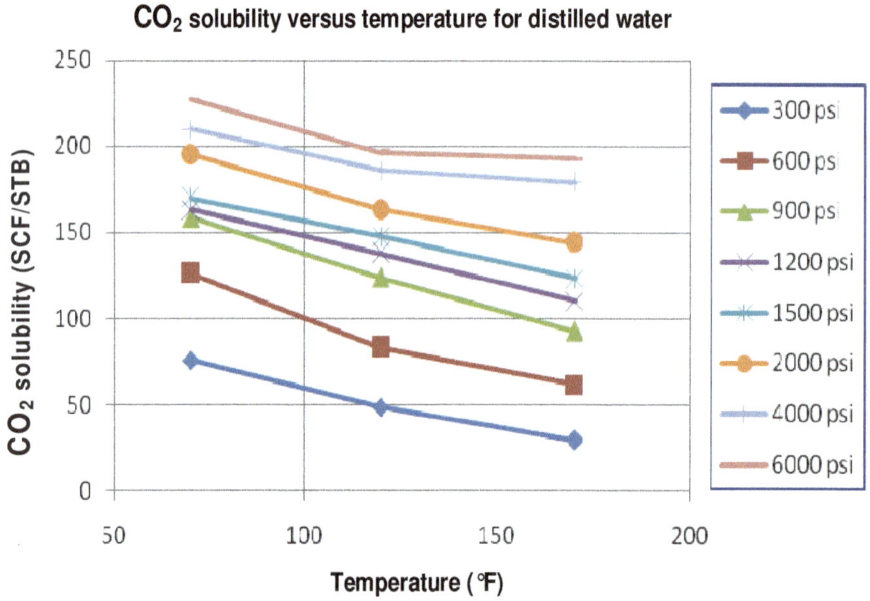

Figure 6. Carbon dioxide solubility versus temperature for a distilled water at pressure range 300 to 6000 psi.

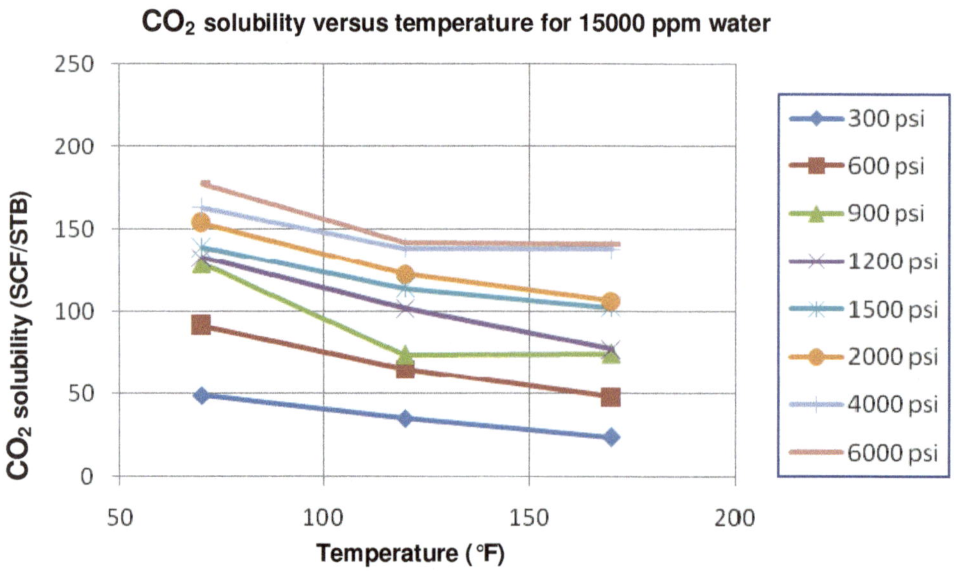

Figure 7. Carbon dioxide solubility versus temperature for a 15000 ppm water salinity at pressure range 300 to 6000 psi.

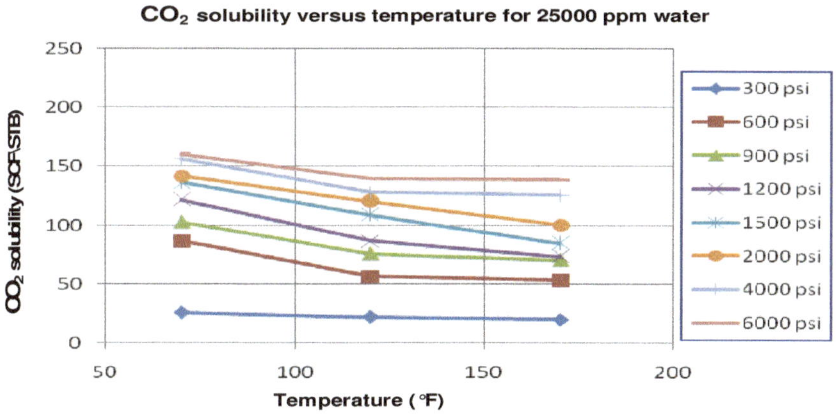

Figure 8. Carbon dioxide solubility versus temperature for a 25000 ppm water salinity at pressure range 300 to 6000 psi.

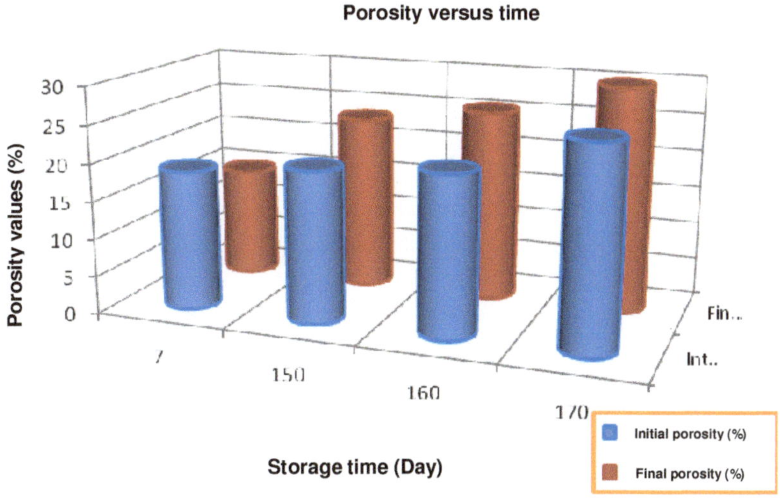

Fiure 9. Initial and final porosity for different storage time intevals.

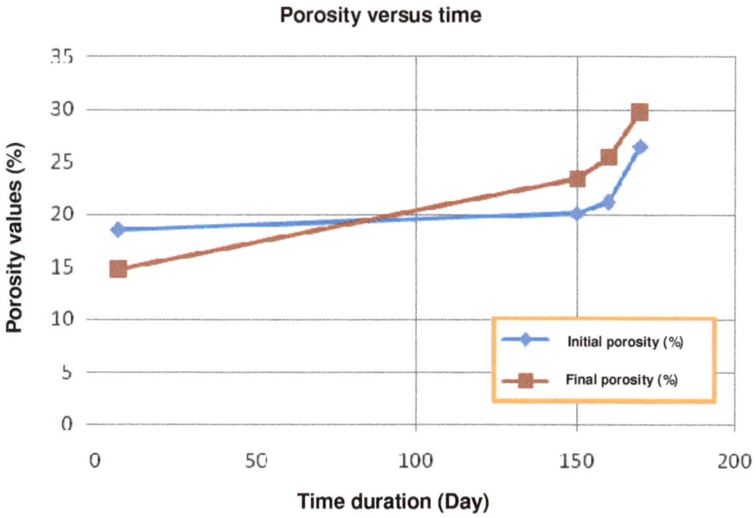

Figure 10. Porosity (initial and final) versus aging time.

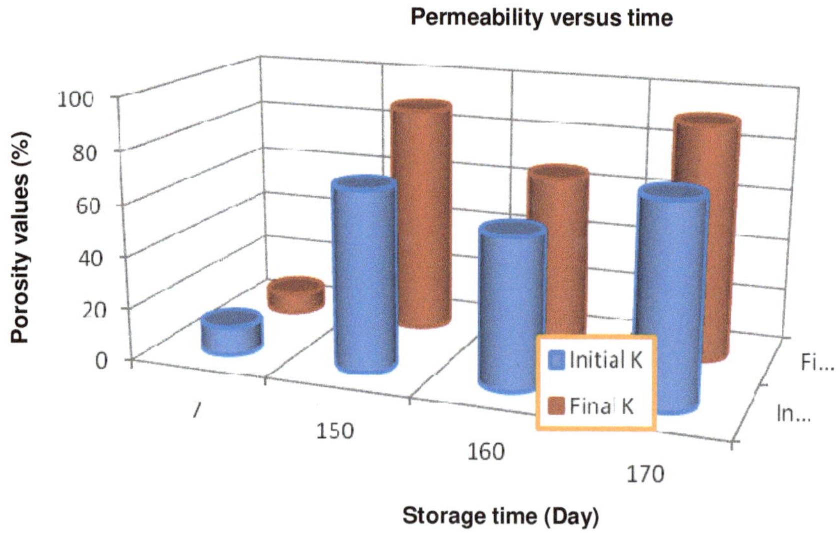

Figure 11. Initial and final permeability for different time intervals.

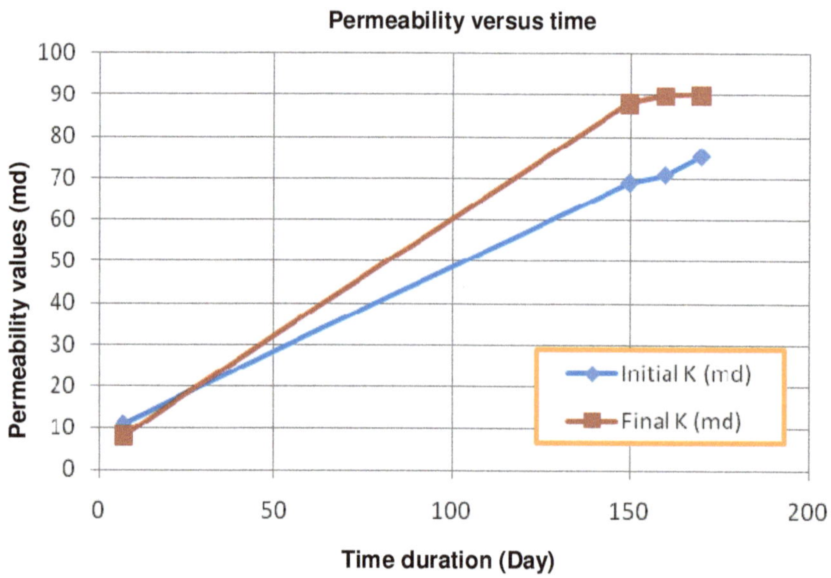

Figure 12. Permeability (Initial and final) versus storing time.

85 days.

For the carbonate rock permeability, the same trend is shown for permeability variation in Figure 11. By investigating this figure, we found that the permeability decreased in the first 27 days of storing CO_2, after that time, the rock permeability has increases with increasing the storing time remarkably as shown in Figure 12. These results can be attributed to the dissolution and precipitation processes that happened inside the pore space due to the geochemical reaction of CO_2 and the limestone and calcite that constitutes our core plugs.

Therefore, it is concluded that the carbon dioxide storage has to be for long time interval to provide enough time for a complete carbonate dissolution process, which leaded to an increase in porosity and permeability. The attained data confirmed the results of Gupta (2010) that the porosity was deceased for some cores and improved for other ones.

Conclusions

This experimental study investigated the effects of brine salinity, pressure and temperature on carbon dioxide solubility plus the variation in porosity and permeability of carbonate reservoirs, the following conclusions can be drawn;

1. The application of CO_2 sequestration and storage in carbonate reservoirs has significant potential to significantly reduce the emission of CO_2 into the atmosphere.
2. The CO_2 solubility in aquifer's brine decreases with an increase in salinity. The temperature effect diminishes and starts to have no effect, especially under pressures equal to or greater than 4000 psia.
3. The CO_2 solubility in brine decreases with increasing reservoir temperature.
4. The increase of pressure increases the CO_2 solubility for different reservoir temperatures and different brine salinities.
5. For reservoirs or aquifers of moderate porosity and good permeability, the porosity and permeability are decreased when CO_2 is stored in carbonate reservoir rocks for short period of time equals to 7 days and increase when stored for longer time intervals equal to or greater than 150 days.

REFERENCES

Alam MM, Hjuler ML, Christensen HF, Fabricius IL (2011). Impact of Supercritical CO_2 Injection on Petrophysical and Rock Mechanics Properties of Chalk: An Experimental Study on Chalk from South Arne Field, North Sea. SPE 147056, presented at the SPE Annual Technical Conference and Exhibition held in Denver, Colorado, USA.

Chang YB, Lim MT, Pope GA, Sepehrnoori K (1994). CO_2 Flow Pattern under Multiphase Flow: Heterogeneous Field-Scale Conditions. J. SPE Reserv. Eng. 9:3.

Chang Y, Coats BK, Nolen JS (1998). A Compositional Model for CO_2 Floods Including CO_2 Solubility in Water. SPE 35164, SPE J. Reserv. Eval. Eng. 1:2.

Cinar Y, Riaz A, Tchelepi HA (2007). Experimental Study of CO_2 Injection into Saline Formations. SPE 110628, presented at the SPE Annual Technical Conference and exhibition, Anaheim, California, USA.

Coats KH, Richardson JG (1967). Calculation of Water Displacement of Aquifer Storage. SPE J. 7(2):105-512.

D'Alesio P, Poloni R, Valente P, Magarini PA (2011). Well-Integrity Assessment and Assurance: The Operational Approach for Three CO_2-Storage Fields in Italy. SPE 133056, presented at the SPE Annual Technical Conference and Exhibition, Florence, Italy, 20–22 September 2010, AND SPE Production and Operations.

Dirik I, Altintas A, Bulbul S, Gumrah F (2044). Analytical Modelling of CO Sequestration in Saline Aquifers.presented at the Canadian International Petroleum Conference, Calgary, Alberta.

Ennis-King J, Paterson L (2002). Engineering Aspects of Geological Storage of Carbon Dioxide.SPE 77809, presented at the S PE Asia Pacific Oil and Gas Conference and Exhibition, Melbourne, Australia.

Gupta A (2010). Capacity and Constraints for Carbon Dioxide Sequestration in Aquifers and Depleted Oil/Gas Reservoirs in Carbonate Environment.SPE 135595, presented at the SPE International Conference on CO_2 Capture, Storage, and Utilization, New Orleans, Louisiana, USA.

Hower TL, Fugate MW, Owens RW (1993). Improved Performance in Aquifers Gas Storage Fields Through Reservoir Management. SPE 26172, Presented at the SPE Gas Technology Symposium, Calgary, Alberta, Canada.

Izgec OD, Bertin HB, Akin S (2006). Experimental and Numerical Modeling of Direct Injection of CO_2 into Carbonate Formations. SPE 100809, presented at the SPE Annual Technical Conference and Exhibition, San Antonio, Texas, USA.

Izgec O, Demiral B, Bertin H, Akin S (2005). Experimental and Numerical Investigation of Carbon Dioxide Sequestration in Saline Aquifers. SPE 94697, presented at SPE/EPA/DOE Exploration and Production Environmental Conference, Galveston, Texas, USA.

Izgec O, Demiral B, Bertin H, France S (2005). CO_2 Injection in Carbonates. SPE 93773, Presented at SPE Western Regional Meeting, Irvine, California.

Kaldi J, Bachu S (2009). Site Selection and Storage Capacity for Geosequestration of Carbon Dioxide.AAPG/SEG/SPE Hedberg Conference "Geological Carbon Sequestration: Prediction and Verification", Vancouver, BC, Canada.

Le Gallo YL, Couillens P, Manai T (2002). CO2 Sequestration in Depleted oil and Gas Reservoirs. SPE 74104, presented at the SPE International Conference on Health, Safety and Environment in Oil and Gas Exploration and Production, Kuala Lumpur, Malaysia.

Mohamed IM, He J, Nasr-El-Din HA (2011). Sulfate Precipitation during CO_2 Sequestration in Carbonate Rock. SPE 139828, presented at the SPE Projects and Facilities Challenges Conference at METS held in Doha, Qatar.

Mohamed IM, He J, Nasr-El-Din HA (2010). Effects of Pressure, CO_2 Volume, and the CO_2 to Water Volumetric Ratio on Permeability Change during CO_2 Sequestration. SPE 136394, presented at the Abu Dhabi International Petroleum Exhibition & Conference held in Abu Dhabi, UAE.

Nghiem L, Yang C, Shrivatava V, Kohse B, Hassam M, Chen D, Card C (2009). Optimization of Residual Gas and Solubility Trapping for CO_2 Storage in Saline Aquifers. SPE 119080, Presented at the SPE Reservoir Simulation Symposium, The Woodlands, Texas, USA.

Nguyen DN (2003). Carbon Dioxide Geological Sequestration: Technical and Economic Reviews. SPE 81199-MS, presented at SPE/EPA/DOE Exploration and Production Environmental Conference, San Antonio, Texas.

Reid BG, Brian JM, Robert KS (2003). Laboratory and Model Tests at Reservoir Conditions for CO2-Brine-Carbonate Rock Systems Interactions.presented at the Second Annual Carbon Sequestration Conference, Washington, D.C.

Ricardo DG (2012). Effects of Dissolution on the Permeability and Porosity of Limestone and Dolomite in High Pressure CO_2/Water Systems. SPE 160911, presented at the SPE International Student Paper Contest at the SPE Annual Technical Conference and Exhibition held in San Antonio, Texas, USA.

Schembre-McCabe JM, Kamath J, Gurton R (2007). Mechanistic Studies of CO_2 Sequestration. IPTC 11391 paper, presented at the International Petroleum Technology Conference held in Dubai, U.A.E.

Shaw J, Bachu S (2002). Screening Evaluation and Ranking of Oil Reservoirs Suitable For CO_2– Flooding EOR and Carbon Dioxide Sequestration. J. Can. Petrol. Technol. 41:9.

Span R, Wagner WJ (1996). "A new equation of state for carbon dioxide covering the fluid region from the triple-point temperature to 1100 K at pressures up to 800 MPa." Phys. Chem. Ref. Data. 25:1509–1596

Taber JJ, Martin FD, Seright RS (1997). EOR Screening Criteria Revisited-Introduction to Screening Criteria and Enhanced Recovery Field Projects. SPE Reserv. Eng. pp. 189-205.

Ülker B, Alkan H, Pusch G (2007). Implications of the Phase-Solubility Behaviour on the Performance Predictions of the CO_2 Trapping in Depleted Gas Reservoir and Aquifers. SPE 107189 presented at EUROPEC/EAGE Conference and Exhibition, London, U.K.

Vanorio T, Nur A, Diaz E (2011). The Rock Physicochemical Basis for Time-Lapse Seismic Reservoir Monitoring of CO_2 Injection. SPE 149031, presented at the SPE/DGS Saudi Arabia Section Technical Symposium and Exhibition held in Al-Khobar, Saudi Arabia.

Wellman PT, Grigg RP, McPherson BJ, Svec RK, Lichtner PC (2003). Evaluation of CO_2-Brine-Reservoir Rock Interaction with Laboratory Flow Tests and Reactive Transport Modeling. SPE 80228, presented at the SPE International Symposium on Oilfield Chemistry held in Houston, Texas, U.S.A.

Zhang G, Taberner C, Cartwright L, Xu T (2009). Injection of Supercritical CO2 Into Deep Saline Carbonate Formations: Predictions From Geochemical Modeling. SPE 121272, SPE Journal, December 2011 AND presented at the EUROPEC/EAGE Conference and Exhibition, Amsterdam.

25

A high temperature N2-foam assisted cyclic steam injection in Fuyv reservoir

Xin Zhao[1*], Panqing Gao[2] and Zhiyong Shao[1]

[1]Daqing Oilfields Lit Co., Daqing, China,
[2]Department of Petroleum Engineering, University of Wyoming, Wyoming, US.

Steam stimulation has already been proved as an important thermal recovery technology. In Daqing, the development and management of high-heterogeneous heavy oil reservoirs plays an important role in stabilizing the oil output. To combine the good features of steam injection and cyclic injection, a cyclic steam flooding has been conducted in the Fuyv reservoir, Daqing. This paper aims to present a detailed process of the cyclic-steam injection. The target Fuyv reservoir in this research is located on the edge of the Songliao Basin. The reservoir characterized by high heterogeneity has a permeability range from 61.5 to 3650 mD. The viscosity of crude oil is averagely 3200 mPa.s. A simulation investigation was performed to instruct the pilot design. The nitrogen co-injected cyclic steam injection is also proposed and analyzed. The mechanisms of EOR by nitrogen assisted cyclic steam stimulation are studied. The pilot injection was fulfilled on the basis of fundamental investigation. The simulation results showed that the heated radius is a function of injection cycles. When the injected steam is only enough to forfeit the heat loss to caprock and beneath formation, the heated radius or heated area will not expand. The limited heated radius and steam overlapping vertically results in a poor sweep efficiency and oil-steam ratio, a large amount of oil untapped. The factors effecting production and injection performance are analyzed, and the injection parameters are optimized. The pilot injection has obtained foreseeable results.

Key words: Cyclic, steam inject, heavy oil, Fuyu reservoir, N_2-foam.

INTRODUCTION

The Fuyv reservoir in Daqing oilfield is a deep heavy oil reservoir (viscosity @ reservoir: 3200 mPa.s), which is located in the edge of Songliao Basin (Harris, 1987; Nguyen et al., 2003; Lin and Yang, 2006; Zhao et al., 2010). Hydrocarbon introduction is from the Fuyu sandstone, which is approximately 1070 m in depth and 21 m in thickness, the permeability is from 62.11 to 3650 mD and the effective porosity is from 23.0 to 39.49% (avg. 34.1%).

As of 2006, the reservoir was put into production by cyclic steam stimulation, which is the major development technology for heavy oil deposits (Zhao et al., 2011). While because of the high oil viscosity, formation heterogeneity and increased formation water due to muti-cycles of steam injection, the cyclic heat loss increases, cyclic production performance declines and oil recovery factor for CSS is limited (Kuehne et al 1990; Martinez et al., 2005; Liu et al., 2008; Liu et al., 2010; Li et al., 2012). Extensive literature investigation indicated that the N_2-foam assisted cyclic steam stimulation was successively applied in many heavy oil reservoirs throughout the world because of profile modification mechanism of nitrogen foam to control fingering and enhance oil recovery rate (Hudgins et al., 1990; Thomas et al., 1991; Bracho and Portillo, 1991; Svrcek et al., 2002; Ma et al., 2005). Therefore, taking the Fuyv reservoir as a research target,

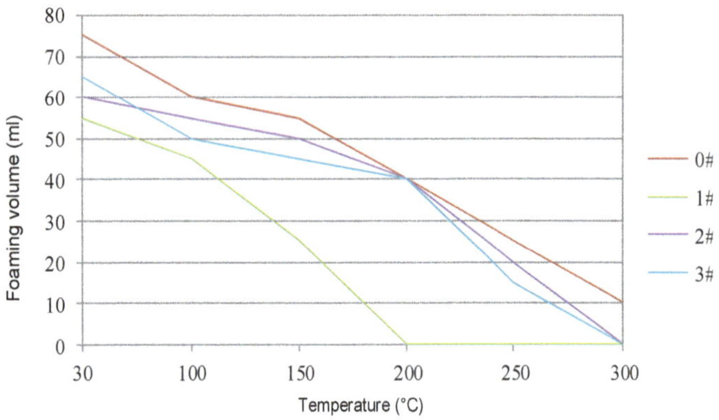

Figure 1. Foaming volume of different sets of N_2-foam system with temperature.

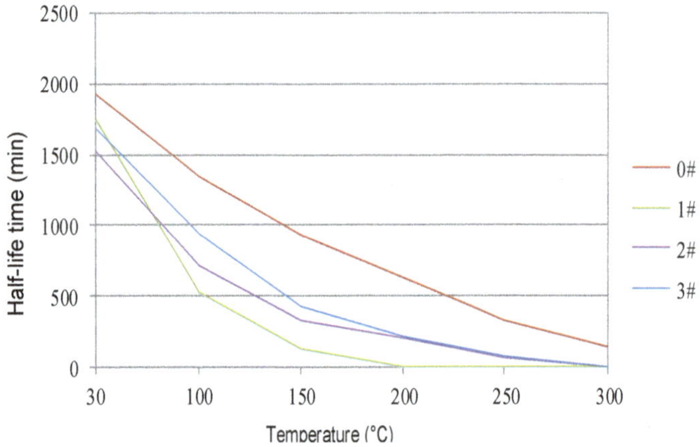

Figure 2. Half-life time of different sets of N_2-foam system with temperature.

four sets of N_2-foam system suitable for the Fuyv reservoi were screened by evaluation experiments and operation strategies of N_2-foam assisted CSS were optimized. The N_2-foam assisted CSS was carried out in two pilot test wells in 2010 and foreseeable results were obtained.

STATIC EVALUATION OF N_2-FOAM SYSTEM

In view of the geologic characteristics and fluid properties in Fuyv reservoir, four sets of N_2-foam system by different foaming agent are evaluated and screened in laboratory.

Foaming volume and half-life time with temperature

The tests of foaming volume and half-life time with temperature are carried out under the same conditions of test pressure (1 MPa) and foaming agent mass concentration (0.5%). The foaming volume test results (Figure 1) indicate that the foaming volumes of four sets of N_2-foam system decreases with increasing temperature. The 0# N_2-foam system can bear with temperature as high as 300°C, while the 1# is more sensitive with temperature and it cannot bear with temperature higher than 200°C. Meanwhile, the half-life time test results (Figure 2) show that the half-life time of 4 sets of N_2-foam system decreases gradually with increasing temperature. Generally, the half-life time of the 0# is higher than the other 3 sets at different temperature; therefore the 0# has the best high-temperature stability.

Adaptability of N_2-foam system to formation water

In the tests of adaptability of N_2-foam system to formation water, the formation water is replicated by sodium bicarbonate, anhydrous calcium chloride, magnesium chloride, sodium chloride, anhydrous sodium sulfate and other chemicals according to the fluid analysis of one well test result in the target area (Table 1). Four sets of N_2-foam system with 1 ml volume is dissolved separately into the replicated formation water to observe whether there occurs phenomenon of turbidity, insolubility or precipitation.

It is observed that the 4 sets of N_2-foam system are well adaptable to the formation water (Figures 3 and 4). Under the normal temperature (30°C), the phenomenon of turbidity, insolubility or precipitation is not observed within 6 days.

Adaptability of N_2-foam system to different formation oil samples

As the structure in Fuyv reservoir is the monocline structure, the formation oil in different position has different oil viscosity and oil composition; therefore it is necessary to test the adaptability of N_2-foam system to different formation oil samples. In view of the well placement in the target area, the oil samples from three different wells are selected to the adaptability test. The mass concentration of each foaming agent is 0.5%, the test pressure is 1 MPa and the test temperature is 30°C. The results (Figures 5 and 6) indicate that the set 0# has the largest foaming volume and longest half-life time in 3 oil samples compared with the other three sets of N_2-foam system. Therefore, the set 0# has a strong adaptability to different types of oil deposits in the target area.

DYNAMIC EVALUATION OF N_2-FOAM SYSTEM

The 1-D coreflooding system is used to carry out the influence of temperature and foaming agent concentration on the foam plugging performance; and

Figure 3. The formation water with dissolved foaming agent on the first day.

Figure 4. The formation water with dissolved foaming agent on the sixth day.

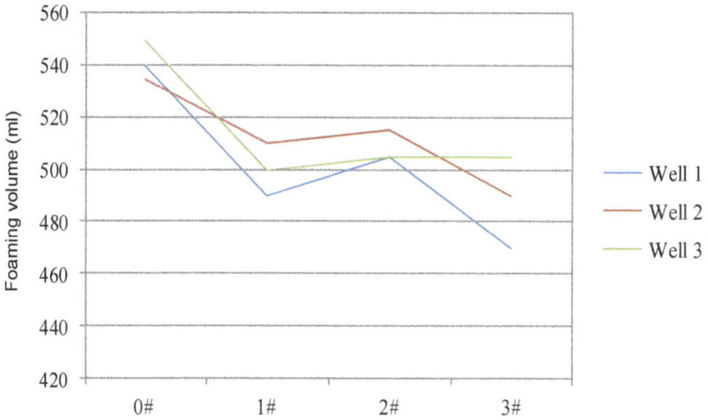

Figure 5. Foaming volume of different sets of N_2-foam system in different oil samples.

then, the oil displacement efficiency of different sets of N_2-foam system is tested and analyzed. The length of the quartz sand filled tube is 60 cm and the diameter is 3.8 cm. The tube is horizontally placed in the constant temperature oven. In the process of the experiment, the first step is to saturate water to test the porosity of the tube and water relative permeability; and then, injecting steam. The pressure difference between the inlet and outlet is recorded as the basic pressure difference; after that, the N_2-foam system is injected at the same volume injection rate with steam. The pressure difference between the inlet and outlet is recorded as the working pressure difference. The ratio between the working pressure difference and basic pressure difference is the foam resistance factor, which is used to evaluate the effectivenss of N_2-foam system. The higher the value of the resistance factor, the better the N_2-foam system performs in the field application.

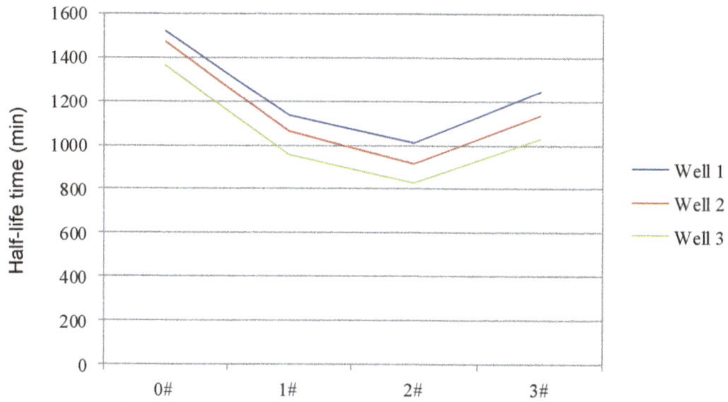

Figure 6. Half-life time of different sets of N_2-foam system in different oil samples.

Effects of temperature on foam plugging performance

The resistance factors of four sets of N_2-foam system under reservoir temperatures of 30, 60, 100, 150, 200, 250 and 300°C are tested. The mass concentration of foaming agent is 0.5%, the test pressure is 1 MPa.

The test results indicate that with increasing temperature, the resistance factor generally decreases, which is the results of smaller foaming volume and shorter half-life time of foam (Figure 7). The foam resistance factor of set 0# performs best at different temperature points, while the set 1# does not have evident resistance effects; therefore, set 1# is excluded from the foam candidates.

Effects of foaming agent concentration on foam plugging performance

The resistance factors of three sets of N_2-foam system (0#, 2# and 3#) are tested with foaming agent mass concentration of 0.3, 0.4, 0.5, 0.6 and 1.0% under experiment pressure of 1 MPa and temperature of 30°C.

The test results indicate that the resistance factor generally increases with increasing foaming agent mass concentration. When the mass concentration increases from 0.3 to 0.5%, the resistance factor increases dramatically. After the mass concentration reaches higher than 0.5%, the incremental resistance factor decreases, which means that when the mass concentration reaches a certain value, the foam resistance capacity becomes stable (Figure 8). Therefore, considering the influence of formation adsorption, the optimum mass concentration of foaming agent is between 0.5 and 0.6%.

Oil displacement efficiency under different injection patterns

In order to evaluate and compare different injection patterns on the performance of oil displacement efficiency,

Table 1. Ions in formation water from a typical well test results.

pH	Na^+	Mg^{2+}	CO_3^{2-}	Cl^-	SO_4^{2-}	HCO_3^-	Total salinity
7.2	1898.3	18.5	66.2	890.2	73.2	3707.6	6654.0

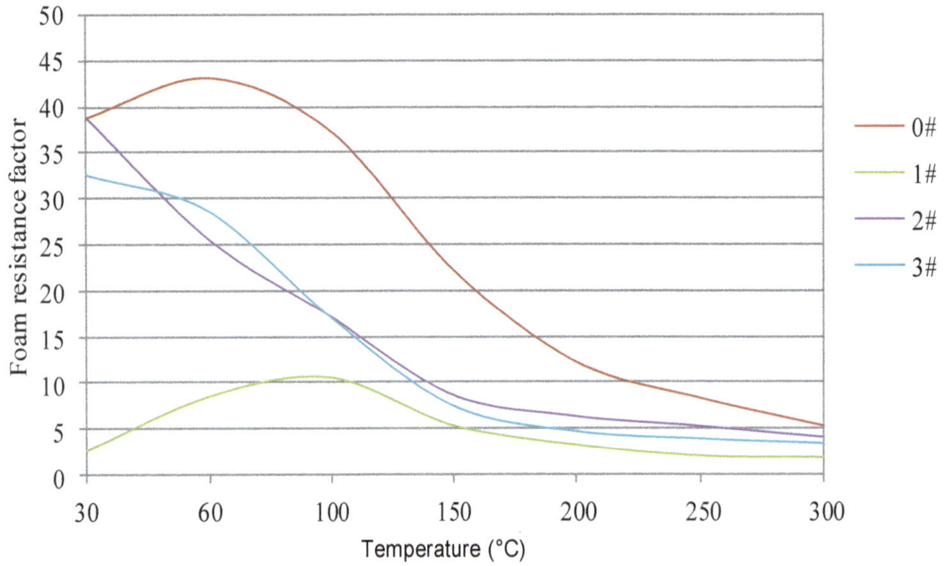

Figure 7. Effects of temperature on foam plugging performance of different sets of N_2-foam system.

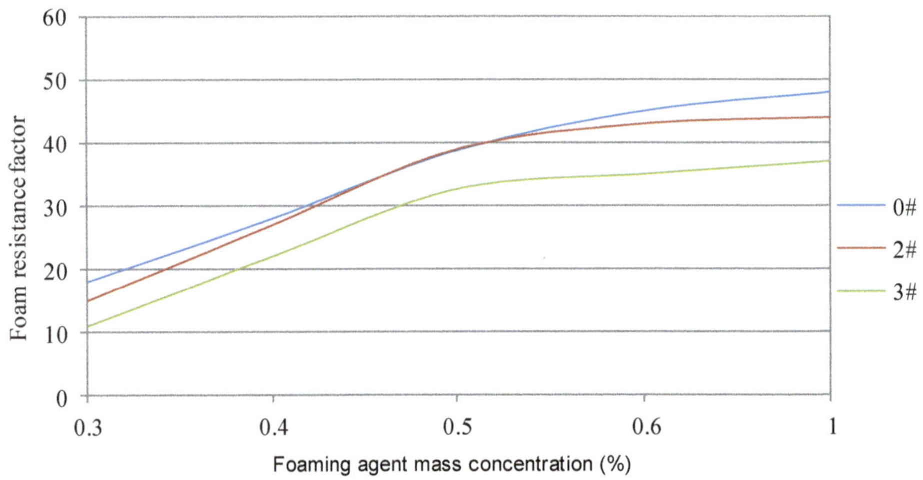

Figure 8. Effects of foaming agent concentration on foam plugging performance of different sets of N_2-foam system.

after reviewing the field conditions of cyclic steam stimulation, four injection patterns are selected and tested by 1-D coreflooding experiments. The sand filled tube has length of 30 cm and diameter of 2.5 cm. The four injection patterns are: steam, steam-nitrogen, steam-foaming agent, steam-nitrogen-foaming agent. The set 0# is selected as the foaming agent in these experiments.

The experiment results (Table 2) indicate that by only injecting steam, the oil displacement efficiency is only 35.4%; meanwhile, the oil displacement efficiency of Steam+Nitrogen injection is slightly higher than only steam injection, which means that the nitrogen itself could not act as a plugging agent to block the high permeability channels in the field application. The case of Steam-Foaming agent also does not yield high oil displacement efficiency, which is because of no resistance

Table 2. Oil displacement efficiency test results under different injection patterns.

Case No.	Injected fluid	Pore volume (ml)	Porosity (%)	Permeability (mD)	HCPV (ml)	So (%)	Oil displacement efficiency (%)
1	Steam	45.0	30.6	334	37.0	82.2	35.4
2	Steam+Nitrogen	43.0	29.2	316	36.0	83.7	38.1
3	Steam+Foaming agent	46.0	31.3	325	37.0	80.4	40.3
4	Steam+Nitrogen+Foaming agent	43.3	29.4	332	36.7	84.8	57.0

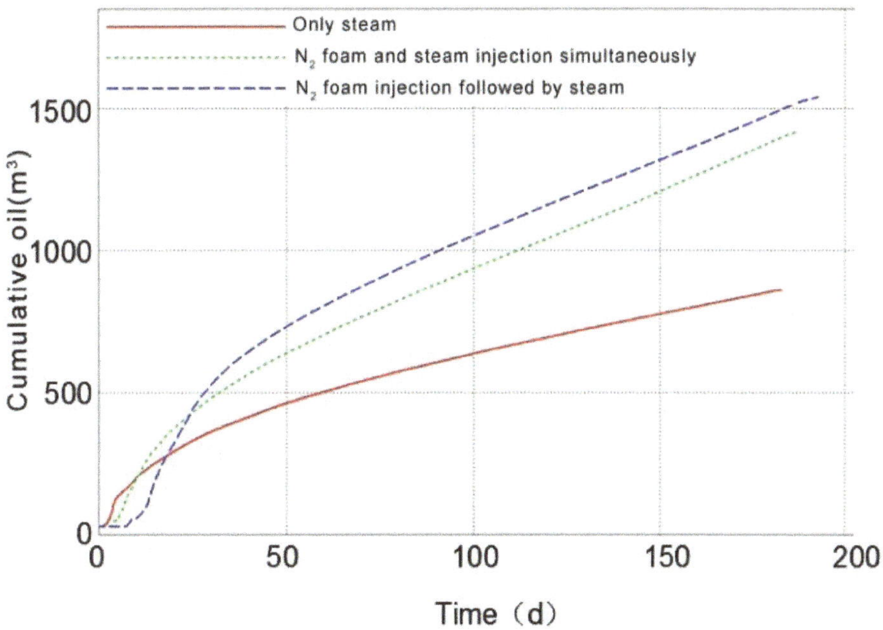

Figure 9. The cumulative oil comparison among different injection strategies.

effect by foaming agent itself; the foaming agent injected along with steam is largely adsorpted by reservoir rock surface. The optimum injection pattern is Steam-Nitrogen-Foaming agent, which can yield highest oil displacement efficiency compared with other injection patterns.

RESERVOIR ENGINEERING DESIGN OF N_2 FOAM ASSISTED CSS

Injection strategy of N_2 foam and steam during N_2 foam assisted CSS

Three injection strategies are simulated: only injecting steam, N_2 foam injection followed by steam injection, injecting N_2 foam and steam simultaneously (Figure 9). The results show that injecting N_2 foam and steam simultaneously is not the best injection strategy. This is because the half-life time of foaming agent is dramatically reduced when it is injected simultaneously with the high-temp steam, therefore initially the N_2 foam is not fully mixed with steam, the nitrogen fingering during injection period occurs and consequently, the CSS performance is poorer than N_2 foam injection followed by steam injection. It is also indicated that the N_2 foam injection followed by steam injection yields cyclic oil 125 m^3 more than by injecting N_2 foam and steam simultaneously, and 679 m^3 more than by conventional CSS. The performance comparison of N_2 foam assisted CSS and conventional CSS is listed in Table 3.

Optimization of the proportion between Nitrogen and steam

The proportion between nitrogen and steam is optimized to determine the optimal nitrogen injection. The results show that the cyclic oil production increases with increase of nitrogen proportion (Figure 10). When the proportion is higher than 35:1, the incremental oil deceases with further increase of nitrogen, and the optimum

Table 3. The CSS performance comparison of different injection strategies.

Injection strategy	Cumulative oil (m^3)	Cumulative liquid (m^3)	Cumulative steam (m^3)	Cumulative Nitrogen (m^3)	Production time (day)	Producers (n)	Average oil (m^3/day)
Only steam	832	4963	2547	0	180	4	1.15
N_2 foam injection followed by steam	1511	8177	2547	101882	180	4	2.10
N_2 foam and steam injection simultaneously	1386	5708	2547	101882	180	4	1.92

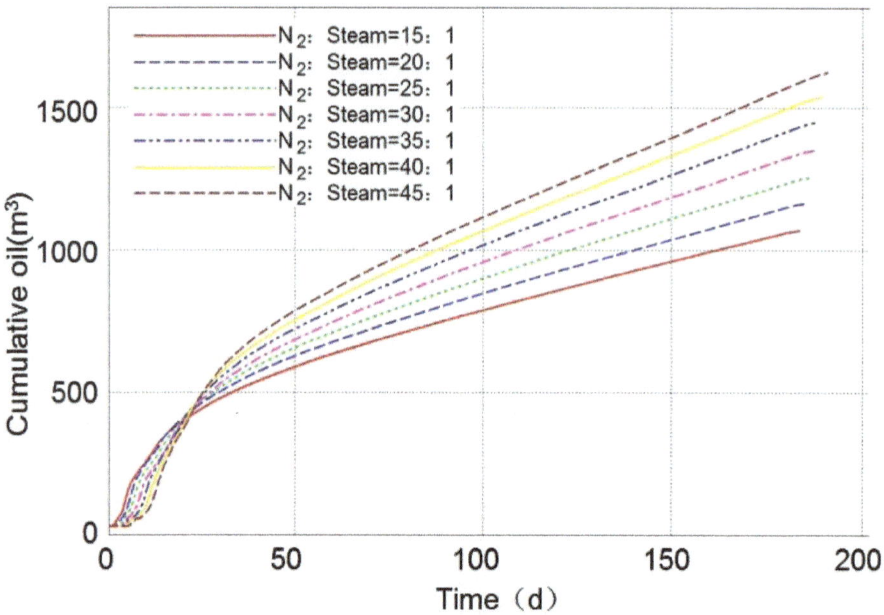

Figure 10. Cumulative cyclic oil production by different proportion between Nitrogen and steam.

proportion is 35:1 (Table 4).

Nitrogen foam injection rate optimization

Four cases of nitrogen foam injection rate are simulated: 2000, 3000, 4000 and 5000 Nm^3/day (Figure 11). The results show that when nitrogen is injected at the rate of 3000 Nm^3/day, the cumulative cyclic oil production is the highest; therefore the optimum nitrogen foam injection rate is 3000 Nm^3/day (Table 5).

FIELD TEST PERFORMANCE ANALYSIS

The field test is carried out based on the foaming agent screening in laboratory and numerical simulation for reservoir engineering parameters. Producers with 5 to 7 cycles of conventional CSS are selected and the injection and production parameters of CSS are optimized according to the reservoir and fluid properties: the proportion between nitrogen and steam is 35:1 and the nitrogen foam injection rate is 3000 N m^3/day as of December 2010; the N_2 foam assisted CSS has been carried out in 16 wells, and the staged performance is encouraging.

Fluid absorption profile improvement

The fluid absorption profiles in 16 wells have been tested and analyzed, which shows an encouraging improvement of the fluid absorption profile vertically. As can been seen from the test results of a typical well that by only steam injection during previous conventional CSS, the zone 1 absorbs most of the steam, which accounts for 83.81% of the total steam injection rate, while for zone 2, the steam absorption percentage is only 16.19%; during steam injection phase of N_2 foam assisted CSS, the fluid absorption profile is improved: for zone 1, the fluid absorption percentage reduces from 83.81 to 43.4%, while for zone 2, the fluid absorption percentage increases from 16.19 to 56.6% (Figure 12). Therefore, it can be seen that by injecting N_2 foam assisted CSS, the vertical oil deposits can be evenly developed. Meanwhile, the pressure data also shows that during injection phase

Table 4. Optimization results of the proportion between Nitrogen and steam.

Proportion (N_2/Steam)	Cumulative oil (m^3)	Cumulative liquid (m^3)	Cumulative steam (m^3)	Cumulative Nitrogen (m^3)	Production time (day)	Producers (n)	Incremental oil (m^3)
15:1	1039	6213	2547	38205	180	4	
20:1	1134	6619	2547	50941	180	4	94.32
25:1	1225	7064	2547	63676	180	4	91.65
30:1	1322	7449	2547	76411	180	4	97.03
35:1	1421	7824	2547	89147	180	4	98.66
40:1	1511	8177	2547	101882	180	4	89.73
45:1	1594	8556	2547	114617	180	4	83.47

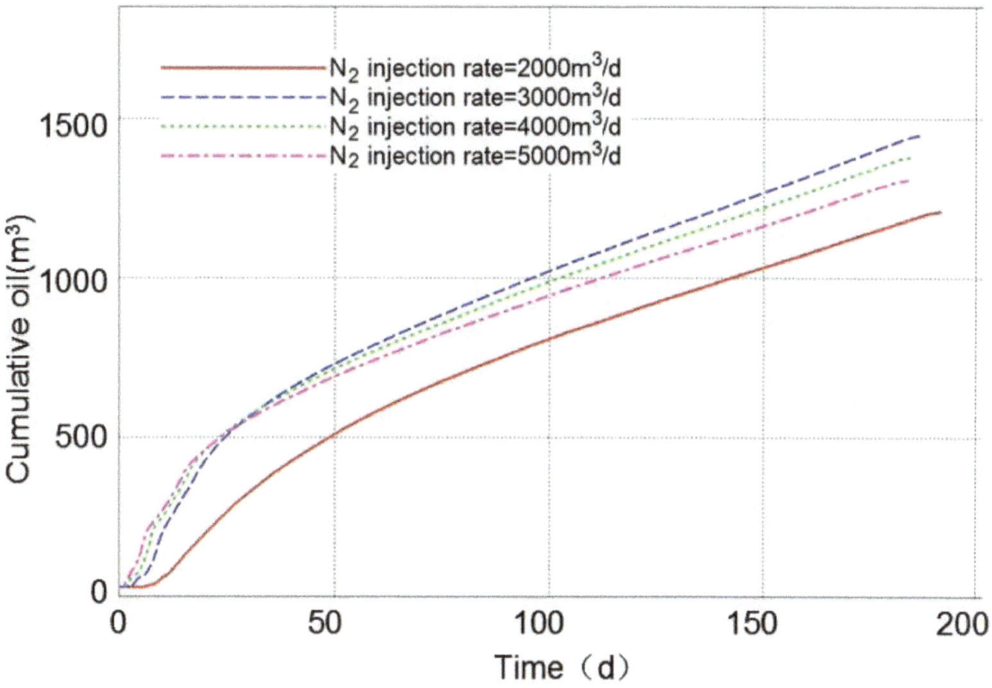

Figure 11. Cumulative cyclic oil production comparison by different nitrogen foam injection rate.

Table 5. Results of nitrogen foam injection rate optimization.

Injection rate (m^3/day)	Cumulative oil (m^3)	Cumulative liquid (m^3)	Cumulative steam (m^3)	Cumulative Nitrogen (m^3)	Production time (day)	Producers (n)
2000	1179	7812	2547	89147	180	4
3000	1421	7824	2547	89147	180	4
4000	1353	7291	2547	89147	180	4
5000	1278	6857	2547	89147	180	4

of N_2 foam assisted CSS, the injection pressure is evidently higher (avg. 3.9 MPa) than those wells injecting only steam due to the resistance effects of the N_2 foam, which could effectively reduce the steam fingering and enhance the cyclic production performance.

Cyclic oil production enhancement

The comparison between the previous conventional CSS and current N_2 foam assisted CSS shows that after nitrogen foam injected, the cyclic oil production and the

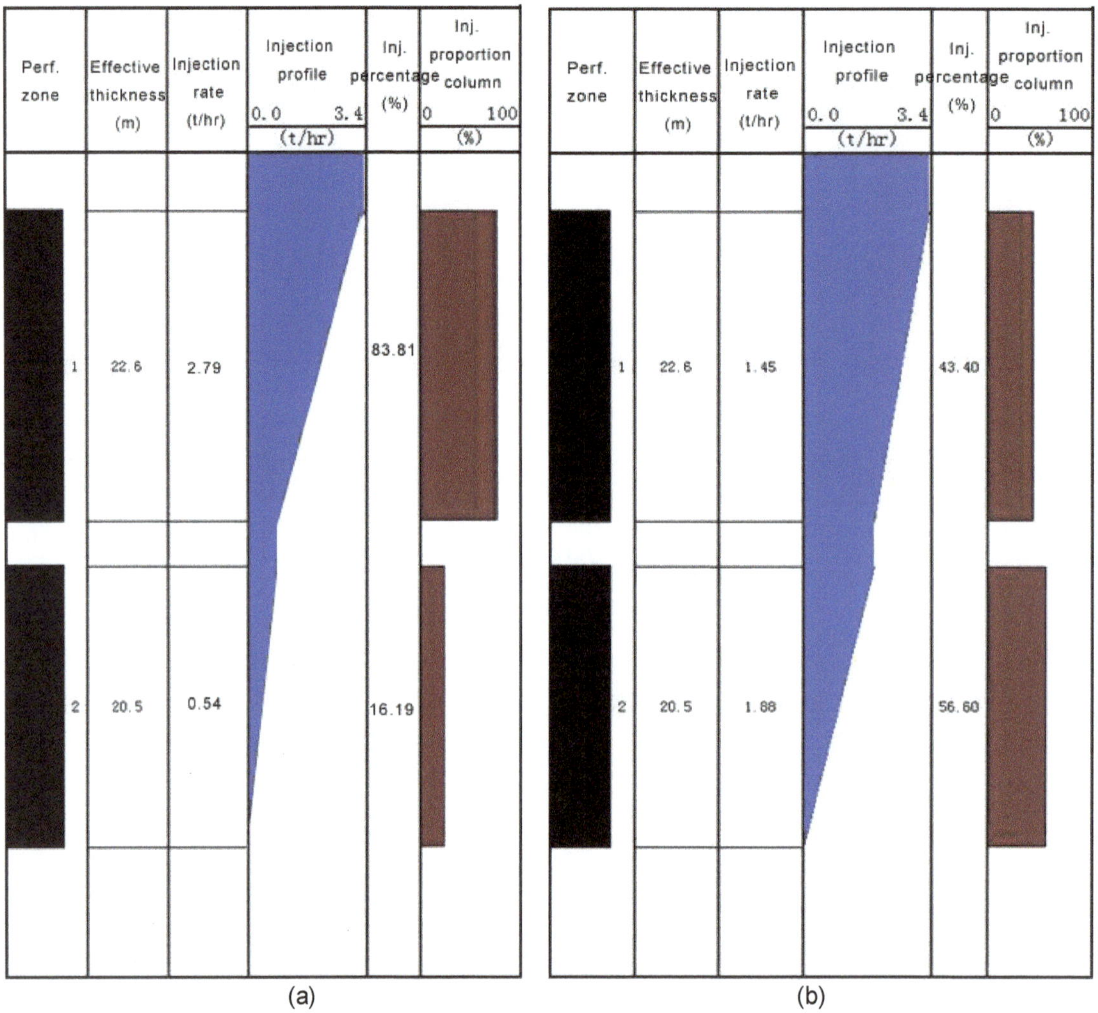

Figure 12. Fluid absorption profile comparison before and after N_2 foam assisted CSS for a typical well. (a) Steam absorption profile (b) Steam and N_2-foam system absorption profile.

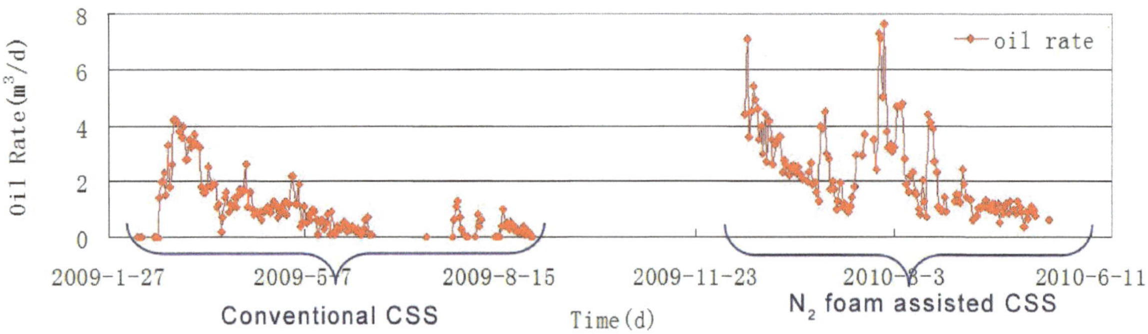

Figure 13. Comparison between conventional CSS and nitrogen assisted CSS in well 1.

cyclic OSR enhanced effectively (Figures 13 and 14). Most wells performed poor and were uneconomic during conventional CSS, while after nitrogen assisted CSS, the performance improved greatly: the cyclic production time increased by 51 days, the OSR enhanced by 0.105 ~ 0.137, the recycling ratio of water enhanced by 12 ~ 141%. Especially, the cyclic oil production in typical well 1 is 114 t and the OSR is 0.144 (Table 6), while after

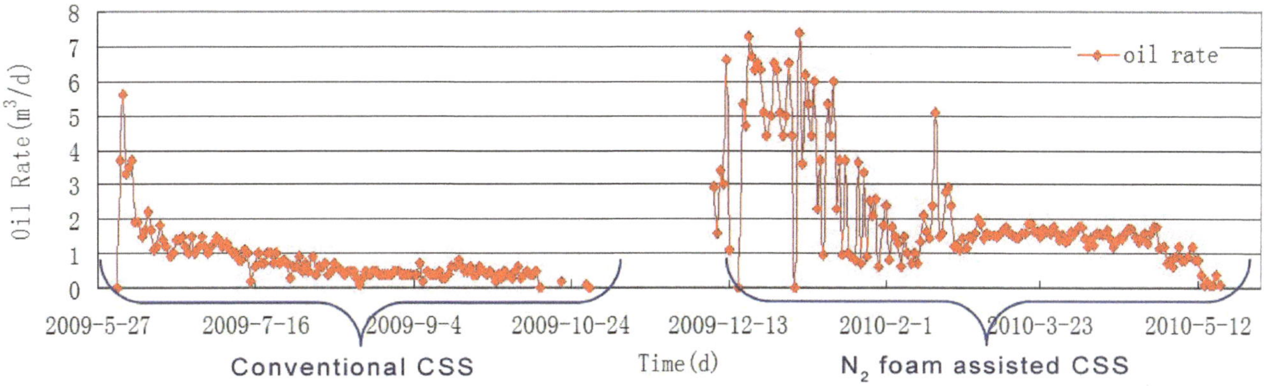

Figure 14. Comparison between conventional CSS and nitrogen assisted CSS in well 2.

Table 6. Incremental oil of well 1 and well 2.

Well No.	CSS method	Cycle	Cyclic oil (m^3)	Incremtnal oil (m^3)	Cyclic liquid (m^3)	Average oil (m^3/day)
Well1	Conventional CSS	6	114	240	284	0.83
	N_2 foam assisted CSS	7	354		655	2.58
Well2	Conventional CSS	6	160	156	613	1.06
	N_2 foam assisted CSS	7	316		902	2.10

nitrogen assisted CSS, the cyclic oil production is enhanced to 354 t and the OSR is enhanced to 0.277. The field tests show that the N_2 foam assisted CSS is an effective technology to enhance the performance of multi-cycle heavy oil wells, and an effective means to enhance the oil recovery factor.

Conclusions

1) Detailed literature investigation is carried out on profile modification mechanism of nitrogen foam to control fingering and enhancing oil recovery rate.
2) Static tests of foam adaptability to temperature, salinity and different oil samples are evaluated using four sets of N_2 foam system.
3) Effects of temperature and foaming agent concentration on foam plugging performance are analyzed and compared, by which the set 0# is selected to carry out the pilot test.
4) The operation parameters regarding injection pattern, injection rate, etc. are obtained by reservoir engineering design, which guided the implementation of N_2 foam assisted CSS successively.
5) Pilot test in 16 wells gave the positive results, which indicate that the fluid absorption profile is improved, consequently the oil deposits could be evenly tapped vertically, and the cyclic production performance is much better than previous conventional CSS, which shows potentials for future tests with larger scale in the Fuyv heavy oil reservoir.

ACKNOWLEDGEMENT

The authors would like to thank Daqing Oilfield E&D Institute, CNPC for permission to publish this paper.

REFERENCES

Bracho L, Portillo F (1991). Steam soak with additives, the maraven experience. 5th UNITAR International Conference on Heavy Crude and Tar Sands, Caracas, Venezuela.
Harris P (1987). Dynamic Fluid-Loss Characteristics of CO_2-Foam Fracturing Fluids. SPE Prod. Eng. *2*(2):89-94.
Hudgins D, Llave F, Chung F (1990). Nitrogen miscible displacement of light crude oil: A laboratory study. SPE Reserv. Eng. *5*(1):100-106.
Kuehne D, Ehman D, Emanuel A, Magnani C (1990). Design and evaluation of a Nitrogen-foam field trial. J. Petrol. Technol. 42(4):504-512.
Li X, Wu Y, Zhao X, Gao P, Qin Q, Yu M (2012). A Cyclic-Steam Injection in Fuyv Reservoir, Daqing. In *SPE EOR* Conference at Oil and Gas West Asia.
Lin Y, Yang G (2006). A successful pilot application for N2 foam flooding in Liaohe Oilfield. In SPE Asia Pacific Oil & Gas Conference and Exhibition.
Liu H, Ye P, li W, Song L, Zhong Y (2010). Application of Nitrogen Foam for Profile Modification in a Heterogeneous Multi-layer Sandstone Oilfield. In SPE Asia Pacific Oil & Gas Conference and Exhibition.
Liu R, Liu H, Li X, Fan Z (2008). The reservoir suitability studies of nitrogen foam flooding in Shengli Oilfield. In *SPE* Asia Pacific Oil and Gas Conference and Exhibition.
Ma C, Wu X, Liu H (2005). Mathematical and physical modeling of heat insulation by injecting nitrogen is steam injection wells. J. Univ. Petrol. China (Edition of Natural Science) 29(5):63-66.
Martinez A (2005). Improving steam soak recovery by using additives. Unitar International Conferenec on Heavy Crude and Tar Sands, Houston, Texas.

Nguyen Q, Currie P, Zitha P (2003). Determination of foam induced fluid partitioning in porous media using X-ray computed tomography. In International Symposium on Oilfield Chemistry.

Svrcek WY, Mehrotra AK (2002). Gas solubility, viscosity and density measurements for athabasca bitumen. J. Can. Pet. Tech. pp. 31-38.

Thomas LK, Dixon TN, Pierson RG, Hermansen H (1991). Ekofisk nitrogen injection. SPE Form. Eval. *6*(2):151-160.

Zhao X, Wu Y, Baiming Z, yu S, xianbao Z, Gao P (2011). Strategies to Conduct Steam Injection in Waterflooded Light Oil Reservoir and a Case in Fuyu Reservoir, Jilin. In SPE Middle East Oil and Gas Show and Conference.

Zhao X, Zhong H, Gao P, Wu Y, Zhong H, Hasi B, Qian Y (2010). Investigation of Steam Injection at Critical Condition and a Pilot Test in 119-52, Chao-601. In International Oil and Gas Conference and Exhibition in China.

Analysis of oil flow in fractured oil reservoir using carbondioxide (CO_2) foam injection

Minhee Choi, Junwoo Seo, Hyemin Park and Wonmo Sung*

Department of Natural Resources and Environmental Engineering, Hanyang University, 222 Wangsimni-ro, Seongdong-gu, Seoul 133-791, Korea.

Most carbonate reservoirs are heterogeneous in terms of pore distribution and the matrix is extremely low permeable, hindering the flow of oil, while highly permeable fractures are the primary flow conduit. This study presents the hybrid discrete fracture network (DFN) approach to represent fracture network, more realistically, which is a continuum approach equipped with local grid refinement (LGR). The LGR is adapted at the cells that fractures are passing through, in order to describe the fracture width less than 0.1 ft. In this approach, control volume of the well lying on the fracture is extremely small and thus, in this study, four-leg horizontal well concept substitutes the vertical well with the use of equivalent well radius for overcoming the numerical convergence problem. The hybrid DFN approach was applied for CO_2-foam injection in fractured carbonate reservoirs to investigate the effect of conformance control. This foam injecting method yields more imbibition of CO_2-foam into low permeable matrix, which contains most of the oil. From the numerical results, it shows that the foam controls CO_2 velocity and this increases the contact time between CO_2 and the rock matrix since CO_2 mobility is controlled by foam, that decreases relative permeability of CO_2 and increases its viscosity.

Key words: Carbonate, fracture, local grid refinement, CO_2-foam, conformance control, discrete fracture network (DFN).

INTRODUCTION

Carbonate reservoirs account for over 40% of total worldwide oil reservoirs, and heavy oil contained in these is estimated to be 1.6 trillion bbl in place (Briggs, 1992; Das, 2007). Most carbonate reservoirs are heterogeneous, comprising fractures, vugs, and cavities. Also, the matrix has very low permeability, hindering the flow of heavy oil, while highly permeable fractures are the primary flow conduit. This renders the recovery process highly ineffective, and a large amount of oil is retained by the tight matrix after the application of primary and secondary recovery methods. Of current enhanced-oil-recovery (EOR) methods, CO_2 flooding has great potential for recovering additional oil after water flooding. The advantage of CO_2 as an injection gas is its relatively low minimum miscibility pressure (MMP), making CO_2 applicable in a broad range of reservoirs (Stalkup, 1983). However, gravity override and viscous fingering caused by geological heterogeneity and unfavorable CO_2 mobility lead to poor macroscopic sweep efficiency and early CO_2 breakthrough.

The application of foam can reduce gas mobility, mitigate the effects of heterogeneity, and increase sweep efficiency (Liu et al., 2011). For the creation of foam, surfactant must be present. Foam can be injected into the

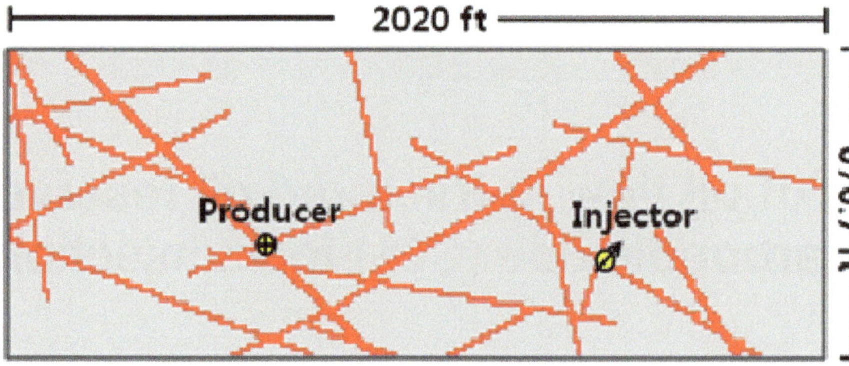

Figure 1. System description.

reservoir by co-injection of surfactant solution and gas, or by surfactant-alternating gas (SAG) injection. Foam moves through the porous media via the film network of bubbles, and gas moves through the system by breaking and by re-forming bubbles throughout the length of the flow path (Hirasaki, 1989). The stability of foam is a function of the surfactant type, surfactant concentration, foam quality, reservoir temperature, salinity, oil saturation, and other variables.

CO_2 foam reduces the mobility of CO_2 within the fracture network and decreases the velocity of CO_2 through the fractures, increasing the contact time between CO_2 and the carbonate rock matrix. This increases the imbibition of CO_2 into the low permeable matrix. Furthermore, with mobility control, foam diverts CO_2 and the surfactant solution to the tight matrix, where it can alter the wettability characteristics of the carbonate matrix from oil-wet to water-wet, causing more CO_2 foam to be imbibed into the matrix. Also, it reduces the interfacial tension (IFT) between the oil and water for release and mobilization of matrix oil, increasing the recovery factor.

Although several field studies and numerous laboratory tests with foam have shown some success (Heller et al., 1985; Hoefner et al., 1995; Holm et al., 1988), foams are a highly complex technology, both chemically and operationally, making them difficult to successfully apply in the oilfield (Sydansk and Romero-Zeron, 2011). Thus, prediction of CO_2 foam behavior in field-scale, especially for fractured carbonate reservoirs with a tight matrix still needs to be studied. In the present study, we conducted field-scale numerical simulations of CO_2 foam injection in fractured carbonate reservoirs with low permeable matrices containing heavy oil, using the hybrid discrete fracture network (DFN) approach. This approach represents fracture network more realistically, which is a continuum approach equipped with local grid refinement (LGR). The LGR is adapted at the cells that fractures are passing through, in order to describe the fracture width less than 0.1 ft. Also, control volume of the well lying on the fracture is extremely small and thus, in this study, four-leg horizontal well concept substitutes the vertical well with the use of equivalent well radius for overcoming the numerical convergence problem. The hybrid DFN approach, which is free of the limitations of the dual porosity/dual permeability (DP/DK) model, allows the investigation of fluid transfer between a fracture and a matrix as well as fluid flow only through the fractures. The foam model in the simulator was used to apply the mobility reduction. In this study, we focused on determining the effect of several operational parameters on CO_2 foam behavior to control fast CO_2 flow through fractures and to improve oil flow in a rock matrix by considering the imbibition of CO_2 foam into the matrix.

SIMULATION SYSTEM

Simulational strategy and scenarios

To analyze CO_2 foam behavior in a fractured carbonate reservoir, a two-dimensional model was constructed, as shown in Figure 1. Generally, a dual-porosity/dual-permeability (DP/DK) model is used to conduct the simulation in fractured reservoirs, including carbonates, but the directional property of each fracture cannot be described using this model (Muhammad, 2011). For this reason, we used the hybrid DFN approach in which it is possible to input permeability into each of the cells in the single-porosity (SP) model. Using this approach, we generated a fracture network, consisting of 19 fractures connected together, with matrix permeability of 1.0 md and fracture permeability of 2000 md.

CO_2 gas and surfactant solution were set up to co-inject and generate CO_2 foam under the reservoir condition. The wells were located on the connected fracture network, and the producer was specified by constant bottom-hole pressure while the injector was specified by constant rate to investigate the effect of injection rate. The total period of injection and production was 3 years, and the other input data in the model are listed in Table 1.

In fluid and rock properties, the oil-wet characteristic of carbonate rock is considered by the oil-water relative permeability curve (Figure 2). Also, because we assumed a dead-oil system without free gas and solution gas, the gas existing in the reservoir repre-

Table 1. Input data for the simulation study.

Parameter		Value
No. of Global Cell (X-Y)		13400 (200 × 67 × 1)
No. of LGR Cell (X-Y)		1569 (3 × 3 × 1)
Permeability [md]	Fracture	2000
	Matrix	1.0
Porosity		0.20
Initial Pressure [psia]		4000
Initial Oil Saturation		0.80
Initial Water Saturation		0.20
Depth [ft]		9000
Production Well BHP [psia]		3000
Injection Rate	Water (STBD)	40.0
	CO_2 (MSCFD)	4.7754
Oil Density [lb/ft^3]		62.5
CO_2 Density [lb/ft^3]		0.117
Surfactant Concentration [wt%]		0.001

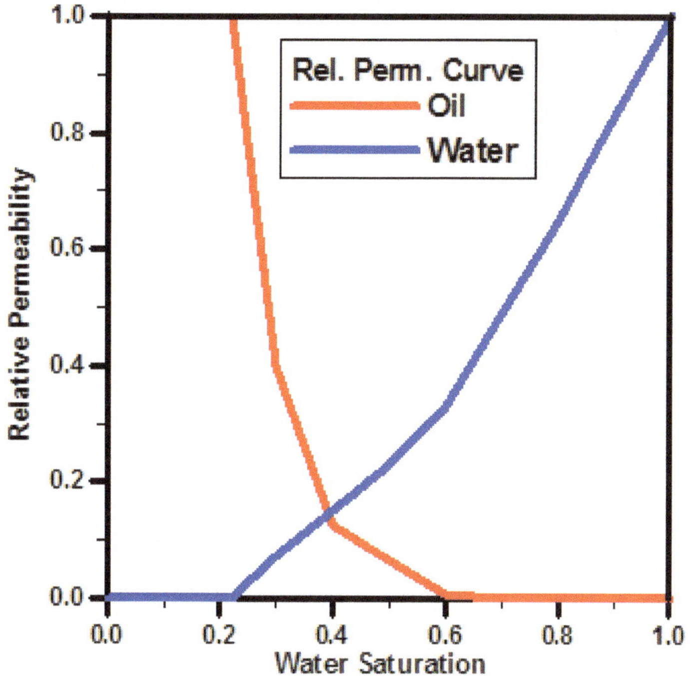

Figure 2. The relative permeability curve for oil-wet carbonate reservoir.

sents only CO_2 gas, which has a surface density of CO_2. Once CO_2 gas is injected into the reservoir, CO_2 becomes miscible or immiscible with oil as it flows (Karaoguz et al., 1989; Shoaib et al., 2009). We assumed an immiscible process because oil has a high molecular weight in the case of heavy oil, making it difficult to develop miscibility with CO_2.

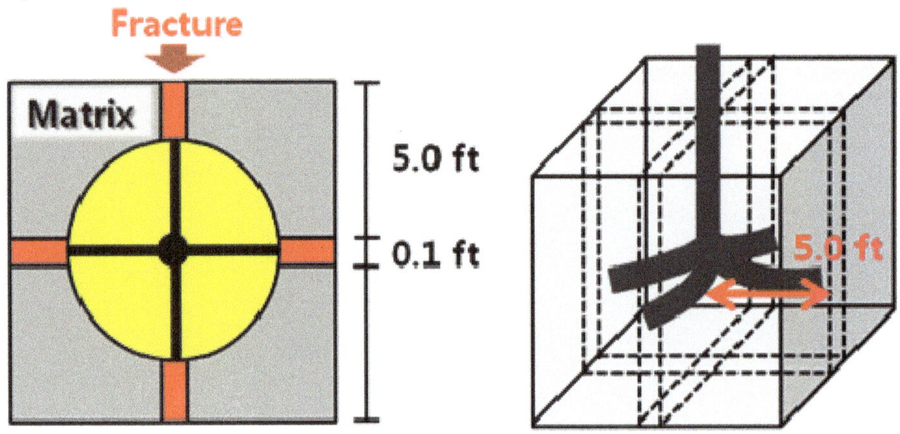

Figure 3. Multi-segment well in LGR fracture cell.

Hybrid discrete fracture network (DFN) approach

In the case of fractured system, local grid refinement (LGR) of the cells passing along the fractures was set up to describe a fracture aperture representing a realistic fracture size. From among the total 13,400 cells, which have dimensions of 200 × 67 × 1, 1596 cells were used in LGR of 3 × 3 cells, and the fracture aperture was set to 0.1 ft (Figure 3). This resulted in increase of the grid cells number to be 26,168 cells, which is increasing by up to twice the original number of cells. Also, It is impossible to locate a vertical well on the fracture cell that can afford a practical flow rate because this cell has an extremely small size. Thus, as shown in Figure 3, a multi-segment well with four legs using the concept of equivalent well radius was adopted instead of vertical well to overcome the numerical convergence problem and to control the flow rate within the small cell volume.

Application of LGR technique is efficient for increasing the degree of accuracy with lower computational time and computer storage to obtain the solution. However, since the long and narrow cells lead to huge computing and numerical errors, extremely small size of time step, in which the initial time step size Δt_i is 0.00001 day (8 s), the maximum time step size $\Delta t_{\max}$ is 0.1 day (2.4 h), and the minimum time step size $\Delta t_{\min}$ is 0.01 day (14.4 min) was applied to control the convergence errors.

Foam model in simulator

In the foam module of the simulator, because flow of foam in a porous medium is highly complex, investigating and modeling the nature of foam generation and coalescence as well as its rheological and physical properties remains a challenge that requires further research. For this reason, the foam model used in this simulation study assumes a simple conservation equation (Eclipse® 100, 2012). Accordingly, the foam generation and breakage mechanism are not considered in detail, and the foam is modeled as a tracer that can be transported with the gas or the water phase, taking into account adsorption onto the rock surface (Emadi et al., 2011).

In this simplified simulation model, the effect of foam on gas mobility is modeled by reducing the gas relative permeability or increasing the gas viscosity. A mobility reduction factor, M_{rf}, is applied to alter both end point of gas relative permeability and trapped gas saturation, and this factor is calculated as follows;

$$M_{rf} = \frac{1}{1+(M_r \cdot F_1 \cdot F_2 \cdot F_3 \cdot F_4)} \tag{1}$$

where, F_i accounts for various effects, such as F_1 for surfactant concentration, F_2 for water saturation, F_3 for oil saturation, and F_4 for gas velocity (Spirove et al., 2012). These dimensionless groups are suitably scaled by exponents e_s, e_o and e_c and weighting factor f_w under varying physical conditions. Each of F_i implies that

$$F_1 = \left(\frac{C_S}{C_S^r}\right)^{e_S}$$

$$F_2 = 0.5 + \frac{a\tan\left[f_w \cdot \left(S_w - S_w^l\right)\right]}{\pi}$$

$$F_3 = \left(\frac{S_o^m - S_o}{S_o^m}\right)^{e_o}$$

$$F_4 = \left(\frac{N_C^r}{N_C}\right)^{e_C}$$

In the above equations, C_s, S_w, S_o and N_c represent surfactant concentration, saturation of water, saturation of oil, and capillary number, respectively. Their corresponding critical values are denoted by C_s^r, S_w^l, S_o^m and N_c^r.

In addition, M_r is the reference mobility reduction factor obtained at maximum surfactant concentration and zero oil saturation. Usually, M_{rf} has the range of 5 to 100. The gas relative permeability is then rescaled from k_{ro} to ($k_{ro} \cdot M_{rf}$) by the foam effect.

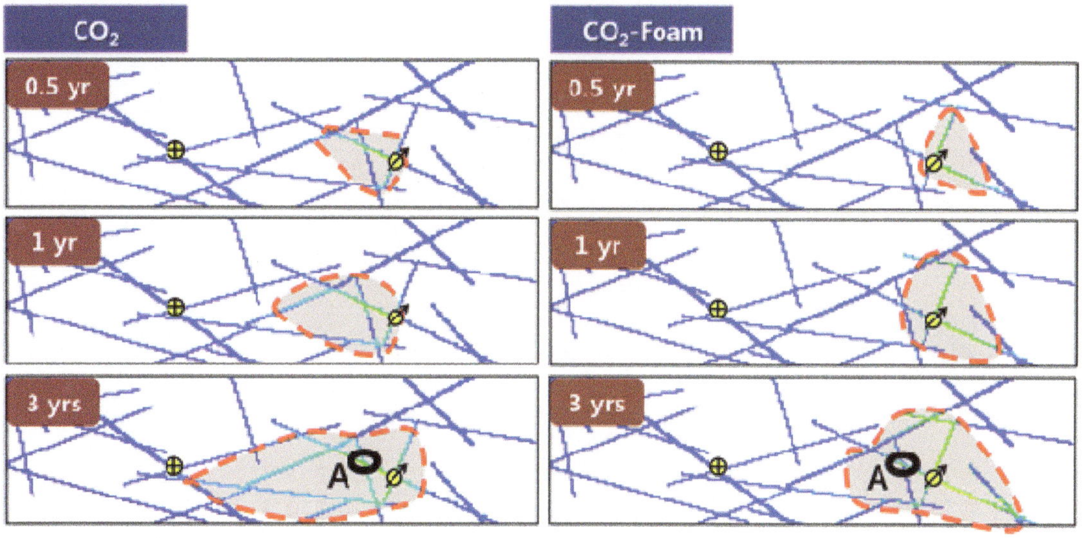

Figure 4. The CO_2 saturation distribution in fractured carbonate system.

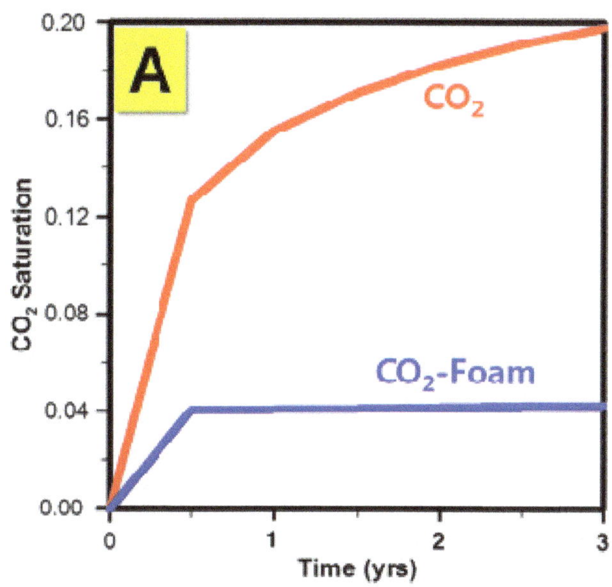

Figure 5. The CO_2 saturation change at location of "A".

RESULTS AND DISCUSSION

Oil flow in the matrix by CO_2 foam

To verify the effect of CO_2 foam injection with application of hybrid DFN approach using LGR on improving oil flow in the tight rock matrix, two simulation studies were conducted. First, we considered injection at 4.7754 MSCFD of CO_2 gas and 40 STBD of water with 0.001 wt% surfactant. A second process was used for comparative analysis of CO_2 flow velocity, which considered continuous injection of only CO_2 gas at 5 MSCFD (Figure 4). Figure 4 shows the CO_2 saturation distribution over 3 years; CO_2 quickly flows through the fractures toward the production well for continuous CO_2 injection compared with the case of CO_2 foam injection. We set point "A" between the two wells to examine the foam effect on flow control by observing the CO_2 saturation change. As shown in Figure 5, CO_2 saturation hardly increases after 0.5 year of CO_2 foam injection,

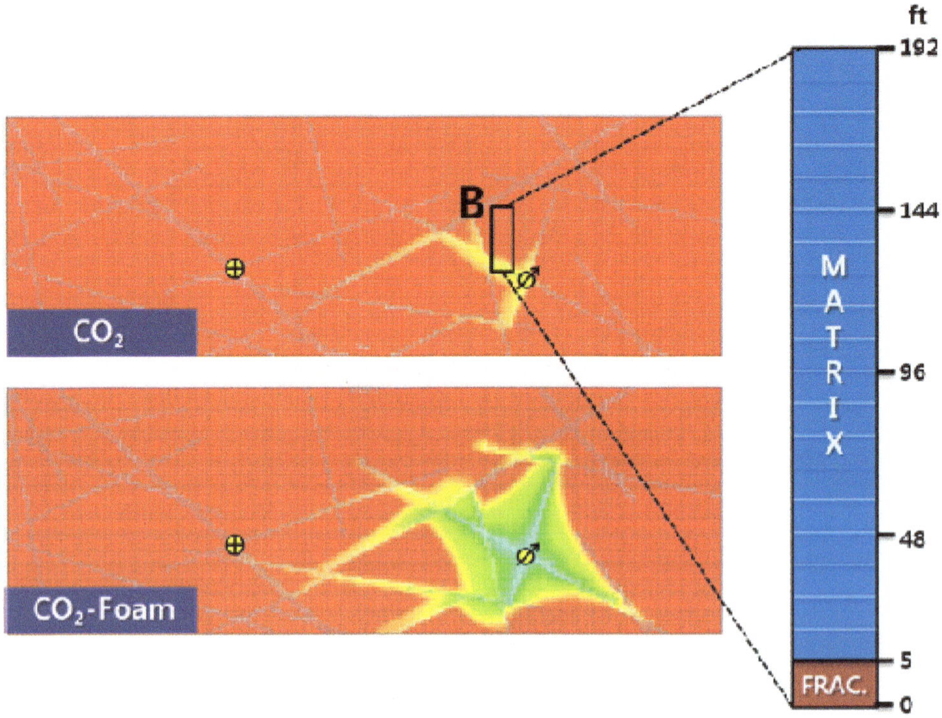

Figure 6. The oil saturation distribution at 1 year of production.

while CO_2 saturation consistently increases for the case of continuous CO_2 injection. This is because the mobility of gas-phase CO_2 is higher than that of liquid-phase foam, producing a greater flow velocity. It was shown that injection pressure for CO_2 foam injection with lower mobility has more than doubled compared to gas-phase CO_2 injection with higher mobility. This trend was also observed in the oil saturation distribution shown in Figure 6. For CO_2 foam injection, the sweep efficiency with oil improved because the liquid-phase foam enables capillary imbibition into the matrix as well as velocity control of fracture flow; for CO_2 injection, CO_2 gas preferentially moves through the fractures. With mobility control, foam diverts CO_2 and the surfactant solution to the tight matrix, where it can alter the wettability characteristics of the carbonate matrix from oil-wet to water-wet, causing more CO_2 foam to be imbibed into the matrix. Also, it reduces the interfacial tension (IFT) between the oil and water for release and mobilization of matrix oil, increasing the recovery factor (Fjelde et al., 2008; Shramm, 1994).

To analyze fluid imbibition into the rock matrix and improvement in oil flow in detail, the saturation change was measured at an arbitrary point "B". Figure 6 shows that section "B" consists of 5 ft of fracture and the remaining matrix, totaling 192 ft long. As indicated in Figure 7, when only CO_2 was injected, oil and CO_2 saturation in the rock matrix started to change at 6 months, and, moreover, the changes reached just 20% of the total distance after 3 years. On the other hand, when CO_2 foam was injected (Figure 8), the change started at 1 month with a rapid increase in foam concentration and, accordingly, a decrease in oil saturation. Because the contact time and area between the injected CO_2 foam and the rock matrix increased, the oil in the rock matrix with a large area was displaced. Thus, the oil saturation change reached the total distance of 192 ft between 12 months and 18 months.

As a consequence of the above results, the curves in Figure 9 for average matrix oil saturation (except for fractures), indicate that both primary oil recovery by natural production and the case of simple CO_2 injection changed slightly, whereas CO_2 foam injection changed 10% from 0.8 to 0.72, confirming that CO_2 foam injection greatly improves oil flow in rock matrix.

Conclusively, as shown in production data at the well (Figure 10), the oil production rate for CO_2 foam injection is enhanced about five times compared to only CO_2 injection. Also, when CO_2 was injected, the CO_2 breakthrough time (BT) was 300 days. However, when CO_2 foam injection was applied, CO_2 breakthrough did not occur during the total production years, indicating that

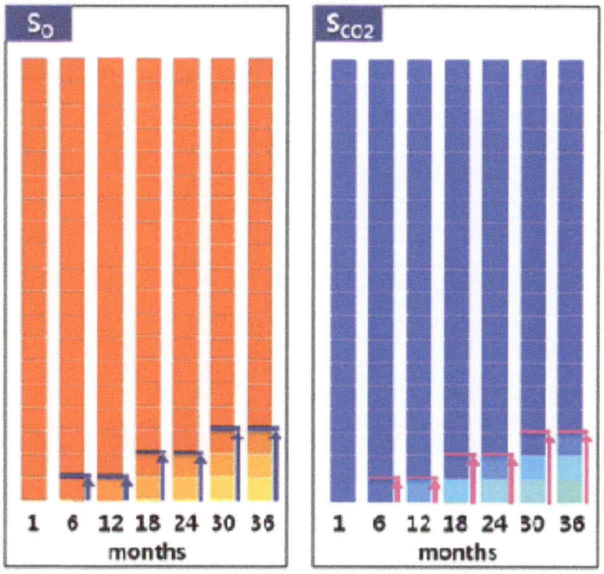

Figure 7. The changes of oil and CO_2 saturations at location of "B" in the case of CO_2 injection.

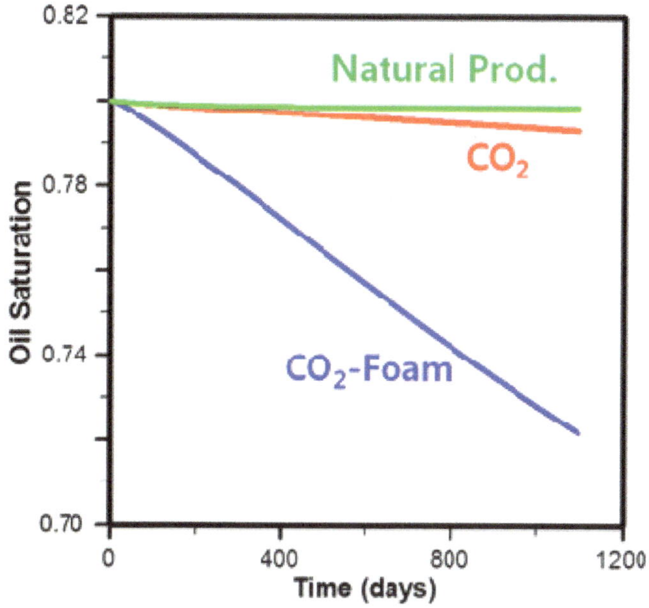

Figure 9. The average matrix oil saturation.

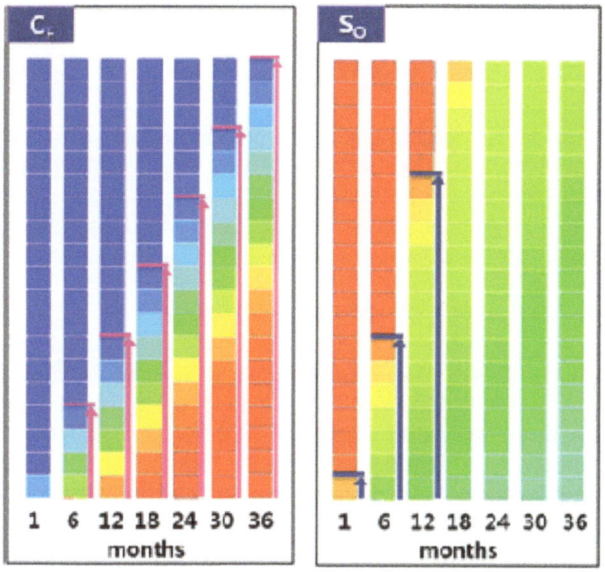

Figure 8. The changes of foam concentration and oil saturation at location of "B" in the case of CO_2 foam injection.

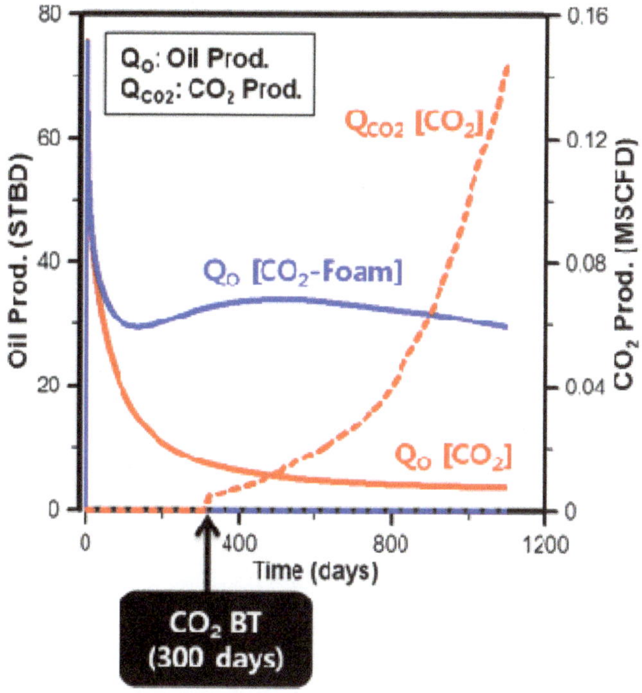

Figure 10. The oil production rate and CO_2 BT for the cases of CO_2 and CO_2 foam injection.

fracture flow was well-controlled by CO_2 foam injection, preventing early BT of CO_2.

Effect of injection fluid

It is important to determine which fluid plays a dominant role in promoting the foam effect when water, including surfactant, and CO_2 gas are co-injected to generate

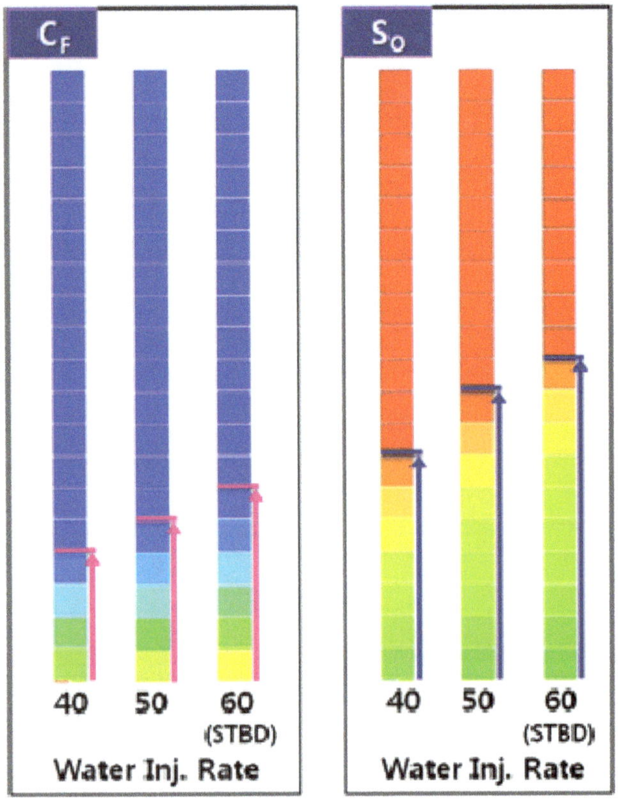

Figure 11. The changes of foam concentration and oil saturation at location of "B" for various cases of water injection rate at 6 months of production.

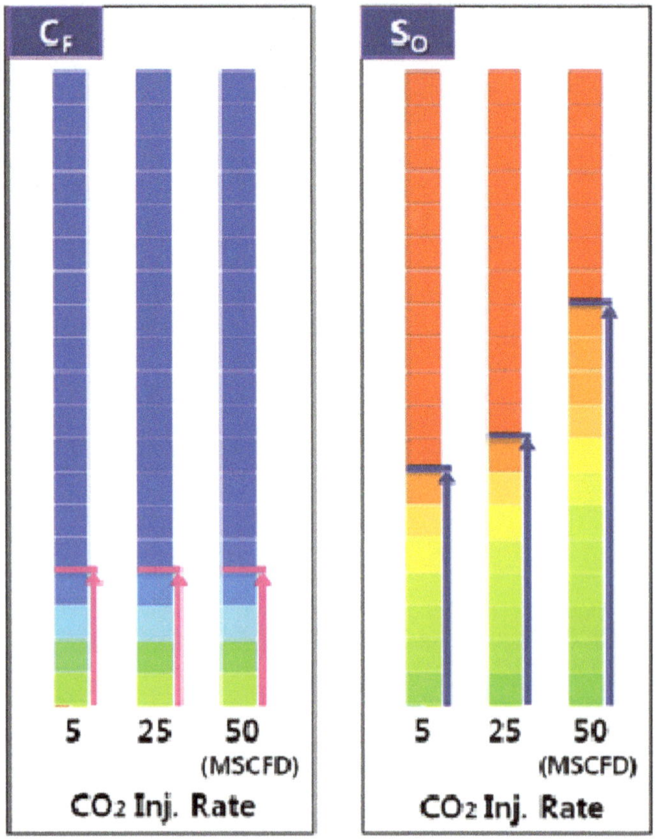

Figure 13. The changes of foam concentration and oil saturation at location of "B" for various cases of CO_2 injection rate at 6 months of production.

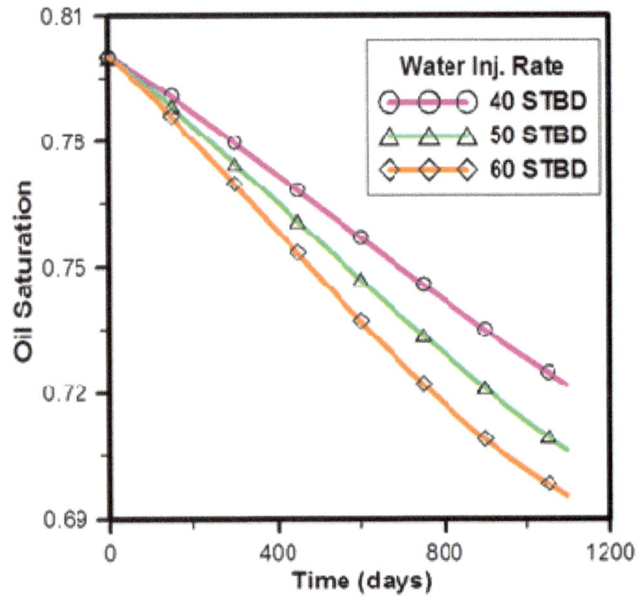

Figure 12. The average matrix oil saturations for various cases of water injection rate.

CO_2foam under the reservoir condition.

For this determination, three simulation runs were performed with water rates ranging from 40 to 50 and to 60 STBD with 0.001 wt% surfactant, while the co-injected CO_2 rate was fixed at 5 MSCFD. Figure 11 shows that higher water injection rates led to higher foam concentrations, decreasing the oil saturation deep in the rock matrix after 6 months. This means that, as the water (including surfactant) injection rate increases, more CO_2 foam is imbibed into the rock matrix, and rapid flow into fractures is controlled, displacing a large portion of oil remaining in the matrix. As a result, the average matrix oil saturation (Figure 12) further decreased when the water injection rate was high.

In sequence, three different CO_2 injection rates were considered: 5, 25, and 50 MSCFD with the same bottom-hole pressure at the injection well for the three simulations above according to water rate. The water injection rate was maintained at 40 STBD with 0.001 wt% surfactant. Figure 13 shows that greater CO_2 injection rates resulted in less oil saturation deep in the rock matrix

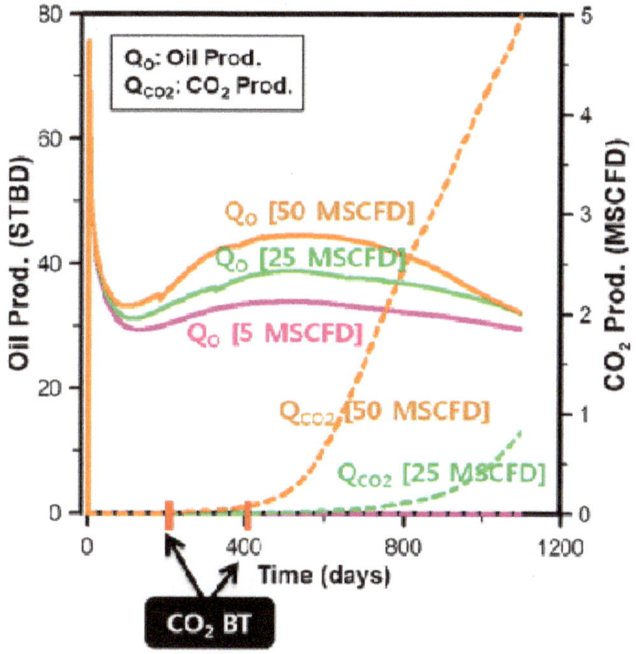

Figure 14. The oil production rate and CO_2 BT for various cases of CO_2 injection rate.

after 6 months, while the foam concentration did not change. Also, the CO_2 BT at the production well occurred early when the CO_2 rate was high (Figure 14). This indicates that an increase in amount of injected CO_2 did not affect fracture flow control via CO_2 foam; the result shown in Figure 13, in which matrix oil saturation was reduced, was likely due to the effect of increased total injection rate. That is, as the CO_2 injection rate increases, the amount of CO_2 imbibed into the rock matrix increases and displaces more oil from the matrix. However, it hardly influences foam flow and thus does not control rapid flow through the fractures on the contrary to the case of increasing water rates.

Effect of surfactant concentration

Since the surfactant is essential for generating foam, it is important to determine the proper surfactant concentration (Zuta et al., 2011). To investigate the effect of surfactant concentration with application of hybrid DFN approach using LGR, two comparative simulation runs were performed with a fixed CO_2 rate of 5 MSCFD and a water rate of 40 STBD with 0.01 and 1.0 wt% surfactant for 6 years.

Figure 15 shows the CO_2 saturation distribution in fractures, showing the effect on fracture flow control. In the case of high surfactant concentration of 1.0 wt%, CO_2 saturation was distributed densely within a narrow radius range compared to the case of low concentration of 0.01 wt%. Thus, CO_2 flow velocity in fractures is effectively slowed by high surfactant concentration, increasing the contact time between CO_2 and the rock matrix. In addition, Figure 16 shows that the CO_2 gas viscosity was held at a higher value for the case of higher surfactant concentration because this facilitated CO_2 foam formation. The viscosity change over time is caused with increases or decreases in pressure and temperature of the reservoir (Hirasaki et al., 1985). Consequently, Figure 17 shows that the CO_2 BT occurred at 1250 days with 11.2 pore volume of injection and 1600 days with 14 pore volume of injection for each case, indicating that BT was delayed with the high surfactant concentration.

Conclusion

A simulation for CO_2 foam injection in fractured carbonate reservoirs was performed to investigate the effect of foam on both rapid flow control in fractures and improved oil flow in a rock matrix. For the simulations, the fracture system was described using LGR and the hybrid DFN approach in which the fracture aperture is represented more realistically in comparison with exsiting works. The main results and conclusions are as follows:

1. Foam controls the CO_2 flow velocity and increases the contact time between CO_2 and the rock matrix. Hence,

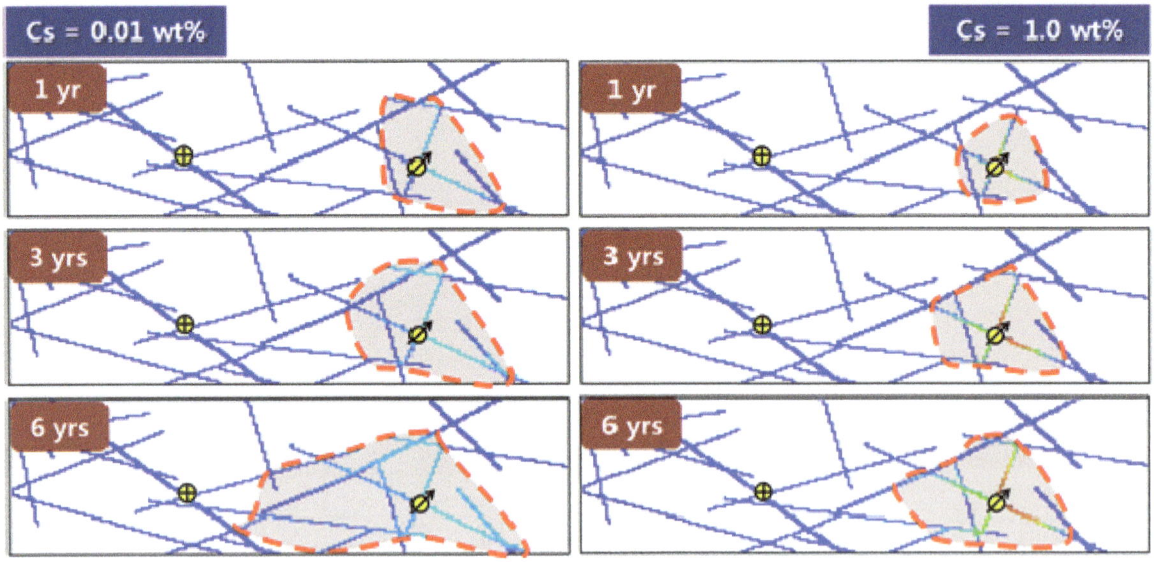

Figure 15. The CO_2 saturation distributions with surfactant concentrations.

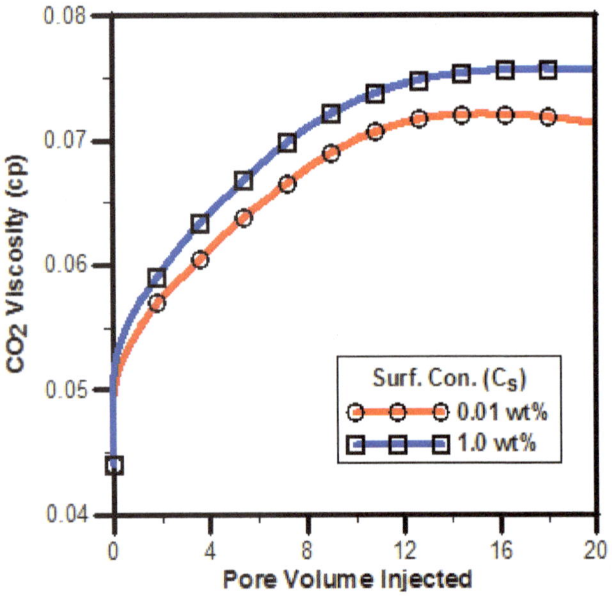

Figure 16. The CO_2 viscosity changes with surfactant concentrations.

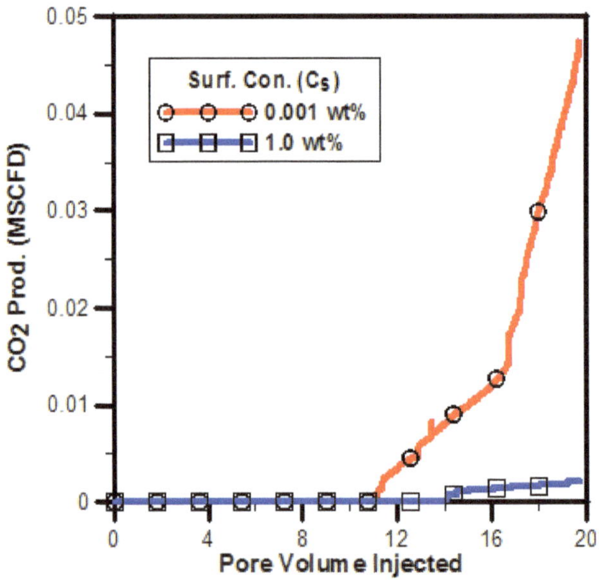

Figure 17. The CO_2 BT with surfactant concentrations.

injected CO_2 foam can be adsorbed and imbibed into the matrix, improving oil flow from the matrix. This is because CO_2 mobility is controlled by foam, which decreases CO_2 relative permeability and increases its viscosity.

2. During 3 years of production, the BT of the injected CO_2 was occurred at 300 days in the case of only CO_2 injection, meanwhile CO_2 breakthrough was not observed at producer when CO_2 foam was applied, indicating that foam prevents early CO_2 breakthrough.

3. The trend of decreasing oil saturation indicated that CO_2 foam injection is more effective for improving oil flow in the matrix; the decrease in oil saturation reached 16% of the total distance after 3 years for CO_2 injection, but it reached the total distance at 1.5 years for CO_2 foam injection.

4. In both cases of increasing water injection rate (with

the same surfactant concentration) and increasing CO_2 injection rate, the oil flow in the rock matrix was enhanced. However, CO_2 BT was occurred earlier, and oil production was lowered with increasing CO_2 injection rate. Therefore, the optimal design for injection rate is essential by considering injecting fluid in advance.

5. Higher surfactant concentration for CO_2 foam resulted in greater CO_2 viscosity. Thus, CO_2 flow velocity can be controlled effectively, resulting in increased oil production and retardation of CO_2 BT. Thus, formulation-related research should be continuously carried out in order to design cost-effective surfactant based EOR which requires an economic viability.

ACKNOWLEDGMENTS

This work was supported by a grand from the Energy Efficiency & Resources of Korea Institute of Energy Technology Evaluation and Planning (KETEP), funded by the Korean Ministry of Knowledge Economy (No.2010201010092A).

REFERENCES

Briggs PG, Beck DI, Black CJJ, Bissel R (1992). Heavy Oil from Fractured Carbonate Reservoirs. SPERE. 7(2): 173-179.

Das S (2007). Application of Thermal Recovery Processes in Heavy Oil Carbonate Reservoirs. Paper SPE 105392 presented at the 15th SPE Middle East Oil and& Gas Show and Conference, Bahrain International Exhibition Centre, Kingdom of Bahrain, 11-14 March.

Eclipse® 100 (2012). Technical Description. Schlumberger.

Emadi A, Sohrabi M, Jamiolahmady M, Ireland S, Rovertson G (2011). Mechanistic Study of Improved Heavy Oil Recovery by CO_2-Foam Injection. Paper SPE 143013 presented at the SPE Enhanced Oil Recovery Conference, Kuala Lumpur, Malaysia, 19-21 July.

Fjelde I, Zuta J, Olawale VD (2008). Oil Recovery from Matrix during CO2-Foam Flooding of Fractured Carbonate Oil Reservoirs. Paper SPE 113880 presented at the 2008 SPE Europec/EAGE Annual Conference and Exhibition, Rome, Italy, 9-12 June.

Heller JP, Boone DA (1985). Field Test of CO_2 Mobility Control at Rock Creek. Paper SPE 14395 presented at the 60th SPE Annual Technical Conference and Exhibition, Las Vegas, USA, 22-25 September.

Hirasaki GJ (1989). Supplement to SPE 19505, The Steam-Foam Process-Review of Steam-Foam Process Mechanisms. Paper SPE 19518, 19 April.

Hirasaki GJ, Lawson JB (1985). Mechanisms of Foam Flow in Porous Media: Apparent Viscosity in Smooth Capillaries. Paper SPE 12129 presented at the 1983 SPE Annual Technical Conference and Exhibition, San Francisco, USA, 5-8 October.

Hoefner ML, Evans EM (1995). CO2 Foam: Results From Four Developmental Field Trials. Paper SPE 22787 presented at the 1994 SPE/DOE Symposium on Improved Oil Recovery, Tulsa, Oklahoma, USA, 17-20 April.

Holm LW, Garrison WH (1988). CO2 Diversion With Foam in an Immiscible CO2 Field Project. SPERE 3(1):112-118.

Karaoguz D, Issever K, Pamir N, Tirek A (1989). Performance of a Heavy-Oil Recovery Process by an Immiscible CO2 Application, Bati Raman Field. Paper SPE 18002 presented at the SPE Middle East Oil Technical Conference and Exhibition, Manama, Bahrain, 11-14 March.

Liu MK, Rossen WR (2011). Sweep Efficiency in CO_2 Foam Simulations With Oil. Paper SPE 142999 presented at the SPE Europec/EAGE Annual Conference and Exhibition, Vienna, Austria, 23-26 May.

Muhammad S (2011). Flow and Transport in Porous Media and Fractured Rock, 2nd Ed. WILEY-VCH, Weinheim.

Shoaib S, Hoffman BT (2009). CO_2 Flooding the Elm Coulee Field. Paper SPE 123176 presented at the 2009 SPE Rocky Mountain Petroleum Technology Conference, Denver, Colorado, USA, 14-16 April.

Shramm LL (1994). Foams: Fundamentals and Applications in the Petroleum Industry. American Chemical Society, Washington DC.

Spirov P, Rudyk SN, Khan AA (2012). Foam Assisted WAG, Snorre Revist with Foam Screening Model. Paper SPE 150829 presented at the SPE North Africa Technical Confernence and Exhibition, Cairo, Egypt, 20-22 February.

Stalkup FI (1983). Miscible Displacement. SPE Monogr. Ser. 8(4):815-826.

Sydansk RD, Romero-Zeron L (2011). Reservoir Conformance Improvement. Society of Petroleum Engineers, USA

Zuta J, Fjelde I (2011). Mechanistic Modeling of CO2-Foam Processes in Fractured Chalk Rock: Effect of Foam Strength and Gravity Forces on Oil Recovery. Paper SPE 144807 presented at the SPE Enhanced Oil Recovery Conference, Kuala Lumpur, Malaysia, 19-21 July.

Minimizing entropy generation for louvered fins in a plate-fin compact heat exchanger

Masoud Asadi[1] and Nasrin Dindar Mehrabani[2]

[1]Department of Mechanical Engineering, Azad Islamic University Science and Research branch ,Tehran, Iran.
[2]Department of Mathematics & Computer Science, Amir Kabir University of Technology, Tehran, Iran.

With distributed power generation market, the most economical solution today is to generate power through small gas turbine systems, arbitrarily categorized as microturbines (5 to 200 kW) orminiturbines (200 to 500 kW). The thermal efficiency of such microturbines is about 20% or less if no recuperator is used in the system but, using a recuperator operating at 87% effectiveness, the efficiency of the gas turbine system increases to about 30%, a substantial performance improvement. However, cost of the recuperator is approximately 30% of the total power plant. This means that the heat exchanger (recuperator) must be designed to get high performance with minimum cost. In this paper, after providing the necessary concise information on compact heat exchangers, the discussion is presented on minimizing entropy generation for a compact heat exchanger with Louvered fins. So, thermodynamical optimization can play a key role in providing better performance for a compact heat exchanger.

Key words: Entropy generation, plate-fin heat exchanger, louvered fin, microturbine.

INTRODUCTION

Heat exchangers play a key role on automobiles, refrigerators and air conditioning system. Nowadays, efficient heat exchangers are required in order to save energy. Aluminum-brazed heat exchanger with multi-louvered fin and flat tube is used widely in industry due to high compactness and excellent heat transfer performance. A number of experimental studies (Davenport, 1980, 1983; Achaichia, 1987; Achaichia and Cowell, 1988; Huihua and Xuesheung, 1989; Aoki et al., 1989; Rugh et al., 1992; Sunden and Svantesson, 1992; Chang et al., 1994, 1996, 1997; Kim and Bullard, 2002; Lyman et al., 2002; Zhang and Tafti, 2003; Qi et al., 2007; Dong et al., 2007) and numerical simulations (Sahnoun and Webb, 1992; Springer and Thole, 1998; Tafti et al., 1999; Leu et al., 2001; Wu and Webb, 2002) have been conducted in recent years, in heat exchangers industry.

A schematic of a louvered fin compact heat exchanger is shown in Figure 1. When such louvered fins were first introduced, it was believed that increased heat transfer was due to added turbulence initiated by flow through the louvered array (Webb, 1991). Early flow visualization studies on scaled up louver designs quickly dispelled this theory. It was observed that as air passed through the louvers, the flow resulted in two distinct flow directions to be classified: axial (or duct) directed flow and louver directed flow. Axial flow occurs when the flow maintains the direction of the inlet flow. Louver directed flow occurs when the flow is aligned parallel to the louvers. These two directions are clearly indicated in the louvered fin cross section in Figure 2. Thus, in a strongly louver directed flow, the individual louvers essentially act as small flat plates aligned parallel to the flow (Hsieh and Jang, 2006). The explanation for increased heat transfer came from the realization that along with the new boundary layer that formed on each of these louvers came a corresponding high heat transfer coefficient (Andrew and Lyman, 2000).

Extensive studies have shown that the flow and heat

*Corresponding author. E-mail: masoud2471@gmail.com.

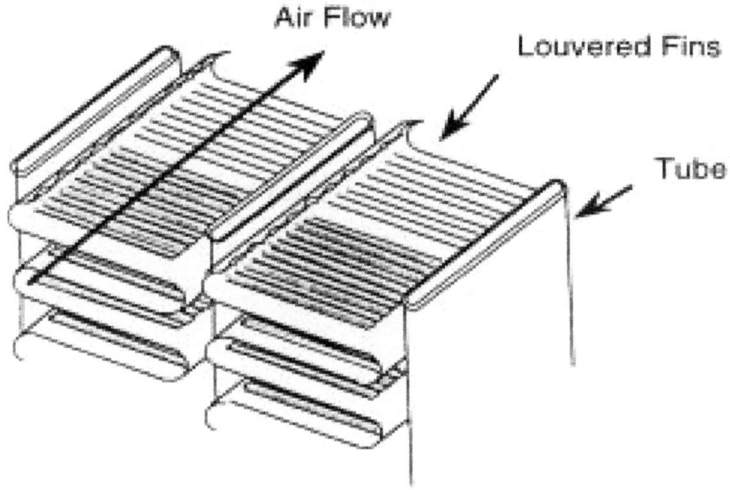

Figure 1. Schematic of a typical louvered fin-and-tube heat exchanger.

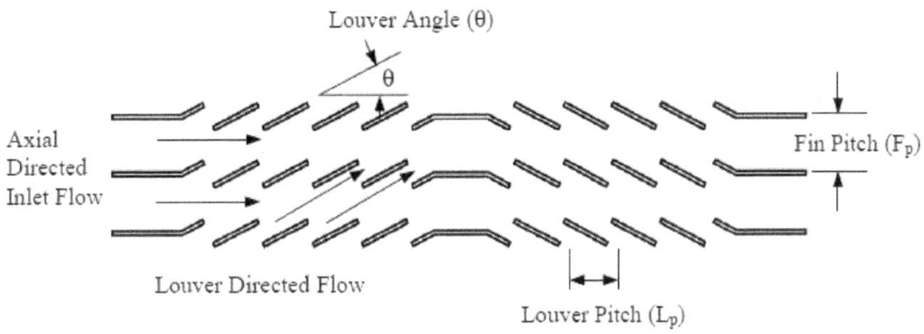

Figure 2. Side view of a typical louvered fin geometry.

transfer resulting in louvered fin arrays is quite complex. It has been shown that the degree to which the flow becomes aligned with the louvers is dependent on geometry and Reynolds number. Vortex shedding, which serves to increase heat transfer, may occur off the leading and trailing edges of the louvers. Heated wakes from louvers continue downstream and interact with downstream louvers creating complex thermal and flow field structures. The attempt to gain a detailed understanding of these mechanisms is what has been done by many researchers such as Davenport (1983) and Achachia (1988). Davenport (1983) performed overall heat transfer experiments on 32 non-standard geometries of louvered fin heat exchangers. After that, Achachia and Cowell (1988) presented experimental data of overall heat transfer and friction factor for 15 variants of louvered fin geometries. Achachia and Cowell (1988) found that at Reynolds numbers below 200, heat transfer performance flattened off considerably.

Even with the significant heat transfer enhancement of louvers, the thermal resistance on the air side of a typical fin-and-tube heat exchanger still dominates the thermal resistance of the system. The thermal resistance of the liquid side is typically 5 to 10 times lower than the thermal resistance of the air side (Yousefi and Darus, 2011: Yousefi et al., 2012). As a result, any modification that result in an increase in the air side efficiency of a heat exchanger is very valuable.

The higher level of the optimization leads to perform the thermal design of the heat exchanger, defining its operating parameters through minimization/maximization of a certain objective function like entropy generation, pressure drop or heat exchanger's costs. So besides calculating the optimized values of volume, pressure and temperature, at this level it is also possible to minimize entropy generation based on constant surface area. The second optimization level involves minimizing entropy generation based on constant volume. Furthermore, when volume and surface area are constant the thermodynamical optimization can be done by modifying Reynolds number.

In this work, the heat exchanger designing process tends to a sizing problem. So using following diagram can be useful.

THERMAL DESIGN

There are two basic types of thermal design problems: rating and sizing. In a rating problem, the geometry and size of the heat exchanger are fully specified. Also, entering flow rates and fluid temperatures are known. So, the job is to compute heat exchanger efficiency and pressure drop for each stream. However, in a sizing problem, the heat exchanger requirement is specified and the designer must calculate the heat exchanger size (Kays and London, 1984). Typically, allowable pressure drops are given for each fluid stream. A number of decisions must be made prior to making the thermal performance calculation (Kays and London, 1984)

1. Heat exchanger flow arrangement, for example, counter flow.
2. Heat exchanger materials, as influenced by fluid temperatures and corrosion potential.
3. Fin geometry and fin thickness, as influenced by design pressure requirements.
4. The type of surface geometry and fin spacing and height. Fouling considerations influence the type of surface geometry and fin spacing that may be selected. Fin height is influenced by the desired fin efficiency.
5. Heat exchanger frontal area. This key decision establishes the Reynolds number for each flow stream. The pressure losses are directly dependent on this decision.

In this work, the heat exchanger designing process tends to a sizing problem. So Figure 3 can be useful.

Since the outlet temperatures are unknown, the $\varepsilon - NTU$ is suitable, where

$$Q = \varepsilon C_{min}\left(T_{h,i} - T_{c,i}\right) = \varepsilon C_{min} \Delta T_{max} \quad (1)$$

Here $\varepsilon, \Delta T_{max}$ and C_{min} are the heat exchanger efficiency, entering temperature difference and the minimum of heat capacity respectively. ε, the heat exchanger efficiency, is a dimensionless number, and is a function of NTU and C^*:

$$\varepsilon = \phi\left(\text{NTU}, C^*, \text{flow arragment}\right) \quad (2)$$

Where C^* is defined by:

$$C^* = \frac{C_{min}}{C_{max}} = \frac{(\dot{m}C_p)_{min}}{(\dot{m}C_p)_{max}} = \begin{cases} \dfrac{\left(T_{c,o} - T_{c,i}\right)}{\left(T_{h,i} - T_{h,o}\right)} & \text{for } C_h = C_{min} \\[2ex] \dfrac{\left(T_{h,i} - T_{h,o}\right)}{\left(T_{c,o} - T_{c,i}\right)} & \text{for } C_c = C_{min} \end{cases} \quad (3)$$

The other important parameter is NTU, Where it is considered as the number of transfer unit.

$$\text{NTU} = \frac{UA}{C_{min}} = \frac{1}{C_{min}} \int_A U dA \quad (4)$$

To calculate the outlet temperatures of both fluids, hot and cold, the – Equations (5) and (6) can be considered.

$$T_{h,o} = T_{h,i} - \varepsilon\left(\frac{C_{min}}{C_h}\right).\left(T_{h,i} - T_{c,i}\right) \quad (5\text{-a})$$

$$T_{c,o} = T_{c,i} + \varepsilon\left(\frac{C_{min}}{C_c}\right).\left(T_{h,i} - T_{c,i}\right) \quad (5\text{-b})$$

One of the key stage in designing a heat exchanger is accurate calculation of fluid properties. But before doing this, it is necessary to compute average temperature of each fluid. Here, according to shah research (Shah, 1981) if $C^* \geq 0.50$ the average temperatures would be

$$T_{h,m} = \frac{T_{h,i} + T_{h,o}}{2} \quad (6\text{-a})$$

$$T_{c,m} = \frac{T_{c,i} + T_{c,o}}{2} \quad (6\text{-b})$$

Else, they would be based on Table 1. Fluid properties are as shown in Table 2.

Heat transfer coefficient is calculated form j, colburn factor. Although there some equations for colburn and friction factors, Kays and London (1984) have introduced these factors versus Reynolds number.

$$h = j\,G\,C_P.\text{Pr}^{\frac{-2}{3}} \quad (7)$$

$$G = \frac{\dot{m}}{A_{fr}} \quad (8)$$

In this formula, G and A_{fr} are mass velocity and free flow cross-sectional area respectively.

Furthermore, frictional pressure drop in both sides is given by:

$$\Delta P_1 = \frac{G_1^2}{2\rho_{in,1}}\left[\left(1 + K_{C,1} - \sigma_1^2\right) + 2\left(\frac{\rho_{in,1}}{\rho_{out,1}} - 1\right) + \left(f_1 \times \frac{S_1}{A_1} \times \frac{\rho_{in,1}}{\rho_{m,1}}\right) - \left(1 - \sigma_1^2 - k_{e,1}\right) \times \frac{\rho_{in,1}}{\rho_{out,1}}\right] \quad (9)$$

Where σ is the ratio of free flow area to the frontal area. Also, K_c and K_e are contraction and expansion coefficients.

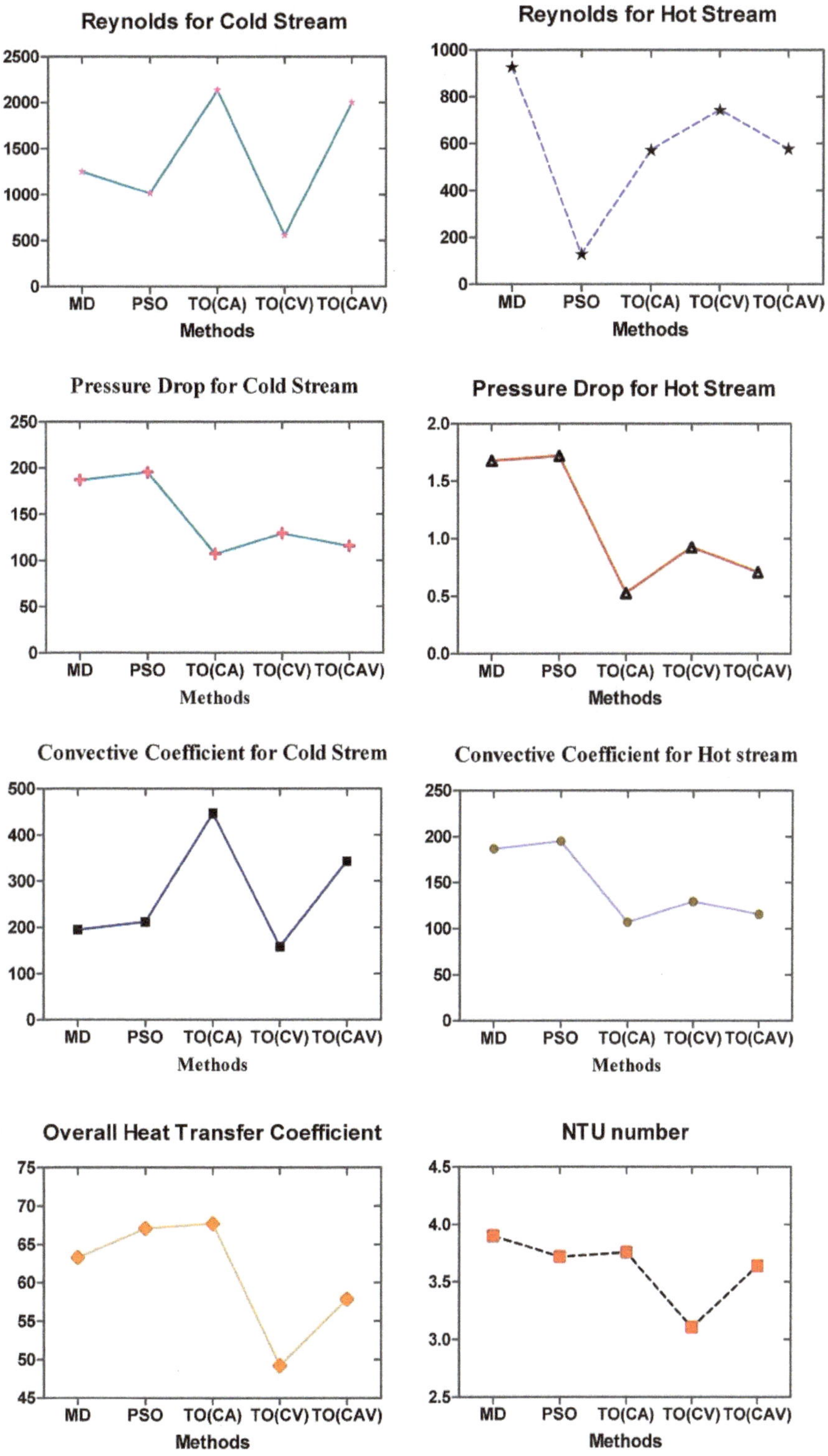

Figure 3. A comparison among the results of different methods.

ENTROPY GENERATION

The irreversibility of any heat exchanger is due to two factors; the transfer of heat across the stream – to – stream temperature difference and the frictional pressure drop. The fluid friction and heat transfer irreversibility can systematically be reduced by showing down the movement of fluid through the heat exchanger. This

Table 1. Approximate bulk mean temperature on hot and cold side of a two-fluid heat exchanger Kuppan (2000).

$C_{max} = hot\ fluid$ $C_{min} = cold\ fluid$	$C_{max} = cold\ fluid$ $C_{min} = hot\ fluid$
$T_{h,m} = \frac{T_{h,i} + T_{h,o}}{2}$	$T_{c,m} = \frac{T_{c,i} + T_{c,o}}{2}$
$T_{c,m} = T_{h,m} - \Delta T_{lm}$	$T_{h,m} = T_{c,m} + \Delta T_{lm}$
$\Delta T_{lm} = \frac{(T_{h,m} - T_{c,i}) - (T_{h,m} - T_{c,o})}{\ln[(T_{h,m} - T_{c,i}) - (T_{h,m} - T_{c,o})]}$	$\Delta T_{lm} = \frac{(T_{h,i} - T_{c,m}) - (T_{h,o} - T_{c,m})}{\ln[(T_{h,i} - T_{c,m}) - (T_{h,o} - T_{c,m})]}$

Table 2. Fluid properties.

	$\rho_i \left(\frac{Kg}{m^3}\right)$	$\rho_0 \left(\frac{Kg}{m^3}\right)$	$K \left(\frac{W}{m.K}\right)$	$\mu \left(\frac{N.S}{m^2}\right)$	$C_P \left(\frac{Kj}{Kg.K}\right)$	$\frac{1}{\rho_m}$	Pr
Gas	0.399	0.491	43.9	2.88×10^{-5}	1.58	0.4182	0.683
Air	2.934	2.017	0.05	3.55×10^{-5}	1.04	2.2684	0.735

technique is synonymous with employing larger heat exchanger, that is, more heat transfer area and more heat exchanger volume (Yousef et al., 2012).

The irreversibility due to heat transfer is

$$\dot{S}_{gen} = \left(\dot{m} C_P\right)_h Ln\frac{T_{h,o}}{T_{n,i}} + \left(\dot{m} C_P\right)_c Ln\frac{T_{c,o}}{T_{c,i}} \tag{10}$$

Where entropy changes associated with the frictional pressure drops $\left(P_{in} - P_{out}\right)_{h,c}$ have not been included. The entropy generation number is defined by Equation (11).

$$N_s = \frac{\dot{S}_{gen}}{\left(\dot{m} C_P\right)_{max}} \tag{11}$$

Here it is assumed ideal gas behavior on either side of the heat exchanger surface. For Simplicity, consider a balanced counter flow arrangement $(C^* = 1)$ in which the stream – to – stream temperature difference and frictional pressure drops are not negligible. So the entropy generation rate in this arrangement is

$$\dot{S}_{gen} = \left(\dot{m} C_P\right)_h Ln\frac{T_{h,o}}{T_{n,i}} + \left(\dot{m} C_P\right)_c Ln\frac{T_{c,o}}{T_{c,i}} - \dot{m}_h R\, Ln\frac{P_{h,o}}{P_{K,i}} - \dot{m}_c R\, Ln\frac{P_{c,O}}{P_{c,i}} \tag{12}$$

Where the first two terms on the right represent the heat transfer irreversibility and the last two terms account for fluid friction. Thus, the entropy generation number becomes.

$$N_S = \frac{\tau}{NTU_h} + \frac{R}{C_P}\left(\frac{\Delta P}{P}\right)_h + \frac{\tau^2}{NTU_c} + \frac{R}{C_p}\left(\frac{\Delta P}{P}\right)_C \tag{13}$$

With

$$\tau = \frac{|T_\circ - T_i|}{(T_i T_o)^{0.5}} \tag{14}$$

The heat transfer irreversibility vanishes when the area is very large $(NTU \to \infty)$ or when the counter flow is isothermal due to end conditions $(\tau = \circ)$. Also, the fluid – friction irreversibility vanishes when the pressure drops on the two sides of the surface are zero.

The one-side irreversibility number depends on six parameters, which can be grouped in two rows:

$$\tau(T_h, T_c), \frac{R}{C_P}, \Pr$$
$$\frac{4L}{D}, \mathrm{Re}_D, G_* \tag{15}$$

The upper row is fixed as soon as the working fluid and two inlet conditions are specified. The second row parameter depends on the geometry of the heat – exchanger duct. For example, Re_D and G_* Can be considered as fixed, while the flow path length $\frac{4L}{D}$ is selected on the basis of minimizing the irreversibility. In actual applications one or more geometric parameters are constrained based on economic considerations. The minimization of irreversibility subject to constant area is important in cases in which the cost of building the heat transfer surface is a major component in the overall cost formula for the heat exchanger system.

The heat transfer area for one side by using the definition of hydraulic diameter is:

$$A = \frac{4L}{D} A_{fr} \tag{16}$$

Where A_{fr} is the flow cross-section. Also, this expression can be put in dimensionless form using Equation (17).

$$G_* = \frac{G}{(2\rho p)^{0.5}} \tag{17-a}$$

$$A_* = \frac{4L}{D} G_*^{-1} \tag{17-b}$$

So, when $N_{s,\min} = \frac{3}{2^{\frac{2}{3}}} \frac{\tau^{\frac{4}{3}} \mathrm{Re}_D^{\frac{1}{3}} f^{\frac{1}{3}} \left(\frac{R}{cp}\right)^{\frac{1}{3}}}{st^{\frac{2}{3}} V_*^{\frac{1}{3}}}$ and Re_D are fixed, the one-side entropy generation number can be minimized by properly selecting G_*:

$$G_{*,opt} = \left\{ \frac{\tau^2}{3A_*^2 \left(\frac{R}{c_p}\right) f\ st} \right\}^{0.25} \tag{18}$$

$$N_{s,\min} = \frac{4}{3^{\frac{3}{4}}} \frac{\tau^{\frac{3}{2}} \left(\frac{R}{C_p}\right)^{\frac{1}{4}} f^{\frac{1}{4}}}{A_*^{\frac{1}{2}} St^{\frac{3}{4}}} \tag{19}$$

These equations demonstrate that low entropy generation levels are possible only when sufficient heat transfer area is available.

Besides, the constant volume constraint governs the design of the heat exchanger for application in which space is on expensive commodity. The volume on one side of the surface is:

$$V = LA_c \tag{20-a}$$

$$V_* = \frac{4L}{D} \mathrm{Re}_D G_*^{-2} \tag{20-b}$$

So, when V_* and Re_D are specified, the minimum irreversibility design is described by:

$$G_{*,opt} = \left\{ \frac{\tau^2 \mathrm{Re}_D^2}{2f\ st \left(\frac{R}{CP}\right) V_*^2} \right\}^{\frac{1}{6}} \tag{21}$$

$$N_{s,\min} = \frac{3}{2^{\frac{2}{3}}} \frac{\tau^{\frac{4}{3}} \mathrm{Re}_D^{\frac{1}{3}} f^{\frac{1}{3}} \left(\frac{R}{cp}\right)^{\frac{1}{3}}}{st^{\frac{2}{3}} V_*^{\frac{1}{3}}} \tag{22}$$

CASE STUDY

To clarify the application of the mentioned optimization procedure, a case study taken from Oil Laboratory in Iran (Mechanical Research Group, 2006 - Report of "Design and Modeling of microturbine heat cycle" in Oil Laboratory of Iran) is considered. A gas–to–air counter flow heat- exchanger having heat duty of 100 kw is needed to be designed for the minimum surface area and volume separately. Gas and air inlet temperatures are 865 k° and 475 k° where gas and air mass flow rates are 1.4676 kg/s and 1.45 kg/s respectively. Pressure drops are set to be limited to 6.36 and 12 kpa at hot and cold side. The heat exchanger material is aluminum with density of 2700 $\frac{kg}{m^3}$. Table 3 presents the operating conditions used in thermal design. Also, louvered fin surface is used on the gas and air sides, where geometric properties of them are according to Table 4.

RESULTS AND DISCUSSION

For the prescribed heat duty and allowable pressure drop, the optimization problem is finding the design variables that minimize the entropy generation based on constant volume and constant surface area. In others word, after mechanical and thermal designing a heat

Table 3. Operating conditions.

Variable	Data
Allowable pressure drop for hot side (%)	6
Allowable pressure drop for hot side (%)	3
Outlet gas temperature(k)	694
Inlet gas temperature(k)	865
Outlet air temperature(k)	670
Inlet air temperature(k)	475
Outlet pressure from compressor(bar)	4
Outlet pressure from turbine(bar)	1.06
Gas mass flow rate (kg/s)	1.4676
Air mass flow rate(kg/s)	1.45

Table 4. Geometric Parameters of Louvered fins.

Fins	$b(\text{mm})$	$D_h(\text{mm})$	$\beta\left(\frac{m^2}{m^3}\right)$	$\frac{S_f}{S}$	$\delta(\text{mm})$
$\frac{1}{2}a-6.06$	6.35	4.453	840	0.640	0.152
$\frac{3}{8}-8.7$	6.35	3.650	1007	0.705	0.152

Table 5. Information about choosing a fin.

Goals	Parameters
volume	1) β 2) b
pressure drop	D_h
heat exchanger efficiency	1) $\frac{S_f}{S}$ 2) D_h
annually overall cos ts	1) D_h 2) β

exchanger, the thermodynamical optimization by modifying mass and air flow rates would be able to improve the heat exchanger performance. However, before doing this, it is necessary computing fluid properties based on average temperatures (Table 2).

When there are some limitations on surface area, the thermodynamical optimization will be doing based on constant surface area. In other words, the width and depth of heat exchanger are specified. Here, the height and fin geometric properties are the only variables. So, pursuant to objective function that can be minimum volume, costs or even weight of the heat exchanger, the height and fin geometric properties can change. To clarify this procedure it is assumed that the objective function be minimum volume, by using Particle Swarm Optimization algorithm (PSO) the total volume of heat exchanger will be optimized. But before optimization process by PSO algorithm, it is essential to select a suitable louvered fin. Table 5 gives useful information about choosing a fin.

As can be seen in this table β and b, which are total transfer area / volume between plates and the height of fin respectively, have direct impact on total volume. Hence, the heat exchanger with smaller b and larger β has smaller volume. On the other hand, when hydraulic diameter and height fin are small there is a logarithmic increase in pressure drop values.

Kays and London (1984) offer good diagrams and tables for different fins. Particle Swarm Algorithm is a heuristic optimization method which was first introduced by Kennedy and Ebehart (1995). In the basic PSO algorithm, particle swarm consists of "n" particles and the position of each particle stands for the potential solution in D-dimensional space.

Each particle can be shown by its current speed and position. So, the speed and position of each particle change according to the following equations:

$$v_{i,j}(t+1)=w\,v_{i,j}(t)+c_1 r_1\cdot\left[p_{i,j}(t)-x_{i,j}(t)\right]+c_2 r_2\cdot\left[p_{g,j}(t)-x_{i,j}(t)\right] \quad (23)$$

$$x_{i,j}(t+1)=x_{i,j}(t)+\Delta t\, v_{i,j}(t+1) \quad (23\text{-a})$$

Here $i=1,2,....,N$ demonstrates the particles of swarm; $t=1,2,...,t_{max}$ indicates the iterations; w is considered as inertia weight factor; $v_{i,j}(t+1)$ is defined as the velocity of the i-th particle with respect to the best previous position of the i-th particle to the j-th dimension. In this formula, $p_{g,j}(t)$ is the best previous position among all the particles along the j-th dimension in iterations. Further information about PSO algorithm can be obtained from several books and papers in this area (Bai, 2010). A chaotic quantum-beha particle swarm approach applied to optimization of heat exchangers (2012); Coelho and Alotto, 2008; Mikki and Kishk, 2006; Clerc and Kennedy, 2002; Patel and Rao, 2010; Rao and Patel, 2010)

At this stage, and when mechanical and thermal designing are completed, optimization by PSO algorithm as well as thermodynamical optimization by modifying mass flow rates minimize the entropy generation.

Thermodynamical optimization tends to decrease gas flow arte and increase air flow rate, which cause to occur a dramatic increase in pressure drop, convective coefficient and Reynolds number for cold side. Although after optimization by PSO algorithm (Table 6) and then thermodynamical optimization, pressure drop have increased, rising NTU as well as decreasing 11% of volume compared with an increase of 9% costs due to pressure change is valuable. As can be observed from

Table 6. Optimization results by PSO algorithm.

Variable	**MD**	**PSO**
Fin pitch	343	343
Plate spacing (fin height) , b	6.35 mm	6.35 mm
Flow passage hydraulic diameter , D_h	3.65 mm	3.65 mm
Fin metal thickness , δ	0.152 mm	0.152 mm
Parting sheet thickness , a	0.152 mm	0.152 mm
Total sheet transfer area/volume between plates , β	1007	1007
Total heat transfer area/total volume , α	491.72	491.72
Fin area/total area , $\frac{S_f}{S}$	0.705	0.705
Contraction coefficient , σ	0.448	0.448
Thermal conductivity of fin , K	18.19	18.19
Height of heat exchanger , H	657 mm	579 mm
Width of heat exchanger , W	700 mm	700 mm
Depth of heat exchanger , D	500 mm	500 mm
Reynolds number of warmer fluid , Re_h	936	1218
Reynolds number of cooler fluid , Re_c	1247	1014
Pressure drop for warmer fluid , ΔP_h	1.68Kpa	1.72 Kpa
Pressure drop for cooler fluid , ΔP_c	0.77 Kpa	0.72 Kpa
Heat transfer coefficient for warmer fluid , h_h	187	195
Heat transfer coefficient for cooler fluid , h_c	195	212
Overall heat transfer coefficient , U	63.29	67.07
Number of transfer unit , NTU	3.9	3.72

Table 7. Optimization outcome based on constant surface area.

	$G_{*,opt}$	$N_{s,\min}$	G	$\dot{m}\left(\frac{kg}{s}\right)$
Hot fluid	12×10^{-3}	58×10^{-3}	3.66	0.78
Cold fluid	13×10^{-3}	282×10^{-3}	20.72	3.04

Table 8, PSO algorithm and thermodynamical optimization are complementary stages because modifying the heat exchanger size and fin geometric properties be done by PSO algorithm, and its thermal designing would improve by thermodynamical optimization.

Furthermore, when the volume of the heat exchanger is constant (Table 9) optimal flow rates can upgrade the heat exchanger performance. Here, similarly to constant surface area (Table 7), thermodynamical optimization is the last stage.

CONCLUSION

Mass flow rate has direct effect on the heat exchanger function, and determining optimal mass flow rate not only would optimize operating costs but also can improve the heat exchanger efficiency and the number of transfer unit. In actual application, and when surface area and volume of the heat exchanger is constant, thermodynamical optimization can be employed. Also, when the heat exchanger has been designed, thermodynamical optimization would determine optimal mass flow rates based on minimum entropy generation number (Tables 10, 11 and 12; Figure 3).

Nomenclature: $A_\circ$ surface area$\left(m^2\right)$; A_* dimensionless form of heat transfer area; b plate width; C_p specific heat $\left(\frac{J}{kg.k}\right)$; D_h hydraulic diameter(mm); E_C efficiency; f friction factor; F correction factor; V_* dimensionless form of mass velocity; A_* dimensionless

Table 8. Comparison among the results of manual designing, PSO algorithm and thermodynamical optimization.

Variable	MD	PSO	TO
Fin pitch	343	343	343
Plate spacing (fin height) , b	6.35 mm	6.35 mm	6.35 mm
Flow passage hydraulic diameter , D_h	3.65 mm	3.65 mm	3.65 mm
Fin metal thickness , δ	0.152 mm	0.152 mm	0.152 mm
Parting sheet thickness , a	0.152 mm	0.152 mm	0.152 mm
Total sheet transfer area/volume between plates , β	1007	1007	1007
Total heat transfer area/total volume , α	491.72	491.72	491.72
Fin area/total area , $\frac{S_f}{S}$	0.705	0.705	0.705
Contraction coefficient , σ	0.448	0.448	0.448
Thermal conductivity of fin , K	18.19	18.19	18.19
Height of heat exchanger , H	657 mm	579 mm	579 mm
Width of heat exchanger , W	700 mm	700 mm	700 mm
Depth of heat exchanger , D	500 mm	500 mm	500 mm
Reynolds number of warmer fluid , Re_h	936	1218	572
Reynolds number of cooler fluid , Re_c	1247	1014	2130
Pressure drop for warmer fluid , ΔP_h	1.68 Kpa	1.72 Kpa	0.53 Kpa
Pressure drop for cooler fluid , ΔP_c	0.77 Kpa	0.72 Kpa	2.55 Kpa
Heat transfer coefficient for warmer fluid , h_h	187	195	107
Heat transfer coefficient for cooler fluid , h_c	195	212	447
Overall heat transfer coefficient , U	63.29	67.07	67.65
Number of transfer unit , NTU	3.9	3.72	3.76

Table 9. Optimization outcome based on constant volume

Fluid	$G_{*,opt}$	$N_{s,\min}$	G	$\dot{m}\left(\frac{kg}{s}\right)$
Hot fluid	16×10^{-3}	67×10^{-3}	4.76	1.018
Cold fluid	3.53×10^{-3}	244×10^{-3}	5.41	0.795

form of surface area; V_* dimensionless form of volume; h convective coefficient $\left(\frac{W}{m^2.k}\right)$; K thermal conductivity $\left(\frac{W}{m^2.k}\right)$; m mass flow rate $\left(\frac{Kg}{s}\right)$; N heat transfer unit; N_S entropy generation number; T temperature of fluids; R_i ideal gas constant; R heat resistance; Re Reynolds number; $\dot{S}_{gen}$ entropy $\left(\frac{Kg}{m^3}\right)$; μ dynamic viscosity (Pas); Subscripts; 1 input; 2

Table 10. The results of thermodynamical optimization based on constant volume.

Variable	MD
Height of heat exchanger , H	657 mm
Width of heat exchanger , W	700 mm
Depth of heat exchanger , D	500 mm
Reynolds number of warmer fluid , Re_h	743.2
Reynolds number of cooler fluid , Re_c	556
Pressure drop for warmer fluid , ΔP_h	0.925 Kpa
Pressure drop for cooler fluid , ΔP_c	0.309 Kpa
Heat transfer coefficient for warmer fluid , h_h	129.23
Heat transfer coefficient for cooler fluid , h_c	158.74
Overall heat transfer coefficient , U	49.19
Number of transfer unit , NTU	3.105

generation; St Stanton number; U_o overall heat transfer coefficient; Greek symbols; ρ fluid density output; h

Table 11. Optimization outcome based on constant surface area and volume.

Fluid	Re_{opt}	$N_{s,\min}$	G	$\dot{m}\left(\frac{kg}{s}\right)$
Hot fluid	577.34	90×10^{-3}	3.69	0.787
Cold fluid	1996	234×10^{-3}	19.41	2.85

Table 12. The results of thermodynamical optimization based on constant surface area and volume.

Variable	MD
Height of heat exchanger , H	657 mm
Width of heat exchanger , W	700 mm
Depth of heat exchanger , D	500 mm
Reynolds number of warmer fluid , Re_h	577.34
Reynolds number of cooler fluid , Re_c	1996
Pressure drop for warmer fluid , ΔP_h	0.709 Kpa
Pressure drop for cooler fluid , ΔP_c	2.553 Kpa
Heat transfer coefficient for warmer fluid , h_h	115.60
Heat transfer coefficient for cooler fluid , h_c	343.53
Overall heat transfer coefficient , U	57.81
Number of transfer unit , NTU	3.64

warmer fluid; C cooler fluid; MD manual design; PSO optimization by PSO algorithm; TA thermodynamical optimization; CA constant surface area; CV constant volume; CAV constant surface area and volume.

REFERENCES

Achaichia A (1987). The performance of louvered tube-and-plate fin heat transfersurface. PhD thesis, Department of Mechanical and Production Engineering, Brighton Polytechnic.

Achaichia A, Cowell TA (1988). Heat transfer and pressure drop characteristics of flat and louvered plate fin surface. Exp. Therm. Fluid Sci. 1:147-157.

Andrew CL (2000). Spatially Resolved Heat Transfer Studies in compact heat exchanger. Master thesis, Virginia Polytechnic and State University.

Aoki H, Shinagawa T, Suga KK (1989). An experimental study of the local heat transfer characteristics in automotive louvered fins. Exp. Therm. Fluid Sci. 2:293-300.

Bai Q (2010). Analysis of particle Swarm optimization algorithm. College of Computer Science and Technology, Inner Mongolia University to nationalities, China, Tongliao, 028043.

Chang Y, Wang C, Chang W (1994). Heat transfer and flow characteristics of automotive brazed aluminum heat exchangers. ASHRAE Trans. 100(2):643-662.

Chang YJ, Wang CC (1996). Air-side performance of brazed aluminum heat exchangers. J. Enhanced Heat Transf. 3(1):15-28.

Chang YJ, Wang CC (1997). A generalized heat transfer correlation for louvered fin geometry. Int. J. Heat Mass Transf. 40(3):533-544.

Clerc M, Kennedy JF (2002). The particle swarm: Explosion, stability and convergence in a multi-dimensional complex space. IEEE Trans. Evolut. Comput. 6:58-73.

Coelho LS, Alotto P (2008). Global optimization of electromagnetic devices using an exponential quantum-behaved particle swarm optimizer. IEEE Trans. Magnetics 44:1074e1077.

Davenport CJ (1980). Heat Transfer and Fluid Flow in Louvered Triangular Ducts. Ph.D. Thesis, Department of Mechanical Engineering, Lanchester Polytechnic.

Davenport CJ (1983). Correlation for heat transfer and friction characteristics of louvered fin. AICHE Symp. Ser. 79(25):19-27.

Dong JQ, Cheng JP, Chen ZJ (2007). Heat transfer and pressure drop correlations for the multi-louvered fin compact heat exchangers. Energy Convers. Manage. 48:1506-1515.

Hsieh CT, Jang JY (2006). 3-D thermal-hydraulic analysis for louver fin heat exchangers with variable louver angle. Appl. Therm. Eng. 26:1629-1639.

Huihua Z, Xuesheung L (1989). The experimental investigation of oblique angles and interrupted plate lengths for louvered fins in compact heat exchangers. Exp. Therm. Fluid Sci. 2:100-106.

Kays WM, London AL (1984). Compact heat exchangers. McGraw Hill, Book Co, New York.

Kennedy JF, Eberhart RC (1995). Particle Swarm Optimization. Proc. IEEE Int. Conf. Neural Netw., Perth, Australia pp. 1942-1948.

Kim MH, Bullard CW (2002). Air-side thermal hydraulic performance of multi-louvered fin aluminum heat exchangers. Int. J. Refrigeration 25:390-400.

Kuppan T (2000). Heat Exchanger Design Handbook. New York ISBN:0-8247-9787-6.

Leu JS, Liu MS, Liaw JS, Wang CC (2001). A numerical investigation of louvered fin-and-tube heat exchangers having circular and oval tube configurations. Int. J. Heat Mass Transf. 44:4235-4243.

Lyman AC, Stephan RA, Thole KA, Zhang LW, Memory SB (2002). Scaling of heat transfer coefficients along louvered fins. Exp. Therm. Fluid Sci. 26:547-563.

Mechanical Research Group (2006). Report of "design and Modeling of microturbine heat cycle" in Oil Laboratory of Iran. March.

Mikki SM, Kishk AA (2006). Quantum particle swarm optimization for electro-magnetics. IEEE Trans. Antennas Propag. 54:2764-2775.

Patel VK, Rao RV (2010). Design optimization of shell-and-tube heat exchanger using particle swarm optimization technique. Appl. Therm. Eng. 30:1417-1425.

Qi ZG, Chen JP, Chen ZJ (2007). Parametric study on the performance of a heat exchanger with corrugated louvered fins. Appl. Therm. Eng. 27:539-544.

Rao RV, Patel VK (2010). Thermodynamic optimization of cross flow plate-fin heat exchanger, using a particle swarm optimization algorithm. Int. J. Therm. Sci. 49:1712-1721.

Rugh JP, Pearson JT, Ramadhyani S (1992). A study of a very compact heat exchanger used for passenger compartment heating in automobiles. ASME Symp. Ser. HTD. 201:15-24.

Sahnoun A, Webb RL (1992). Prediction of heat transfer and friction for the louver fin geometry. Int. J. Heat Mass Transf. 114:893-900.

Shah RK (1981). Compact heat exchanger design procedures. In: S. Kakac, A.E. Bergles, F. Mayinger, Heat exchangers: Thermal-hydraulic fundamentals and design pp. 495-536.

Springer ME, Thole KA (1998). Experimental design for flow field studies of louvered fins. Exp. Therm. Fluid Sci. 18:258-269.

Sunden B, Svantesson J (1992). Correlation of j- and f factors for multi louvered heat transfer surfaces. In: Proc. 3rd UK Natl. Conf. Heat Transf. pp. 805-811.

Tafti DK, Zhang LW, Wang G (1999). Time-dependent calculation procedure for fully developed and developing flow and heat transfer in louvered fin geometries. Numer. Heat Transf. Part A 35:225-249.

Wu XM, Webb RL (2002). Thermal and hydraulic analysis of a brazed aluminium evaporator. Appl. Therm. Eng. 22:1369-1390.

Yousef M, Darus AN, Mohammadi H (2012). An imperialist competitive algorithm for optimal design of plate- fin heat exchangers. 55:3178-3185.

Yousefi M, Darus AN (2011). Optimal design of plate-fin heat exchangers by particle swarm optimization. In: Fourth International Conference on Machine Vision (ICMV 2011): Computer Vision and Image Analysis; Pattern Recognition and Basic Technologies.

Yousefi M, Enayatifar R, Darus AN (2012). Optimal design of plate-fin heat exchangers by a hybrid evolutionary algorithm. Int. Commun. Heat Mass Transf. 39(2):258-263.

Zhang X, Tafti DK (2003). Flow efficiency in multi-louvered fins. Int. J. Heat Mass Transf. 46:1137-1150.

A new simulator for surfactant flooding

Yongwei Li

E&P Institute, Daqing Oil Company, Heilongjiang, China 163318.

Surfactant as a successful enhanced oil recovery (EOR) agent has been widely used in many mature reservoirs. This research focuses on the description of surfactant solution at low permeability condition. A new three-dimensional, two-phase, three-component surfactant simulator is presented. The simulator is based on the non-Darcy flow characteristics of surfactant flooding in the low permeability formations. The change of threshold pressure and influences of surfactant on convection, diffusion, adsorption, and retention, are all considered. A new equation for the calculation of surfactant adsorption is employed, which can significantly promote the matching degree between the mathematical model and field practice. This mathematical model was adopted to conduct a field scale simulation modeling for a surfactant flooding pilot of the Chao-522 field, Daqing.

Key words: Surfactant flooding, simulation, threshold pressure, pilot test.

INTRODUCTION

Oil-production from enhanced oil recovery projects continues to supply an increasing percentage of the world's oil. Taber et al. (1997) estimated that more and more of the worldwide oil production would come from EOR. That number has continued to increase and in the future it is expected that EOR will eventually produce the majority of the world's oil. In China, the use of polymer in Daqing has already obtained a great success. Surfactant functions work by adding certain concentrations of surfactants to injection water to reduce the interfacial tension (IFT) between displacing and displaced phases (Hornof et al., 1983; Gao et al., 2010). Flow of the surfactant formulation through the reservoir allows the trapped oil droplets to deform and become mobile oil. These oil droplets then coalesce and form a new flowing oil bank.In the process of surfactant flooding, the surfactant adsorbs onto the oil-water interface and surface of rock which may also make a wettability change of rock (Kang and Liu 1996). The experiment shows that the oil drops are becoming easier to deform when the oil-water interfacial tension reduces, so the resistant force lowers when the oil drops flow through the pore throat. Increase of dispersion of crude oil in the water. With the decrease of IFT, the crude oil can disperse in the surfactant solution, meantime, the surface of oil drops are charged after adsorption, so the oil drops are not easy to stick onto the surface of rock particles.

Low permeability reservoirs account for a considerable reserve and play an important role in the stable output in Daqing. The pilot test in Daqing shows surfactant flood can be a good candidate of the enhanced oil recovery for the low permeability reservoirs. The indoor experiments show that the surfactant flooding can lower the threshold pressure and increase the oil recovery efficiency of low permeability oilfield (Liu et al., 1987). Several pilot tests of surfactant flooding were carried out in Daqing's low permeability oilfields, such as Yushulin and Chaoyanggou oilfields; the objectives of pilot tests are to reduce the injection pressure, to increase the injection rate, and to enhance the oil recovery (Feng et al., 2010; Zhao et al., 2010, 2011; Ge and Ghassemi, 2007 and 2008). The experimental screening of surfactant was finished, but the theoretical study on surfactant flooding in low permeability reservoirs is limited (Sun and Li, 1996);

The reservoir simulation software that includes the threshold pressure is not reported. Because of the existence of threshold pressure, current commercial simulators cannot exactly reflect the flow of surfactant solution in the low-permformations. In order to overcome this obstacle, the investigation of a simulator for surfactant floods with threshold pressure was conducted.

Because the cost of surfactant is comparatively high, the amount of surfactant used is one of biggest concerns and should be first determined for the field application to obtain the maximum economic benefit. Thus, it is very important to conduct reservoir numerical simulation study. On the basis of compositional model, a mathematical model of surfactant flooding was established, in which the changes of threshold pressure and relative permeability which are caused by surfactant flooding (Yin et al., 2010), and influence of adsorption and retention of surfactant in the reservoir are included in the model. This model was used to optimize the injection plan for pilot test of surfactant flooding in Chaoyanggou oilfield of Daqing to provide a theoretical basis of decision for the development of oilfield.

MODELING

Mathematical Model

Feature and assumption

(a) Two phases, three components (oil, water, and surfactant);
(b) Non-Darcy flow of oil and water;
(c) The reservoir rock and fluids compressible;
(d) Anisotropic and heterogeneous reservoir;
(e) Description of threshold pressure change;
(f) Consideration of convection, diffusion and adsorption on rock;
(g) Influences of capillary force and gravity effect;
(h) No salinity considered (Low salinity in Chaoyanggou reservoirs).

Continuity equations

The flow equation of two-phase, three-component is used to describe the process of surfactant flooding. The downward direction is positive direction of z-axis, then the mathematical model of each component is as follows:

$$\nabla\cdot[\frac{KK_{ro}\rho_o}{\mu_o}C_{o1}(\nabla p_o-\rho_o g\nabla z\pm D_p\vec{e})+\frac{KK_w\rho_w}{\mu_w}C_{w1}(\nabla p_w-\rho_w g\nabla z\pm D_p\vec{e})]=\frac{\partial}{\partial t}\phi(\rho_o s_o C_{o1}+\rho_w s_w C_{w1})-\rho_o q_o C_{o1}-\rho_w q_w C_{w1} \tag{1}$$

$$\nabla\cdot[\frac{KK_{ro}\rho_o}{\mu_o}C_{o2}(\nabla p_o-\rho_o g\nabla z\pm D_p\vec{e})+\frac{KK_{rw}\rho_w}{\mu_w}C_{w2}(\nabla p_w-\rho_w g\nabla z\pm D_p\vec{e})]=\frac{\partial}{\partial t}\phi(\rho_o s_o C_{o2}+\rho_w s_w C_{w2})-\rho_o q_o C_{o2}-\rho_w q_w C_{w2} \tag{2}$$

$$\nabla\cdot\phi[D_{o3}s_o\nabla(\rho_3 C_{o3})+D_{w3}s_w\nabla(\rho_3 C_{w3})]+\nabla\cdot K[\frac{K_{ro}\rho_3}{\mu_o}C_{o3}(\nabla p_o-\rho_o g\nabla z\pm D_p\vec{e})+\frac{K_{rw}\rho_3}{\mu_w}C_{w3}(\nabla p_w-\rho_w g\nabla z\pm D_p\vec{e})]=\frac{\partial}{\partial t}(\phi\rho_3 s_o C_{o3}+\rho_3 a_{o3}+\phi\rho_3 s_w C_{w3}+\rho_3 a_{w3})-\rho_3 q_o C_{o3}-\rho_3 q_w C_{w3} \tag{3}$$

$$p_c=p_o-p_w \tag{4}$$

$$s_o+s_w=1 \tag{5}$$

$$M=C_{o3}/C_{w3} \tag{6}$$

Where subscript 1 denotes the oil component, subscript 2 stands for the water, and sub 3 represents the surfactant.

For the process of surfactant flooding, it is assumed that the volume percentage of water component in oleic phase, C_{o2}, is zero, and the volume percentage of oil component in aqueous phase, C_{w1}, is zero too, and $C_{o1}+C_{o3}=1$, $C_{w2}+C_{w3}=1$. Then the above model can be simplified, and the block-center difference can be used to solve equation.

Numerical solution of mathematical model

Differential equations

$$\Delta\cdot[\frac{KK_{ro}\rho_o}{\mu_o}C_{o1}(\Delta p_o-\rho_o g\vec{k}\pm D_p\vec{e})=\Delta_t\phi(\rho_o s_o C_{o1})-\rho_o q_o C_{o1} \tag{7}$$

$$\Delta\cdot[\frac{KK_{rw}\rho_w}{\mu_w}C_{w2}(\Delta p_w-\rho_w g\vec{k}\pm D_p\vec{e})]=\Delta_t\phi(\rho_w s_w C_{w2})-\rho_w q_w C_{w2} \tag{8}$$

$$\Delta\cdot\phi[D_{o3}s_o\Delta(\rho_3 MC_{w3})+D_{w3}s_w\Delta(\rho_3 C_{w3})]+\Delta\cdot K[\frac{K_{ro}\rho_3}{\mu_o}MC_{w3}(\Delta p_o-\rho_o g\vec{k}\pm D_p\vec{e})+\frac{K_{rw}\rho_3}{\mu_w}C_{w3}(\Delta p_w-\rho_w g\vec{k}\pm D_p\vec{e})]=\Delta_t(\phi\rho_3 s_o MC_{w3}+\rho_3 a_{o3}+\phi\rho_3 s_w C_{w3}+\rho_3 a_{w3})-\rho_3 q_o MC_{w3}-\rho_3 q_w C_{w3} \tag{9}$$

Where Δ is spatial difference operator,

$$\Delta\varphi = \frac{\varphi_{i+\frac{1}{2}} - \varphi_{i-\frac{1}{2}}}{\Delta x_i}\vec{i} + \frac{\varphi_{j+\frac{1}{2}} - \varphi_{j-\frac{1}{2}}}{\Delta y_j}\vec{j} + \frac{\varphi_{k+\frac{1}{2}} - \varphi_{k-\frac{1}{2}}}{\Delta z_k}\vec{k}$$

Δ_t is time difference operator, $\Delta_t\varphi = \frac{\varphi^{n+1}_{i,j,k} - \varphi^{n}_{i,j,k}}{\Delta t^n}$; $\vec{e}$ is unit vector, $\vec{e} = \vec{i} + \vec{j} + \vec{k}$.

The choice of positive or negative sign for threshold pressure is determined by the signs of $\Delta_l p^n$ (l=x, y, z), the signs between them are always opposite to guarantee that the threshold pressure gradient is always the resistance force.

Model solution

The method of implicit pressure, explicit saturation, andimplicit concentration is used to solve equation, the steps are as follows: Difference Equations (7), (8), (4) and (5) are used to implicitly calculate pressure; The values of pressure are substituted into difference equation (8) to explicitly calculate values of water saturation; then substituting pressure value p_o and value of saturation s_w into Equations (6) and (9) to implicitly compute values of concentration C_{o3}, C_{w3}.

Adsorption treatment

The isothermal equation of Langmuir adsorptionis used to describe the adsorption process. The effects of salinity, concentration of surfactant and rock permeability on adsorption are considered. The isothermal equation is (Liao et al., 1999):

$$C_r = bC_{rm}C/(1+bC) \quad (10)$$

The concentration C in isothermal equation of Langmuir adsorption is the average concentration of surfactant in different phases of liquids in pores. This will cause major error in calculation. Because the saturations of oleic, and aqueous phases are always changing in the process of surfactant flooding, the concentration of surfactant in each phase is different and changes with the time. Therefore, in this study, the adsorption from oil phase and adsorption from water phase are calculated separately. A function $B_i(s_i)$ is used to adjust the adsorption calculation due to the partial contact between solid particles and phase. The calculation equation of surfactant adsorption in oleic phase and aqueous phase are:

$$a_{o3} = B_o(s_o)C_{or}(C_{o3}) \quad (11)$$

$$a_{w3} = B_w(s_w)C_{wr}(C_{w3}) \quad (12)$$

where: b_o, C_{orm}, b_w, C_{wrm} and dimensionless function $B_i(s_i)$ are determined by experiments, and they meet the following conditions: $B_i(0) = 0$, $B_i(1) = 1$.

Relative permeability curve

The mechanism of surfactant flooding is to reduce interfacial tension (IFT) between oil phase and water phase. Macroscopically, this mechanism reflects that relative permeability curve of oleic phase is moved up. Therefore, correct representation of relationship between the oleic and aqueous phases' relative permeability and surfactant concentrations is very important for accurate simulation of surfactant flooding. In the process of simulator development, two methods to determine the relative permeability of oil and water phases at different surfactant concentrations are provided. The first is obtained through the interpolation of relative permeability curves which are measured at the various surfactant concentrations. The second method is the theoretical calculation equation. The interpolation method is based on the experimental data, and can represent the real situation more accurately. Therefore, this method is recommended. However the interpolation method also needs large amount of experimental data. When the enough experimental data are not available, the theoretical formula can be used. The following is the introduction of interpolation method.

Inputting N different relative permeability curves determined at different surfactant concentrations into the simulator, assuming that the surfactant concentration corresponding to i^{th} relative permeability curve is C_{w3i}. The curve is treated with saturation normalization:

$$s_{oNi} = \frac{s_o - s_{ori}}{1 - s_{wc} - s_{ori}} \quad (13)$$

When the surfactant concentration in some place of oil reservoir is C_{w3x}, which is between C_{w3i} and C_{w3i+1}, the interpolation method is applied to calculate relative permeability at C_{w3x}. First, the residual oil saturation s_{orx} corresponding to this concentration is calculated:

$$s_{orx} = s_{ori} + \frac{C_{w3x} - C_{w3i}}{C_{w3i+1} - C_{w3i}}(s_{ori+1} - s_{ori}) \quad (14)$$

Table 1. Comparison of water injection history for Chao-522.

Year	Major oil layers		Non-major layers		Non-connected layers		Total	
	Permeability of water absorbing thickness (%)	Permeability of water absorbing layers (%)	Permeability of water absorbing thickness (%)	Permeability of water absorbing layers (%)	Permeability of water absorbing thickness (%)	Permeability of water absorbing layers (%)	Permeability of water absorbing thickness (%)	Permeability of water absorbing layers (%)
1995	88.5	82.2	91.0	85.7	48.1	50.0	87.8	81.4
1996	90.1	87.1	59.7	61.1	52.9	33.3	82.2	76.6
1997	81.7	78.5	61.1	52.1	85.1	75.0	76.0	69.5
1998	74.6	75.2	55.9	47.3			67.1	61.1
1999	70.1	66.6	60.3	56.5	45.2	40.0	63.8	61.4
2000	55.7	52.6	64.1	57.1	40.9	30.0	59.3	54.6
2001	62.6	58.5	61.3	58.5	56.6	50.0	60.9	57.7

Table 2. Classification of water injection for Chao-522.

Water injection (m^3/d)	No. of wells	Effective thickness (m)	Thickness of connected layers (m)	Fracturing pressure (MPa)	Injection pressure (MPa)	Allocated injection (m^3)	Actual injection (m^3)	Water injection intensity (m^3/d·m)
<10	8	8.5	6.5	14.1	14.6	110	60	0.88
10-20	15	9.2	7.6	14.1	14.2	310	267	1.93
20-30	14	9.3	7.1	13.9	13.5	420	401	3.08
>30	9	10.1	7.8	14.0	13.9	480	442	4.86
total	46	9.3	7.3	14.0	14.0	1320	1170	2.73

The value of s_{oNx} is used to consult the ith and (i+1)th relative permeability curves to obtain the relative permeability of oleic phase, K_{roxi} and K_{roxi+1}, the oil phase relative permeability corresponding to concentration C_{w3x} and saturation s_o is:

$$K_{rox} = K_{roxi} + \frac{C_{w3x} - C_{w3i}}{C_{w3i+1} - C_{w3i}}(K_{roxi+1} - K_{roxi}) \quad (15)$$

The interpolation method of relative permeability of water phase is the same as the oil phase.

Integrated study in Chao-522

The pilot is located in the Chao 82 - 152 well block. There are 4 injecting and 10 production wells in the pilot area. The average permeability is 18.5 × 10^{-3} μm^2. The geological reserve is 24.6 × 10^4 t. The comprehensive water cut is 82.9% at the end of history match. The average daily oil production rate per well is 2.95/day. Four injection wells startedinjecting water since January of 2002. In January of 2005, just before the start of pilot test, the designed water injection rate was 122.5 m^3/day; the actual water injection was 98.3 m^3/day. The designed injection rate was still not reached even the bottom hole flowing pressure of water injection wells was up to 23.4 MPa which was close to the fracturing pressure (Tables 1 and 2).The shortage of injection made serious problems on the pressure maintenance and stable oil output. Thus, this block was selected as the pilot test of surfactant flooding, aiming to reduce the injection pressure and to increase injection.

Core flooding tests

In order to fully understand the non-Darcy flow behaviors of surfactant solution in low permeability reservoirs, the core flooding tests of surfactant injection on natural cores were conducted. The chosen displacing-liquid was solution of nonionic alkanol acid amide surfactant and auxiliary agent, which aimed to reduce the interfacial tension between crude oil and water of Chaoyanggou oilfield to reach ultra low interfacial tension (IFT). The experimental results are shown in Table 1. As shown in Table 3, after injecting the displacing liquid of surfactant, the pressure of chase water injection was reduced by 40% compared with that of fresh water injection before

Table 3. Experimental results of core flooding.

Core No.	Ø (%)	K ($10^{-3}\mu m^2$)	S_{oi} (%)	Injection rate (%)	Injected PV	Initial injection pressure (MPa)	Final injection pressure (MPa)	Decrease of pressure (%)	Increase of recovery (%)
1	20.0	21.6	64.1	0.5	2.8	2.2	0.91	53.6	5.4
2	17.1	15.7	66.2	0.5	3.5	2.6	1.28	51.5	5.3
3	14.2	9.8	59.5	0.5	4.1	2.1	1.20	42.8	4.8
4	12.5	4.3	63.6	0.5	2.7	2.7	1.55	42.6	5.0
5	10.7	2.1	58.6	0.5	3.9	2.7	1.60	40.7	4.6

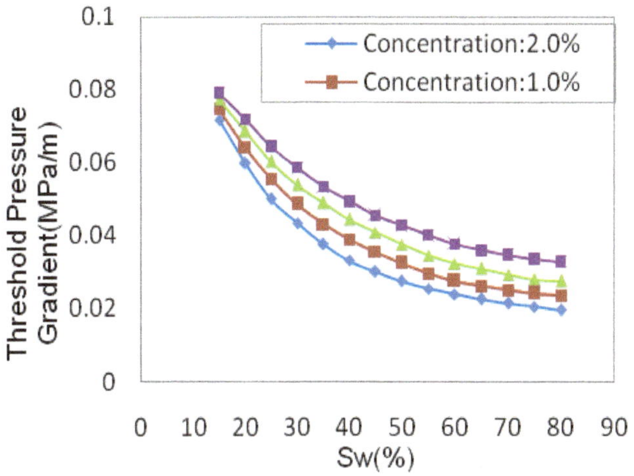

Figure 1. Relation between threshold pressure and surfactant.

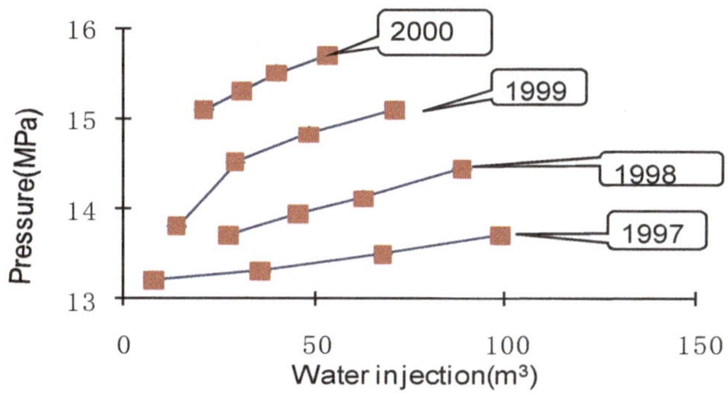

Figure 2. Water injection curve for Well Chao 84-144 concentration, water saturation.

the surfactant slug. Thus the injection pressure can be significantly dropped by the participation of surfactant, compared with secondary water injection.The results also showed the recovery efficiency was enhanced by 5.12%. In order to further estimate the effects of surfactant concentration on the threshold pressure gradient, the relationship between surfactant concentration, water saturation and threshold pressure gradient were measured, shown in Figure 1. From Figure 1, we can find that, with the increase of surfactant concentration and water saturation, the threshold pressure gradient reduces gradually, but the decrease extent becomes smaller.

Simulation study

A simulator was developed based on the above theoretical model, and then it is applied to pilot test of surfactant flooding in Chao 522 Block of Chaoyanggou low permeability oilfield. Before numerical simulation of surfactant injection, the history match of water flooding was first conducted. The calculated water saturation and pressure field are initial values for the surfactant flooding research; the predicted ultimate recovery efficiency of conventional water flooding is 32.34% of OOIP.

The pilot test is located in the Chao 82 - 152 well block; there are four injecting wells and 10 production wells, the average permeability is 18.5 × 10^{-3} μm^2, the geological reserve is 24.6 × 104 t, the comprehensive water cut is 82.9% and the average daily oil production rate per well is 2.95/day. Four injection wells were put into production in January of 2002, as of January of 2005, just before the start of pilot test, the allocated water injection rate is 122.5 m3/day, the actual water injection is 98.3 m3/day, the bottom hole flowing pressure of water injection wells is 23.4 MPa, which is close to the fracturing pressure, but the allocated injection rate cannot be achieved (Figure 2), therefore, this block was selected as the pilot test of surfactant flooding, aiming to reduce the injection pressure and to increase injection.

Determination of surfactant concentration

The calculation results of development indexes and economic indexes of surfactant injection schemes are shown in Table 4. With the increase of surfactant concentration, the bottom hole flowing pressure of injection wells decreases, the water injection rate increases, the recovery efficiency increases, but the extent of increase becomes smaller. Because the production cost increases due to the increase of amount of surfactant used, based on the economic evaluation, the reasonable slug concentration is 1.0%.

Table 4. Net benefit of different schemes in condition of different slug concentration.

Scheme	Concentration (%)	Slug volume (PV)	Cost increase (10^4 RMB)	Injection increase (10^4m^3)	Recovery Increase (%)	Cumulative oil increase (10^4t)	Revenue increase (10^4 RMB)	Net benefit increase (10^4RMB)
1	0.5	0.02	16.9	0.974	0.88	0.216	245.1	228.2
2	1.0	0.02	33.8	1.331	1.37	0.336	381.6	347.8
3	1.5	0.02	50.7	1.428	1.42	0.348	395.5	344.7
4	2.0	0.02	67.6	1.504	1.47	0.360	409.5	341.8

Table 5. Development indexes and economic indexes of schemes at the different slug volumes.

Scheme	Concentration (%)	Slug volume (PV)	Cost increase (10^4 RMB)	Injection increase (10^4m^3)	Recovery Increase (%)	Cumulative oil increase (10^4t)	Revenue increase (10^4 RMB)	Net benefit increase (10^4 RMB)
5	1.0	0.05	84.6	1.822	1.94	0.476	540.3	455.7
6	1.0	0.10	169.2	2.716	3.17	0.778	882.8	713.6
7	1.0	0.20	338.4	3.451	3.69	0.905	1027.6	689.2
8	1.0	0.40	676.8	3.874	4.05	1.017	1126.9	450.1

Table 6. Net benefit of schemes with different slugs.

Scheme	Concentration (%)	Slug volume (PV)	Slug mode	Cost increase (10^4RMB)	Recovery increase (%)	Cumulative oil increase (10^4t)	Revenue increase (10^4RMB)	Net benefit increase (10^4RMB)
9	1.0	0.10	2	172.5	3.23	0.793	899.5	727.8
10	1.0	0.10	3	179.0	3.27	0.802	911.8	732.4
11	1.0	0.10	4	188.0	3.30	0.810	919.0	631.0

Optimization of surfactant slug size

As the surfactant concentration is kept at 1.0% of Scheme 2, the slug volumes of Plans 5~8 are 0.05, 0.10, 0.20 and 0.30 PV, respectively, the calculation results of development indexes and economic indexes are summarized in Table 5. As shown in Table 5, with the increase of slug volume, recovery efficiency also increases, but the increase becomes smaller, and the cost goes up. Through the economic evaluation, the reasonable slug volume is 0.10 PV of Scheme 6, the net benefit is the largest.

Injection process

Scheme 8 and 9 are based on the Scheme 6, the injection slugs are 2, 3 and 4, respectively, the calculation results are shown in Table 6. The results show that the effect of multiple slugs is better than that of one slug injection, this is because the multi-slug injection mode can prolong the displacement time of surfactant, reducing the inefficient flow of surfactant in the reservoir. Comparing multi-slug injection mode with the one slug injection mode, the former has higher recovery efficiency, this is because the multi-slug injection mode can prolong the displacement time of surfactant, reducing the ineffective flow of surfactant in the reservoir. But the multiple slugs injection mode has complicated operation process, which increases the operation cost, considering that there is no obvious difference in economic benefit between Scheme 9 and Scheme 10, in order to have an easier field operation, the Scheme 9 is chosen.

Pilot testing

Based on the results of the field scale simulation, the pilot test was performed. According to the optimized surfactant injection process, the most favorable surfactant concentration was 1.0%. Volume of each and every slug was 0.10 PV. The injection procedure in the pilot testing can be seen as follow:

(a) Main surfactant slug
(b) Water slug
(c) Supplementary surfactant slug
(d) Chase water

Results comparison between simulation and pilot

The injection of surfactant solution lowered the threshold pressure and increased the water injectivity. The pilot test of surfactant flooding started in January of 2005, the effect of surfactant flooding was seen in injectors after one month of surfactant injection, the average bottom hole flowing pressure of injectors was decreased to 22.9 MPa, and the daily water injection rate increased to 124.8 m^3/day, the allocated injection rate was met, the extent of increase of water injection was 27.3%, so the effect of pressure-reducing and increase of injection rate was significantly obvious. For June of 2006 to June of 2007, the predicted value of average bottom hole flowing pressure was 22.3 MPa, the actual value is 22.7 MPa, the relative error was 2.3%; the predicted average daily water injection rate was 122.5 m^3/day, the actual value was 128.9 m^3/day, the relative error was 5.2%. The comparison of water injection profile before and after surfactant for well Chao 82 - 152 showed that the water absorbing thickness was increased by 2 m, the daily water injection increased from 14 to 22 m^3 which was increased by 57.1%. Especially, the water intake of FI 72 layers increased from 0 to 2 m^3 (Figure 3).

The displacement efficiency was improved, especially in low permeability zones. In the condition of 300 m well pattern, the displacement pressure difference was 0.073 MPa/m, the zones, which had higher threshold pressure than 0.073 MPa/m could not be displaced in the condition of conventional water flood. On average, the 18 zones were drilled through by wells, 11 zones could be displaced by conventional water flood, accounting for 60%. The simulation predicted results showed that 14 zones can be displaced by surfactant flooding; actually, 15 pay zones were displaced, increased by 22%, indicating that water was injected into some low permeable zones after surfactant flooding because of decrease of threshold pressure.

The prediction showed significant decline of water cut and obvious increase of oil production. At the end of the injection of first slug, the predicted water cut was 75.5%, the actual water cut was 74.2%, the absolute error was 1.3%, the water cut was decreased by 8.7% over the pre-pilot test; the simulation predicted daily oil production was 3.72 t/day, the actual value was 3.89 t/day, and the relative error was 4.2%, the increase of daily oil production was 0.94 t/day compared with that of pre-pilot test, the average increase of oil for single well was 31.5%, and the cumulative increase of oil is 3479.0 t.

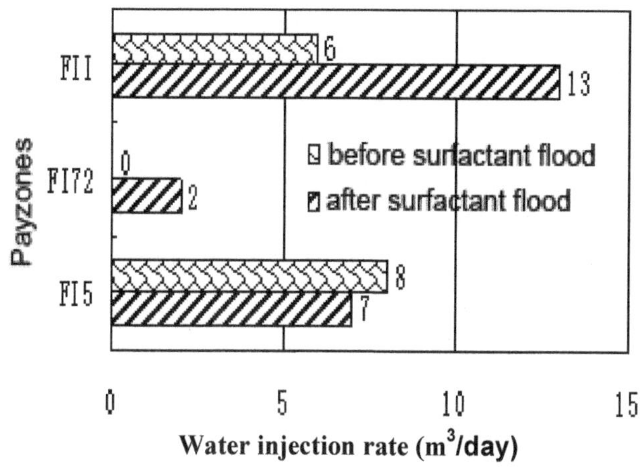

Figure 3. Comparison of water injection for different payzones for well Chao 82 - 152.

It was assumed that price of crude oil was 1450.0 RMB/t. The operation cost was 550.0 RMB /t. Deducting 86.0 × 10^4 RMB of investment, as of December 2006, the net economic benefit was 261.9 × 10^4 RMB, the ratio of cost to revenue was 1/3, so the economic benefit was significant.

The comparison between simulation results and pilot test show that the calculation error of main development indexes is within 6%, meeting the requirement of engineering calculation. The mathematical model presented in this paper includes the threshold pressure gradient, and can be used to simulate such characteristics of surfactant flooding in low permeability reservoir as effect of pressure-reducing and injection-increase, and increase of displaced payzones; the conventional chemical flooding simulators do not have this function.

Conclusions

1. The laboratory experiments showed that the surfactant flooding lowered the threshold pressure by more than 40% and increase the oil recovery efficiency of low permeability by 5.0% of OOIP at Chao-522 reservoir conditions;
2. A three-dimensional, two-phase, three-component mathematical model for surfactant flooding is established, which is able to reflect the changes of threshold pressure, influence of convection, diffusion and adsorption of surfactant on rock;
3. The method to calculate surfactant adsorption quantity and treatment of relative permeability curve are improved, which increase the matching between simulation results and field application, the calculation error of main development indexes is within 6.0%;
4. The pilot test shows that surfactant flooding can

decrease water injection pressure, increase the water injectivity in low permeable layers, and increase number of displaced zones, increase the oil production rate, the extent of increase reaches 3.0%;

5. This investigation of surfactant flooding simulator is an excellent opportunity to manage the surfactant flood in the Chao-522 field and show that this should be of reference value to improve the surfactant floods in the low permeability reservoirs.

Nomenclature: K, Permeability; K_{ro}, oil phase relative permeability; K_{rg}, gas phase relative permeability; K_{rw}, water phase relative permeability; ϕ, porosity; g, gravity coefficient; μ_{w}, water viscosity; μ_{o}, oil viscosity; μ_{g}, gas viscosity; S_{w}, water saturation; S_{o}, oil saturation; S_{g}, gas saturation p_{o}, oil phase pressure; C, surfactant concentration; Δ_{t}, time difference operator; $\vec{e}$, unit vector; s_{orx}, residual oil saturation; $C_{\mathrm{w}3x}$, surfactant concentration; $C_{\mathrm{w}3i}$, i^{th} relative permeability curve; a,b, adsorption coefficients; C_{r}, adsorption concentration.

REFERENCES

Feng Y, Ji B, Gao P (2010). "An Improved Grey Relation Analysis Method and Its Application in Dynamic Description for a Polymer Flooding Pilot of Xingshugang Field, Daqing." SPE 128510, presented at the North Africa Technical Conference and Exhibition, 14-17 February 2010, Cairo, Egypt

Gao P, Towler B, Zhang X (2010). "Integrated Evaluation of Surfactant-Polymer Floods". SPE-129590, presented at the 2010 SPE EOR Conference at Oil & Gas West Asia, 11-13 April 2010 in Muscat, Oman.

Ge J, Ghassemi A (2007). "Pore Pressure and Stress Distributions around an Injection-Induced Fracture." GRC Annual Conf., Sparks, Nevada.

Ge J, Ghassemi A (2008). "Analysis of Failure Potential around a Hydraulic Fracture in Jointed Rock." Presented in 42nd U.S. Rock Mechanics Symposium, San Francisco, California.

Hornof V, Chaaraoui A (1983). "Viscosity of Surfactant-polymer Solution." SPE 11775; the paper was presented at the International Symposium on Oilfield and Geothermal Chemistry held in Denver Co.

Kang W, Liu Y (1996). "Mechanism of Tertiary Oil Recovery." Beijing, Chemistry Industry Press, pp. 101-122.

Liao GZ, Wang QM, Wang DM (1999). "Principles of Chemical Combination Flooding ant its Application." Beijing, Petroleum Industry Press, pp. 132-166.

Liu Q, Jiang Y (1987). "The Application of Surfactant in Oilfield Development." Beijing, Petroleum Industry Press, pp. 22-37.

Sun KP Li X (1996). "Theory Basis of Reservoir Numerical Simulation." Beijing, Petroleum Industry Press, pp. 198-212.

Yin D, Gao P, Pu H (2010). "Investigation of a New Simulator for Surfactant Floods in Low Permeability Reservoirs and Its Application in Chao-522 Field, 2010 Update." SPE 128645, presented at the 2010 SPE EOR Conference at Oil & Gas West Asia, 11-13 April 2010 in Muscat, Oman.

Zhao X, Wu Y, Baiming Z, yu S, xianbao Z, Gao P (2011). Strategies to Conduct Steam Injection in Waterflooded Light Oil Reservoir and a Case in Fuyu Reservoir, Jilin. In SPE Middle East Oil and Gas Show and Conference.

Zhao X, Zhong H, Gao P, Wu Y, Zhong H, Hasi B, Qian Y (2010). Investigation of Steam Injection at Critical Condition and a Pilot Test in 119-52, Chao-601. In International Oil and Gas Conference and Exhibition in China.

Permissions

All chapters in this book were first published in JPGE, by Academic Journals; hereby published with permission under the Creative Commons Attribution License or equivalent. Every chapter published in this book has been scrutinized by our experts. Their significance has been extensively debated. The topics covered herein carry significant findings which will fuel the growth of the discipline. They may even be implemented as practical applications or may be referred to as a beginning point for another development.

The contributors of this book come from diverse backgrounds, making this book a truly international effort. This book will bring forth new frontiers with its revolutionizing research information and detailed analysis of the nascent developments around the world.

We would like to thank all the contributing authors for lending their expertise to make the book truly unique. They have played a crucial role in the development of this book. Without their invaluable contributions this book wouldn't have been possible. They have made vital efforts to compile up to date information on the varied aspects of this subject to make this book a valuable addition to the collection of many professionals and students.

This book was conceptualized with the vision of imparting up-to-date information and advanced data in this field. To ensure the same, a matchless editorial board was set up. Every individual on the board went through rigorous rounds of assessment to prove their worth. After which they invested a large part of their time researching and compiling the most relevant data for our readers.

The editorial board has been involved in producing this book since its inception. They have spent rigorous hours researching and exploring the diverse topics which have resulted in the successful publishing of this book. They have passed on their knowledge of decades through this book. To expedite this challenging task, the publisher supported the team at every step. A small team of assistant editors was also appointed to further simplify the editing procedure and attain best results for the readers.

Apart from the editorial board, the designing team has also invested a significant amount of their time in understanding the subject and creating the most relevant covers. They scrutinized every image to scout for the most suitable representation of the subject and create an appropriate cover for the book.

The publishing team has been an ardent support to the editorial, designing and production team. Their endless efforts to recruit the best for this project, has resulted in the accomplishment of this book. They are a veteran in the field of academics and their pool of knowledge is as vast as their experience in printing. Their expertise and guidance has proved useful at every step. Their uncompromising quality standards have made this book an exceptional effort. Their encouragement from time to time has been an inspiration for everyone.

The publisher and the editorial board hope that this book will prove to be a valuable piece of knowledge for researchers, students, practitioners and scholars across the globe.

List of Contributors

I. M. Kutasov
Pajarito Enterprises, 640 Alta Vista, Suite 124, Santa Fe, New Mexico 87505, USA

L.V. Eppelbaum
Department of Geophysics and Planetary Sciences, Raymond and Beverly Sackler Faculty of Exact Sciences, Tel Aviv University, Ramat Aviv 69978, Tel Aviv, Israel

A. A. Sulaimon
Department of Petroleum Engineering, University of Ibadan, Oyo State, Nigeria

G. K. Falade
Department of Petroleum Engineering, University of Ibadan, Oyo State, Nigeria

W. DeLandro
Shell Petroleum Development Company, Nigeria

S. Adsavakulchai
School of Engineering, University of the Thai Chamber of Commerce 126/1 Vibphavadee Rangsit Road, Bangkok, Thailand

C. Huntula_
School of Engineering, University of the Thai Chamber of Commerce 126/1 Vibphavadee Rangsit Road, Bangkok, Thailand

Gbenga Folorunso Oluyemi
School of Engineering, Robert Gordon University, Schoolhill, Aberdeen, AB10 1FR, UK

M. Babs Oyeneyin
School of Engineering, Robert Gordon University, Schoolhill, Aberdeen, AB10 1FR, UK

Siddharth Arora
Department of Mechanical Engineering, Bits Pilani Dubai

Shobhit Chitkara
Department of Mechanical Engineering, Bits Pilani Dubai

R. Udayakumar
Department of Mechanical Engineering, Bits Pilani Dubai

Muhammad Ali
ADC Energy Systems. Dubai

Andisheh Alimoradi
Department of Mining, Petroleum and Geophysics Engineering, Shahrood University of Technology, Shahrood, Iran

Ali Moradzadeh
Department of Mining, Petroleum and Geophysics Engineering, Shahrood University of Technology, Shahrood, Iran

Mohammad Reza Bakhtiari
Department of Petroleum Engineering, Amir Kabir University of Technology, Tehran, Iran
National Iranian Oil Company, Exploration Directorate, Tehran, Iran

Ahmmed S. Ibrehem
Department of Chemical and Petroleum Engineering, UCSI University, Kuala Lumpur, 56000, Malaysia

Hikmat S. Al-Salim
Department of Chemical and Petroleum Engineering, UCSI University, Kuala Lumpur, 56000, Malaysia

Huang Wei
Liaoning Shihua University, Fushun, Liaoning, P.R. China

You Hongjun
SAIT Polytechnic, Calgary AB, Canada

Ariffin Samsuri
Department of Petroleum Engineering, Faculty of Petroleum and Renewable Energy Engineering, Universiti Teknologi, Malaysia

Amir Hamzah
Department of Petroleum Engineering, Faculty of Petroleum and Renewable Energy Engineering, Universiti Teknologi, Malaysia

Mandana Akia
Chemical Engineering Department, Iran University of Science and Technology, Tehran, Iran
Chemical Engineering Department, Kermanshah University of Technology, Kermanshah, Iran
State Key Laboratory for Heavy Oil Processing, Key Laboratory of Catalysis, CNPC, China University of Petroleum, Dongying 257061, China

Seyed Mahdi Alavi
Chemical Engineering Department, Iran University of Science and Technology, Tehran, Iran

Zi-Feng Yan
State Key Laboratory for Heavy Oil Processing, Key Laboratory of Catalysis, CNPC, China University of Petroleum, Dongying 257061, China

Abdelkader Khiari
Laboratoire des Ressources Naturelles et Aménagement des Milieux Sensibles, Larbi Ben M'hidi University, Oum el Bouaghi, 04000 Algeria

Djemoi Merzoug
Laboratoire des Ressources Naturelles et Aménagement des Milieux Sensibles, Larbi Ben M'hidi University, Oum el Bouaghi, 04000 Algeria.

Noureddine Gherraf
Laboratoire des Ressources Naturelles et Aménagement des Milieux Sensibles, Larbi Ben M'hidi University, Oum el Bouaghi, 04000 Algeria

Mourad Kourichi
Laboratoire de génie de procédés, UKMO Ouargla University 30000 Algeria

Noureddine Maza
Laboratoire des Ressources Naturelles et Aménagement des Milieux Sensibles, Larbi Ben M'hidi University, Oum el Bouaghi, 04000 Algeria

Zainal Zakaria
Department of Gas Engineering, Faculty of Petroleum and Renewable Energy Engineering, Universiti Teknologi Malaysia, 81310 Johor Baru, Johor, Malaysia

Nor Ismail Hashim
Department of Gas Engineering, Faculty of Petroleum and Renewable Energy Engineering, Universiti Teknologi Malaysia, 81310 Johor Baru, Johor, Malaysia

Bhupendra Kumar Sharma
Department of Mathematics, Birla Institute of Technology and Science- Pilani, -333031, India

Tara Chand
Department of Mathematics, University of Rajasthan, Jaipur-302004, India

R. C. Chaudhary
Department of Mathematics, University of Rajasthan, Jaipur-302004, India

Ahmmed S Ibrehem
Department of Chemical Engineering, University of Technology Mara (UiTM), 40450, Shah Alam Malaysia

Hikmat S. Al-Salim
Chemical and Petroleum Engineering Department, UCSI - University, Cheras, Kuala Lumpur, 56000, Malaysia

Mojtaba Tafaoli-Masoule
Department of Mechanical Engineering, Fuel Cell Research Technology Group, Babol University of Technology, P. O. Box 484, Babol, Iran

Mohsen Shakeri
Department of Mechanical Engineering, Fuel Cell Research Technology Group, Babol University of Technology, P. O. Box 484, Babol, Iran

Arian Bahrami
Department of Mechanical Engineering, Fuel Cell Research Technology Group, Babol University of Technology, P. O. Box 484, Babol, Iran

Mehrdad Alemi
Department of Petroleum Engineering, Amirkabir University of Technology, Tehran, Iran
Department of Petroleum Engineering, Shahid Bahonar University of Kerman, Iran
Environmental and Energy Research Center, Shahid Bahonar University of Kerman, Iran

Hossein Jalalifar
Department of Petroleum Engineering, Shahid Bahonar University of Kerman, Iran
Environmental and Energy Research Center, Shahid Bahonar University of Kerman, Iran

Gholamreza Kamali
Department of Mining Engineering, Shahid Bahonar University of Kerman, Iran

Mansour Kalbasi
Department of Petroleum Engineering, Amirkabir University of Technology, Tehran, Iran
Department of Chemical Engineering, Amirkabir University of Technology, Tehran, Iran

A. C. Pacheco
Departamento de Engenharia Química, Escola de Química da Universidade Federal do Rio de Janeiro - UFRJ, Rio de Janeiro –RJ, Brazil

A. F. Souza
Departamento de Engenharia Química, Escola de Química da Universidade Federal do Rio de Janeiro - UFRJ, Rio de Janeiro –RJ, Brazil

D. P. Costa
Departamento de Engenharia Mecânica e Energia -DEMEC, Instituto Politécnico da Universidade do Estado do Rio de Janeiro - IPRJ-UERJ, Nova Friburgo-RJ, Brazil

D. A. G. Aranda
Departamento de Engenharia Química, Escola de Química da Universidade Federal do Rio de Janeiro - UFRJ, Rio de Janeiro –RJ, Brazil

L. D. T. Câmara
Departamento de Engenharia Mecânica e Energia -DEMEC, Instituto Politécnico da Universidade do Estado do Rio de Janeiro - IPRJ-UERJ, Nova Friburgo-RJ, Brazil

E. G. Ladopoulos
Interpaper Research Organization, 8, Dimaki Str., Athens, GR - 106 72, Greece

Khulud M. Rahuma
University of Tripoli, Tripoli-Libya, P. O. Box 13126, Libya

F. Sondi
University of Tripoli, Tripoli-Libya, P. O. Box 13126, Libya

M. Milad
University of Tripoli, Tripoli-Libya, P. O. Box 13126, Libya

M. AbuGheit
University of Tripoli, Tripoli-Libya, P. O. Box 13126, Libya

H. S. Liang
Exploration and Development Research Institute, Chinese Petroleum Corporation (CPC), Miaoli, Taiwan

S. L. Lin
Exploration and Development Research Institute, Chinese Petroleum Corporation (CPC), Miaoli, Taiwan

T. K. Huang
Exploration and Development Research Institute, Chinese Petroleum Corporation (CPC), Miaoli, Taiwan

D. L. Chen
Exploration and Development Research Institute, Chinese Petroleum Corporation (CPC), Miaoli, Taiwan

M. C. Tom Kuo
Department of Mineral and Petroleum Engineering, National Cheng Kung University, Tainan, Taiwan

Morteza Elahi Naraghi
Petroleum and Geosystems Engineering Department, University of Texas at Austin, Austin, TX

Peyman Ezzatyar
Science and Research Branch, Islamic Azad University, Tehran, Iran

Saeid Jamshidi
Chemical and Petroleum Engineering Department, Sharif University of Technology, Tehran, Iran

Adel M. Salem Ragab
Department of Petroleum Engineering, School of Sciences and Engineering (SSE), American University in Cairo (AUC), Egypt
Department of Petroleum Engineering, Faculty of Petroleum and Mining Engineering, Suez University, Egypt

Sepehr Sadighi
Project Manager, Catalysis and Nanotechnology Division, Catalytic Reaction Engineering Department, Research Institute of Petroleum Industry (RIPI) Iran

S. Reza. Seif Mohaddecy
Project Manager, Catalysis and Nanotechnology Division, Catalytic Reaction Engineering Department, Research Institute of Petroleum Industry (RIPI) Iran

Adel M. Salem
American University in Cairo (AUC), and Suez University, Egypt

Shedid A. Shedid
British University in Egypt (BUE), Cairo, Egypt

Xin Zhao
Daqing Oilfields Lit Co., Daqing, China

Panqing Gao
Department of Petroleum Engineering, University of Wyoming, Wyoming, US

Zhiyong Shao
Daqing Oilfields Lit Co., Daqing, China

Minhee Choi
Department of Natural Resources and Environmental Engineering, Hanyang University, 222 Wangsimni-ro, Seongdonggu, Seoul 133-791, Korea

Junwoo Seo
Department of Natural Resources and Environmental Engineering, Hanyang University, 222 Wangsimni-ro, Seongdonggu, Seoul 133-791, Korea

Hyemin Park
Department of Natural Resources and Environmental Engineering, Hanyang University, 222 Wangsimni-ro, Seongdonggu, Seoul 133-791, Korea

Wonmo Sung
Department of Natural Resources and Environmental Engineering, Hanyang University, 222 Wangsimni-ro, Seongdonggu, Seoul 133-791, Korea

Masoud Asadi
Department of Mechanical Engineering, Azad Islamic University Science and Research branch ,Tehran, Iran

Nasrin Dindar Mehrabani
Department of Mathematics & Computer Science, Amir Kabir University of Technology, Tehran, Iran

Xin Zhao
Daqing Oilfields Lit Co., Daqing, China

Panqing Gao
Department of Petroleum Engineering, University of Wyoming, Wyoming, US

Zhiyong Shao
Daqing Oilfields Lit Co., Daqing, China

Yongwei Li
E&P Institute, Daqing Oil Company, Heilongjiang, China 163318